THE ORION COMPLEX:
A CASE STUDY OF INTERSTELLAR MATTER

ASTROPHYSICS AND
SPACE SCIENCE LIBRARY

A SERIES OF BOOKS ON THE RECENT DEVELOPMENTS
OF SPACE SCIENCE AND OF GENERAL GEOPHYSICS AND
ASTROPHYSICS PUBLISHED IN CONNECTION WITH
THE JOURNAL SPACE SCIENCE REVIEWS

VOLUME 90

THE ORION COMPLEX:
A CASE STUDY
OF INTERSTELLAR
MATTER

by

C. GOUDIS

Max-Planck-Institut für Astronomie, Heidelberg,
Federal Republic of Germany

D. REIDEL PUBLISHING COMPANY

DORDRECHT : HOLLAND / BOSTON : U.S.A.

LONDON : ENGLAND

Library of Congress Cataloging in Publication Data

Goudis, C., 1947–
The Orion complex.

(Astrophysics and space science library ; v. 90)
Bibliography: p.
Includes index.
1. Interstellar matter. 2. Orion (Constellation). I. Title. II. Series.
QB790.G68   1982   523.1'2   82–7659
ISBN-13: 978-94-009-7714-3   e-ISBN-13: 978-94-009-7712-9
DOI: 10.1007/ 978-94-009-7712-9

Published by D. Reidel Publishing Company,
P.O. Box 17, 3300 AA Dordrecht, Holland.

Sold and distributed in the U.S.A. and Canada
by Kluwer Boston Inc.,
190 Old Derby Street, Hingham, MA 02043, U.S.A.

In all other countries, sold and distributed
by Kluwer Academic Publishers Group,
P.O. Box 322, 3300 AH Dordrecht, Holland.

D. Reidel Publishing Company is a member of the Kluwer Group.

To my Daughters Ismini and Alexandra

Κ' ἦρθαν καὶ οἱ γύφτοι ποὺ γνωρίζουν
τῶν πλανητῶν τὰ κατατόπια
κι ὅλα τὰ μυστικὰ τῶν ἄστρων,
καὶ ποὺ μιλᾶνε μὲ τ' ἀστέρια,
καὶ ποὺ θωρώντας τα μαντεύουν
ζωές, ἀγάπες, μοῖρες, χάρους.

Κ' ἦρθαν κ' οἱ γύφτοι οἱ διαβασμένοι
κ' οἱ σκεφτικοὶ κ' οἱ βυθισμένοι
στ' ἀξήγητου τὸ ξήγημα...

Ο ΔΩΔΕΚΑΛΟΓΟΣ ΤΟΥ ΓΥΦΤΟΥ
Κωστῆς Παλαμᾶς

And the gypsies came, those who know the whereabouts of the planets and all the secrets of the stars, those who talk to the stars, and who by looking at them foresee lives, loves, destinies, deaths.

And the gypsies came, the learned, the thoughtful, the absorbed in the explanation of the unexplained...

THE DODECALOGUE OF THE GYPSY
Kostis Palamas

TABLE OF CONTENTS

INTRODUCTION

This work deals with some of the most typical complexes of interstellar matter and is intended to serve both as a reference book for the specialist and as an introduction for the newcomer to the field. It is hoped to meet the first aim by presenting a holistic view of the well studied complexes in Orion, built on information derived from various branches of modern Astrophysics. The wealth of published data is presented in the form of photographs, contour maps, diagrams and numerous heavily annotated tables. The second aim is pursued by providing an outline of the complexes, the physical problems associated with them, the empirical models describing their behaviour and, in addition, by including an extended Appendix section summarizing the numerous methods employed to derive the physical parameters of an H II region and the dust and molecular cloud physically associated with it. The book consists of five chapters and four Appendix sections.

Chapter 1, which is concerned with the large scale view of the Orion region, outlines the morphology of the area and examines in particular the nature of Barnard's Loop and the associated filamentary structure in addition to the origin of the I Orion OB association.

Chapter 2 focuses on the ionized gas of the Orion H II/molecular complex i.e. the Great Orion Nebula (M42 or NGC 1976) and the small H II region to the north (M43 or NGC 1982) (these two nebulae are also referred to in earlier radio astronomical studies as Orion A or W10). The chapter deals with:

(1) The optical and radio structure of both nebulae as derived from photographic, photoelectric and radio astronomical observations of both line and continuum radiation.

(2) The physical parameters of the nebulae, and the parameters of the most prominent stars associated with them.

(3) The distinct ionized objects i.e. the small stellar-size condensations ($\lesssim 1\overset{\prime\prime}{.}5$) and knots ($\simeq 10''$) present in the core of the Orion Nebula (M42).

(4) The dust contained in M42, it's spatial distribution, energy spectrum and chemical composition.

Chapter 3 examines the Orion Complex as a whole, i.e. the H II regions M42 and M43, the associated molecular clouds OMC 1 and OMC 2 and their interrelations. The chapter includes an investigation of:

(1) The infrared structure of the H II regions and the corresponding molecular clouds i.e. the structure of the Ney–Allen (Trapezium) infrared nebula associated with M42 and the Kleinmann–Low(KL) nebula/IR cluster of stars which is embedded in the molecular cloud OMC 1 lying behind M42, and the structure of the IR cluster associated with OMC 2.

(2) The physical parameters of the dust derived from infrared observations and relating to both the extended and the compact sources of the complex.

(3) The morphology and the physical parameters of the molecular clouds OMC 1 and OMC 2.

(4) The distribution of masers in OMC 1 and OMC 2 and the physical association of the numerous masering sources in OMC 1 with the compact sources of the IR cluster.

(5) The magnetic field of the complex.

Chapter 4 contains a discussion of the empirical models introduced to attempt to explain certain aspects of this very complex region.

Chapter 5 investigates the second prominent H II region and molecular cloud complex of the Orion region, NGC 2024 (Orion B, W12), in a manner similar to that employed for the Orion complex.

The Appendix sections of the book include:

(1) A brief discussion concerning the equations of radiative transfer in an H II region, a dust cloud and a molecular cloud (Appendix I).

(2) A summary of the various methods employed to derive the physical parameters of an H II region with comments concerning their merits and limitations, and analytical formulae expressing the physical parameters in terms of observable quantities (Appendix II).

(3) A summary of methods used to derive (a) the extinction suffered by the optical radiation in the presence of dust and (b) the physical parameters of the dust associated with an H II region (Appendix III).

(4) A brief summary of the methods used to derive the physical parameters of a molecular cloud (Appendix IV).

The aim of the Appendix sections is to present in a coherent manner the most important methods leading to the determination of the physical parameters of an H II region, a dust and a molecular cloud respectively. It should be kept in mind that more emphasis has been given to the use of the methods, their limitations and merits and to the conditions under which they can be legitimately used than to the fundamental physics underlying them. Therefore the treatment of certain topics is necessarily limited. The Appendices constitute in a sense a practical guide for the observational astronomer, intending to bridge the gap between the theory and observation of H II/Dust/Molecular complexes. The reader who strives for a deeper understanding of the principles involved is strongly encouraged to consult the classical works of: L. Spitzer, *Physical Processes in the Interstellar Medium*, Wiley Interscience, New York, 1978 (theoretical); S. A. Kaplan and S. B. Pikelner, *The Interstellar Medium*, Harvard University Press, Cambridge, Massachusetts, 1970 (general); D. E. Osterbrock, *Astrophysics of Gaseous Nebulae*, W. H. Freeman, San Francisco, 1974 (mainly optical); and the outstanding articles of E. J. Chaisson (general) and G. G. Fazio (infrared) from the *Frontiers of Astrophysics* Ed. by E. H. Avrett, Harvard University Press, Cambridge, Massachussets, 1976; P. G. Mezger (radioastronomical) from the *Interstellar Matter*, Swiss Society of Astronomy and Astrophysics, 1972; M. W.

Werner, E. E. Becklin, and G. Neugebauer (infrared) published in *Science* **197**, 723, 1977; and P. M. Solomon (molecular) published in *Physics Today* **26**, 32, 1973. The topic of Interstellar Masers associated with the H II complexes has not been touched at all, since it is fully covered by the excellent articles of J. M. Moran (*Frontiers of Astrophysics*, see reference above, and, more recently, in the *CRC Handbook on Laser Science and Technology*, 1981) and M. J. Reid and J. M. Moran (*Annual Review of Astronomy and Astrophysics*, 1981). The reader who wants to elaborate on the optical instrumentation and the variety of observational techniques employed in this field is strongly advised to consult the unique monograph of J. Meaburn, *Detection and Spectrometry of Faint Light*, D. Reidel, Dordrecht, 1976.

The present work is related to a number of publications by the same author concerned with the classification of the published data for certain well known H II regions (Goudis, 1975a, b, e, 1976a, b, c, e, 1979; also Walsh, 1980). The favourable reception of these articles by the astronomical community encouraged the author to undertake this study, with the hope that it will prove useful to a widerange of students and workers in the field of interstellar matter.

ACKNOWLEDGEMENTS

I wish to express my thanks to the numerous authors who gave me their permission to reproduce certain figures from their works. I also wish to extend my thanks to the publishers and editors of many scientific journals and monographs who consented to the reproduction of these figures. Analytically: Figure 2.1.6 of the present book is from the journal *Annales d'Astrophysique*; Figures 1.2.2, 2.1.2, and 5.1.8 are from the *Astronomical Journal*; Figures 1.1.5, 1.1.7, 1.2.3, 2.1.10, 2.1.15, 2.1.16b, 2.1.18, 2.2.4, 2.3.2, 2.3.3, 2.3.5, 2.3.6, 2.3.7, 2.3.9, 2.3.10, 2.3.11, 2.3.12, 2.4.1, 2.4.3, 2.4.6, 2.4.9, 3.1.3, 3.2.6, 3.2.7, 3.3.4, 3.3.10, 3.3.11, 3.4.4, 3.4.6, 3.4.7, 3.4.8, 4.1.2, 4.1.5, 5.1.2, 5.1.3, 5.1.9, 5.1.11, 5.1.12, 5.1.13, 5.2.4, 5.3.1, 5.3.2, II.3.2, III.2.1a, and III.2.1b are from the journal *Astronomy and Astrophysics*; Figure 2.2.5 is from the journal *Astrophysical Letters*; Figures 1.1.8, 1.1.9, 1.1.10, 1.2.4, 1.2.5, 1.2.6, 1.3.2, 1.3.3, 2.1.8, 2.1.9, 2.1.16c, 2.1.21, 2.3.8, 2.4.7, 2.4.8, 2.4.11, 3.1.1. 3.1.4, 3.1.5, 3.1.6, 3.1.7, 3.2.1, 3.2.2, 3.2.3, 3.2.5, 3.2.8, 3.2.9, 3.3.1, 3.3.2, 3.3.3, 3.3.5, 3.3.6, 3.3.7, 3.3.8, 3.4.9, 3.4.11, 4.1.1, 5.1.5, 5.2.1, and 5.2.2 are from the *Astrophysical Journal*; Figures 2.1.5, 2.1.7, 2.1.13, 2.1.17, 2.2.2, 2.3.4, 5.1.7, and II.3.3 are from the journal *Astrophysics and Space Science*; Figure 2.1.16a is from the *Australian Journal of Physics*; Figures 3.4.1 and II.2.1 are from the book *Frontiers of Astrophysics*, published by the Harvard University Press; Figures 3.3.9, 3.4.2, and 3.4.3 are from the book *Giant Molecular Clouds in the Galaxy*, published by the Pergamon Press; Figures 2.4.10, 3.2.4, and III.2.2 are from the book *Infrared and Submillimeter Astronomy*, published by D. Reidel; Figures 1.1.4 and 1.2.1 are from the book *Interstellar Dust and Related Topics*, published by D. Reidel on behalf of the IAU; Figure 3.4.10 is from the book *Interstellar Molecules*, published by D. Reidel on behalf of the IAU; Figures 2.1.14, 2.3.1, 5.1.1, and 5.1.4 are from the journal *Mercury*; Figures 2.1.19, 2.1.20, 2.2.1, 2.4.4, 2.4.5, 3.4.5, 5.1.6a, 5.1.6b, 5.1.10, and 5.2.3 are from the *Monthly Notices of the Royal Astronomical Society*; Figure 2.4.12 is from the journal *Nature*; Figures 4.1.1, 5.3.1, and III.2.1 are from the journal *Naturwissenschaften*; Figure 1.3.1 is from the book *Protostars and Planets*, published by the University of Arizona Press; Figure 1.1.2 is from the *Publications of the Astronomical Society of Japan*; Figures 2.1.1, 3.1.2, and 4.1.3 are from the *Publications of the Astronomical Society of the Pacific*; Figure 2.2.3 is from the *Review of Modern Physics*; and Figures 1.1.6 and 2.1.3 are from the *Zeitschrift für Astrophysik*. I also wish to thank the Asiago, Anglo-Australian, Hale, Haute Provence, Kitt Peak and Lick Observatories and the NASA/Ames Research Center for permitting me to reproduce photographs and material obtained at their establishments.

I would specifically like to thank Professor H. Elsässer, who has given me invaluable support and Professor G. Münch for his kindness to read the manuscript and discuss it with me. Professor Münch has made a number of very constructive comments for which I am grateful. I would also like to extend my thanks to Prof. T. de Jong who has refereed the present work. I also like to acknowledge many helpful discussions with Dr A. Harris. I am also indebted to Mrs E. Behme and C. Thome for the efficient typing of the manuscript and to Mr W. Neumann and Mrs D. Gayer for their excellent photographic printing. Last, but not least, to my wife Elpis, for her continuous psychological support and encouragement during the writing of this work, I express my sincere gratitude.

C. GOUDIS

Heidelberg, October 1980

THE LARGE SCALE VIEW OF THE ORION REGION

1.1. General Morphology of Barnard's Loop
and the Neighbouring Region

The general structure of the photographically visible part of Barnard's Loop (BL; Barnard, 1894a, b, 1903, 1927) is shown in Figures 1.1.1, 1.1.2, and 1.1.3. These are narrow filter photographs taken in the light of $H\alpha$, $H\beta$ and [N II] lines respectively.

The general morphology of the area is also shown in Figure 1.1.4 which is a contour map of the $H\alpha$ emission from the loop and the λ Orionis Nebula; this is a spherically shaped Nebula around the star λ Orionis which lies to the north of BL. Located in the inner region of the loop are two main visible complexes: the Orion Nebula (around the star θ^1 Orionis) and a complex of numerous H II regions the most prominent of which is NGC 2024 (around the star ζ Orionis). The complicated filamentary structure of the area around BL is shown in a widefield photograph taken in $H\alpha$ light (Figure 1.1.5). The faint U-shaped filamentary nebulosity at the upper right corner of Figure 1.1.5 which is probably associated with BL, is shown separately in Fig 1.1.6. A compilation of recent photographic and photometric work on Barnard's Loop and the surrounding area is presented in Table 1.1.I.

The radio continuum emission from the loop is shown in Figure 1.1.7 in which a 1420 MHz contour map of the Orion Region is presented. A compilation of the numerous radio continuum maps of the area is presented in Table 1.1.II.

The molecular structure of the region is presented in Figure 1.1.8, where the extent of the two main molecular complexes lying within BL, as derived from CO observations, is shown. The two molecular complexes coincide roughly with the various dark clouds (Lynds clouds) which have been detected in the area. The Northern Molecular Complex is associated with the cloud L1630, the nebulae around the star ζ Orionis (NGC 2024, NGC 2023, IC 434 (Horsehead Nebula, etc.), and the small nebulae NGC 2068 and NGC 2071 further north. The Southern Molecular Complex coincides mainly with the dark cloud L1641 and is associated with the Orion Nebula (NGC 1976 or M42 and NGC 1982 or M43) and various smaller H II sources (NGC 1981, NGC 1972, NGC 1973, NGC 1977 to the north of the Orion Nebula and NGC 1980, NGC 1999 to the South, etc.). The correlation of these features is shown in Figure 1.1.9. Relevant information relating to published CO contour maps is given in Table 1.1.III.

The stellar association which dominates the Orion Region is the I Orion OB association which is subdivided into four subgroups Ia, Ib, Ic and Id. The rough boundaries of these subgroups and their location with respect to the two molecular complexes and BL are shown in Figure 1.1.10.

Details for some main stars which are in or near the I Orion association are given in Table 1.1.IV.

Information relating to the extent of BL and other features which are thought to be physically associated with the loop are presented in Table 1.1.V. These features are: Barnard's Loop, the Orion association of OB stars (I Ori), the Orion Reflection Nebula (ORN) which is centred at the I Ori association and extends far beyond the BL, the "Orion's cloak" which is a high velocity ionized shell enveloping BL, the nearby Eridanus region which is active in soft X-ray emission and the extended shell of ionized and neutral hydrogen which contains BL and the Eridanus faint filaments as part of a more extended shell structure.

1.2. Nature of Barnard's Loop and the Associated Filamentary Structure

The nature, origin and evolution of Barnard's Loop is still a subject of considerable research. Various theories have been put forward all of which accomodate a particular set of data, but a generally acceptable consistent explanation is still lacking.

In recent years, however, the wealth of data accumulated by various workers in the field, the technological advances in instrumentation and the newly acquired ability to carry out astronomical observations from space platforms, have assisted in the clarification and understanding of certain aspects of the phenomena involved.

A first step towards such an understanding is the comparison of the various morphological features of BL. Such a comparison, is shown in Figure 1.2.1 and involves the spatial correlation of the brightest part of BL (ridge of maximum intensity) as it appears in Hα, H I (21 cm) and UV emission. The picture which arises is one of a neutral hydrogen (H I) ring, the inner part of which is ionized (H II ring). It should be noted that BL appears much more amorphous in the UV than in the visible although in both cases the pattern is similar. In addition BL has a greater surface brightness than expected from pure atomic emission. This was first discovered by O'Dell *et al.* (1967) and further corroborated by Carruthers and Opal (1977). These authors attributed the excess surface brightness in the UV to the presence of a dust cloud scattering the starlight of the stars of the Orion association. Many of these stars are strong UV emitters. This fact, and the presense of dust particles with relatively high albedo in the UV, makes the above explanation plausible. Witt and Lillie (1978) have carried out further surface brightness measurements (in the range of 4250–1550 Å) and have confirmed the existence of an extended ($\sim 33°$) ultraviolet reflection nebulosity (Orion Reflection Nebulosity, ORN) much more extended than BL ($\sim 15°$, see Table 1.1.V). The explanation of the observed brightnesses is consistent with a model involving a combination of (a) back-scattering of the light from the early type stars of the I Ori association by the molecular cloud/dust complex lying behind them (Figures 1.1.8 and 1.1.9, Table 1.1.III) and (b) forward-scatteering by optically thin dust lying along the line of sight between the Earth and the Orion stars. Also included in the correlation of Figure 1.2.1 is the distribution of stars in the area, derived from star counts on the Palomar Sky Survey Prints. (The stars seem to be distributed in the shape of an elliptical ring, in the inner part of the H II ring, with a gap SE from θ^1 Orionis).

Early studies of the kinematics of BL based on observations of the 21 cm H I emission led Menon (1958), to conclude that the ring is a result of an explosive event which occured in the Orion association about 3×10^6 yr ago (see Table 1.2.I). Gordon (1970), however, interpreted her observations of neutral hydrogen emission in terms of rotation (Figure 1.2.2). Another mechanism proposed by O'Dell *et al.* (1967) to explain the origin and support of the loop is stellar radiation pressure exerted upon the grains which drag along them the associated gas. This "matter sweeping" mechanism results in the creation of the observed shell. Another line of reasoning was followed by Appenzeller (1974) who made observations of the linear interstellar polarization of starlight in the region in order to study the effects of the magnetic field on the structure of the area. His polarization map of the Orion Region is shown in Figure 1.2.3. His suggestion that the interstellar gas in Orion is suspended in an interstellar "magnetic pocket" provides a plausible explanation for the equilibrium of BL. The magnetic field in the area may also be responsible for the elliptical structure of BL.

The structure of BL implied in the above discussion is that of an isolated elliptical shell of interstellar matter, a form which was derived from the extrapolation of the prominent features in the optical and radio domain of BL (Table 1.1.V, Nos. 1 and 2). However recent widefield mosaic photographs capable of registering the faint Hα emission from the area, photometric diffuse Hα surveys and detailed study of the neutral emission have shown that the whole area in the vicinity exhibits a very complicated shell structure (Figure 1.1.5 and Table 1.1.V, Nos. 7 and 8). BL is part of this very extended interstellar shell or system of shells. This suggestion was further strengthened by the study of the kinematics of the whole Orion–Eridanus area by Reynolds and Ogden (1979). Their main conclusion was that BL is part of a very large (280 pc) expanding shell which dominates the Orion–Eridanus area of the sky (Table 1.1.V, No. 7 and Table 1.2.I, No. 3). The structure of this area, as derived from the Hα emission, is shown in Figure 1.2.4, while a schematic representation of the shell with respect to the galactic plane, the I Orion association, the Orion Molecular Cloud and the Sun is shown in Figure 1.2.5. The ionizing source of the optically emitting inner part of the shell is probably the UV radiation from the stars of the Orion association. The shell is the result of a supernova explosion (or series of explosions) with an initial energy of the order of $E_0 \sim 10^{52}$ ergs, which occured in the Orion association about 2×10^6 yr ago (Reynolds and Ogden, 1979). According to the estimates of Cowie *et al.* (1979), however, this may have been as early as 3×10^5 yr ago.

The assymetric shape of the shell (Figure 1.2.5) can be explained by considering Barnards' Loop as a fragment of the shell which has been dramatically decelerated by the denser gas near the galactic plane, and hence could not expand with the rest of the shell. An interaction of BL with a molecular cloud (L1638) can be inferred from Figure 1.1.9.

An alternative scenario proposed by Cowie *et al.* (1979) based on the detection of a high velocity (~ -100 to -120 km s^{-1}) low ionization shell with a diameter of 120 pc surrounding the Orion area is shown in Figure 1.2.6. The high velocity feature is attributed to the radiative shock from the explosion of a less powerful supernova ($E_0 \sim 10^{50}$ ergs) which occured as early as 3×10^5 yr ago. An intermediate velocity feature ($35 < |v| < 100$ km s^{-1}) arises from a denser fragmented interstellar shell which expands with 20 km s^{-1}. BL, lying in the inner edge of this denser shell, is ionized by the UV radiation of the OB stars of the I Ori association. The observed denser shell, the age of which is $(2–4) \times 10^6$ yr, constitutes the 'memory' of an older super-nova explosion (or explosions). According to this model BL belongs to a less extended shell ($15°$ or 120 pc for a distance of 450 pc) than the one proposed by Reynolds and Ogden (1978) ($\sim 35°$ or 280 pc for a distance of 400 pc). If, however, the Eridanus 'hot spot' (Naranan *et al.*, 1976, Figure 1.2.4) which lies inside the area of the extended shell is a supernova remnant (Table 1.2.II) at the same distance as the 'Orion supernova (e)' then a connection between their remnant shells is possible according to the tunnel theory of Cox and Smith (1974). In that case the end result may be similar to the extended shell observed by Reynolds and Ogden (1978). The circumstantial evidence supporting the case for one or more supernova explosions in the area is presented in Table 1.2.II.

An additional source of energy which play a part (still unclear) in the energetics of the shell is stellar winds originating from the hot evolved stars of the I Ori OB as-sociation. These according to Castor *et al.* (1975), can transfer significant amounts of energy to the interstellar medium (see also Goudis and Meaburn (1978) and Goudis (1978)). The estimated total stellar wind luminosity of the stars of the Ori association is, according to Abbott (1978), 2×10^{37} ergs s^{-1} which over the expected lifetime of these stars ($\sim 10^6$ yr) makes the total energy transferred to the interstellar medium equal to 6×10^{50} ergs. However it should be remembered that by the time this energy has been transferred to the ISM, one or more of the stars would be expected to have exploded as SN providing instantaneously energy $\gtrsim 10^{51}$ ergs. In fact the rate of occurence of supernova explosions in an OB association is 1 per $10^5 - 2 \times 10^5$ yr (Cowie *et al.*, 1979).

Details concerning the age of the I Orion association, the age of the remnants from the proposed supernova explosions and the energetics of the various processes oc-curing in the area are presented in Table 1.2.III. Information about the masses of the various complexes contained in the Orion region is given in Table 1.2.IV.

1.3. Origin of the I Orion OB Association

The distribution of the stars of the I Ori OB association (and other associations) in distinct groups characterised by different ages led Blaauw (1964) to conclude that the formation of such an association does not occur at one time but in progressive steps. In the case of the Orion association four distinct groups are discernible (Figure 1.3.1). The older subgroup, Ia, has a characteristic size of 50 pc while that of the younger, Id, is only 2 pc. Information concerning the age and size of these four subgroups of the Orion association is presented in Tables 1.3.I and 1.3.II. Figures 1.3.1 and 1.1.10 show that only the younger subgroups Ic and Id are associated with the molecular complexes contained within BL. This is in accordance with the idea that the subgroups have been created out of the fragments of a parent molecular cloud, the older parts of which were previously exhausted in the formation of the older subgroups (Ia and Ib). The initiation of such a process is not well understood. One of the current ideas favours a trigger mechanism caused by a SN explosion such as that already discussed. Once the formation of the first subgroup of OB stars has occurred the sequential formation of the rest can be explained either as a result of further SN explosions originating from quickly evolved massive stars of the first subgroup (supernova cascade) or as a consequence of the ionization driven shock fronts propagating from the OB stars of the first subgroup into the rest of the molecular cloud.

According to the first view, which was originally suggested by Öpik (1953) and has recently been studied in detail by Ögelman and Maran (1976), the initial supernova blast compresses the molecular cloud (consisting mainly of hydrogen) into a high density shell composed of inhomogeneous condensing fragments from which the first subgroup of stars is created. Some of the massive new stars evolve quickly into new SN, and a repetition of the first step occurs, resulting in the formation of a second subgroup of stars. A schematic representation of this idea is shown in Figure 1.3.2. Further details concerning the proposed mechanism are included in the legend of the figure.

According to the alternative view, originally suggested by Oort (1954) and recently further elaborated by Elmegreen and Lada (1977), an ionization front (I) driven by the OB stars of the first subgroup, and preceded by a shock front (S) propagates into the molecular cloud. The layer of dense material sandwiched between the $I - S$ front becomes gravitationally unstable and breaks into condensations from which a new subgroup of OB stars is formed. The process repeats itself until the molecular cloud has been exhausted, thus having been transformed into an association of OB stars consisting of subgroups displaced both in place and time. A schematic representation with further details of this view is shown in Figure 1.3.3. A necessary prerequisite for this mechanism is an already existing subgroup of OB stars lying on the edge of a molecular cloud. The formation of this subgroup could have been the result of a supernova explosion (as mentioned above) or other mechanisms such as density wave shocks, cloud-cloud collisions or stellar winds (Lada et al., 1978).

TABLE 1.1.I.

Photographs and contour maps of Barnard's Loop and the neighbouring region

No.	Photographs (P)/ Contour maps (CM)	Wavelength range (λ = peak wavelength $\Delta\lambda$ = width of the filter)	Remarks		References
1	Hα (P)	$\lambda=6563$ Å $\Delta\lambda=8$ Å	(1)		Georgelin (unpublished)
2	Hα (P)	$\lambda=6563$ Å	(1)		Courtès (1967)
3	Hα (P)	$\lambda=6563$ Å $\Delta\lambda=40$ Å	(2),	(3)	Meaburn (1965)
4	Hα (P)	$\lambda=6563$ Å $\Delta\lambda=297$ Å	(1),	(3)	Isobe (1973)
5	Hα (P)	$\lambda=6563$ Å $\Delta\lambda=10$ Å	(4)		Sivan (1974)
6	Hα (P)	$\lambda=6563$ Å $\Delta\lambda=9$ Å	(1)		Johnson (1979)
7	Hβ (P)	$\lambda=4881$ Å $\Delta\lambda=15$ Å	(1)		Isobe (1978)
8	[N$\textsc{ii}$] (P)	$\lambda=6584$ Å $\Delta\lambda=10$ Å	(1)		Goudis and Meaburn (1974)
9	UV (P)	in the range 2200–4900 Å	(1),	(9)	O'Dell *et al.* (1967)
10	UV (P)	in the range 1230–2000 Å	(1),	(9)	Carruthers and Opal (1977a)
11	Hα (CM)		(1),	(5)	Isobe (1973,1978)
12	Hα (CM)		(6)		Reynolds *et al.* (1974)
13	Hα (CM)		(7)		Reynolds and Ogden (1979)
14	Hβ (CM)		(1),	(8)	Isobe (1978)
15	UV (CM)		(1),	(10)	Carruthers and Opal (1977a)

Remarks to Table 1.1.I:

(1) Barnard's Loop.

(2) A mosaic of photographs covering the Eridanus Nebulocity in the proximity of Barnard's Loop.

(3) Hα + [N $\textsc{ii}$] photograph.

(4) Mosaic from wide field photographs covering Barnard's Loop and the surrounding area.

(5) Isodensity contour map (Isobe, 1973) and relative intensity map (Isobe, 1978) constructed from an Hα photograph.

(6) Contour map of diffuse galactic Hα emission from $l^{\textsc{ii}} = 0$ to $l^{\textsc{ii}} = 240°$ and from $b^{\textsc{ii}} = -50°$ to $+30°$, made with a spatial resolution of $5°-10°$. A ring like structure of enhanced Hα emission $\sim 30°$ in diameter and centred on $l^{\textsc{ii}} = 200°$ and $b^{\textsc{ii}} = -25°$ outlines an extended shell of gas, part of which is BL.

(7) Absolute intensity contour map of the extended Orion–Eridanus region; BL is part of this structure.

(8) Relative intensity contour map of BL produced from an Hβ photograph.

(9) Photograph showing an amorphous appearance of BL, probably due to starlight scattered by interstellar grains in the nebula.

(10) Isodensity contour map produced from a UV photograph taken in the range of 1230–2000 Å.

TABLE 1.1.II.

Radio continuum contour maps of Barnard's Loop and the Orion region

No.	HPBW (arc min)	Frequency ν (MHz)	Wavelength λ	Remarks	References
1	85×100	19.7	15.2 m	(1)	Risbeth (1958)
2	72	820	36.6 cm	(1)	Berkhuijsen (1972)
3	66×72	240	1.25 m	(1)	Salter (1970)
4	50×60	85.5	3.5 m	(1)	Risbeth (1958)
5	45	408	73.5 cm	(1)	Haslam et al. (1970)
6	34	1420	21 cm	(1)	Reich (1978)
7	8.2	2700	11 cm	(2)	Caswell and Goss (1974)

Remarks to Table 1.1.II:

(1) Barnard's Loop and neighbouring region.

(2) Contour map of the radio continuum emission in the vicinity of Orion A (M42 + M43) and Orion B (NGC 2024).

TABLE 1.1.III

Contour maps of molecular complexes associated with the Orion region

No.	Molecule	HPBW (arc min)	Frequency ν (GHz)	Wavelength λ (mm)	Remarks	References
1	^{12}CO ($J=1-0$)	8	115.2712	2.6	(1)	Kutner et al. (1977)
2	^{13}CO ($J=1-0$)	8	110.2014	2.7	(1)	
3	^{12}CO ($J=1-0$)	2.6	115.2712	2.6	(1)	

Remarks to Table 1.1.III:

(1) Two main molecular complexes were detected: (1) the Northern Molecular Complex which coincides with the dark cloud L1630 and (2) the Southern Molecular Complex which is associated with the dark cloud L1641; see Figures 1.1.8 and 1.1.9.

TABLE 1.1.IV
Prominent stars in the Orion region

No.	Star	HD number [1]	α(1950) [1]	δ(1950) [1]	l^{II}	b^{II}	Spectral type [2]	Apparent visual magnitude [2] (V)	Colour index [2] ($B-V$)	Annual proper motion [2] α(")	δ(")	Radial velocity [2] (km s⁻¹)	Distance [3] (pc)	Velocity components of [4] UV absorption lines V_{LSR} (km s⁻¹)
1	π^5 Ori	31237	4h51m38.s661	+ 2°21′37″.24	196.25	−24.60	B2 III	3.72	−0.19	−0.001	+0.000	+23	230	−43 (<60)
2	λ Eri	33328	5 06 45.092	− 8 49 00.32	209.20	−26.70	B2 IV	4.27	−0.20	+0.003	+0.000	+3	263	—
3	λ Lep	34816	5 17 16.190	−13 13 37.19	214.80	−26.25	B0.5 IV	4.28	−0.28	−0.003	−0.005	+20	457	—
4	23 Ori	35149	5 20 12.177	+ 3 29 52.42	199.20	−18.00	B1 V	4.99	−0.15	−0.003	−0.001	+18	363	−118, −102, −60
5	25 Ori	35439	5 22 08.980	+ 1 48 07.84	201.00	−18.30	B1 V?pe	4.94	−0.21	+0.000	+0.000	+19	363	—
6	τ Ori	35468	5 22 26.828	+ 6 18 21.17	196.90	−16.00	B2 III	1.64	−0.23	−0.006	−0.014	+18	93	− (<80)
7	33 Ori	36351	5 28 36.990	+ 3 15 20.92	200.50	−16.20	B1.5 V	5.44	−0.20	+0.001	+0.004	+20	363	−(100−120), −77, −50
8	δ OriA	36486	5 29 27.017	− 0 20 04.41	203.85	−17.75	O9.5 II	2.20	−0.21	+0.001	−0.001	+16	380	Several blended components, $V> -100$
9	ϕ^1 Ori	36822	5 32 04.419	+ 9 27 26.67	195.40	−12.30	B0 IV	4.41	−0.17	+0.004	−0.004	+33	400	−116(?), −97(?), −55
10	λ Ori	36861	5 32 22.913	+ 9 54 08.35	195.10	−12.00	O8 III	3.66	—	+0.001	−0.006	+34	400	−115(?), −79, −59
11	42 Ori	37018	5 32 55.055	− 4 52 10.72	208.50	−19.10	B2 III	4.60	−0.20	+0.003	+0.001	+30	500	−83, −60 (<100)
12	θ^1 OriC	37022	5 32 48.995	− 5 25 16.22	209.00	−19.40	O6 p	5.16	+0.06	+0.003	+0.003	+33	478	Several blended components, $V> -80$
13	θ^2 OriA	37041	5 32 55.449	− 5 26 50.72	209.05	−18.40	O9.5 Vp	5.07	−0.10	+0.000	+0.006	+36	478	−74, −59, −38
14	ι Ori	37043	5 32 59.126	− 5 56 28.26	209.50	−19.60	O9 III	2.77	−0.25	+0.003	+0.004	+22	478	−106, −89, −68, −40
15	ε Ori	37128	5 33 40.476	− 1 13 56.30	205.20	−17.25	B0 Ia	1.70	−0.19	+0.000	+0.000	+26	398	−50 (<100)
16	σ Ori	37468	5 36 14.048	− 2 37 38.49	206.80	−17.30	O9.5 V	3.75	−0.24	+0.000	+0.004	+29	436	−119, −97, −84, −66
17	ζ Ori	37742	5 38 14.043	− 1 58 02.98	206.45	−16.60	O9.5 Ib	2.05	—	+0.004	−0.002	+18	436	−107, −88, −69, −54, −35
18	κ Ori	38771	5 45 23.013	− 9 41 09.45	214.50	−18.50	B0.5 Ie	2.04	−0.18	+0.004	−0.002	+21	417	—

Remarks to Table 1.1.IV:

(1) From the 'Smithsonian Astrophysical Observatory (SAO) Star Catalog' (Smithsonian Institution, Washington, D.C., 1966).

(2) From the 'Catalogue of Bright Stars' (Yale University Observatory, 1964).

(3) From Warren and Hesser (1977), Murdin and Penston (1977), Lesh (1968, 1972).

(4) From Cowie et al. (1979).

TABLE 1.1.V
Position and size of various large scale structures, seen in the direction of the Orion region

No.	Feature	Aproximate coordinates of the centre				Angular size ($°$)	Distance (pc)	Linear size (pc)	Remarks	References
		α (1950)	δ (1950)	l^{II} ($°$)	b^{II} ($°$)					
1	Barnard's Loop	5^h34^m	$-3°$	207	-18	17×11	450	134×86	(1)	Menon (1958)
2	Barnard's Loop	5 34	-3	207	-18	14×10	450	110×78	(1)	O'Dell et al. (1967)
3	I Orion Association	5 34	-3	207	-18	11×10	460	100	(2)	Blaaw (1964)
4	Orion Reflection Nebula	5 34	-3	207	-18	~ 33	450	250	(3)	Witt and Lillie (1978)
5	Orion's Cloak	5 34	-3	207	-18	~ 15	450	120	(4)	Cowie et al. (1979)
6	Eridanus 'Hot Spot'	4 15	-12	205	-40	15	200	52	(5)	Naranan et al. (1976)
7	Extended Shell	4 58	0	200	-25	35	400	280	(6)	Reynolds et al. (1974) Reynolds and Ogden(1979)
8	Extended Shell			198	-40	~ 31	150?	90	(7)	Heiles and Jenkins (1976) Heiles (1976)

Remarks to Table 1.1.V:

(1) Approximate size of the ellipse inferred from optical photographs of Barnard's Loop.

(2) Approximate position: from $l^{II} = 199°$ to $l^{II} = 210°$ and $b^{II} = -12°$ to $b^{II} = -21°$, $z = -130$ pc; the quoted linear size is the largest overall diameter. Number of O–B2 stars: 56 The I Orion association is subdivided into four subgroups Ia, Ib, Ic, and Id; for details see Tables 1.3.I and 1.3.II.

(3) Position and size of the Orion Reflection Nebula (ORN) inferred from surface brightness measurements in the range of 4250–1500 Å. A combination of stellar light back-scattered by the molecular complex (lying behind the Orion stars), and of stellar light forward-scattered by the dust lying between the Earth and Orion seems to be an adequate explanation of the observed spectral characteristics of the ORN.

(4) Shell of low ionization, low density material surrounding the I Ori and λ Orion is associations and expanding with 100–120 km s^{-1}. This may be the radiative shock from a SN exploded 3×10^5 yr ago.

(5) Active soft X-ray emission region, probably a SNR.

(6) Ringlike structure of enhanced Hα emission probably associated with H I filaments and covering substantial part of the Orion-Eridanus region. The ionization of the region is due to the UV radiation emitted from the stars of the I Ori association.

(7) Elliptically shaped expanding H I ring probably associated with Hα filaments (see note (6) above). The distance adopted is rather arbitrary.

TABLE 1.2.I
Expansion of Barnard's Loop and the extended shell associated with it

No.	Line	Velocity of expansion km s^{-1}	Remarks	References
1	H I (21 cm)	10	(1)	Menon (1958)
2	H I (21 cm)	23	(2)	Heiles (1976)
3	Hα, [N II]	15	(3)	Reynolds and Ogden (1979)

Remarks to Table 1.2.1:

(1) The observed velocities were explained as due to an expanding shell containing both neutral and ionized hydrogen around a quite H I sphere. Centre of expansion $l^{II} = 173°.5$ $b^{II} = -18°$. Inner shell radius $= 38$ pc, thickness $= 29$ pc. Mass of shell $\sim 5.8 \times 10^4 M_\odot$. Kinetic energy of shell $E_k = 5.8 \times 10^{49}$ ergs. Expansion time: 2.4×10^6 yr.

(2) Observed neutral hydrogen velocity of extended H I ring probably associated with BL (Table 1.1.V, No. 8).

(3) Mean velocity of expansion of the extended ring of Hα emission (Table 1.1.V, No. 7) associated with the extended neutral hydrogen ring (Table 1.1.V. No. 8); BL is probably the bright optical part of the Hα ring.

CHAPTER 1

TABLE 1.2.II

Circumstantial evidence supporting the occurence of one or more supernovae
in the Orion–Eridanus region

No.	Evidence	Remarks
1	Expanding shell ($V\sim 20$ km s^{-1}) and filamentary structure associated with it	(1)
2	Runaway stars originally belonging to the I Ori OB association	(2)
3	Soft X-ray emission from the interior of the shell	(3)
4	High ($V\sim 100$ km s^{-1}) and intermediate ($V = 35$–100 km s^{-1}) velocity shells	(4)

Remarks to Table 1.2.II:

(1) Shell consisting of BL and various Hα and H I filaments (Sivan, 1974; Reynolds *et al.*, 1974; Heiles and Jenkins, 1976; Heiles, 1976; Reynolds and Ogden, 1979); the real extent of the shell is not well asserted; it may cover a substantial part of the Orion–Eridanus region ($\sim 30°$ or 280 pc for a distance of 400 pc (Reynolds and Ogden (1979)) or it may be limited to the elliptical ring around BL ($\sim 15°$ or 120 pc for a distance of 450 pc).

(2) Runaway stars are believed to originate in a SN explosion which occurred in a binary system ('sling' effect; Shklovsky, (1968)). The stars AE Aurigae and μ Columbae are moving away and in opposite directions from the centre of the Orion region with $V = 127$ km s^{-1} (Blaauw and Morgan, 1954); the star 53 Arietis is also moving away with $V = 70$ km s^{-1} (Blaauw, 1956). Estimated kinematic ages of these stars (from the moment of their 'ejection') are: 2.2×10^6 yr, 2.7×10^6 yr, and 4.9×10^6 yr, respectively (Shklovsky, 1968); therefore, the occurrence of this explosive event happened $\sim 3 \times 10^6$ yr ago. This is consistent with the age of the I Ori OB association (Table 1.2.III). Some of the massive stars of the older subgroup (Ia) have probably passed through the SN stage.

(3) The emission is probably due to the very hot, low density gas occupying the interior of a Supernova Remnant (Chevallier, 1974). An enhancement of X-ray emission all over the extended Ori-Eri shell has been observed in the 0.5–1.2 keV band by Nousek (1978) ($T \sim 3 \times 10^6$ K). A similar enhancement in the 0.2–0.3 keV band has been observed around the 'Eridanus hot spot' (Figure 1.2.4) by Williamson *et al.* (1974) and Naranan *et al.* (1976). The 'Eridanus hot spot' which extends over a field of 15° (Table 1.1.V) may be a separate SNR; the circumstantial evidence supporting this view (Naranan *et al.*, 1976), is based on the association of the 'spot' with: (1) optical filaments; (2) a weak hard X-ray source (3U–0431–10, Giacconi *et al.* (1974)); (3) a runaway pulsar 10° from the centre (MP 0450, Vaughn *et al.* (1969)); (4) an expanding H I shell inferred from the 21 cm profiles (Tolbert, 1971; Heiles and Habing, 1974); and (5) O VI seen in absorption (York, 1974). For a distance of 200 pc, the X-ray diameter is 52 pc and the estimated kinematic age is $(2$–$4) \times 10^6$ yr. Alternatively (Reynolds and Ogden, 1979) the 'Eridanus hot spot' belongs to the extended Ori-Eri shell. The lack of X-ray emission in the 0.2–0.3 keV band around the I Ori association (energetic centre) may be due to higher interstellar absorption in the lower X-ray energies.

(4) The shells were detected by Cowie *et al.* (1979) from observations of UV absorption lines of various ionization stages of C, N, Si, and S, for stars lying towards the direction of the I Ori OB association. The high velocity shell (V expansion ~ 100 km s^{-1}) enveloping the Orion region ($\sim 15°$; Table 1.1.V) is believed to represent the radiative shock of a SNR. The occurrence of the explosive event happened $\sim 3 \times 10^5$ yr ago. Total mass of the shell ~ 100 $M_\odot$ and kinetic energy $E_k \sim 10^{49}$ ergs. Pre-shock density (density of the ambient interstellar matter before the passing of the disturbance) $N_e \sim 3 \times 10^{-3}$ cm^{-3}. The lower velocity component ($35 < V_{obs} < 100$ km s^{-1}) originates from a denser fragmented shell ($N_e > 3$ cm^{-3}) which expands with $V_{exp} \sim 20$ km s^{-1}. The existence of this component has also been verified by Shull (1979) ($V_{obs} -50$ to -43 km s^{-1}). The lower velocity shell probably indicates an older SN explosion which occurred 2×10^6 yr ago.

TABLE 1.2.III

Ages and energetics of the Orion Supernovae and the I Ori Association

No. Region	Angular size (°)	Linear size (pc)	Age (yr)	E_k (ergs) (kinetic energy of the shell)	E_0 (ergs) (initial energy of the explosion)	Remarks	References
1 I Ori OB association	11×10	100	3×10^6		6×10^{50}	(1), (2)	Blaauw and Morgan(1954), Blaauw(1964), Walker(1969)
Orion Supernovae:							
2 explosive event			2.4×10^6	6×10^{49}	6×10^{50}	(3)	Menon (1958)
3 young supernova	15	120	3×10^5	10^{49}	10^{50}	(4)	Cowie et al. (1979)
4 Old supernova	15	120	$(2-4) \times 10^6$			(5)	Cowie et al. (1979)
5 Eridanus supernova	19	?	1×10^5 2×10^6	3.8×10^{50}	3.6×10^{51}	(6)	Naranan et al.(1976)
6 Orion+Eridanus supernova	35	280	2×10^6	2×10^{50}	$\gtrsim 10^{52}$	(7)	Reynolds and Ogden (1979)

Remarks to Table 1.2.III:

(1) The quoted size is the largest projected dimension; the linear size was estimated for a distance of 460 pc (Table 1.1.V). For further information concerning the four subgroups of the I Orion association see Tables 1.3.I and 1.3.II.

(2) The quoted energy is the total energy of stellar winds emitted from the stars of the I Ori association which is transferred to the interstellar medium (ISM) over their lifetime ($\sim 10^6$ yr).

(3) The quoted total energy was adopted on the assumption that the kinetic energy of the shell is 10% of the initial energy of explosion.

(4) The occurrence of such a supernova was inferred from the detected high velocity shell around BL which expands with a velocity of 100 km s^{-1} (Table 1.2.II, note (4)).

(5) The occurrence of such an old supernova (or supernovae) was inferred from the detection of a lower velocity shell expanding with a velocity of 20 km s^{-1} (Table 1.2.II; note (4)).

(6) Energy estimates made by Heiles (1976) for a distance of 150 pc. For a distance of 450 pc the energetic requirements are: $E_{kin} = 3.4 \times 10^{51}$ ergs, $E_0 = 3.2 \times 10^{52}$ ergs.

(7) Energy estimates made for a distance of 400 pc.

TABLE 1.2.IV

Mass of various complexes in the Orion region

No.	Complex	Mass ($M_\odot$)	Remarks	References
1	M total (H I+H II+Dust)	1.1×10^5	(1)	Menon (1958)
2	M H$_2$ (Northern Molecular Complex/L1630)	$2.5 \times 10^4 - 1 \times 10^5$	(2)	Tucker et al. (1973)
3	M H$_2$ (Northern Molecular Complex/L1630)	1×10^5		Kutner et al. (1977)
4	M H$_2$ (Southern Molecular Complex/L1641)	$\sim 6 \times 10^4$		Kutner et al. (1977)
5	M total of molecular gas	$\sim 2 \times 10^5$		Kutner et al. (1977)
6	M I Ori association	7.6×10^3	(7)	Blaauw (1964)
7	M H I (atomic hydrogen)	7×10^4	(3)	Gordon (1970)
8	M extended ionized shell	8×10^4	(4)	Reynolds and Ogden (1979)
9	M extended H I component	5.2×10^5	(5)	Reynolds and Ogden (1979)
10	M high velocity envelope	10^2	(6)	Cowie et al. (1979)

Remarks to Table 1.2.IV:

(1) Total mass of neutral and ionized hydrogen and dust inside the ring. A spherical distribution was assumed with a radius of 60 pc, an average density of neutral and ionized hydrogen atoms of ~ 5 cm^{-3} and a dust to gas ratio of 1/100 by mass.

(2) Mass estimated for an area of $1°.5 \times 4°$ (13×35 pc for a distance of 500 pc) and a path length of 13 pc.

(3) H I in the Orion Region extends up to 30° in longitude.

(4) Mass of an extended ionized hydrogen shell, 280 pc in diameter, expanding with a velocity of 15 km s^{-1}. BL is part of this structure. Average density within the shell ~ 1.1 cm^{-3}.

(5) Mass of neutral component enveloping the extended ionized shell, estimated for a distance of 400 pc and an angular extend of 19°. Mass estimated for a distance of 150 pc (Heiles, 1976): $7.3 \times 10^4 \, M_\odot$. Velocity of expansion ~ 23 km s^{-1}.

(6) Mass of shell enveloping the Orion region and expanding with velocity of the order of 100 km s^{-1}.

(7) Mass of the I Ori association estimated for 56 O–B2 stars (see howerver Table 1.3.I including more than 80 O–B2 stars).

TABLE 1.3.I

Extent and age of the four subgroups of the I Orion association

No.	Subgroup	Largest projected dimension[1]		Age (10^6 yr)		
		Angular size (°)	Linear size[2] (pc)	Strand (1958)	Blaauw (1964)	Warren and Hesser (1978)
1	Ia (Northwest)	6.20	50		12	7.9
2	Ib (Belt)	2.50	25		8	5.1
3	Ic (Outer Sword)	1.90	15		6	3.7
4	Id (Orion Nebula/ Trapezium cluster)	0.25	2	0.3	~ 2	< 0.5

(1) sizes taken from Blaauw (1964).
(2) for a distance of 460 pc.

TABLE 1.3.II

Stellar statistics of the four subgroups of the I Orion Association[1]

No.	Spectral type	Number of stars belonging to the subgroup:			
		Ia	Ib	Ic	Id
1	O6	0	0	0	1
2	O9	0	1	1	1
3	B0	0	4	2	2
4	B1	4	6	8	6
5	B2	13	12	4	8

(1) Table originally constructed by Reeves (1978) from data taken from Levato and Abt (1976) and Abt and Levato (1977).

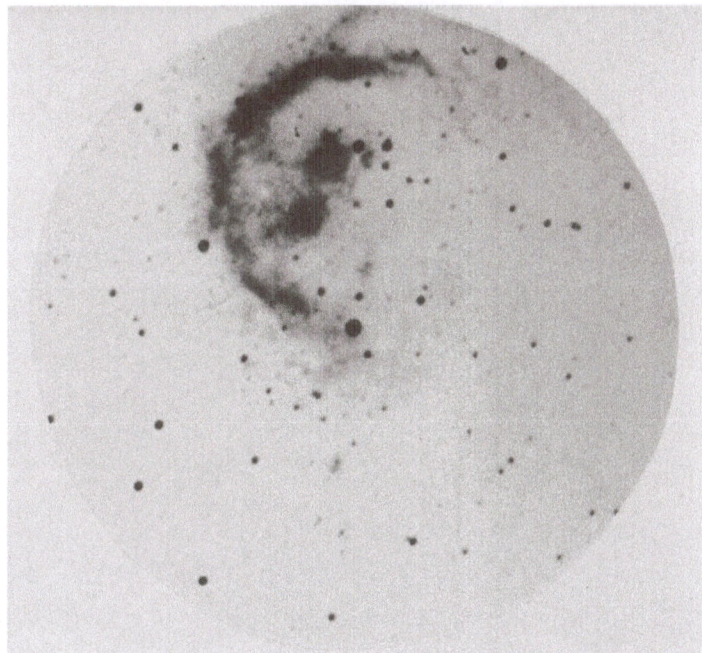

Fig. 1.1.1.　Hα photograph of Barnard's Loop (BL) taken with a Westinghouse WL–30677 S25 image tube through a 9 Å interference filter centred on 6563 Å. The emulsion of the plate was Kodak IIIa-J and the exposure time was 2 hr (Courtesy of Dr. P. G. Johnson.)

Fig. 1.1.2.　Hβ photograph of Barnard's Loop (BL) taken with a Nikon F camera at the Tokyo Astronomical Observatory through a 15 Å interference filter centred on 4881 Å. The emulsion of the film was Kodak 103a-O and the exposure time was 5 hr and 25 min (after Isobe, 1978).

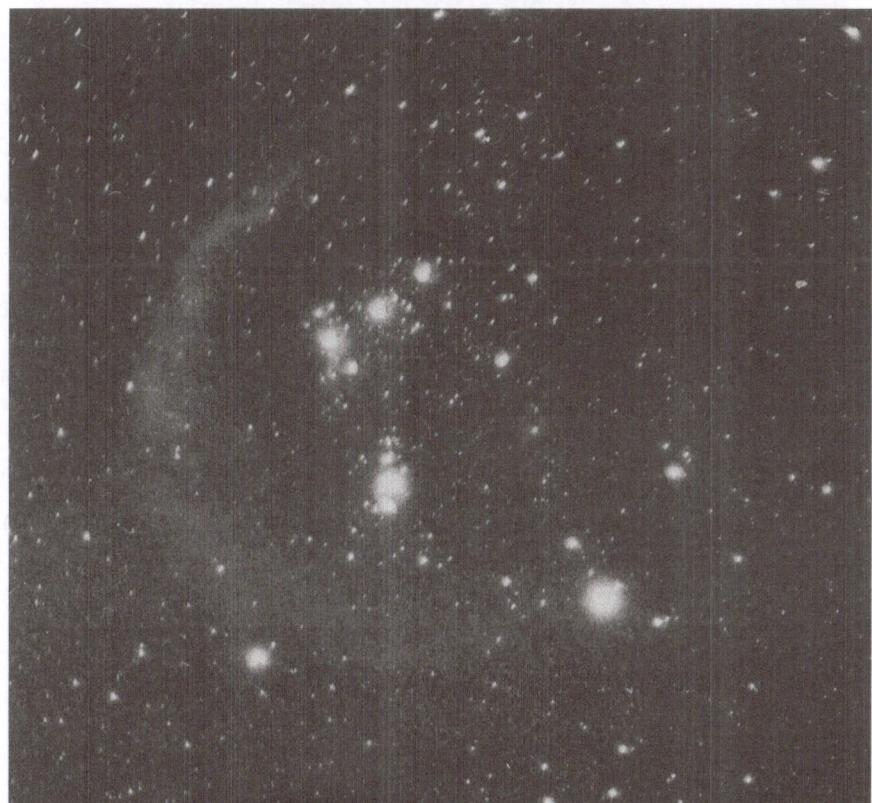

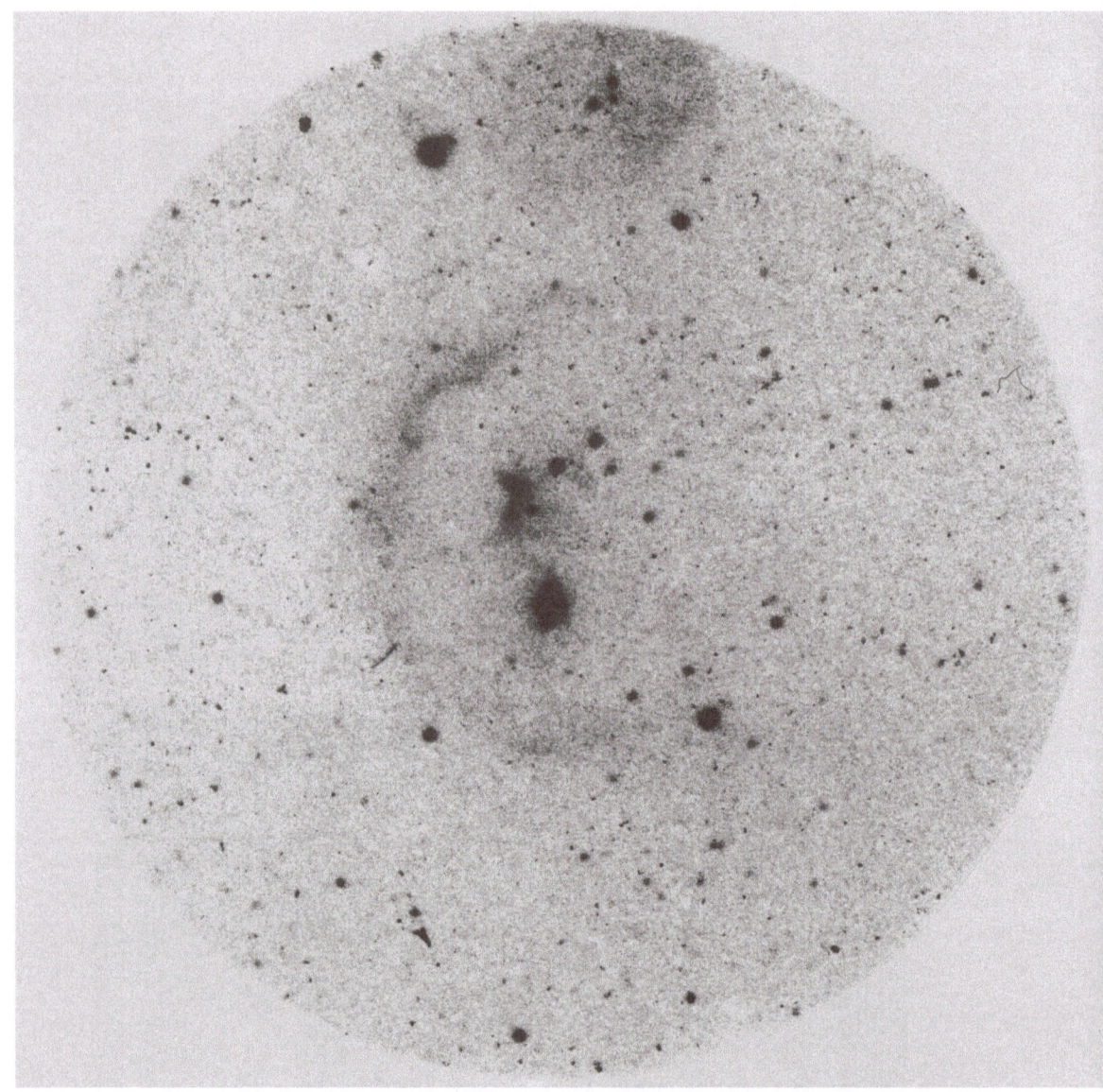

Fig. 1.1.3. [N ɪɪ] photograph of Barnard's Loop (BL) taken with a Westinghouse WL-30677 S25 image tube through a 9 Å interference filter centred on 6584 Å. The emulsion of the plate was Kodak IIIa-J and the exposure time was 2 hr (after Goudis and Meaburn, 1974).

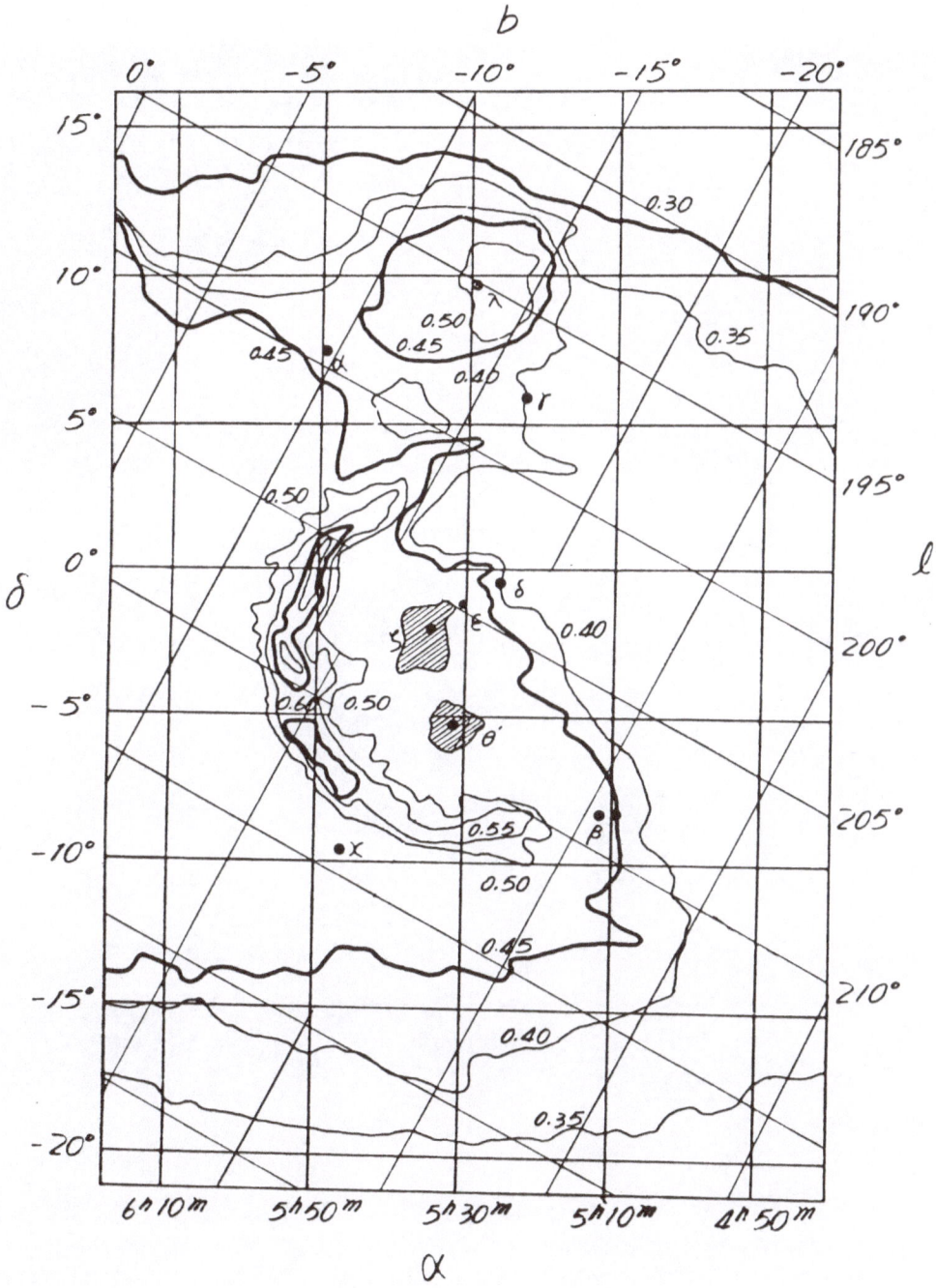

Fig. 1.1.4. Hα contour map of Barnard's Loop (after Isobe, 1973). The nebulosity ∼5° to the north of BL is the λ-Orionis nebula. The region around the star θ¹ Orionis contains the Orion Nebula (M42 + M43) whereas the region around the star ζ Orionis contains numerous H II regions the most prominent of which is NGC 2024.

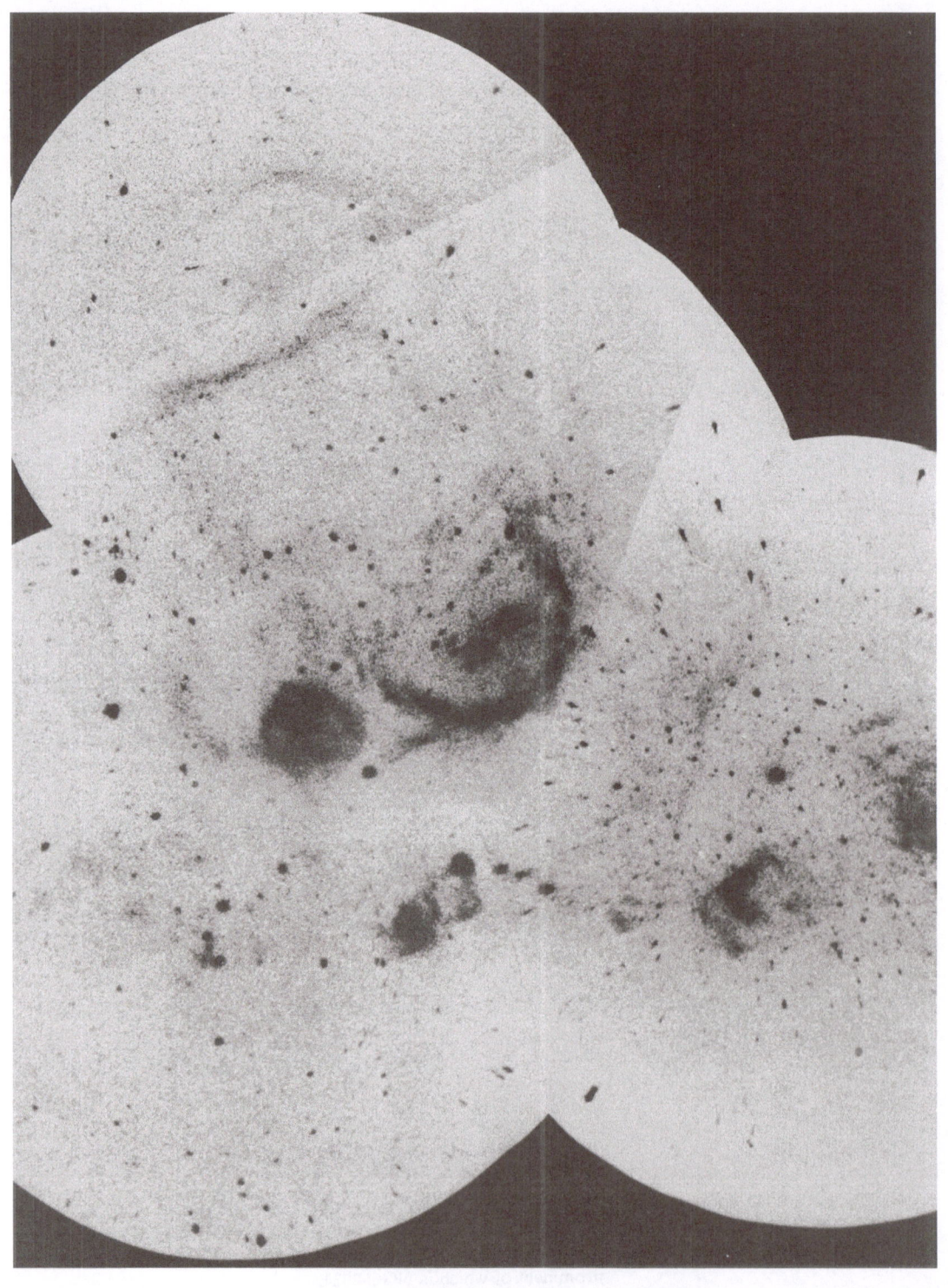

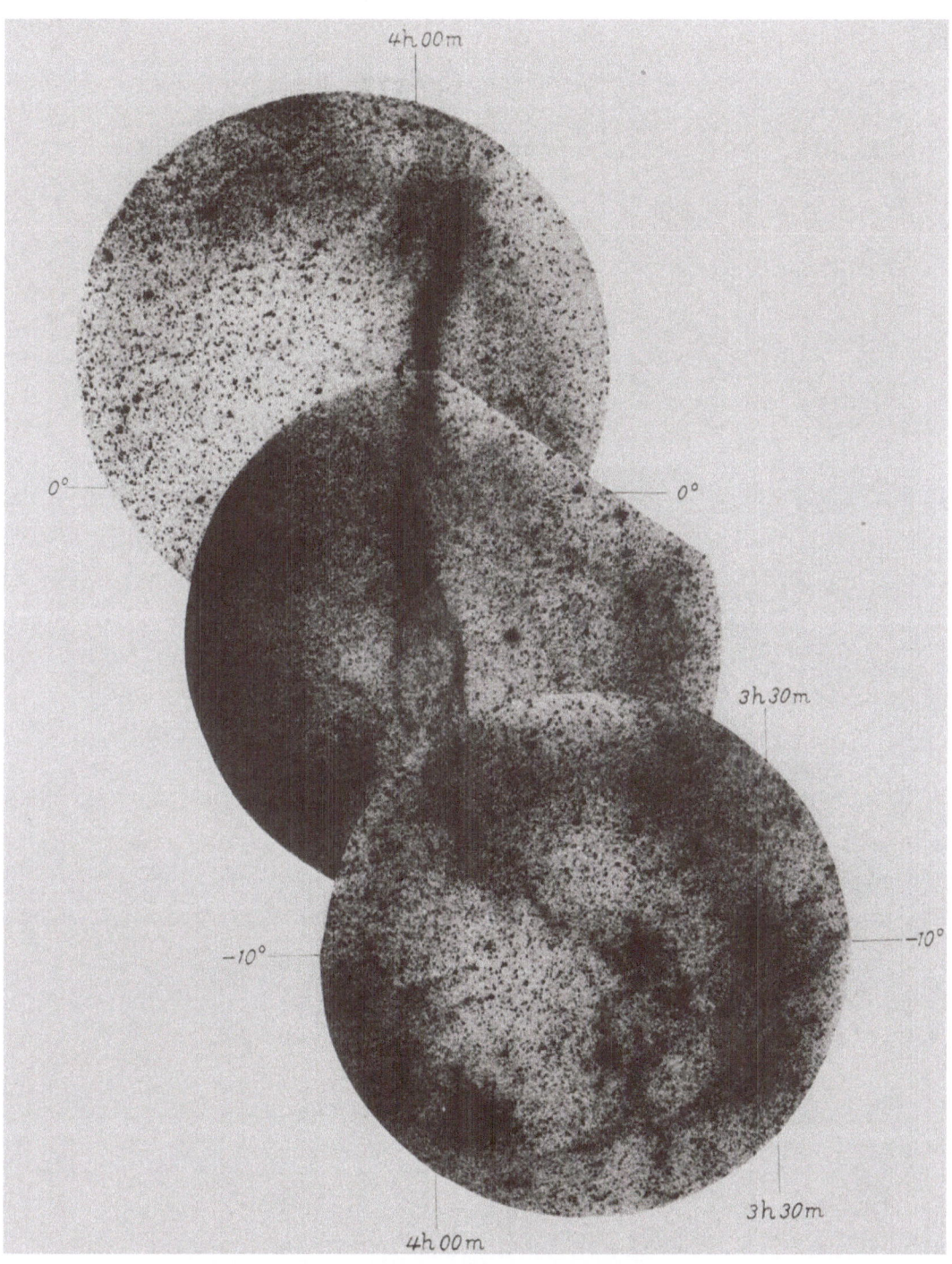

Fig. 1.1.6.   Hα mosaic of the U-shaped faint filamentary structure in the vicinity of Barnard's Loop (see previous Figure 1.1.5) taken with an $f/1$ Schmidt camera through a 40 Å interference filter centred on 6563 Å (after Meaburn, 1967).

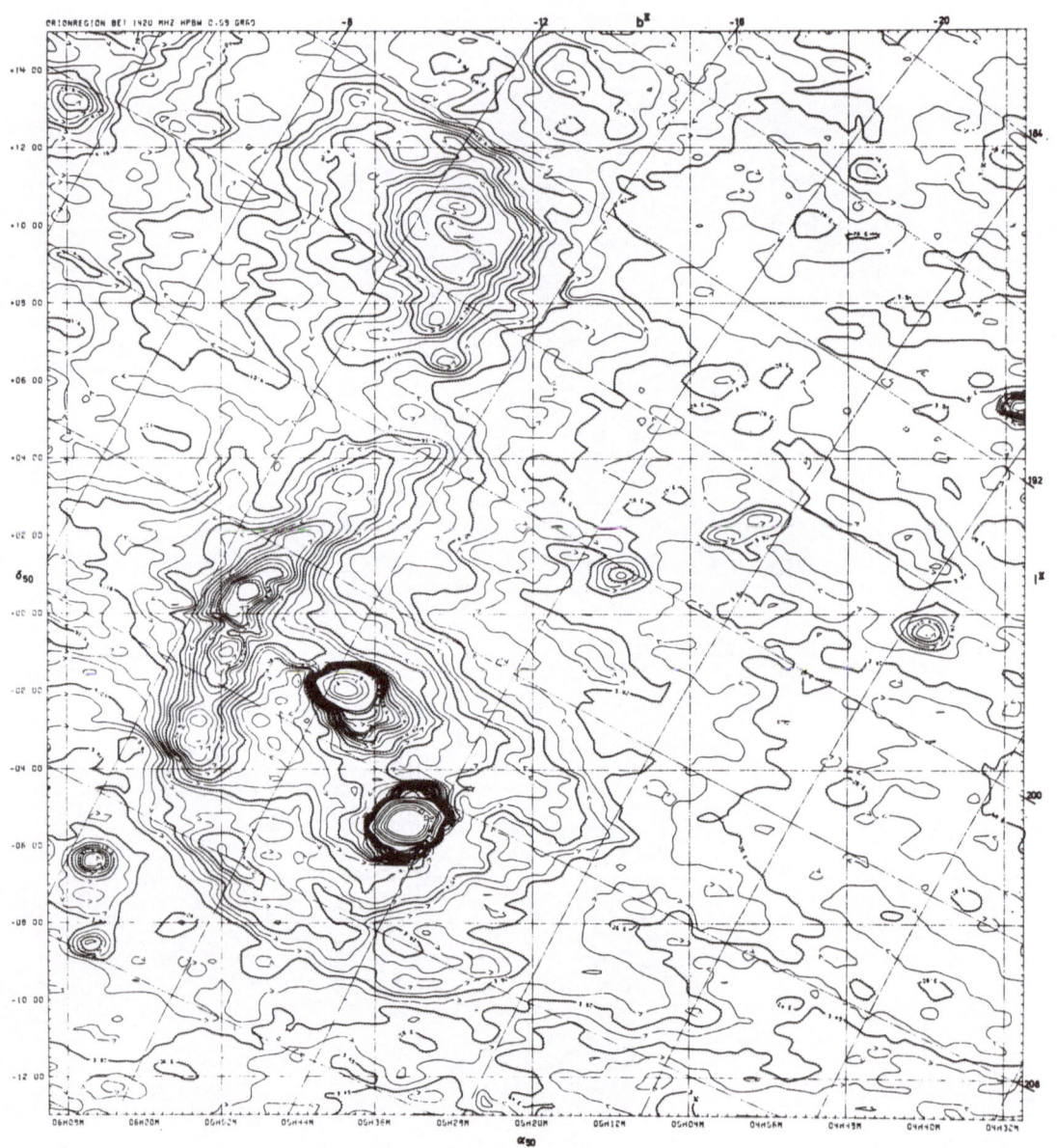

Fig. 1.1.7.   Radio continuum contour map of the Orion Region at 1420 MHz with an angular resolution of 34′ (after Reich, 1978). Note the similarity with the Hα contour map of Figure 1.1.4.

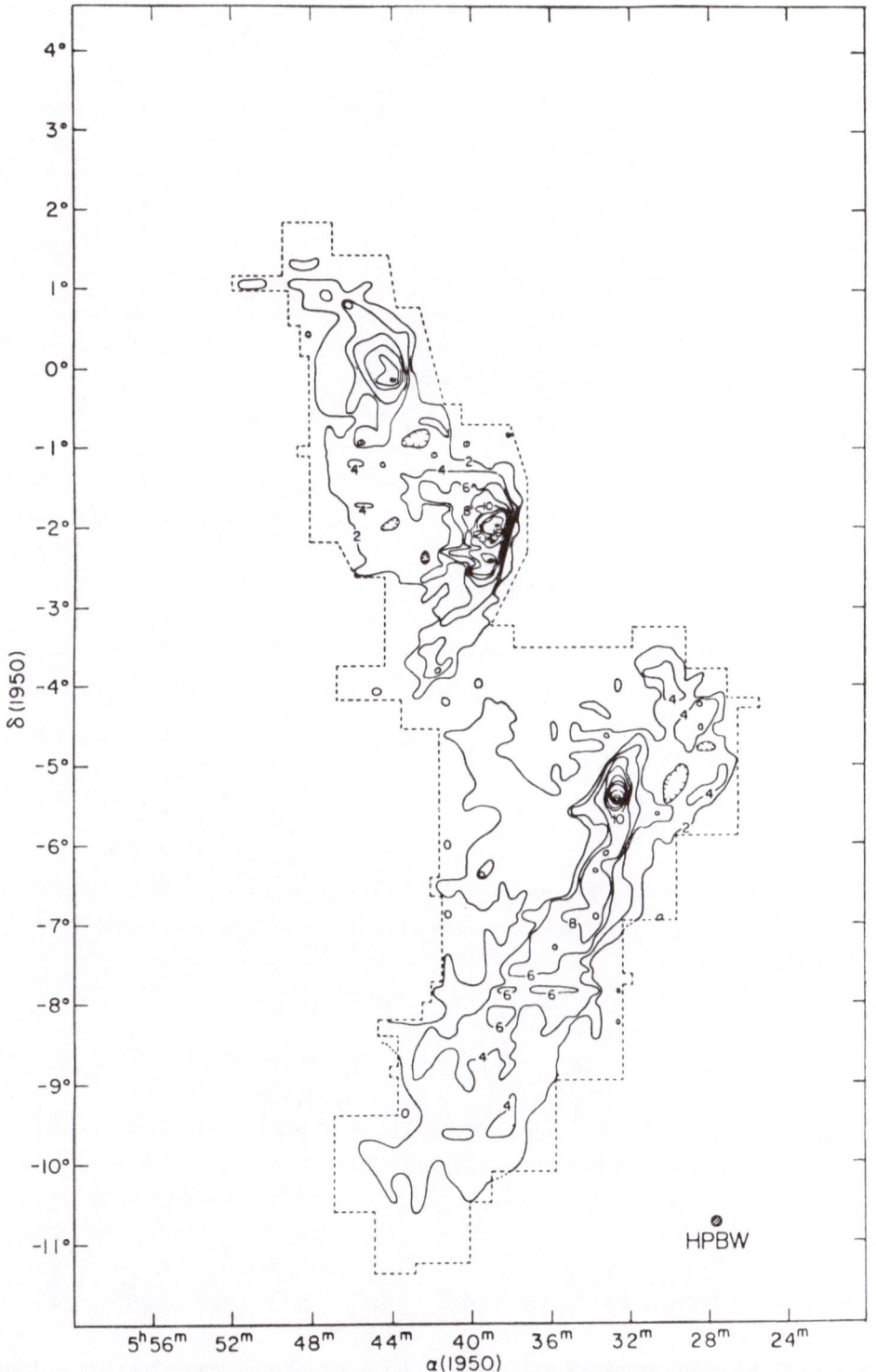

Fig. 1.1.8. ¹²CO contour map of the two main molecular complexes enveloped by Barnard's Loop (after Kutner *et al.*, 1977). The angular resolution is 8′.

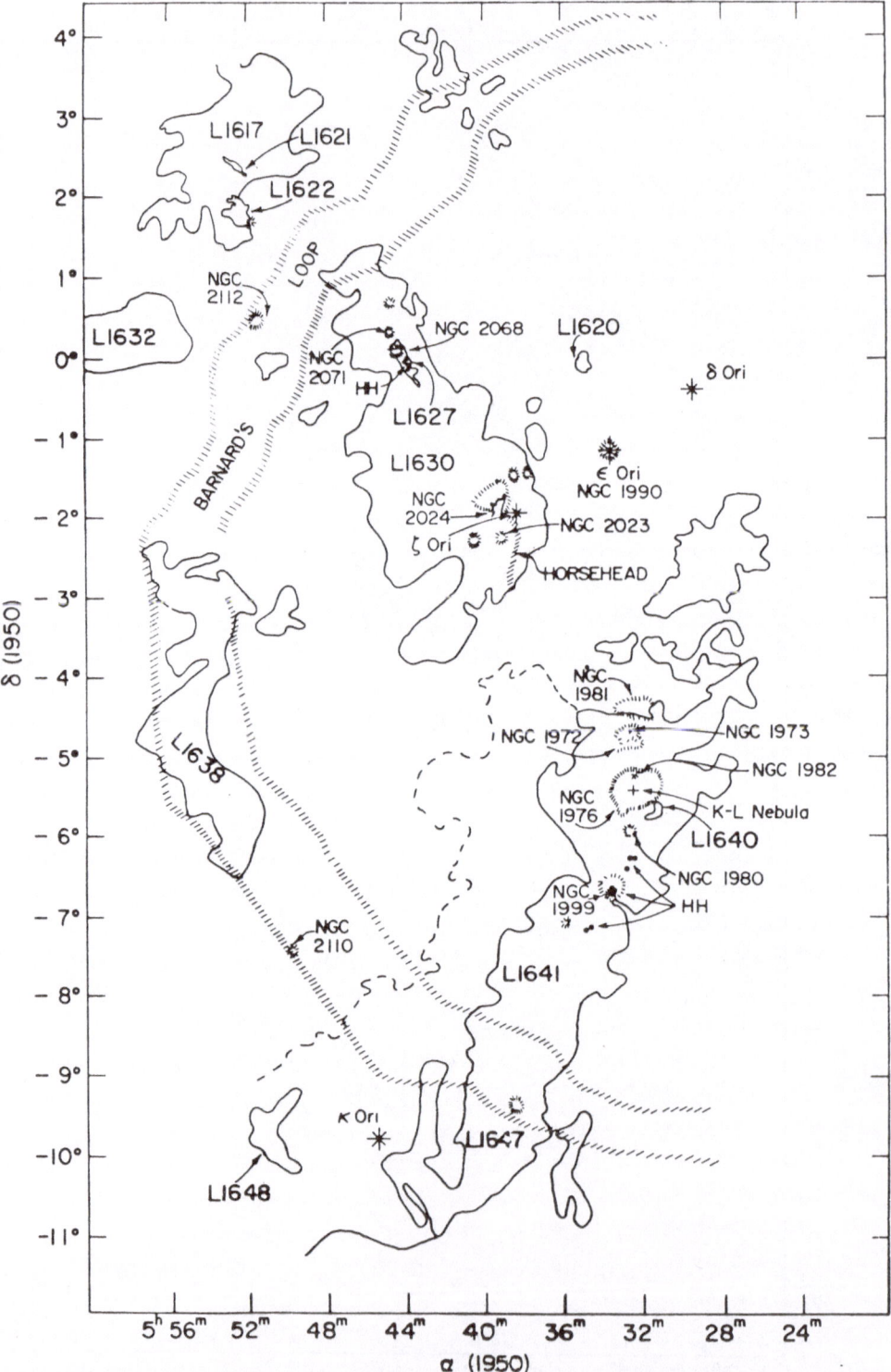

Fig. 1.1.9. Schematic representation of Barnard's Loop and the neighbouring region (after Kutner *et al.*, 1977). Dark clouds are designated by their L number (after Lynds, 1962). Herbig-Haro objects are designated by HH.

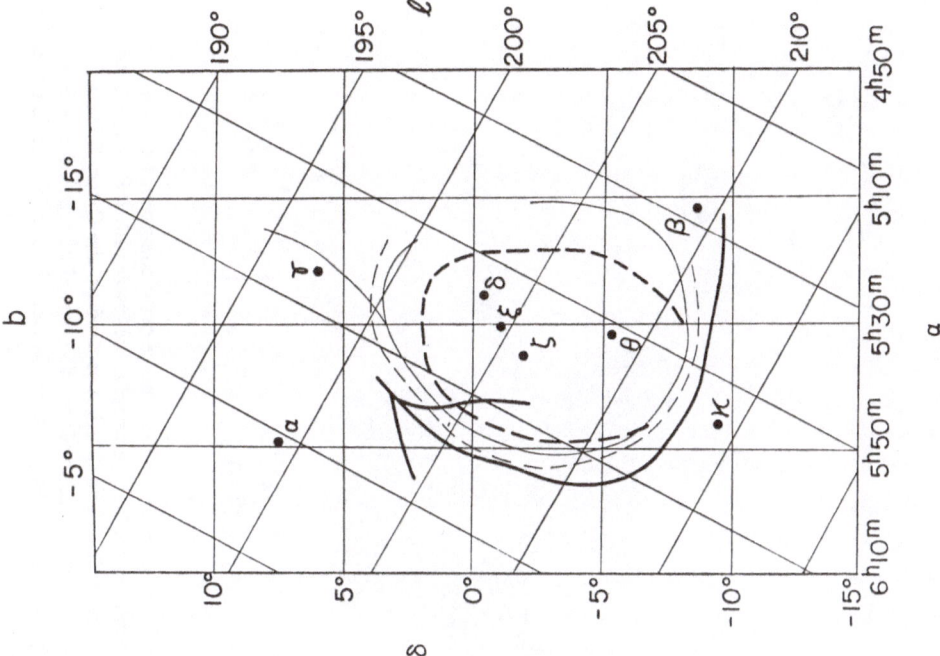

Fig. 1.2.1. Comparative positions of Barnard's Loop as inferred from Hα (thin dashed line), UV (thin solid line), and H I (21 cm) (thick solid line) observations made by Isobe (1973), O'Dell et al. (1967), and van Woerden (1967), respectively (after Isobe, 1973). Only the contours of maximum intensity were used. The picture emerging is that of a neutral hydrogen (H I) shell the inner part of which is ionized (H II loop). The thick broken line indicates the ridge of maximum star density.

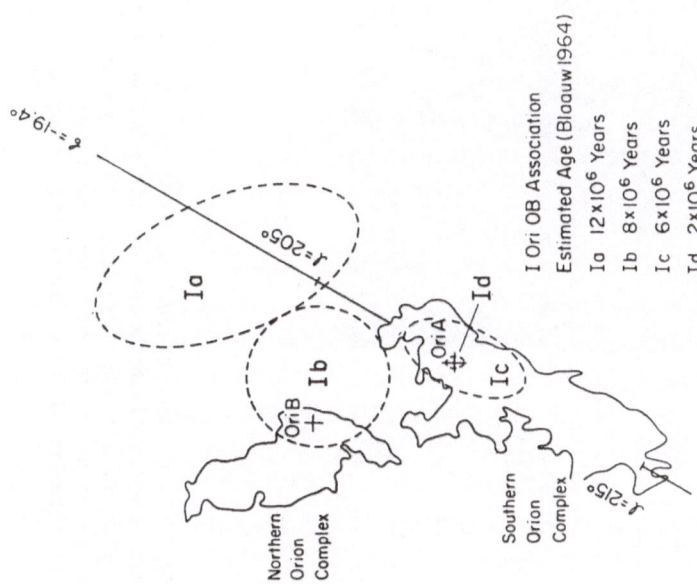

Fig. 1.1.10. Schematic representation of the two Orion molecular complexes and the approximate boundaries of the four subgroups Ia, Ib, Ic, Id of the I Orion OB association (after Kutner et al, 1977).

I Ori OB Association

Estimated Age (Blaauw 1964)

Ia 12×10⁶ Years
Ib 8×10⁶ Years
Ic 6×10⁶ Years
Id 2×10⁶ Years

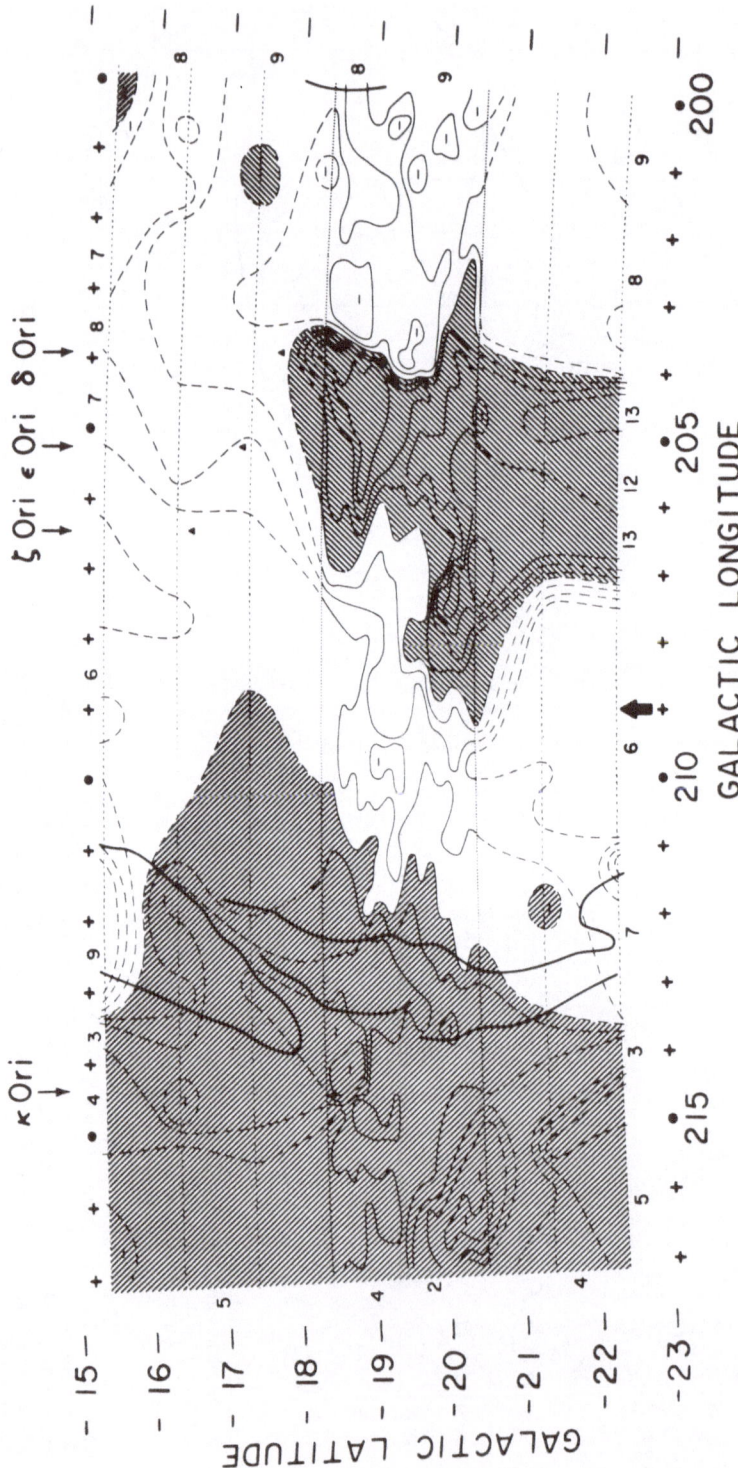

Fig. 1.2.2.   Contours of equal radial velocity of H I associated with the Orion Region (after Gordon, 1970). The velocity component of ~ +12 km s⁻¹ which is associated with the area around $l^{II} \simeq 205°$ indicates recession with respect to the centre of the Nebula whereas the lower velocity component around $l^{II} \simeq 215°$ indicates a velocity of approach. This kinematic pattern was interpreted by Gordon (1970) as rotation centred on the Orion Nebula. The axis of rotation in running east of north and is tilted ~ 35° with respect to the perpendicular to the galactic plane. The period of rotation is ~ 5 × 10⁷ yr. An H I cloud at −10 km s⁻¹ is also present along the line of sight. This may be physically associated with the Orion Region as defined by these H I observations extends at least over 30°.

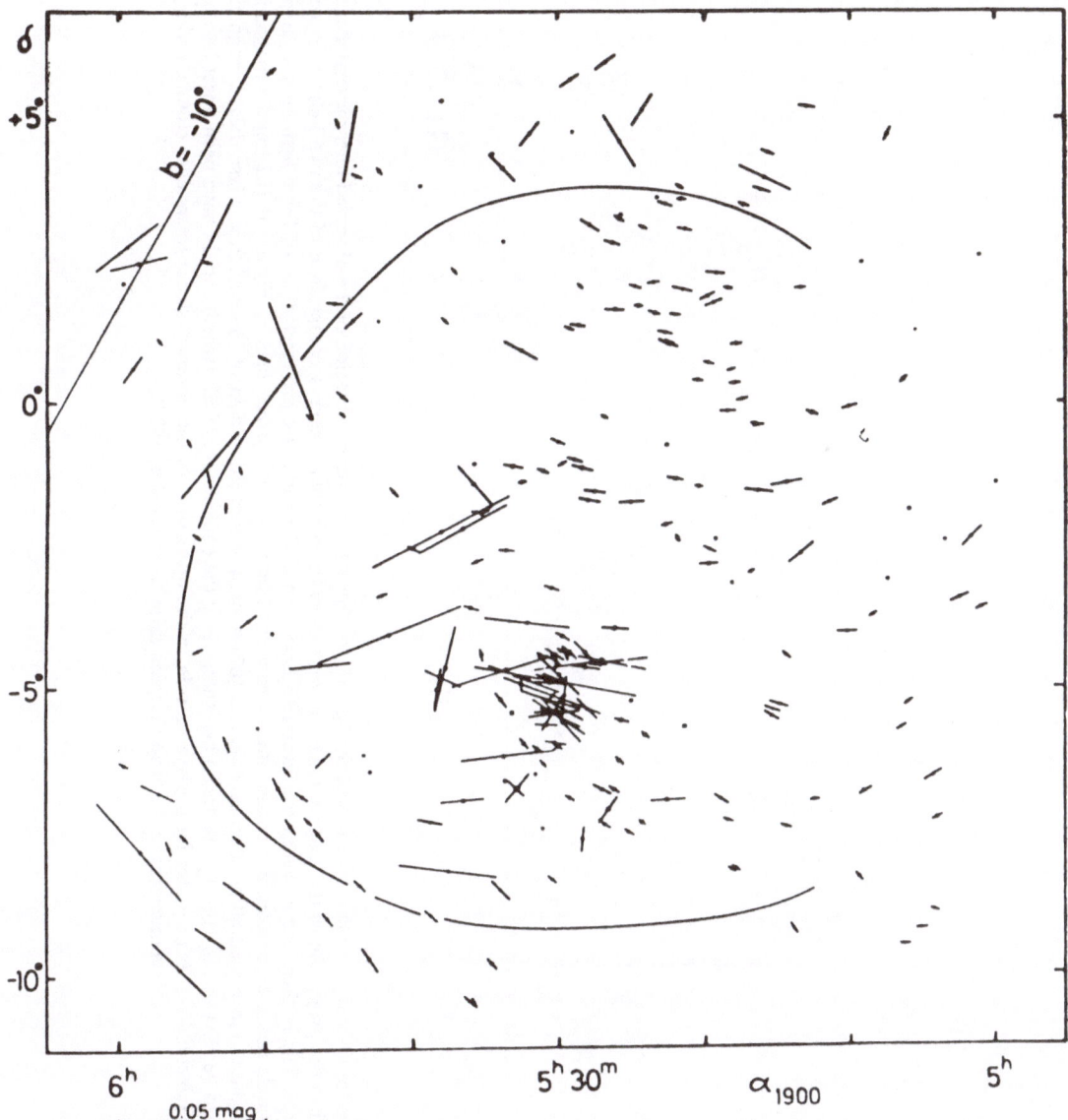

Fig. 1.2.3a. Linear polarization map of the Orion Region (after Appenzeller, 1974). The solid-curve indicates the position of Barnard's Loop.

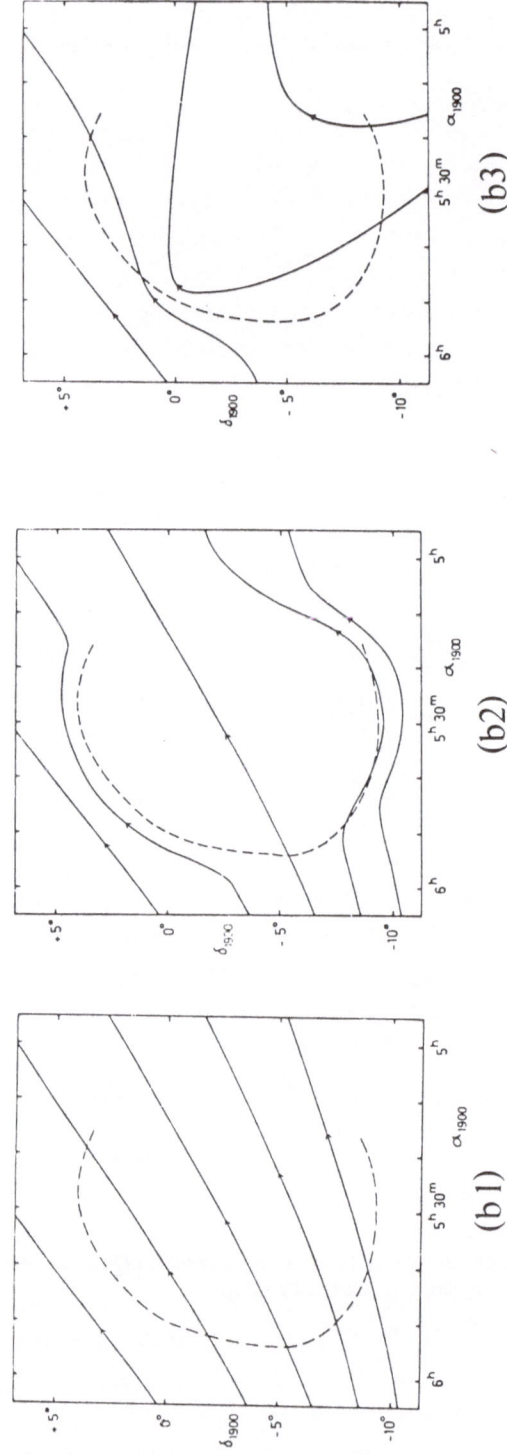

Fig. 1.2.3b.   Projected field lines for several possible magnetic field configurations. Since the planes of vibration of the observed linear interstellar polariza-tion (Figure. 1.2.3a) are parallel to the projection of the magnetic lines of force onto the celestial sphere, a comparison of the two helps to reject certain assumed magnetic field structures. The broken line indicates the Barnard's Loop. Solid lines indicate projected lines of force assuming an undisturbed parallel magnetic field (b1), a simple radial expansion of a conducting gaseous shell into the undisturbed field of Figure 1.2.3a. Case (b3), a magnetic pocket (b3) (after Appenzeller, 1974). Note the incompatibility of model (b1) and (b2) with the polarisation data of Figure 1.2.3a. Case (b3), a magnetic pocket containing Barnard's Loop, might be relevant. If true, it provides an explanation for the equilibrium of BL (which should have been upset because of the gravitational attraction of the Galactic disk). It may also have played a role in the formation of the observed dense interstellar clouds, protostars and young stars in the Orion Region (after Appenzeller, 1974).

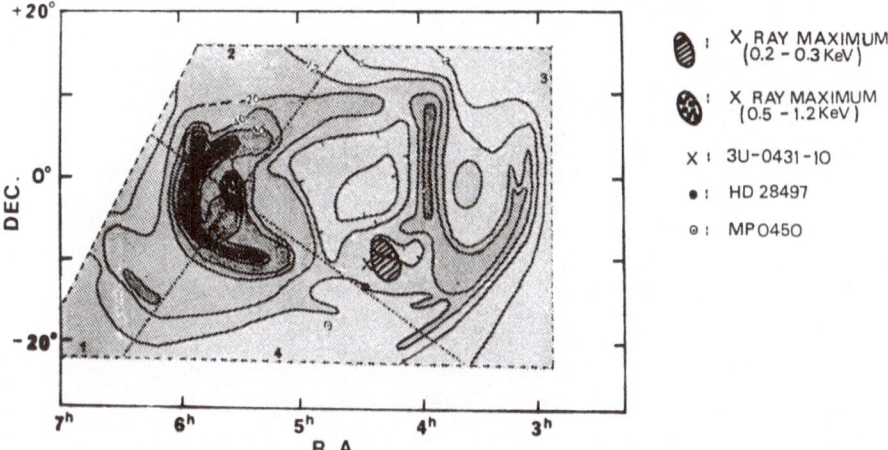

Fig. 1.2.4. Absolute Hα intensity map of the extended Orion-Eridanus region (after Reynolds and Ogden, 1979). The contour units are in Rayleighs (R) ($1R = 2.4 \times 10^{-7}$ ergs cm^{-2} s^{-1} sr^{-1} at Hα). Shown on the map are the soft X-Ray Eridanus 'hot spot' as detected in the 0.2–0.3 keV band (Naranan *et al.*, 1976), the mean hard X-ray Source 3U–0431–10 (Giacconi *et al.*, 1974), the runaway pulsar MP 0450 (Vaughn *et al.*, 1969) which is 10° from the centre and the star HD 28497 (G208.8 + 37.4) against which the O VI line is seen in absorption (York, 1974). Also shown is the X-Ray maximum in the 0.5–1.2 keV band (Nousek, 1978).

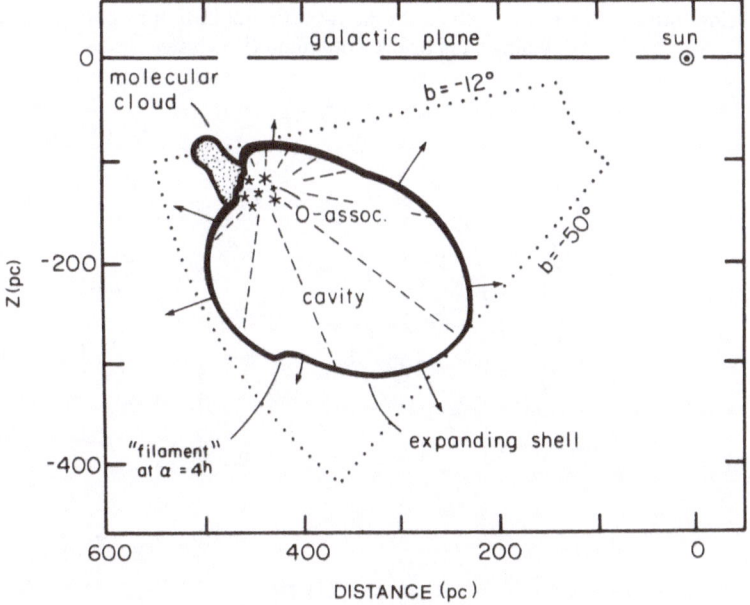

Fig. 1.2.5. Schematic representation of the extended Orion-Eridanus shell (after Reynolds and Ogden, 1979). *Z* is the distance away from the galactic plane. A highly energetic SN explosion (or series of explosions) which occured in the Orion Association may be the cause of the observed shell. Parts of the shell moving towards the galactic centre have encountered denser matter (neutral/molecular interstellar clouds) and have been decelerated, whereas parts of the shell moving to less dense regions away from the galactic plane are still expanding with velocities as high as 30 km s^{-1}. The energetic stellar winds originating from the evolved hot stars of the I Orion OB Association may also have played an important role in shaping the shell. Barnard's Loop is the ionized inner part of the decelerated portion of the shell. The cause of the ionization is probably the hard UV radiation emitted from the OB stars of the I Orion Assocation.

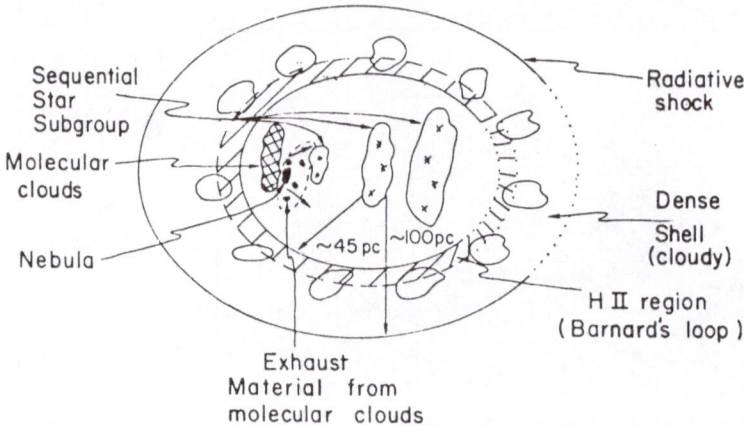

Fig. 1.2.6. Schematic representation of a model of the Orion Region (after Cowie *et al.*, 1979). The radiative shock, which envelopes BL, moves at 100 km s⁻¹ with respect to the centre of BL and is probably the result of a SN explosion which occured as early as 3×10^5 yr ago. Within it, is a denser shell, fragmented into clouds with typical velocities of 10–20 km s⁻¹. This can be the result of older SN explosions occured $(2-4) \times 10^6$ yr ago. The inner part of this shell, becomes ionized by the UV radiation emitted from the OB stars of the I Ori Association (Barnard's Loop). The explosion of the initial SN may have triggered the creation of the older subgroup Ia. The other subgroups were formed in a sequential manner (Figures 1.3.1. and 1.3.2.) by exhausting the material of a vast molecular cloud. The younger subgroups Ic and Id are located near the remnants of the cloud from which new stars are still being formed. The Orion Nebula results from the ionization of parts of this cloud by the young stars of the Trapezium (Id) cluster.

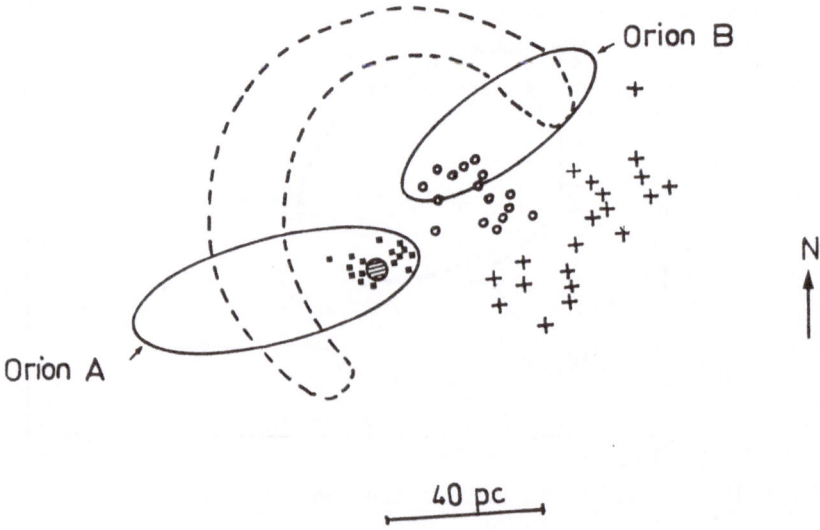

Fig. 1.3.1. Schematic representation of the four subgroups of the I Orion OB Association (after Reeves, 1978). The stars of the subgroups Ia (North–West) Ib (Belt), and Ic (Outer Sword) are indicated by crosses, open circles and filled squares respectively. The stars of subgroup Id (Orion Nebula or Trapezium Cluster) are all within the hatched circle which also contains the H II Region M42 (Orion Nebula) the core of the associated molecular cloud (OMC 1) and the young infrared cluster (IR cluster) which is embedded within it (see Chapters 3 and 4). Barnards' Loop (BL) is marked by the dashed line. The two CO clouds (Northern and Southern Molecular Complex; Figure 1.1.10) are marked as Orion B and Orion A, respectively.

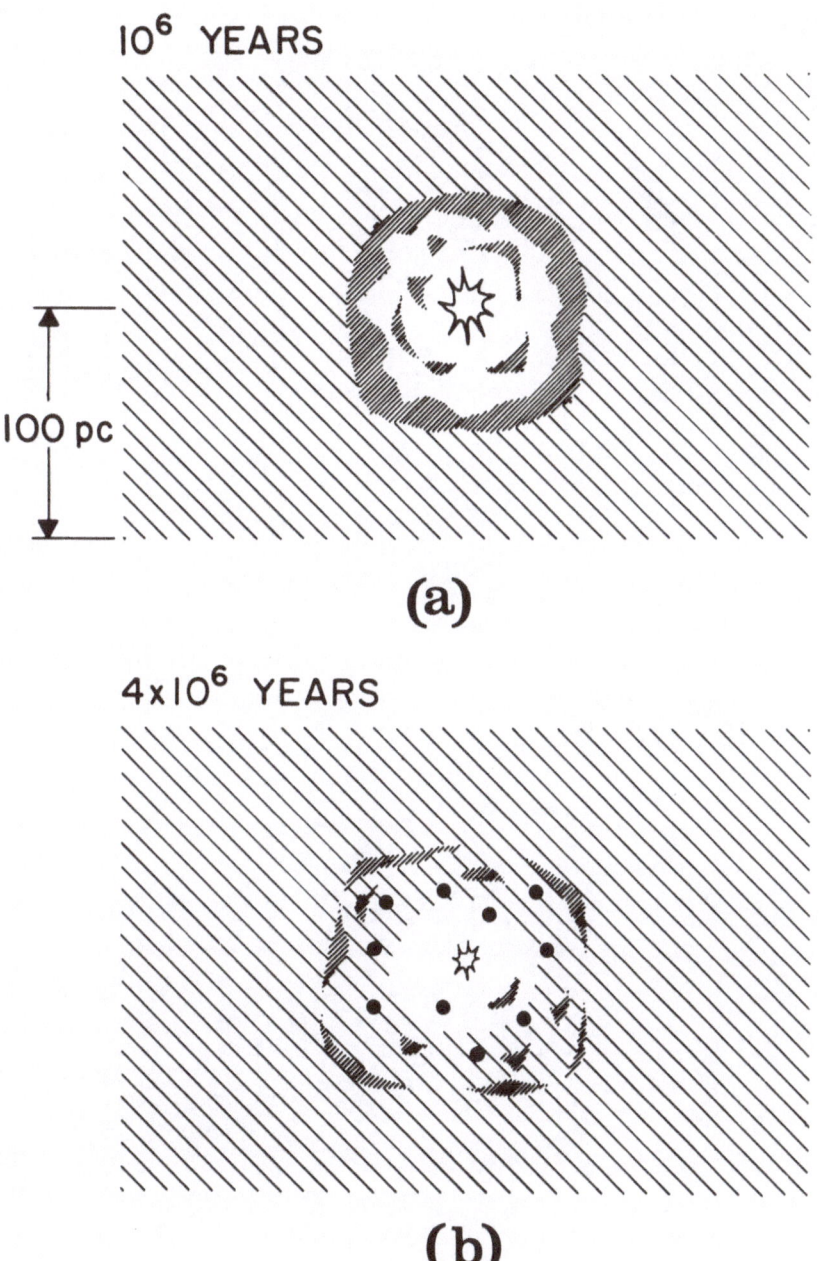

Fig. 1.3.2a–b.

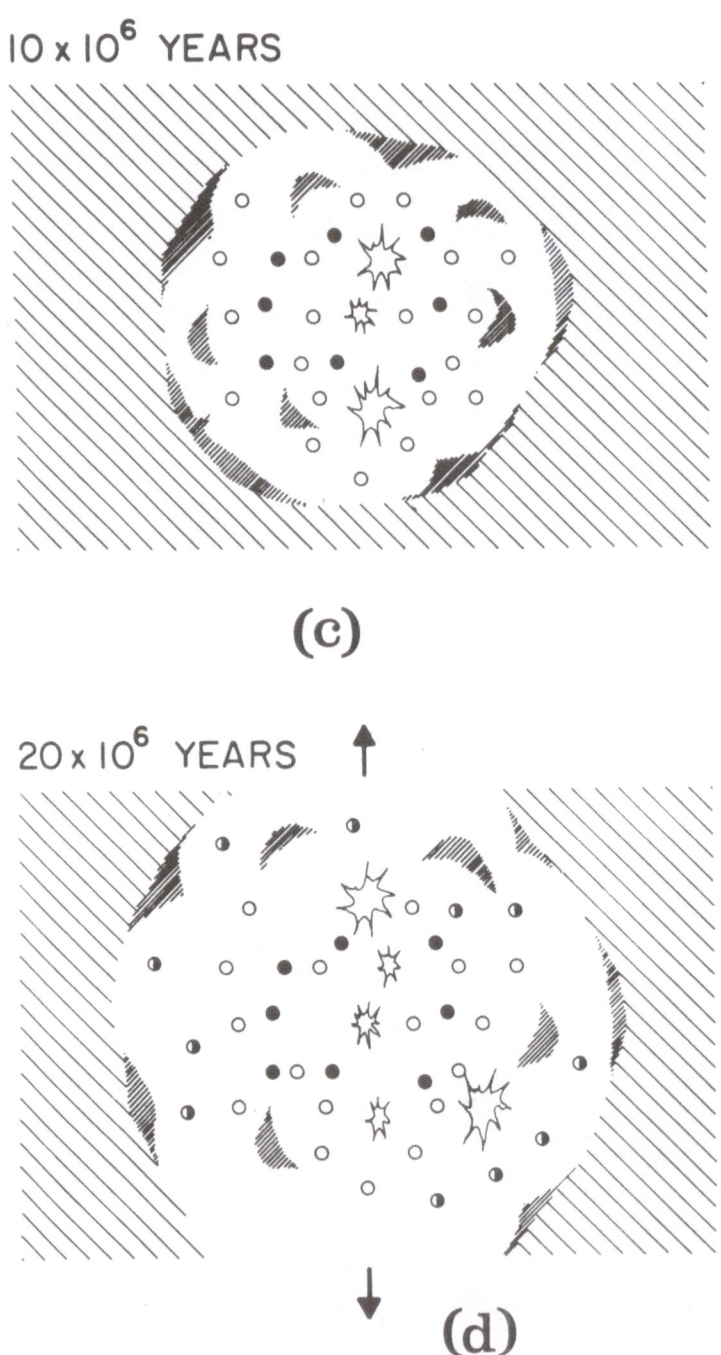

Fig.   1.3.2c–d.

Figs. 1.3.2a–d. Schematic representation of a 'supernova cascade' leading to the formation of an OB Association (after Ögelman and Maran, 1976). The z-direction (distance away from the galactic plane) is vertical in this view. The hatched area represents a 200 pc layer of neutral hydrogen of the galactic disk. A supernova explosion triggers a sequence of events which can be briefly described in the following four stages:

(a) 10^6 yr *after the initial explosion:* a 100 pc in diameter, dense, inhomogeneous, fragmented shell has been created by the explosion. The cavity contains a hot ($\sim 10^6$ K) tenuous (10^{-2} atoms cm^{-3}) gas.

(b) 4×10^6 yr *after the initial explosion:* new stars are created from the fragments of the shell. Ten of them with $M > 10 M_\odot$ have reached the main sequence (filled circles).

(c) 10×10^6 yr *after the initial explosion:* two stars with $M > 20 M_\odot$ have already reached the stage of SN at $t = 6 \times 10^6$ yr; their explosions have expanded the cavity and formed approximately 20 new OB stars (empty circles).

(d) 20×10^6 yr *after the initial explosion:* two second generation OB stars with $M > 20 M_\odot$ have reached the stage of SN at $t = 1.2 \times 10^7$ yr and have produced a third generation of OB stars (half filled circles). The cavity (~ 200 pc in diameter) breaks out of the galactic disk ejecting high velocity clouds and loops.

A modification of this scenario, by assuming an initial supernova explosion on the edge of a neutral molecular cloud, can possibly account for the events leading to the formation of the subgroups of the I Orion OB Association and the extended shell observed around them. Some evidence supporting the occurrence of more than one supernova explosions in the Orion Region is available (e.g. Cowie *et al.*, 1979). The filamentary structure of this part of the shell which moves away from the denser areas of the galactic centre (Figures 1.1.5 and 1.2.5) can also be explained along these lines (ejected loops in phase (d)).

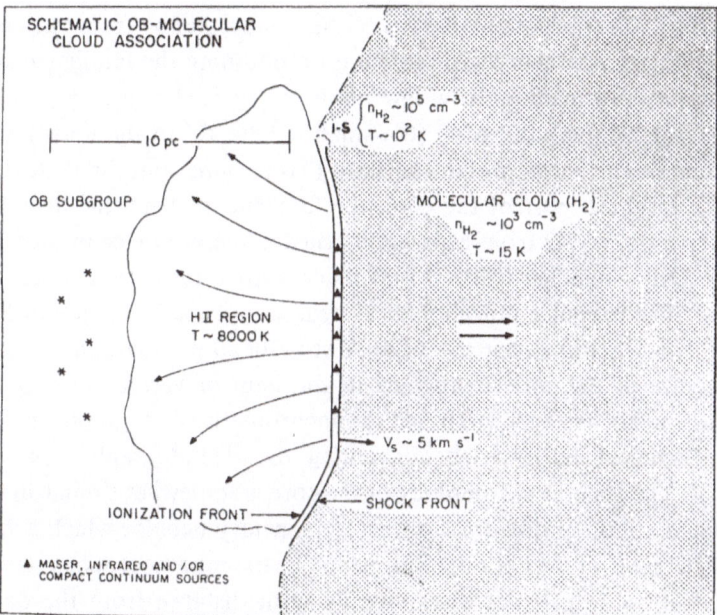

Fig. 1.3.3. Formation of an OB subgroup of stars from a molecular cloud (after Elmegreen and Lada, 1977). An OB cluster on the edge of a large molecular cloud emits ionizing radiation (Lyman continuum) which drives an ionization front (*I*-front) into the cloud, leaving behind an H II region. The *I*-front is preceded by a shock front (*S*-front). The dense neutral material sandwiched in the thin layer between the $I - S$ fronts becomes gravitationally unstable, a situation which eventually leads to the formation of an OB subgroup of stars. These stars, although initially obscured from our view, become apparent from the presence of H_2O and OH masers, infrared and compact radio continuum sources. In the subsequent evolution, the newly formed OB stars move out of the thin layer between the $I - S$ fronts and eventually destroy their birthplace. UV radiation emitted by the newly formed OB stars creates a new H II region; consequently a new $I - S$ front propagates into the molecular cloud initiating the formation of a new OB subgroup.

THE H II REGIONS M42 AND M43

2.1. Optical and Radio Structure

The photographic investigation of the Orion Nebula has a history of almost 100 years: the first photograph of the object was obtained by Anslie Common in 1883. Since then numerous photographs have been obtained by both professional and amateur astronomers all over the world. A compilation of the most recent photographs taken in the light of various emission lines and the continuum are presented in Table 2.1.I. Some relevant details concerning the ionization state of the most important elements from which the emission lines originate are presented in Table 2.1.II. The relative and absolute strengths of the most important lines are presented in Table 2.1.IIIa and 2.1.IIIb, respectively.

In addition a selection of the best published photographs from various observers are presented here. A schematic representation outlining the familiar features of the nebula as appears in many narrow line photographs is shown in Figure 2.1.1. Both M42 (the Great Nebula) and M43 (the small nebula 10' to the north) are included. The positions of the most prominent stars (Trapezium stars or θ^1 A, B, C, D and θ^2 A, B), the brightest part of the core, the filament (or bar), the dark bay and the dark lane separating M42 from M43 are marked. An absolute calibrated Hα contour map of the Orion nebulae (M42, M43) made with a resolution of 1.'5 is given in Figure 2.1.2. The irregular morphological structure of the 3'–5' core of M42 is highlighted in a Hα + [N II] $\lambda\lambda$ 6563 + 6584 Å broadband photograph in Figure 2.1.3.

The structure of the core as appears in the light of various lines is shown in a sequence of some of the best published monochromatic photographs in Figures 2.1.4 to 2.1.10. Technical information concerning these photographs can be found in Table 2.1.IV. The general morphology of the core is somewhat similar in the different monochromatic photographs (in contrast to planetary nebulae which exhibit various well stratified states of excitation). This is due to the inhomogeneous structure of the gas, the density of which decreases sharply with distance from the density centre (Wurm and Rosino, 1959). Subtle differences between certain features are however discernible, and these play an important role in the understanding of the physical processes governing this complex region. In this respect the bar appears stronger in the low excitation [N II], [O II], [S II] lines than in the higher excitation [O III] line (compare Figure 2.1.5 and Figure 2.1.7). This in relation with other factors ([O III] zone and velocity splitting lying on the side towards the Trapezium, association of the bar with dust/neutral material), indicate that the bar is the manifestation of an ionization front powered by the θ^1 Orionis stars (Courtès, 1968; Elliott and Meaburn, 1974). Another difference is apparent from the comparison of the photographs taken in the light of the permitted O I λ 8446 Å and the forbidden [O I] λ 6300 Å lines. The O I λ 8446 Å emission exhibits a unique filamentary structure not seen in the [O I] λ 6300 Å emission (see Figure 2.1.8 and 2.1.9). In addition, both lines exhibit a velocity redshifted by about 10 km s^{-1} with respect to the other nebular lines. This implies that the gas emitting these lines do not originate in the same region as that emitting the other lines. The filamentary structure of the O I λ 8446 Å emission is probably due to density fluctuations in the transition zones between the H II and H I regions where the gas and grain content is thought to be higher than average (Münch and Taylor, 1974). A morphological difference is also exhibited by the emission line He I λ 10830 Å. This can be seen in Figure 2.1.11 and resembles to a certain extent the [O III] structure of the core of the nebula (Figure 2.1.7). The observed He I λ 10830 Å line is unique in the sense that it behaves like a resonanse line and, therefore, is subject to radiative transfer effects.

The implications of the morphological investigation of the core of Orion are summarized in the sketch of Figure 2.1.12. The optical manifestations of the physical phenomena are thought to be viewed tangentially by the observer. The bar is probably an ionization front viewed edge on, powered by the UV radiation of the Trapezium stars and propagating into the neutral/molecular complex surrounding the area. The higher energy photons emitted by the Trapezium stars are capable of ionizing trace elements such as O^0 and S^0 to the higher ionization states O^{++} and S^{++}, from which the [O III], [S III] emission originates. At greater distances the remaining lower energy photons are only capable of ionizing such elements to the first ionization state (O^+, S^+, N^+), from which the [O II], [S II], [N II] forbidden line emission originates. This emission delineates more distinctly the ionization front since its presence is predominantly confined in and near the H I–H II interface, and not in the immediate surroundings of the exciting stars which are dominated by the emission of the highly ionized element. This interpretation is supported by the predicted variation of the volume emissivities of the [O II], [N II], and [O III] lines as a function of the distance from the ionization front, shown in Figure 2.1.13.

The morphology of the outer regions of the Orion Nebula is dominated by a complicated filamentary structure which is beautifully illustrated in Figure 2.1.14. This is a red photograph of both M42 and M43, printed through an unsharp mask in order to enhance the very faint filamentary structure surrounding the objects.

The less spectacular M43 nebula resembles a typical spherical H II region and exhibits no striking morphological features. A photograph of the core of M43 in the light of Hα + [N II] is shown in Figure 2.1.15.

The radio continuum structure of M42 and M43 has been the subject of many radioastronomical investigations in the last decade. A compilation of the numerous contour maps published by various workers, at different wavelengths and with ever increasing angular resolution, is presented in Table 2.1.V. The derived structure is always smeared by the antenna pattern. Earlier results obtained with low angular resolution gave the impression of smooth spherical structures for both M42 and M43. A selection of radio continuum contour maps of both H II regions made with intermediate angular resolutions (2'–3') are shown in Figures 2.1.16a, b, and c. The emission from both objects is thermal as can be seen from their radio continuum spectra shown in Figures 2.1.17 and 2.1.18, respectively. Recent radio maps made with high angular resolution have resolved the core of M42 into smaller components as shown in Figure 2.1.19. The derived structure shows strong similarities with the complicated optical structure of the core of the source. This is clearly demonstrated in Figure 2.1.20.

A different structure, however, appears from the study of the carbon recombination lines which probably originate in a shell or a sheet of predominantly neutral matter (C II zone) partially surrounding the H II region. The carbon becomes ionized by 912–1100 Å photons leaking out of the H II region i.e. photons whose energy is less than the ionization potential of hydrogen (13.6 eV) but greater than the first ionization potential of carbon (11.3 eV; see also Table 2.1.II). The distribution of the C II emission in the central part of the C II region, as derived from the study of the C85a recombination line is shown together with the H II distribution in Figure 2.1.21. The form of the C II emission is nearly elliptical (8' × 6') and is similar to that of the molecular ridge lying behind the H II region (see for example Figure 3.3.5 later in the text). It is in contrast, however, to the nearly circular distribution of the H II emission originating in the less extended 3' core of M42 as seen in the radio continuum contour maps of similar angular resolution (2') The distribution of the C II emission supports the view that the C recombination lines originate in a shell of partially ionized matter which defines the near edge of the dense molecular cloud lying behind the H II region (see Chapter 3.3). Kinematic considerations also support this view (see Section 2.2.).

Finally the X-Ray emission associated with the Orion Nebula region should be mentioned. This originates from the weak X-ray source 4U 0531–05 ≡ 2A 0531–05 (Giacconi et al., 1972). Its position, obtained from observations made from the Uhuru (Forman et al., 1978), Ariel-5 (White and Ricketts, 1977; Cooke et al., 1978), and ANS (den Boggende et al., 1978) satellites suggests a physical association with the nebula. The observed X-Ray emission originates in a rectangular region of 50' × 10' in the centre of the nebula and was considered to be coronal emission ($T \sim 4 \times 10^6$ K) around T Tauri stars within the nebula (den Boggende et al., 1978). Recently, a more concentrated X-Ray source ($< 2' = 0.3$ pc in diameter) located at the Trapezium system of stars was discovered by Bradt and Kelly (1979) at $\alpha = 5^h32^m49\overset{s}{.}2$, $\delta = -5°25'08''$ (with an error radius of 35''). These authors have suggested that both the extended and compact sources may originate in high temperature regions ($T > 10^6$ K) shocked by stellar winds; for detailed discussion of the various probable mechanisms see Ku and Chanan (1979) and references therein.

2.2. Physical Parameters Derived from Radio And Optical Observations

A brief outline of the methods currently used to derive the physical parameters of the ionized gas from observations of the radio continuum emission is presented in the Appendix, Section (A.II.1), of the present work. The physical parameters of the Orion Nebula (M42) and the nearby M43 nebula derived from radio continuum observations are presented in Tables 2.2.I and 2.2.II, respectively. These include the angular (θ) and linear ($2R$) diameter of each source at distance D from the Earth, the root mean square electron density (rms N_e), the mass of the ionized gas (M H II), the emission measure (E), the excitation parameter (u), and the rate of the Lyman continuum photons (Lc) necessary to the ionization of the observed source. The angular resolution of the original observations from which these parameters were derived can be found for the majority of the references, in Table 2.1.IV together with the respective radio continuum map produced.

The relevant information concerning the stars thought to play a significant role in the ionization of M42 (θ^1, A, B, C, D; θ^2 A, B) and M43 (NU Orionis) is presented in Tables 2.2.III, 2.2.IV, and 2.2.V.

The excitation parameters of these stars can be derived from the information concerning their spectral type and luminosity class given in these tables (see Appendix A.II.1, Figure II.1.2). From such an estimate it can be concluded that the energy requirements for the ionization of M42 and M43 are matched by the radiation output of stars θ^1 Ori C and NU Orionis, respectively. This is however a somewhat crude method since the function of stellar excitation parameter versus stellar type and luminosity class is not accurately known (see Figure II.1.2 in Appendix A.II.1). In addition, the assumption of an ionization bounded nebula, which is necessary for the application of the method, is not fulfilled in the case of M42 whose front face is density bounded (see comments in Appendix A.II.1).

Estimates of the local electron density of M42 as derived from optical observations are presented in Table 2.2.VI. The relevant methods are briefly discussed in Appendix A.II.2. The variation of density with distance from the centre as derived from optical and radio measurements is shown in Figure 2.2.1. The density centre lies about 30″ to the west of the Trapezium group of exciting stars, as can be seen from the distribution of the radio continuum emission (e.g. Gordon, 1969; Schraml and Mezger, 1969) and more clearly from the study of the distribution of the [O II] ratio (Elliott and Meaburn, 1973; denstity peak at 15″S, 2^s W from θ^1 Ori C). The observed density distribution in comparison with the predicted density distribution, as derived from a spherically symmetric variable density model, is also shown in Figure 2.2.1. The density gradient is an aspect of the model introduced by Brocklehurst and Seaton (1972) to interpet the observed shape of the radio recombination lines. According to previous studies by Hjellming and Churchwell (1969); Hjellming *et al.* (1969), Hjellming and Davies (1970), and Hjellming and Gordon (1971), the recombination lines are produced in compact knots (condensations, clumps) with $N_e \sim 10^4$ cm^{-3},

$T_e \sim 10^4$ K and a linear size of ~ 0.1 pc which are embedded in a more tenuous extended gas. This idea was in line with the model suggested a decade earlier by Oster-brock and Flather (1959) to account for differences in the values of electron density derived from optical methods (giving local densities of optically visible inhomogene-ous features or condensations of ionized matter) and the root mean square density derived from radio observations (giving the average density of the whole volume occupied by these condensations). According to this concept only part of the volume which appears as an H II region is occupied by ionized matter, in the form of condensa-tions. A crude estimate of the fraction of the volume occupied by these ionized con-densations is expressed by the 'condensation' or 'filling factor' $\alpha = ($rms N_e/local $N_e)$ (see Appendix A.II.2) which for the Orion Nebula lies in the range of 3–5 % (Dopita *et al.*, 1974). This interpretation however has not been accepted by all. Chaisson (1976) and Chaisson and Dopita (1977) have pointed out that in Orion only 4 % of the total radio continuum flux originates in the two small clumps which are the only ones detected with $N_e > 10^4$ cm^{-3} by the high resolution observations of Webster and Altenhoff (1970). The bulk of the continuum seems to originate in a reasonably uniform medium with $N_e \sim 10^{3.3}$ cm^{-3}. This is an argument also used to question the validity of a subsequent model introduced by Lockman and Brown (1975), which adopts a very compact central core $(N_e \sim 10^{4.5}$ cm^{-3}; Region I) a Gaussian core source $(N_e \sim 10^{3.5}$ cm^{-3}; Region II), and an extended envelope of emission $(N_e \sim 10^{2.3}$ cm^{-3}; Region III). The parameters adopted are included in Table 2.2.I. The model predicts a temperature increase towards the edge of the nebula which is itself a controversial issue (see Table 2.2.XI). More recently another variable density model has been introduced by Shaver (1980). This is an isothermal model $(T_e = 8200$ K) consisting of a 0.13 pc core with rms $N_e = 7 \times 10^3$ enveloped by four lower density outer regions. It should be noted that despite their inadequacies these models have managed to mimic progressively more and more of the observed characteristics of both the radio recombination lines and continuum (i.e. flux density, T_L/T_C ratios, shape of the line profiles, variations of line width with the principal quantum number n). In addition the presence of inhomogeneities in Orion cannot be doubted (see for example Figure 2.1.3 where structures down to 1″ or 0.002 pc are visible) and a density versus distance gradient is not inconsistent with the available observational evidence (see Figure 2.2.1). It is the role these inhomogeneities play and the theoretical problems they impose which still remain unsolved. For instance, their lifetime should be $\sim 10^{3.5}$ yr which is far less than the average nebular lifetime of 10^5 yr (Chaisson, 1976). A summary of the sizes and densities of the most typical features associated with M42 is presented in Table 2.2.VII. The numbers are rounded and the termino-logy used is necessarily descriptive since the exact physical nature of some of the features is not known. Additionally the small radio structures could well be artifacts of the relatively low angular resolution, and finer structure cannot be ruled out. The optically visible minute condensations and knots will be discussed in detail in Section 2.3.

Estimates of the temperatures of M42 and M43 from various optical and radio observations are presented in Table 2.2.VIII, IX, and X. The relevant methods are briefly outlined in the Appendix, Section A.II.3. The problem of the possible existence of a temperature gradient over the nebula is still open. This is better demonstrated in Table 2.2.XI, where the most important results relevant to the debate are presented.

Information concerning the chemical composition of M42 and M43 is presented in Tables 2.2.XII, XIII, and XIV (singly ionized helium abundance derived from optical and radio methods), and Tables 2.2.XV, XVI, and XVII (ionic and total abundance of various elements present in M42). For further details see the captions to the tables. The main methods employed are briefly described in Appendix A.II.4.

Information concerning the kinematic structure of M42 and M43 is presented in Tables 2.2.XVIII to 2.2.XXI. Tables 2.2.XVIII and 2.2.XIX give measurements of the radial velocities of M42 derived from optical and radio recombination lines respectively. From Table 2.2.XIX it can be seen that the C II radial velocity ($\sim +9$ km s^{-1}) differs distinctly from that of H II and He II (~ -5 km s^{-1}) but is similar to that derived from the various molecular lines originating from the molecular cloud lying behind the H II region (see Section 3.2 and Table 3.3.VII later in the text). This is additional evidence supporting the view that the C recombination lines originate in a zone of partially ionized matter which is the interface between the H II region and the neutral/molecular complex lying behind (see previous Section 2.1). Table 2.2.XIX also includes kinematic information derived from the H I (21 cm) line seen in absorption in the direction of the Orion nebula. The line is probably due to the presence of foreground H I clouds along the line of sight. Part of the H I is believed to be physically associated with the Orion H II region/molecular complex (see further, Section 2.4 and Figure 2.4.9). Further points concerning the kinematic information derived from studies of the various H, He, C, and Z (unknown heavy element) recombination lines and the H I (21 cm) absorption line are included in the commentary accompanying Table 2.2.XIX. Information about the widths (full widths at half maximum intensity) of the various lines are included in Table 3.3.IX later in the text (Chapter 3) together with the widths of the various molecular lines. From this table it can be seen that the width of the carbon recombination lines is very narrow (~ 4 km s^{-1}) compared to that of the hydrogen lines (total width due to thermal and turbulent broadening ~ 25 km s^{-1}, width due to turbulence ~ 15 km s^{-1}) but similar to the width of the narrow molecular lines which are characteristic of the extended molecular ridge ($\sim 3' \times 6'$) lying behind the H II region. This implies that the C II lines originate in a cold region ($T_e \simeq 50$–100 K; Boughton (1978)), which is consistent with the idea of an interface zone between the hot H II region ($T_e \simeq 10^4$ K) and the cold neutral/molecular complex ($T \simeq 50$–70 K; see Chapter 3) associated with it. In the optical regions [C I] lines have been recently observed, which if excited by electron collisions require typical H II conditions ($T_e \simeq 10^4$ K, $N_e \simeq 10^{3.5}$ cm^{-3}) (Hippelein and Münch, 1978). The relation between C II recombination lines and [C I] is not yet clear. Radial velocity measurements of M43 derived from radio recombination lines are given in Table 2.2.XX. Information concerning the turbulence in M42 and M43 and the line splitting and high velocity features observed in M42 is presented in Table 2.2.XXI. The methods for estimating the radial velocity and turbulence of an H II region are briefly outlined in the Appendix, Section A.II.5. The several small regions (0.02 to 0.07 pc across) in the dense ($\sim 10^4$ cm^{-3}) core of M42 in which line splitting of the order of 25 km s^{-1} occurs (Wilson *et al.* (1959) from high resolution 3″ observations of the [O III] line) are schematically shown in Figure 2.2.2. The splitting of the [O III] and [O II] lines across the core of M42 (for an E–W line 25″ south of θ^1 Ori C) is shown in Figure 2.2.3. Splitting of the same order has also been observed in the [N II] line on a larger scale (0.5 pc across) in the lower density (2×10^2 cm^{-3}) outer areas of M42 (Deharveng, 1973).

The distribution of the velocities of the ionized hydrogen over the core of M42 as derived from observations of the H76α line is shown in Figure 2.2.4. The velocity distribution over a more extended area covering M42 and M43 as derived from observations of the H109α line is shown in Figure 2.2.5.

Information concerning the age of M42 is open to discussion. Estimates made on the basis of dynamic considerations are of the order of 10^5 yr (Vandervort, 1964).

2.3. Ionized Stellar-Size Condensations (ISC) and Ionized Knots (IK) Associated with the Core of M42

The complex nature of the $\sim 4'$ core of M42 is again demonstrated in Figure 2.3.1 which is a red photograph printed through an unsharp mask. A variety of condensations, faint stars and other objects not seen clearly or not seen at all in normal prints are visible. The existence of ionized condensations in the core of Orion was first suggested by Osterbrock and Flather (1959) to explain the discrepancies in the electron density distribution derived from radio and optical methods (see Section 2.2 and Appendix A.II.2). Direct photographic evidence of the clumpy nature of the core of M42 was produced by many workers (e.g., Münch and Wilson, 1962; Elliott and Meaburn, 1974; Figures 2.1.3, 2.1.6). These photographs show distinct intensity variations over the core of M42 which are attributed to the presence of condensations of ionized matter within the tenuous ionized core. It should be noted however that such an interpretation, although quite plausible, is not necessarily unique. Indeed Münch and Persson (1971) have argued that alternative causes of the observed intensity variations such as fluctuations in the dust concentration or in the effective thickness of the nebula cannot be ruled out.

The most convincing photographic evidence for the existence of small highly ionized stellar-size condensations ($\leqslant 1.''5$ in diameter) probably comes from the work of Lacques and Vidal (1977) who succeeded in detecting 6 such condensations by photographing the core of M42 in the light of Hα and [O III] lines. They concluded that these condensations which will be referred to here as ISC 1 to 6 (Ionized Stellar-Size Condensations) are highly ionized since they emit in the Hα, Hβ, and [O III] lines but not in the [N II] or the [S II] lines. Their Hα, [O III], and [N II] photographs of the core of M42 and an identification chart of ISC 1 to 6 are shown in Figures 2.3.2a–c and 2.3.3, respectively. All of the condensations are within or very near the Trapezium stars, have sizes $\leqslant 1\overset{''}{.}5$ and probably as small as $0\overset{''}{.}2$–$0\overset{''}{.}4$ and a minimum electron density in the range of $1.5 \times 10^5\,cm^{-3}$–$4.1 \times 10^5\,cm^{-3}$ (their actual density might be considerably higher, i.e. in the range of 2.2×10^6–$3.5 \times 10^6\,cm^{-3}$).

The same authors have suggested that the nature of the ISCs is similar to that of the partially ionized globules (PIGs). The presence of PIGs in the core of M42 was first suggested by Dyson (1968) and later successfully employed by Dopita et al. (1974) to explain the multitude of phenomena occuring within the core, including the line splitting observed by Wilson et al. (1959) (see Figures 2.2.2 and 2.2.3) and the dependence of the observed radial velocity of various ions on their ionization potential (see note (8) in Table 2.2.XVIII). A schematic model of such a partially ionized condensation (< 0.01 pc $= 4''$ in size), neutral inside and enveloped by an ionization front with the ionized material flowing away into the less dense surroundings, is shown in Figure 2.3.4. The physical parameters and probable nature of ISC 1 to 6 are also summarized in Table 2.3.II.

Another, more extended ($\sim 10''$) type of object present in the core of M42 which is probably of a different nature will be referred to as IK (ionized knots). This nomenclature avoids identifications which imply definite knowledge of the nature of the objects (e.g. HH, Herbig Haro objects). This discrimination between ISCs (ionized stellar-size condensations with $1.5''$ diameters) and IKs (ionized knots with diameters of the order of $10''$) is purely descriptive and is employed for convenience. Two of the ionized knots IK 1 and IK 2 are shown in Figure 2.3.5, which is a negative masked print in the light of [S II] $\lambda\,6731$ Å of the core of M42. One of them IK 2 is also present in Figure 2.3.6, which is a positive masked print in the light of [N II] $\lambda\,6584$ Å of the same region of the core of M42 (IK 1 is virtually absent). Three more ionized knots IK 3, 4, and 5 are shown in Figure 2.3.7. Information concerning the photographs shown in Figures 2.3.1 to 2.3.7 is presented in Table 2.3.I.

Many questions concerning the exact nature of these knots still remain. One of them, IK 1, is most certainly a Herbig Haro object, M42 HH 1, as suggested by Gull et al (1973) and definetely confirmed by the spectroscopic studies of Münch (1977). Various spectra of IK 1 demonstrating its HH nature are shown in Figure 2.3.8. The objects IK 2, 3, and 4 have also been identified as Herbig Haro objects by Cantó et al. (1980) although the classification of the last two is admittedly debatable (Taylor and Münch, 1978). Studies of the kinematics of all these knots have revealed the existence of supersonic radial velocities within them which are always negative with respect to the adjacent ionized nebular material. It should be noted that supersonic velocity features have been known to occur in M42 (see Table 2.2.XXI). There is no compelling evidence that these features are truly nebular and in fact they may well originate from objects similar to the ones discussed above (Münch, 1977). Information concerning the kinematic structure of IK 1, 2, 3, 4, and 5 is presented in Figures 2.3.9 (IK 1), 2.3.10 (IK 2), and 2.3.11 (IK 3, 4, 5), and in Table 2.3.II which also gives information about their electron densities. A model relevant mainly to objects

IK 1 and 2, which seem to be Herbig Haro objects embedded in neutral filaments in M42, is shown in Figure 2.3.12. This model is based on the idea that HH objects are the products of shocks from energetic stellar winds originating from stars embedded in the surrounding neutral/dusty material (Schwartz, 1975; Münch, 1977; Dopita, 1978). The physical nature of the remaining objects (IK 3, 4, 5) is more problematic. Two of them, IK 3 and 4, are thought to be HH objects (Cantó *et al.*, 1980). Alternatively, together with IK 5, they may be parts of an expanding spherical shell of interstellar matter driven by the energetic stellar wind from θ^2 Orionis (Taylor and Münch, 1978).

Finally, another interesting structure near the bar of M42 should be mentioned. This is a $\sim 1'$ (0.15 pc) cartwheel-like filamentary object, the centre of which is at $\alpha = 5^h32^m47^s.1$ and $\delta = -5°28'17''$ (Taylor and Axon, 1979). The structure is clearly visible in Figure 2.3.13 which is a red photograph taken by Münch and printed through an unsharp mask.

According to Taylor and Axon (1979), the object may be the result of the interaction between a proto-stellar wind and the ambient H II gas. Further investigations are needed to establish the physical parameters and the kinematic structure of this peculiar object.

2.4. Dust in the Orion Nebula (M42)

The extent of the obscuration of light in the direction of the Orion nebula is shown in Figure 2.4.1. The mapping of the 'obscuration' or 'extinction' of light has been obtained by comparing the attenuated optical radiation of the nebula, measured in the Hβ line, with the unattenuated radio continuum emission measured at 24 GHz. The general method is briefly outlined in the Appendix, Section A.III.1. The presence of dust in the Orion Nebula is also demonstrated by the different appearance of the nebula in the continuum (Figure 2.4.2) compared to that in the various spectral lines (see for example Figures 2.1.6, 2.1.7, 2.1.8, and 2.1.9). This difference was attributed (first by Wurm and Rosino, 1956) to stellar light scattered by dust (non-atomic component) which is mixed with the nebular continuum (atomic component). In fact the atomic component constitutes roughly 20–30% of the observed continuum, the rest being stellar continuum scattered by the dust. The main illuminating star of the nebula is θ^1 Ori C and a comparison of the light received directly from this star with the dust scattered continuum shows excellent agreement (Simpson, 1973). The contribution of Thomson, or electron scattering, is insignificant.

There is strong evidence supporting the view that the bulk of the dust detected along the line of sight towards the Orion Nebula is in fact physically associated with the gas and does not lie in the foreground of the complex. This is presented in Table 2.4.I and in Figures 2.4.3, 2.4.4, 2.4.6, 2.4.7, and 2.4.8 which are relevant to this table. Early photoelectric studies of the nebula, corroborated by subsequent work (O'Dell and Hubbard, 1965; O'Dell *et al.*, 1966; Simpson, 1973; Dopita *et al.*, 1975; Barbieri *et al.*, 1976), have demonstrated the existence of a dust to gas ratio increasing with distance from the Trapezium area. These results have shown the existence of a 'hole' in the dust within 3' of the θ^1 Ori stars, where little scattering by dust occurs. This can clearly be seen in the contour map of the absorbing matter in Orion shown in Figure 2.4.3 (Barbieri *et al.*, 1976). However, the very high gas to dust ratio, reported in early studies for the central part of the nebula (O'Dell *et al.*, 1966), has not been corroborated by recent investigations (Perinotto and Patriarchi, 1980). Estimates of the gas to dust ratio and relevant comments are presented in Table 2.4.II.

Two main mechanisms were proposed to explain the dust depletion in the central part of the nebula: expulsion of dust grains by radiative pressure (e.g. O'Dell and Hubbard, 1965) and evaporation of grains by the intense stellar radiation (e.g. Isobe, 1970).

In the first mechanism the dust grains are accelerated by the pressure exerted on them by the stellar radiation and are eventually pushed out of the area around the ionizing stars together with the ionized gas with which they are strongly coupled. The result would be a hole around the excitation centre of the nebula like that seen in the Rossette Nebula (Mathews, 1967). The detailed treatment of such a mechanism is complicated and depends on various assumptions. For example, the radiation pressure exerted on the grains is counteracted by the drag produced by their mutual attraction. The importance of the drag depends on the electric charge of the grains which is also an important protective screen against their destruction by ion and photon sputtering (bombardment of grains by ions with energies ~ 1 eV and photons with energies ~ 15 eV present in the ionized gas). The destruction of grains by sputtering is thought to occur on short time scales if $N_e > 10^3$ cm^{-3} and the grains are electrically neutral. Charged grains however will be protected against sputtering for $\gtrsim 5 \times 10^5$ yr, except in regions with $N_e > 10^4$ cm^{-3} (Mathews, 1967, 1969). Acceleration of grains by radiation pressure will theoretically alter the size distribution since larger particles will be subjected to smaller acceleration and will tend to dominate the area around the source of radiation. The time scale of such a modification of the size distribution depends on the charge of the grains. For electrically neutral grains such a modification will occur in 10^5 yr (Krishna Swamy and O'Dell, 1967); for positively charged grains, however, no substantial alteration will occur within 2×10^6 yr (Mathews, 1967).

In the second mechanism, the grains are evaporated by absorbing UV radiation from the exciting stars. The radiation absorbed by the grains is: (1) Lyman continuum directly emitted from the stars and (2) Lα radiation emitted by the recombination of hydrogen and resonantly trapped within the H II region. The bulk of the Lα photons is thought to be absorbed by the dust. This can result in the evaporation of grains as stable as H_2O ice (Krishna Swamy and O'Dell, 1967). Part of the Lyman continuum is also absorbed by the dust and is responsible for the 'infrared excess' observed in the energy distribution of H II regions (see Appendix A.III.2). Evaporation of grains such as ice particles will alter their size distribution. This occurs because as soon as evaporation starts (at a temperature depending on the chemical composition of the dust e.g. $T \sim 100$ K for H_2O ice particles) it proceeds faster for grains of smaller size. More complicated 'dirty ice' particles (i.e. grains composed of a graphite core and an 'ice mantle') can survive theoretically in an H II region but must lose their ice mantles within 0.1 pc of an O5 star (Greenberg and Hogg, 1974). The subject of the formation and destruction of dust grains has been recently reviewed by Salpeter (1977).

In the case of M42 the end result of the above processes is the alteration of the scattering properties of the dust around θ^1 Orionis. This was thought to be the reason for the observed anomalous extinction law, found by Baade and Minkowski (1937), for the central part (3′) of the Orion Nebula [the ratio of total to selective absorption $R = A_v/(E(B - V)) \sim 6$ estimated by Johnson (1965, 1968) and Lee (1969) becoming normal, i.e. $R \sim 3$, for areas 10′ from the Trapezium (Isobe, 1970)]. The existence of such an abnormality is, however, strongly challenged by recent investigations (Penston et al., 1975; also Walker, 1969; Penston, 1973) which indicate a normal reddening law for the whole of the Orion Nebula cluster. This result together with recent investigations indicating a gas to dust ratio in Orion not much different from that of the interstellar medium (Perinotto and Patriarchi, 1980; Table 2.4.II) leave the whole subject in need of further investigation.

The most realistic model dealing with the distribution of dust in Orion is probably that introduced by Schiffer and Mathis (1974), which takes into account the dust contained in the H I/Molecular Complex lying behind the ionized gas. The physical association of at least part of the H I cloud with the dust can be inferred from the similarity of the distribution of the H I optical depth (Figure 2.4.9, angular resolution 2′) with that of the visual extinction of light (Figure 2.4.1 by Chaisson and Dopita (1977) made with comparable angular resolution of 1′.7 and the similar map by Isobe and Kurihara (1970) made with a 2′ angular resolution). It should be noted here that most of the apparent extinction ($A_v \sim 1^m5$) occurs in the 'bay' area where the H I optical depth is higher. Such a correlation provides extra support to the hypothesis (Balick et al., 1974, and others) that the 'bay' is in fact a 'tongue' or 'folding' of an optically thick feature of neutral gas and dust protruding in front of the nebula from the massive Orion Molecular Cloud which lies behind the ionized gas. This idea was further strengthened by Münch and Persson (1971) who have shown that the reddening in the bay is caused by a patch of dust lying in front of the emitting gas. The model proposed by Shiffer and Mathis (1974) consists of a slab of dust at the front surface of the H I/Molecular cloud and a shell of dust in front of it, enveloping the Trapezium stars which lie 1′–2′ (~ 0.2 pc) in front of the slab. The model indicates that most of the scattered light in the visible arises from the dust shell within the nebula which contains grains of albedo 0.6–0.7 (at Hβ λ 4861 Å). According to this model the 'hole' in the dust around θ^1 Ori C could be explained if the neutral condensations in the vicinity are either optically thick or very small.

Condensations containing dust grains were proposed by Dopita et al. (1975) as a possible source capable of supplying Orion with dust and of replenishing the losses caused by the destruction of grains (occuring within 10^3–10^5 yr).

An aspect of the dust which is constantly under review is its chemical composition.

The earlier assumptions of solid H_2O particles or ice mantle particles with graphite core (dirty ice particles) by Wickramasinghe (1963), Isobe (1973), Dopita et al. (1975), and others, were recently supplemented by silicates. This was a result of the infrared investigations of the Trapezium Nebula. The observed infrared energy distribution which represents the thermal reradiation by dust of the absorbed stellar radiation can be explained if grains of silicate composition are present. The findings of the infrared studies of the Trapezium or Ney–Allen Nebula (which is the IR nebula associated with M42 as opposed to the Kleinman–Low Nebula $\sim 1'$ NW of the Trapezium which lies in the core of the massive molecular cloud on the backside of M42 (see Figure 3.1.2) are presented in Table 2.4.III. It should, however, be noted that the observed emission band in the Trapezium Nebula shows a profile far broader than any known terrestrial or cosmic silicate. This is demonstrated in Figure 2.4.10 (Panagia, 1977) in which the profile of the 10μ silicate band observed in Trapezium (Forrest et al., 1975) has been compared with that computed for the lunar silicate No. 14321 (Bussoletti and Zambetta, 1976). The Trapezium profile shows an excess particularly pronounced around $\lambda = 8.5$ and $\lambda = 11.8\mu$.

Similar unidentified features at 8.7μ and 11.3μ characterize the emission from the bar (Aitken et al., 1979). The narrow emission features in the bar occur in a layer of neutral gas which is adjacent to the ionization front. These features may not be of thermal origin but are probably due to infrared fluorescence caused by the strong θ^1 Ori C UV irradiation of the neutral region longwards of the Lyman limit (Aitken et al., 1979). The presence of such unknown emission features complicates the estimation of the dust temperature which is thought to be ~ 75 K (Werner et al., 1976). The total energy distribution of the bar in the range of 5–1000μ, shown in Figure 2.4.11, has led Becklin et al., (1976) to propose the presence of two dust components with temperatures of 60 K and 300 K and grain sizes of 0.1 and 0.01μ, respectively (see also Table 2.4.III).

An effort made recently by Panagia (1977) to present a more coherent view of the problem concerning the population and chemical composition of dust associated with H II regions, has led to the suggestion of a dust population consisting of four different components. Details of these components and relevant information on this subject are presented in Table 2.4.IV. It should be noted that according to Panagia (1977) the 'hole' in the dust, observed in Orion may be due to the evaporation of the first of these components which probably consists of frozen molecules. This is a very interesting point and might well be related to the finding of Elliott and Meaburn (1974), that the denser part of M42 coincides with one of the peaks of the HCHO molecular ridge (Figure 2.4.12). In view of the generic relation of M42 with the associated molecular cloud, a mechanism that infuses molecular dust from the neutral/molecular cloud into the ionized gas (M42) could be in operation. Investigations of this possibility would be useful. Such an infusion of dust seems possible in view of the lower temperature of the dust (and gas) contained in the molecular cloud ($T_{gas} \simeq T_{dust} \simeq 80$ K; see next chapter) compared to the sublimation temperature of the 'frozen molecules' component of Panagia (1977) ($T \sim 100$ K).

More information concerning the dust in M42 and its relation to the dust detected in the Molecular Cloud, out of which the Orion Nebula, was born, will be discussed in the next chapter dealing with the infrared structure of the Orion complex (H I regions in association with the Molecular Cloud).

TABLE 2.1.I

Line and continuum photographs of the Orion Nebula

No.	Photograph	Wavelength range (λ = peak wavelength of the filter, $\Delta\lambda$ = full width of the filter	Region	References
1	Hα+[N II] $\lambda\lambda$ 6563, 6584 Å	$\lambda = \sim 6650$ Å $\Delta\lambda = \sim 300$ Å	M42, M43	Münch (unpublished), Osterbrock (1974), Malin (1979)
2	Hα+[N II] $\lambda\lambda$ 6563, 6584 Å	$\lambda = \sim 6600$ Å $\Delta\lambda = \sim 1000$ Å	M42, M43	Gull (1974a)
3	Hα+[N II] $\lambda\lambda$ 6563. 6584 Å	$\lambda = \sim 6650$ Å $\Delta\lambda = \sim 700$ Å	M42 (core)	Münch and Wilson (1962)
4	Hα+[N II] $\lambda\lambda$ 6563, 6584 Å	$\lambda = 6563$ Å $\Delta\lambda = 140$ Å	M42, M43	Wurm and Rosino (1959, 1965)
5	Hα+[N II] $\lambda\lambda$ 6563, 6584 Å	$\lambda = 6563$ Å $\Delta\lambda = 90$ Å	M42, M43	Gull (1974a)
6	Hα+[N II] $\lambda\lambda$ 6563, 6584 Å	$\lambda = 6563$ Å $\Delta\lambda = 150$ Å	M43	Thum et al. (1978)
7	Hα+[N II] $\lambda\lambda$ 6563, 6584 Å	$\lambda = 6563$ Å $\Delta\lambda = 30$ Å	M42 (core)	Gow et al. (1978)
8	Hα λ 6563 Å	$\lambda = 6563$ Å $\Delta\lambda = \sim 10$ Å	M42	Gull (1974b)
9	Hα λ 6563 Å	$\lambda = 6566$ Å $\Delta\lambda = 8.4$	M42 (core)	Goudis et al. (1977)
10	Hα λ 6563 Å	$\lambda = 6562$ Å $\Delta\lambda = 10$	M42 (core)	Laques and Vidal (1979)
11	Hβ λ 4861 Å	$\lambda = 4861$ Å $\Delta\lambda = 90$ Å	M42 (core)	Wurm and Rosino (1959)
12	Hβ λ 4861 Å	$\lambda = 4862$ Å $\Delta\lambda = 60$ Å	M42, M43	Gull (1974a)
13	Hβ λ 4861 Å	$\lambda = 4861$ Å $\Delta\lambda = 30$ Å	M42 (core)	Gow et al. (1978)
14	Hβ λ 4861 Å	$\lambda = 4861$ Å $\Delta\lambda = \sim 10$ Å	M42	Gull (1974b)
15	Hβ λ 4861 Å	$\lambda = 4863$ Å $\Delta\lambda = 9.2$ Å	M42 (core)	Goudis et al. (1977)
16	Hβ λ 4861 Å	$\lambda = 4862.5$ Å $\Delta\lambda = 11$ Å	M42 (core)	Laques and Vidal (1979)
17	He I λ 10830 Å	$\lambda = 10830$ Å $\Delta\lambda = 100$ Å	M42 (core)	Münch (unpublished)
18	He I λ 10830 Å	$\lambda = 10830$ Å $\Delta\lambda = 100$ Å	M42 (core)	Andrillat and Duchesne (1974)
19	He I λ 10830 Å	$\lambda = 10830$ Å $\Delta\lambda = 40$ Å	M42 (core)	Gow et al. (1978)
20	[N II] λ 6584 Å	$\lambda = 6584$ Å $\Delta\lambda = 8$ Å	M42 (core)	Elliott and Meaburn (1974)
21	[N II] λ 6584 Å	$\lambda = 6584$ Å $\Delta\lambda \sim 10$ Å	M42 (core)	Gull (1974b)
22	[N II] λ 6584 Å	$\lambda = 6584$ Å $\Delta\lambda = 10$ Å	M42 (core)	Laques and Vidal (1979)
23	[N II] λ 6584 Å	$\lambda = 6584$ Å $\Delta\lambda = 10$ Å	M42 (core)	Cantò et al. (1980)
24	O I λ 8446 Å	$\lambda = 8450$ Å $\Delta\lambda = 20$ Å	M42 (core)	Münch and Taylor (1974)
25	[O I] λ 6300 Å	$\lambda = 6300$ Å $\Delta\lambda = 12$ Å	M42 (core)	Münch and Taylor (1974)
26	[O I] λ 6300 Å	$\lambda = 6300$ Å $\Delta\lambda \sim 10$ Å	M42 (core)	Gull (1974b)
27	[O I] λ 6300 Å	$\lambda = 6300$ Å $\Delta\lambda = 10$ Å	M42(core)	Goudis et al. (1977)
28	[O II] $\lambda\lambda$ 3726, 3729 Å	$\lambda = 3727$ Å $\Delta\lambda = 40$ Å	M42, M43	Wurm and Rosino (1965)
29	[O II] $\lambda\lambda$ 3726, 3729 Å	$\lambda = 3730$ Å $\Delta\lambda = 32$ Å	M42 (core)	Courtès and Viton (1967)
30	[O II] $\lambda\lambda$ 3726, 3729 Å	$\lambda = 3726$ Å $\Delta\lambda \sim 10$ Å	M42	Gull (1974b)
31	[O III] $\lambda\lambda$ 4959, 5007 Å	$\lambda = 5007$ Å $\Delta\lambda = 130$ Å	M42 (core)	Wurm and Rosino (1959, 1965)
32	[O III] λ 5007 Å	$\lambda = 5009$ Å $\Delta\lambda = 10$ Å	M42 (core)	Elliott and Meaburn (1974)
33	[O III] λ 5007 Å	$\lambda = 5007$ Å $\Delta\lambda \sim 10$ Å	M42	Gull (1974b)
34	[O III] λ 5007 Å	$\lambda = 5008$ Å $\Delta\lambda = 8.4$ Å	M42 (core)	Laques and Vidal (1979)
35	[S II] $\lambda\lambda$ 6717, 6731 Å	$\lambda = 6730$ Å $\Delta\lambda = 180$ Å	M42, M43	Wurm and Rosino (1959)

(continued)

Table 2.1.I (continued)

No.	Phorograph	Wavelength range (λ=peak wavelength of the filter, $\Delta\lambda$=full width of the filter)		Region	References
36	[S II] $\lambda\lambda$ 6717, 6731 Å	$\lambda=6725$ Å	$\Delta\lambda=90$ Å	M42, M43	Gull (1974a)
37	[S II] λ 6717 Å	$\lambda=6716$ Å	$\Delta\lambda=9$ Å	M42 (core)	Goudis *et al.* (1977)
38	[S II] λ 6717 Å	$\lambda=6717$ Å	$\Delta\lambda=9.2$ Å	M42 (core)	Laques and Vidal (1979)
39	[S II] λ 6731 Å	$\lambda=6732.5$ Å	$\Delta\lambda=10$ Å	M42 (core)	Laques and Vidal (1979)
40	[S II] λ 6731 Å	$\lambda=6731$ Å	$\Delta\lambda=9$ Å	M42 (core)	Cantó *et al.* (1980)
41	[S II] $\lambda\lambda$ 10284, 10318, 10336 Å	$\lambda=10330$ Å	$\Delta\lambda=200$ Å	M42 (core)	Andrillat and Duchesne (1974)
42	[S III] λ 9069 Å	$\lambda=9060$ Å	$\Delta\lambda=45$ Å	M42 (core)	Andrillat and Duchesne (1974)
43	[A III] λ 7751 Å	$\lambda=7751$ Å	$\Delta\lambda=70$ Å	M42 (core)	Andrillat and Duchesne (1974)
44	Continuum	$\lambda\sim8500$ Å	$\Delta\lambda>7400$ Å	M42 (core)	Addrillat and Duchesne (1974)
45	Continuum	$\lambda=6900$ Å	$\Delta\lambda=90$ Å	M42, M43	Gull (1974a)
46	Continuum	$\lambda=6470$ Å	$\Delta\lambda=90$ Å	M42, M43	Gull (1974a)
47	Continuum	$\lambda=6441$ Å	$\Delta\lambda=11$ Å	M42 (core)	Laques and Vidal (1979)
48	Continuum	$\lambda=6000$ Å	$\Delta\lambda=260$ Å	M42, M43	Wurm and Rosino (1959)
49	Continuum	$\lambda=5672$ Å	$\Delta\lambda=70$ Å	M42 (core)	Goudis *et al.* (1977)
50	Continuum	$\lambda=5300$ Å	$\Delta\lambda=400$ Å	M42, M43	Osterbrock (1974)
51	Continuum	$\lambda=5200$ Å	$\Delta\lambda=240$ Å	M42, M43	Wurm and Rosino (1959, 1965)
52	Continuum	$\lambda=4770$ Å	$\Delta\lambda=60$ Å	M42, M43	Gull (1974a)
53	Continuum	$\lambda=4600$ Å	$\Delta\lambda=49$ Å	M42 (core)	Courtès and Viton (1967)
54	Continuum	$\lambda=4200$ Å	$\Delta\lambda=32$ Å	M42 (core)	Courtès and Viton (1967)
55	Continuum	$\lambda=3920$ Å	$\Delta\lambda=32$ Å	M42 (core)	Courtès and Viton (1967)
56	Continuum	$\lambda=3500$ Å	$\Delta\lambda=300$ Å	M42, M43	Wurm and Rosino (1965)
57	Continuum	$\lambda=3425$ Å	$\Delta\lambda=150$ Å	M42, M43	Wurm and Rosino (1965)
58	Continuum (UV)	$\lambda\sim1450$ Å	$\lambda\lambda1230$–2000 Å	M42, M43	Carruthers and Opal (1977b)

Remarks to Table 2.1.I:

(1) The exact width of the narrow band filters used by Gull (1974, unpublished) is not known to the author; an approximate $\Delta\lambda \simeq 10$ Å is cited, based on the information by Gull (1974a) that the bandpasses are in the range of 6–20 Å.

TABLE 2.1.II

Ionization potentials and relevant emission lines

Atomic number (Z)	Element (X)	Ionization Potential (in eV) (State of ionization)			Important emission lines in M42
		I–II X^+	II–III X^{++}	III–IV X^{+++}	
1	H	13.6			Hα λ 6563 Å, Hβ λ 4861 Å
2	He	24.5	54.4		He I λ10830 Å
6	C	11.3	24.4	47.9	C I λ 9850 Å, C I λ 8727 Å
7	N	14.5	29.6	47.4	[N II] λ 6584 Å
8	O	13.6	35.1	54.9	O I λ 8446 Å, [O I] λ6300 Å, [O II] $\lambda\lambda$ 3726, 3729 Å, [O III] λ 5007 Å
16	S	10.4	23.4	34.8	[S II] $\lambda\lambda$ 6717, 6731 Å, [S III] λ 9069 Å, [S III] λ 9531 Å

TABLE 2.1.IIIa

Relative line intensities in the Orion Nebula

λ (Å)	Identification	Transitions	Relative intensity ($I_{H\beta}=100$)			
			Not corrected for interstellar reddening		Corrected for interstellar reddening[8]	
			Aller and Liller (1959)[2,3,4]	Mendez (1967)[3,5,6,7]	Aller and Liller (1959)[2,3,4]	Mendez (1967)[3,5,6,7]
3725.99	[O II]	$2p^3\ ^4S - 2p^3\ ^2D$ ⎱	127	66.3	181.61	96.79
3728.76	–	– ⎰				
3734.37	H I	$2\ ^2P - 13\ ^2D$	3.5	–	4.99	–
3750.19	H I	$2\ ^2P - 12\ ^2D$	4.4	2.4	6.24	3.47
3770.66	H I	$2\ ^2P - 11\ ^2D$	5.4	3.2	7.61	4.60
3797.91	H I	$2\ ^2P - 10\ ^2D$	7.8	4.1	10.89	5.84
3835.41	H I	$2\ ^2P - 9\ ^2D$	10.9	6.3	15.02	8.84
3868.74	[Ne III]	$2p^4\ ^3P - 2p^4\ ^1D$	19.7	15.5	26.82	21.48
3889.91	{H I, He I}	{$2^3P - 8\ ^2D$, $2^3S - 3\ ^3P$}	}18.1	}10.9	}24.462	}14.99
3967.51	[Ne III]	$2p^4\ ^3P - 2p^4\ ^1D$	}24.4	–	}32.18	–
3970.06	H I	$2\ ^2P - 7\ ^2D$		17.4		23.30
4026.12	He I	$2\ ^3P - 5\ ^3D$	2.7	1.8	3.5	2.37
4068.62	[S II]	$2p^3\ ^4S - 3p^3\ ^2P$	2.7	–	3.45	–
4101.74	H I(Hδ)	$2\ ^2P - 6\ ^2D$	25	20.9	31.56	26.75
4267.13	C II	$3\ ^2D - 4\ ^2F$	0.36	–	0.43	–
4340.44	H I(Hγ)	$2\ ^2P - 5\ ^2D$	41	40.8	48.08	48.28
4363.21	[O III]	$2p^2\ ^1D - 2p^2\ ^1S$	1.55	–	1.81	–
4387.90	He I	$2p\ ^1P - 5d\ ^1D$	0.6	–	0.69	–
4471.50	He I	$2\ ^3P - 4\ ^3D$	4.6	3.5	5.20	3.98
4658.20	[Fe III]	$3d^6\ ^5D_4 - 3d^6\ ^3F_4$	1.2	–	1.28	–
4712	{He I, [A IV]}	{$2p\ ^3P - 4s\ ^3S$, $3p^3\ ^4S - 3p^3\ ^2D$}	}1.1	–	}1.15	–
4861.30	H I (Hβ)	$2\ ^2P - 4\ ^2D$	100	100	100	100
4921.90	He I	$2p\ ^1P - 4d\ ^1D$	1.5	–	1.48	–
4958.91	[O III]	$2p^2\ ^3P - 2p^2\ ^1D$	113	117.1	110.18	114.01

(continued)

Table 2.1. IIIa (continued)

λ (Å)	Identi-fication	Transitions	Relative intensity ($I_{H\beta}=100$)			
			Not corrected for interstellar reddening		Corrected for interstellar reddening[8]	
			Aller and Liller (1959)[2,3,4]	Mendez (1967)[3,5,6,7]	Aller and Liller (1959)[2,3,4]	Mendez (1967)[3,5,6,7]
5006.84	[O III]	$2p^2\ {}^3P-2p^2\ {}^1D$	342	360	329.27	345.84
5875.63	He I	$2\ {}^3P-3\ {}^3D$	31	12	23.98	9.14
6548.09	[N II]	$2p^2\ {}^3P-2p^2\ {}^1D$	18	–	12.4	–
6562.90	H I (Hα)	$2\ {}^2P-3\ {}^2D$	350	346.6	235.5	227.94
6583.36	[N II]	$2p^2\ {}^3P-2p^2\ {}^1D$	55	–	36.86	–
6716.40	[S II]	$3p^3\ {}^4S-3p^3\ {}^2D$	} 14	} 4.9	} 9.16	} 3.13
6730.80	[S II]	$3p^3\ {}^4S-3p^3\ {}^2D$				
7065.30	He I	$2\ {}^3P-3\ {}^3S$	} 30	11.3	} 18.60	6.82
7135.80	[A III]	$3p^4\ {}^3P_2-3p^4\ {}^1D$		11.4		9.80
7318.60						
7319.90	[O II]	$2p^3\ {}^2D-2p^3\ {}^2P^o$	} 17.3	} 10.3	} 10.33	} 5.96
7329.90						
7330.70						
9069.40	[S III]	$3p^2\ {}^3P-3p^2\ {}^1D$	72	49.5	35.02	23.10
9229.02	H	$3\ {}^2D-9\ {}^2F$	5.8	4.7	2.78	2.16
9545.97	H	$3\ {}^2D-8\ {}^2F$	8	–	3.74	–
9532.10	[S III]	$3p^2\ {}^3P-3p^2\ {}^1D$	181	125	84.63	55.94
10049.38	H	$3\ {}^2D-7\ {}^2F$	10	10.1	4.51	4.35
10830.00	He I	$2\ {}^3S-2\ {}^3P$	70	80	29.96	32.61
10938.09	H	$3\ D^2-6\ {}^2F$	20	17.8	8.5	7.20

Remarks to Table 2.1.IIIa:

(1) For a comprehensive compilation of numerous lines observed by various workers see Kaler (1976).

(2) 4′ × 4′ area centred 1.2 W of the Trapezium (brightest position of M42).

(3) Corrected for atmospheric extinction.

(4) Average accuracy for all except the weakest lines: ±1.2%.

(5) 45″ SW of Trapezium (brightest position of M42).

(6) 14″ diaphragm.

(7) Average accuracy better than ±5%.

(8) Observed intensity corrected for interstellar reddening by Kaler (1976) according to the relation $I_{corrected} = I_{observed} \times 10^{cf(\lambda)}$ where $c = \log[F_c(H\beta)/F_0(H\beta)]$ where $F_c(H\beta)$ and $F_0(H\beta)$ are the true (unattenuated) and observed (attenuated) Hβ flux, respectively. The factor c is derived from the ratio of the radio flux density to the Hβ flux (see general method in Appendix III.1). The reddening function $f(\lambda)$ is derived from the curve deduced by Whitford (1958) ($R = 3.2$) in such a way as $f(\infty) = -1$ and $f(H\beta) = 0$.

TABLE 2.1.IIIb

Absolute line intensities in the Orion Nebula[1,2,3,6]

No.	Beam (arc sec)	Line	λ (Å)	Flux 10^{-12} ergs cm^{-2} s^{-1} or 10^{-19} W cm^{-2}	10^{-9} W cm^{-2} sr^{-1}	Remarks	References
1	8×10	[O II]	3727	481	32	(4), (5)	
2	8×10	[Ne III]	3868	86.8	5.8	(4), (5)	
3	8×10	Hγ	4340	217	14.5	(4), (5)	
4	8×10	[O III]	4363	4.6	0.3	(4), (5)	
5	8×10	He II	4686	<2	<0.13	(4), (5)	
6	8×10	Hβ	4861	470	31.3	(4), (5)	
7	8×10	[O III]	4959	429	28.6	(4), (5)	
8	8×10	[O III]	5007	1394	93	(4), (5)	Lester et al.
9	8×10	[N II]	5755	2.6	0.17	(4), (5)	(1979)
10	8×10	[S III]	6312	3.6	0.24	(4), (5)	
11	8×10	[N II]	6548	51.4	3.4	(4), (5)	
12	8×10	Hα	6563	895	59.6	(4), (5)	
13	8×10	[N II]	6583	151	10.1	(4), (5)	
14	8×10	[S II]	6716	5.0	0.33	(4), (5)	
15	8×10	[S II]	6731	10.1	0.67	(4), (5)	
16	8×10	[Ar III]	7136	51.9	3.5	(4), (5)	
17	8×10	[O II]	7326	50.7	3.4	(4), (5)	
18	38	Pα12	8750	11.1	0.04	(7)	Barbieri et al. (1976)
19	38	[S III]	9069	100.5	0.39	(8), (9)	Olthof et al. (1978)
20	120	[S III]	9069	8800	3.3	(7)	Lowe et al. (1977)
21	10	[S III]	9531	363	19.6	(4)	Lester et al. (1979)
22	120	[S III]	9531	20000	7.5	(7)	Lowe et al. (1977)
23	120	Pε	9545	800	0.3	(7)	Lowe et al. (1977)
24	10	Pδ	10049	17.8	0.96	(4)	Lester et al. (1979)
25	120	Pδ	10049	1400	0.51	(7)	
26	120	He I	10830	13300	5.0	(7)	
27	120	Pγ	10938	2700	1.03	(7)	
28	120	Pβ	12818	4800	1.80	(7)	Lowe et al.(1977)

Remarks to Table 2.1.IIIb:

(1) Many permitted lines have also been detected in the bright inner part of M42 ($\sim 15''$ S of the Trapezium stars) in the range of 6500–9100 Å; these have been identified with N I and O I; they are probably due to direct starlight excitation (Grandi, 1975).

(2) Polarisation measurements of the Hα, Hβ, [O III] λ 5007 Å and [O II] λ 3727 Å made by Gull (1974a) were found to be smaller or comparable to the polarization of the continuum which was $\sim 1\%$ near the Trapezium increasing to $\sim 10\%$ at $10'$ distance.

(3) Other forbidden lines identified in the Orion spectrum include the [Fe II], [Fe III] (Kaler et al., 1965), [Ni II], [Cu II] (Thackeray, 1975), and [Ni III] (Flather and Osterbrock, 1960).

(4) The line intensities were estimated from observations made at the position of the electron density peak of M42 ($1'$ W and $10''$ S of θ^1 Ori C; Elliot and Meaburn (1973), Martin and Gull (1976)); the quoted fluxes have been corrected for absorption by using the extinction law of Miller and Mathews (1972).

(5) Flux calibration based on stars HD 162487 and HD 46150; the accuracy is $\sim \pm 20\%$ for the brightest and $\pm 40\%$ for the faintest lines.

(6) The strength of infrared lines with $\lambda > 1.3\mu$ can be found in Table 3.2.VIII.

(7) Not corrected for interstellar extinction.

(8) At the brightest region $38''$ W to the Trapezium.

(9) Not corrected for interstellar extinction; flux corrected for extinction: 0.85×10^{-9} W cm^{-2} sr^{-1} or 225×10^{-19} W cm^{-2}.

TABLE 2.1.IV

Technical information related to the photographs of the core of M42 shown in Section 2.1

No.	Figure	Line	Telescope	Auxilliary device
1	2.1.2	$H\alpha$ + [N II] $\lambda\lambda$ 6563, 6584 Å	3 m (120 inch) Hale; f: 3.67 prime focus	
2	2.1.4	$H\beta$ λ 4861 Å	1.22 m (50 inch) Asiago; f: 5 prime focus	image tube
3	2.1.5	[N II] λ 6584 Å	2.5 m (98inch) Isaac Newton; f: 14.7 Cassegrain focus	image tube
4	2.1.6	[O II] $\lambda\lambda$ 3726, 3729 Å	0.8 m (31 inch) Haute Provence	image tube
5	2.1.7	[O III] λ 5007 Å	2.5 m (98 inch) Isaac Newton; f: 14.7 Cassegrain focus	image tube
6	2.1.8	O I λ 8446 Å	1.5 m (60 inch) Palomar; f: 8.75 Ritchey–Chrétien focus	image tube
7	2.1.9	[O I] λ 6300 Å	1.5 m (60 inch) Palomar; f: 8.75 Ritchey–Chrétien focus	electronic camera
8	2.1.10	[S III] λ 9069 Å	1.93 m (76 inch) Haute Provence; f: 5 Newtonian focus	image tube
9	2.1.11	He I λ 10830 Å	1.5 m (60 inch) Palomar; f: 8.76 Ritchey–Chrétien focus	image tube

No.	Filter Type	Peak Wavelength λ(Å)	Full width $\Delta\lambda$ (Å)	Photo cathode	Plate	Exposure time (min)	Print P=Positive N=Negative	Remarks	References
1	Schott RG2	6650	~300		Kodak III-E	1	P	(1)	Münch and Wilson (1962)
2	Dye filter	4861	90		103a-O		N	(2)	Wurm and Rosino (1959)
3	Interference filter	6584	8	S25	Kodak IIIα-J	5	N	(3)	Elliott and Meaburn (1974)
4	BPM filter	3730	32				N	(4)	Courtés and Viton (1967)
5	Interference filter	5009	10	S25	Kodak IIIα-J	2.5	N	(3)	Elliott and Meaburn (1974)
6	Interference filter	8450	20	ERMA-4	Kodak IIα-D		P	(5)	Münch and Taylor (1974)
7	Interference filter	6300	12	ERMA-4	Kodak IIα-D		P	(6)	Münch and Taylor (1974)
8	Interference filter	9060	45	S1	Kodak Definix	12	N	(7)	Andrillat and Duchesne (1974)
9	Interference filter	10830		S1	Kodak IIα-D	30	P	(8)	Münch (unpublished)

Remarks to Table 2.1.IV:

(1) Filter transmitting λ > 6300 Å; effective bandwidth $\Delta\lambda$ ~300 Å; resolution of the plate set by the seeing conditions ~0".5.

(2) Filter transmits substantial amount of continuum.

(3) Westinghouse WL-30677, S25 image tube with phosphor output; IIIaJ emulsion (baked); field of view: 4'; Resolution set by the seeing conditions ~1"–2".

(4) Multiple Band Pass monochromatic filter.

(5) Transmission of P17, λ 8467 Å and P18, λ 8438 Å emission lines contribute 30% to the O I image. Photocathode 40 mm in diameter; images recorded on 35 mm film.

(6) Transmission of [S III] λ 6312 Å line contribute 15% to the [O I] image; photocathode 40 mm in diameter; images recorded on 35 mm film.

(7) Photocathode 20 mm in diameter; field of view 5'.

(8) Filter transmitting λ > 10600 Å.

TABLE 2.1.V

Radio continuum contour maps of the Orion Nebula (Orion A ≡ M42 + M43)

No.	HPBW (arc min or arc sec)	Frequency ν (MHz)	Wavelength λ (mm or cm)		Remarks	References
1	7′.2	35 000	8.6	mm	(1)	Tolbert (1965)
2	6′.4	5 000	6	cm		Mezger and Henderson (1967)
3	4′.2	7 800	3.9	cm		Gordon and Meeks (1968)
4	4′.2	5 000	6	cm	(1)	Gardner and Morimoto (1968)
5	4′.0	5 000	6	cm		Goss and Shaver (1970)
6	3′.35	14 500	2	cm	(1)	Baars et al. (1965)
7	3′.0	408	73.5	cm		Mills and Shaver (1968) Shaver and Goss (1970α)
8	2′.8	10 700	2.8	cm		MacLeod and Doherty (1968)
9	2′.4	72 800	4.1	mm	(1)	Kaifu et al. (1973)
10	2′.3	15 800	1.9	cm	(1)	Zisk (1966)
11	2′.2	15 460	1.94	cm		Gordon (1969)
12	2′.0	15 370	1.95	cm		Schraml and Mezger (1969)
13	2′.0	87 000	3.5	mm		Fukui and Iguchi (1977)
14	2′.0	1 420	21	cm		Lockhart and Goss (1978)
15	1′.9	36 500	8.2	mm		Sorochenko and Berulis (1970)
16	1′.6	31 400	9.55	mm	(1)	Johnston and Hobbs (1969)
17	1′.37	10 700	2.8	cm	(2)	Thum et al. (1978)
18	1′.33	23 400	1.3	cm		Rodriguez and Chaisson (1978)
19	1′.3	24 000	1.25	cm		Chaisson and Dopita (1977)
20	21″ × 60″	2 700	11	cm	(3)	Altenhoff (1972)
21	7″ × 35″	26 950	1.1	cm	(3)	Webster and Altenhoff (1970)
22	7″ × 20″	5 000	6	cm	(3)	Martin and Gull (1976)

(1) M42 only.

(2) M43 only.

(3) core of M42 only.

TABLE 2.2.I

Physical parameters of M42 (G209.0-19.4) derived from radio continuum measurements

No.	Source	Coordinates (1950) α	δ	Distance D (kpc)	Angular size $\theta\alpha \times \theta\delta$	Angular diameter $\bar\theta$ ($G = \theta_G = \sqrt{\theta\alpha \times \theta\delta}$, Sph $= \theta$ spherical)
1	M42	$5^h31^m51^s.8$	$-5°25'16''.9$	0.45		3' (Sph)
2	M42 (core)			0.45		10' (Sph)
3	M42 (total)			0.45		
4	M42			0.53		
5	M42	5 31 51.8	-5 25 16.9	0.50		2'.33 (G)
6	M42 (core)			0.50		16'.3 (G)
7	M42 (total)			0.50		
8	M42 (core)			0.50		
9	M42	5 32 44±4	-5 24 54±60	0.50	3'.7 × 4'.2	3'.9 (G)
10	M42	5 32 50.3	-5 25 06		3' × 3'	3' (G)
11	M42			0.43		
12	M42	5 32 51.1±0.7	-5 25 16±11			
13	M42	5 32 50	-5 25 24		3'.28 × 3'.45	
14	M42	5 33 42	-5 24 40			
15	M42	5 32 50.2±2	-5 25 12±25	0.50	2'.6 × 3'.5	
16	M42	5 32 48.9±0.9	-5 24 48±14		3'.2 × 4'	
17	M42	5 32 50	-5 25 36		3'.1 × 3'.1	
18	M42	5 32 48.8±1	-5 25 2±10	0.50	3'.4 × 3'.4	
19	M42			0.50		6'.2 (G)
20	M42	5 32 51	-5 25 39	0.50	3'.8 × 4'.3	
21	M42 component 1	5 32 48.9±0.5	-5 25 27±8		6'.4 × 1'.9	
	component 2	5 32 56.3±0.2	-5 26 06±3		37''.5 × 35''	
	component 3	5 32 51.2±0.2	-5 25 25±3		33'' × 16''	
	component 4	5 32 51.7±0.2	-5 26 01±3		18'' × 15''	
	component 5	5 32 48.5±0.2	-5 25 36±3		60'' × 27''	
	component 6	5 32 46.9±0.2	-5 25 16±3		55'' × 22''	
22	M42 (core)					
23	M42 (core)					
24	M42	5 32 51	-5 25 39	0.50	3'.2 × 3'.3	
25	M42			0.43		3'.9 (Sph)
26	M42 Region I					0'.33 (G)
	Region II					4'.3 (G)
	Region III					19'.1 (G)
27	M42 (core)			0.50		1' (G)
28	M42 (core)			0.50		4' (G)
29	M42	5 32 50	-5 24 48	0.50	2'.9 × 3'.5	
30	M42			0.50	2'.9 × 3'	
31	M42 (core)			0.50		3'.5 (G)

Table 2.2.1 (continued)

No.	Linear diameter $2R$ (pc)	Root mean square density rms N_e[9] (cm^{-3})	Mass of ionized hydrogen M_{HII} ($M_\odot$)	Emission measure E (pc cm^{-6})	Excitation parameter u (pc cm^{-2})	Rate of Lyman continuum photons Lc[1] (photons s^{-1})	Remarks	References
1								Sharpless (1952)
2			6	4×10^6				Pariiskii (1961b)
3			116	0.04×10^6				Pariiskii (1961b)
4								Becker and Fenkart (1963)
5	4.4	4.1×10^2	450					Pottasch (1965)
6	0.37	4.95×10^3	2.44	33.3×10^6	54	5.7×10^{48}		Gol'nev et al. (1965)
7	2.60	2.74×10^2	50	0.71×10^6	55	6.0×10^{48}		Gol'nev et al. (1965)
8	0.30			3×10^6				Zisk (1966)
9	0.83	1.65×10^3	12	2.3×10^6				Mezger and Henderson (1967)
10	0.70	2.6×10^3	11.3	4.7×10^6				MacLeod and Doherty (1968)
11								Miller (1968)
12				$(3.1\pm0.6) \times 10^6$				Mills and Shaver (1968), Gordon (1969)
13				3.76×10^6				Gardner and Morimoto (1968)
14	0.65	2.2×10^3		3.9×10^6				Gordon and Meeks (1968)
15		2.24×10^3	7	3.2×10^6	55.2	6.1×10^{48}		Schraml and Mezger (1969)
16	0.70	2.12×10^3	8.8	3×10^6				Johnston and Hobbs (1969)
17		2.47×10^3	8.6	4.1×10^6	60	7.8×10^{48}		Wilson et al. (1970)
18		2.7×10^3		3.65×10^6				Sorochenko and Berulis (1970)
19	0.70	2.34×10^3	9.0	3.8×10^6	60.6	8.1×10^{48}		Shaver and Goss (1970b)
20	0.60	1.7×10^3	13	2.5×10^6	62	8.6×10^{48}		Reifenstein et al. (1970)
21		$(3.3\pm0.6) \times 10^3$		$(3.1\pm0.6) \times 10^6$				Webster and Altenhoff (1970)
		$(3.4\pm0.6) \times 10^3$		$(6.3\pm1.3) \times 10^6$				
		$(40.3\pm8) \times 10^3$		$(6.7\pm1.3) \times 10^6$				
		$(47.9\pm9) \times 10^3$		$(8.0\pm1.6) \times 10^6$				
		$(2.3\pm0.4) \times 10^3$		$(3.0\pm1.6) \times 10^6$				
		$(2.9\pm0.5) \times 10^3$		$(4.4\pm0.9) \times 10^6$				
22		2×10^4	9.1					Hjellming Gordon (1971)
23		1.6×10^4	5.5				(7)	Brocklehurst and Seaton (1972)
24		2.3×10^3		3.8×10^6			(7), (8)	Berulis and Sorochenko (1973)
25	0.50	3.45×10^3		6×10^6	56	6.4×10^{48}		Goudis (1975)
26	0.043	3.16×10^4					(4)	Lockman and Brown (1975)
	0.56	3.16×10^3					(4)	
	2.50	2×10^2					(4)	
27	0.15	6×10^3		5.2×10^6				Martin and Gull (1976)
28	0.66	2.5×10^3	13	5×10^6	68	$\sim 1 \times 10^{49}$		Chaisson and Dopita (1977)
29		2.4×10^3	12	4.3×10^6		1.1×10^{49}		Rodriguez and Chaisson (1978)
30		5.04×10^3					(5)	Churchwell et al. (1978)
31	0.13	7×10^3		7.24×10^6			(10)	Shaver (1980)

TABLE 2.2.II
Physical parameters of M43 (G208.9–19.3) derived from radio continuum measurements

No.	Source	Coordinates (1950)		Distance D (kpc)	Angular size $\theta\alpha \times \theta\delta$	Angular diameter θ_G $(G=\theta_G= \sqrt{\theta\alpha \times \theta\delta})$	Linear diameter $2R$ (pc)
		α	δ				
1	M43	$5^h 33^m 04\overset{s}{.}2$	$-5° 17' 43''$		$1\overset{'}{.}8 \times 1\overset{'}{.}8$		0.46
2	M43	5 33 55	-5 17 39				
3	M43	5 33 03.8$\pm$3	-5 17 50$\pm$40	0.50	$2\overset{'}{.}7 \times 2\overset{'}{.}5$		0.56
4	M43			0.50		$2'(G)$	0.50
5	M43	5 33 03	-5 18 18	0.70$\pm$0.70	$2\overset{'}{.}5 \times 2\overset{'}{.}5$		0.70
6	M43			0.50	$2\overset{'}{.}2 \times 2\overset{'}{.}5$		
7	M43	5 33 04	-5 17 45	0.50	$2\overset{'}{.}7 \times 2\overset{'}{.}7$	$2\overset{'}{.}7(G)$	
8	M43			0.40	$2'8 \times 2\overset{'}{.}8$		0.50
9	M43			0.50	$2\overset{'}{.}8 \times 2\overset{'}{.}9$		

Remarks to Table 2.2.I and 2.2.II:

(1) Calculated from relation (II.1.51) (see Appendix II.1) for $T_e = 8500$ K.
(2) Calculated from relation (II.1.51) (see Appendix II.1) for $T_e = 7000$ K.
(3) LTE assumed; Non LTE values: rms $N_e = 2.3 \times 10^3$ cm^{-3}, $E = 4.4 \times 10^6$ pc cm^{-6}.
(4) Model: I = very compact central source; II = Gaussian core source; III = extended envelope. Electron temperatures: $T_e(\text{I}) = 7500$ K; $T_e(\text{II}) = 10\,000$ K; $T_e(\text{III}) = 12\,500$ K.
(5) $\theta e \times \theta b$ instead of $\theta\alpha \times \theta\delta$.
(6) LTE assumed; Non-LTE values: rms $N_e = 2.7 \times 10^3$ cm^{-3}, $E = 1.5 \times 10^6$ pc cm^{-6}.
(7) Peak value.
(8) From an analysis of radio line and continuum observations; assuming a density distribution consisting of uniform spherically symmetric shells centred on the core, they find a peak rms $N_e = 1.6 \times 10^4$ cm^{-3}. The density falls to 5×10^3 cm^{-3} at $1'$ away and to 10^3 cm^{-3} at $3'$. The T_e is taken to be constant and equal to 9500 K.
(9) The condensation factor (see Appendix II.2; relation II.2.4) is in the range of 0.025–0.070 (Osterbrock and Flather, 1959; Pariiskii, 1961; Dopita *et al.*, 1974).
(10) Isothermal model ($T_e = 8200$ K) consisting of a 0.13 pc core with rms $N_e \simeq 7 \times 10^3$ enveloped by four lower density outer regions.

TABLE 2.2.III
Positions of prominent stars associated with M42 and M43

Nebula	Stars	Catalogue number			Coordinates (1950)*		Remarks
		Henry Draper	Bonner Durchmus- terung	Smithson- ian Astro- physical Observatory	α	δ	
M42	θ^1 Ori A	HD 37020	BD $-5°1315$				(1), (7), (10)
	θ^1 Ori B	HD 37021	BD $-5°1315$				(1), (2), (3), (5), (10)
	θ^1 Ori C	HD 37022	BD $-5°1315$	SAO 132314	$5^h 32^m 48\overset{s}{.}995$	$-5°25' 16\overset{''}{.}22$	(3), (9), (10), (10)
	θ^1 Ori D	HD 37023	BD $-5°1315$				
	θ^2 Ori A	HD 37041	BD $-5°1319$	SAO 132321	5 32 55.449	-5 26 50.72	(1), (2), (4), (6), (10), (11)
	θ^2 Ori B	HD 37042	BD $-5°1320$	SAO 132322	5 32 58.911	-5 26 52.21	(10)
M43	NU Ori	HD 37061	BD $-5°1325$	SAO 132328	5 33 3.747	-5 17 54.89	(8), (10)

* SAO catalog of stars.

Table 2.2.II (continued)

No.	Root mean square density rms $N_e^{(9)}$ (cm^{-3})	Mass of ionized hydrogen $M_{H\,II}$ ($M_\odot$)	Emission measure (pc cm^{-6})	Excitation parameter u (pc cm^{-2})	Rate of Lyman continuum photons Lc$^{(2)}$ (photons s^{-1})	Remarks	References
1	9.1×10^2	1.1	3.8×10^5				MacLeod and Doherty (1968)
2	2.5×10^3		12×10^5			(6)	Gordon and Meeks (1968)
3	6.1×10^2	1.3	2.1×10^5	20	3.4×10^{47}		Schraml and Mezger (1969)
4	7.13×10^2	1.3	2.8×10^5	21.6	4.3×10^{47}		Shaver and Goss · (1970b)
5	6.13×10^2	3.1	2.8×10^5	26	7.4×10^{47}		Wilson et al. (1970)
6	6.6×10^2	0.98	2.2×10^5				Berulis and Sorochenko (1973)
7	7×10^2	2	3.1×10^5	24	5.9×10^{47}		Rodriguez and Chaisson (1978)
8	6.1×10^2	1	1.8×10^5	18	2.5×10^{47}		Thum et al. (1978)
9	1.1×10^3		3.3×10^5			(5)	Churchwell et al. (1978)

Remarks to Table 2.2.III:

(1) Binary star.

(2) The secondary component may be a black hole (Trimble and Thorne, 1969; Wilson, 1972).

(3) Possible X-Ray emitter (Bradt and Kelley, 1979).

(4) Possible X-Ray emitter (Barbon et al., 1972).

(5) The primary component is a normal B3 star of mass $\sim 6M_\odot$, lying on the ZAMS (hotter component); the secondary component is a pre-main sequence star exhibiting an A5–F2 spectrum and has a mass of $\sim 1.8M_\odot$ (cooler component); for a detailed study see Popper and Plavec (1976).

(6) Observed absorption lines probably originate in the secondary component (Aikman and Goldberg, 1974).

(7) Possibly a pre-main sequence object (Lohsen, 1975).

(8) Bolometric luminosity $L_* = 2.4 \times 10^4 L_\odot$ (Thum et al., 1978).

(9) The θ^1 Ori C star exhibits an inverse P Cygni profile (Conti and Leep, 1974) which may imply an extremely young star, if interpreted as due to infalling material. Such an interpretation is however debatable (Strom et al., 1975).

(10) Information concerning the spectral type, luminosity class, and apparent (V) and absolute (M_V) visual magnitude of the stars, together with information concerning the visual extinction A_v, is presented in Table 2.2.IV.

(11) Becklin et al. (1976) have argued that θ^2 Ori A is probably a foreground star; they based their suggestion on the 5–20μ spectrum observed near the θ^2 Ori A region (exhibiting a normal silicate feature in emission) which is different from the peculiar spectrum originating from regions along the barlike ionization front in Orion; Taylor and Münch (1978) have however pointed out several facts indicating a physical association of this star with the nebular material (strong He I λ 3888 Å absorption line in the spectrum of θ^2 Ori A indicating the presence of a dense H II region around the star, kinematic considerations indicating association of the star with the nebular material and the star cluster contained in M42, and polarimetric measurements showing that the continuum seen in projection near θ^2 Ori is light originating from the star and scattered from the dust grains contained in the nebular material).

TABLE 2.2.IV

Spectral type and magnitudes of some prominent stars in M42

No.	Star	Spectral type	Apparent visual magnitude (V)	Absolute visual magnitude (M_V)	Total visual extinction (A_v)	References
1	θ^1 Ori A	B 0.5 V	6.82			Lee (1968) Carruthers (1969)
		B 0.5 Vp	6.72	−4.0		Underhill (1966)
				−2.8	1.52^m	Johnson (1967)
2	θ^1 Ori B	B 3	8.10			Ney *et al.* (1973)
		B 3 V	8.40			Hall and Garrison (1969)
		B 3		$-0.8 \pm 0.3^{(1)}$		Popper and Plavec (1976)
3	θ^1 Ori C	O 6p	5.12			Lee (1968), Carruthers (1969)
			5.40			Conti (1972)
				−4.5	1.62^m	Johnson (1967)
		O 7	5.16	−4.2		Conti and Aschuler (1971)
		O 7 Vp				Snow and Morton (1976)
4	θ^1 Ori D	O 9.5 V	6.70			Lee (1968), Carruthers (1969)
		B 0.5 Vp	6.70	−4.0		Underhill (1966)
				−2.8	1.50^m	Johnson (1967)
5	θ^2 Ori A	O 9.5 Vp	5.08			Lee (1968), Carruthers (1969)
		O 9 V	5.07	−4.8		Underhill (1966)
				−3.6	0.68^m	Johnson (1967)
		O 9 V	5.08	−4.1		Conti and Aschuler (1971)
		O 9.5				Warren and Hesser (1977)
6	θ^2 Ori B	B 1 V	6.38	−3.6		Underhill (1966)
		B 0.5 Vp	6.41			Lee (1968), Carruthers (1969)

(1) Secondary conponent: $M_V = +0.2 \pm 0.4$.

TABLE 2.2.V

Spectral type and magnitude of the exciting star of M43

No.	Star	Spectral type	Apparent visual magnitude (V)	Absolute visual magnitude (M_V)	Total visual extinction (A_v)	References
1	NU Ori	B 0.5 V				Schild and Chaffee (1969), Penston *et al.* (1975)
			6.5		$1.74^{m(1)}$	Thum *et al.* (1978)
		B 1 V	6.83			Penston (1972) cited in Ney *et al.* (1973)

(1) Foreground extinction $\sim 0.2^m$; A_v due to material associated with the star itself: 1.5^m.

TABLE 2.2.VI
Electron density of M42 derived from optical emission lines

No. Region	Method	Electron density $N_e(\text{cm}^{-3})$	Remarks	References
1 M42	[O II] ratio	3.9×10^3	(3)	Aller and Liller (1956)
2 M42 (core)	[O II] ratio	1.8×10^4	(1), (3)	Osterbrock and Flather (1959)
3 M42 (outer part)	[O II] ratio	10^3	(1), (3)	Osterbrock and Flather (1959)
4 M42 (core)	[O II] ratio	10^4	(3), (14)	Cruvellier (1967)
5 M42 (outer part)	[O II] ratio	10^2	(3), (14)	Cruvellier (1967)
6 M42 (core)	[S II] ratio	2×10^4	(2), (4), (13)	Danks (1970)
7 M42 (outer part)	[S II] ratio	6×10^2	(2), (4)	Danks (1970)
8 M42 (core)	[S II] ratio	4.5×10^3	(2), (4)	Danks and Meaburn (1971)
9 M42 (outer part)	[S II] ratio	1.8×10^2	(2), (4)	Danks and Meaburn (1971)
10 M42 (ionized globules)	[O II] ratio	$> 1.5 \times 10^4$	(3), (14), (15)	Elliott and Meaburn (1973b)
11 M42 (material around the globules)	[O II] ratio	$< 3 \times 10^3$	(3), (14), (15)	Elliott and Meaburn (1973b)
12 M42 (core)	[S II] ratio	2.5×10^4	(4)	Perinotto and Patriarchi (1974)
13 M42 (bright filament 14′ W of Trapezium)	[S II] ratio	$(3.6 \pm 1.5) \times 10^2$	(4)	Dufour and Mathis (1975)
14 M42 (45″ N of θ^1 Ori C)	[O II] ratio	4.6×10^3	(3), (8)	
15 M42 (45″ N of θ^1 Ori C)	[S II] ratio	6.3×10^3	(4), (8)	
16 M42 (45″ N of θ^1 Ori C)	[Cl III] ratio	5.9×10^3	(5), (8)	
17 M42 (45″ N of θ^1 Ori C)	[O II] ratio	6.9×10^3	(6), (8)	
18 M42 (45″ N of θ^1 Ori C)	[S II] ratio	1×10^4	(7), (8)	Peimbert and Torres-Peimbert (1977)
19 M42 (25″ S of θ^1 Ori C)	[O II] ratio	4.8×10^3	(3), (8)	
20 M42 (25″ S of θ^1 Ori C)	[O II] ratio	9.3×10^3	(6), (8)	
21 M42 (outer part: 3′ S of θ^1 Ori C)	[O II] ratio	1.2×10^3	(3), (8)	
22 M42 (position 1)	[S II] ratio	8×10^3	(4), (9)	
23 M42 (position 2)	[S II] ratio	6×10^3	(4), (10)	Hawley (1978)
24 M42 (position 3)	[S II] ratio	1.7×10^3	(4), (11)	
25 M42 (30″ W of θ^1 Ori C)	[S II] ratio	$(6 \pm 2) \times 10^3$	(4)	
26 M42 (outer part: 60″ E of Brun 405)	[S II] ratio	$(4 \pm 3) \times 10^3$	(4), (12)	
27 M42 (7″ S, 357″ W of θ^1 Ori C)	[S II] ratio	1×10^3	(4)	
28 M42 (7″ S, 305″ W of θ^1 Ori C)	[S II] ratio	9×10^2	(4)	McCall (1979)
29 M42 (365″ N of θ^1 Ori C)	[S II] ratio	$< 1.2 \times 10^3$	(4)	
30 M42 (365″ N, 52″ E of θ^1 Ori C)	[S II] ratio	2.4×10^3	(4)	
31 M42 (1.0 W, 10″ S of θ^1 Ori C)	[S II] ratio	1.2×10^4	(4)	Lester et al. (1979)
32 M42 (condensations)	model	$> 1.5 \times 10^5$ -4.1×10^5	(16), (17)	Laques and Vidal (1979)
33 M42 (condensations)	$H\alpha$/[O III] ratio	2.2×10^6 -3.5×10^6	(16), (18)	Laques and Vidal (1979)
34 M42 (ionized knots)	[S II] ratio	4.8×10^3 -1.3×10^4	(4), (19)	Cantó et al. (1980)

Remarks to Table 2.2.VI:

(1) For the distribution of N_e over the nebula see Figure 2.2.1.
(2) For the distribution of N_e over the nebula see the original paper.
(3) [O II] λ 3729 Å/λ 3726 Å.
(4) [S II] λ 6716 Å/λ 6731 Å.
(5) [Cl III] λ 5518 Å/λ 5538 Å.
(6) [O II] λ 7325 Å/λ 3727 Å.
(7) [S II] λ 4072 Å/λ 6724 Å.
(8) For values of N_e at various positions over the nebula see the original paper.
(9) $\Delta\alpha = +0''$, $\Delta\delta = -120''$ from Bond 628.
(10) $\Delta\alpha = +0''$, $\Delta\delta = +120''$ from Bond 628.
(11) $\Delta\alpha = -59''$, $\Delta\delta = +300''$ from Bond 628.
(12) $\sim 7\overset{'}{.}3$ away of θ^1 Ori C.
(13) From [Cl III] ratio: $N_e = 1.8 \times 10^4$ cm^{-3} and from [Ar IV] ratio $N_e = 10^4$ cm^{-3} (based on observational results by Aller and Liller (1959) and density curves derived by Saraph and Seaton (1970)).
(14) For values of the [O II] ratio versus distance from the centre of the Trapezium see the original paper; for a similar distribution covering a more extended area see Caplan (1972).
(15) The density maximum is situated $\sim 30''$ W of the Trapezium stars.
(16) Range of densities for 6 condensations.
(17) Diameter of each condensation $\leq 1\overset{''}{.}4$.
(18) Diameter of condensation in the range of $0\overset{''}{.}19 - 0\overset{''}{.}45$.
(19) Range of densities for various velocity components originating within four HH objects detected in the core of Orion; the density of the surrounding medium is much lower (1.7×10^3 cm^{-3}– 2.9×10^3 cm^{-3}).

TABLE 2.2.VII

Typical sizes and densities of various features seen in the core of M42

Feature	Angular size	Linear size (for a distance of 500 pc)	Electron density N_e (cm^{-3})	Remarks	References
Gaussian core	$\sim 3'$	~ 0.6 pc	$\sim 2 \times 10^3$	(1)	Schraml and Mezger (1969)
Small structures	$20''$	0.05 pc	$\sim 5 \times 10^4$	(2)	Webster and Altenhoff (1970)
Ionized knots	$\sim 10''$	0.02 pc	$2 \times 10^3 - 1.5 \times 10^4$	(3)	Cantó *et al.* (1980)
Ionized stellar-size condensations	$< 1''.5$	< 0.004 pc	$> 1.5 \times 10^5 - 4 \times 10^5$	(4)	Laques and Vidal (1979)

Remarks to Table 2.2.VII:

(1) Typical parameters derived from radio continuum maps made with a $2'$ angular resolution. Centre of Gaussian core, i.e. density peak: $\sim 30''$ W of the Trapezium stars (Gordon, 1969; Elliott and Meaburn, 1973; Martin and Gull, 1976). Core embedded in a tenuous envelope extending up to $10'$ from the centre.
(2) Two compact structures.
(3) Four ionized knots, probably HH objects produced by energetic stellar winds from obscured stars; for further details see Table 2.3.I.
(4) Six stellar-size ionized condensations, probably as small as $0''.20 - 0''.40$ ($0.0005 - 0.001$ pc) with densities in the range of $2 \times 10^6 - 3.5 \times 10^6$ cm^{-3}; probably partially ionized globules; for further details see Table 2.3.I.

TABLE 2.2.VIII

Electron temperature of M42 derived from optical emission lines and
optical continuum measurements

No.	Method	Electron temperature T_e (K)	Remarks	References
1	[O II], [O II] ratios	8 500	(32)	Aller and Liller (1956)
2	O, Hγ profiles	9 700		Münch (1958)
3	[O III] ratio	10 000 ± 300	(24)	Aller and Liller (1959)
4	[O III]/Hα, [O II]/Hα	9 900	(30), (31)	Pottasch (1965)
5	Balmer Continuum/Hβ	4 000		Kaler et al. (1965)
6	[O III] ratio	8 900 ± 300	(1), (24)	
7	[O III] ratio	9 450 ± 300	(2), (24)	
8	[O III] ratio	9 450 ± 300	(3), (24)	
9	[O II], [N II] ratios	13 700 ± 800	(1), (25)	
10	[O II], [N II] ratios	12 400 ± 800	(2), (25)	
11	[O II], [N II] ratios	10 500 ± 800	(3), (25)	Peimbert (1967)
12	Balmer Continuum/Hβ	7 100$^{+1600}_{-1200}$	(1)	
13	Balmer Continuum/Hβ	6 500$^{+1500}_{-1100}$		
14	Balmer Continuum/Hβ	6 100$^{+1400}_{-1000}$	(3)	
15	Balmer Decreement	2 000		Kaler (1967b)
16	Hα, [N II] profiles	7 400 ± 1500		Courtès et al. (1968)
17		9 600		Mendez (1968)
18	Hα/[N II]	7 200	(4)	Foukal (1969)
19	Hα/[N II]	7 000—9000	(33)	Baudel (1970)
20	Hα, [N II] profiles	6 500—11700	(4)	Dopita (1972)
21	Hα, [N II] profiles	4 000—13600	(4)	
22	Hα, [N II] profiles	12 400	(5)	Dopita et al. (1973)
23	Hβ, [O III] profiles	8 000—11850	(4)	
24	Hα, [N II]; Hβ, [O III]	9 700	(5)	Dopita (1973)
25	[N II] ratio	10 520 ± 800	(6), (26)	
26	[N II] ratio	9 390 ± 550	(7), (26)	
27	[O III] ratio	8 840 ± 500	(6), (24)	
28	[O III] ratio	8 970 ± 500	(7), (24)	Simpson (1973)
29	Balmer Continuum/Hβ	8 000 ± 800	(6)	
30	Balmer Continuum/Hβ	7 700 ± 800	(7)	
31	Continuum 3000–5000 Å	8 300 ± 500		Hua (1974)
32	[O III] ratio	9 200 ± 100	(4), (5), (24)	Bohuski et al. (1974)
33	[S III] ratio	9 700 ± 1000	(14)	Foukal (1974)
34	[N II] ratio	9 560	(8), (26)	Dufour and Mathis (1975)
35	Hβ, [O III] profiles	9 000 ± 350	(4)	Gibbons (1976)
36	Hα, He I profiles	10 410 ± 800	(5), (4)	Chaisson and Dopita (1977)
37	[N II] ratio	8 600	(9), (4), (26)	
38	[N II] ratio	9 750	(10), (4), (26)	
39	[O III] ratio	8 500	(9), (4), (24)	
40	[O III] ratio	8 500	(10), (4), (24)	Peimbert and Torres-Peimbert (1977)
41	[O II] ratio	9 800	(9), (4), (27)	
42	[O II] ratio	11 300	(10), (4), (27)	
43	[S II] ratio	10 900	(9), (4), (28)	
44	[N II] ratio	9 100	(11), (26)	
45	[N II] ratio	8 900	(12), (26)	
46	[N II] ratio	8 700	(13), (26)	Hawley (1978)
47	[O III] ratio	8 500	(11), (24)	
48	[O III] ratio	9 200	(12), (24)	

(continued)

Table 2.2.VIII (continued)

No.	Method	Electron temperature T_e (K)	Remarks	References
49	[S III] ratio	5 000	(15)	Olthof *et al.* (1978)
50	[N II] ratio	9 100 ± 600	(16), (26)	
51	[N II] ratio	8 400 ± 900	(17), (26)	
52	Balmer Continuum	$6\,600^{+1500}_{-1100}$		Barker (1979)
53	Paschen Continuum	$11\,200^{+1300}_{-1200}$	(16)	
54	Paschen Continuum	$12\,500^{+1400}_{-1300}$	(17)	
55	[N II] ratio	7 600 ± 900	(18), (26)	
56	[N II] ratio	8 000 ± 1200	(19), (26)	
57	[N II] ratio	< 8 000	(20), (26)	
58	[N II] ratio	7 900 ± 1200	(21), (26)	McCall (1979)
59	[O III] ratio	8 430 ± 80	(22), (24)	
60	[O III] ratio	9 000 ± 400	(23), (24)	
61	[N II] ratio	9 500 ± < 1000	(26), (29)	
62	[O III] ratio	8 100 ± 1000	(24), (29)	Lester *et al.* (1979)
63	[S III] ratio	8 100 ± < 1000	(14), (29)	

Remarks to Table 2.2.VIII:

(1) Position 35″ N of θ^1 Ori C.

(2) Position 35″ S of θ^1 Ori C.

(3) Position 35″ N of HD 37042.

(4) For values of T_e over the entire nebula see the original paper.

(5) $\alpha = 5^h32^m49^s$, $\delta = -5°25'20''$.

(6) Central part of M42.

(7) Bright bar near θ^2 Ori A.

(8) Bright filament 14′ W of Trapezium.

(9) 45″ N of θ^1 Ori C.

(10) 25″ S of θ^1 Ori C.

(11) Position: $\Delta\alpha = 0''$, $\Delta\delta = -120''$ from Bond 628.

(12) Position: $\Delta\alpha = 0''$, $\Delta\delta = +120''$ from Bond 628.

(13) Position: $\Delta\alpha = -59''$, $\Delta\delta = +300''$ from Bond 628.

(14) [S III]: $I(\lambda\,9532\,\text{Å} + \lambda\,9069\,\text{Å})/I(\lambda\,6312\,\text{Å}) = 4\exp(2.3 \times 10^4\,T_e^{-1}) + \tfrac{2}{3}$.

(15) S III: $\lambda\,9069\,\text{Å}/\lambda\,18.7\mu$.

(16) 28″ S of θ^1 Ori C.

(17) 37″ N of θ^1 Ori C.

(18) 7″ S, 375″ W of θ^1 Ori C.

(19) 7″ S, 305″ W of θ^1 Ori C.

(20) 365″ N of θ^1 Ori C.

(21) 365″ N, 52″ E of θ^1 Ori C.

(22) 30″ W of θ^1 Ori C.

(23) 60″ E of Brun 405 (7.3 from θ^1 Ori C).

(24) [O III] $(\lambda\,4363\,\text{Å})/(\lambda\,5007\,\text{Å} + \lambda\,4959\,\text{Å})$, and N_e.

(25) [O II] $\lambda\,3727\,\text{Å}/\lambda\,7325\,\text{Å}$, [N II] $\lambda\,5755\,\text{Å}/(\lambda\,6584\,\text{Å} + \lambda6548\,\text{Å})$.

(26) [N II] $(\lambda\,5755\,\text{Å})/(\lambda\,6584\,\text{Å} + \lambda\,6548\,\text{Å})$, and N_e.

(27) [O II] $\lambda\,7325\,\text{Å}/\lambda\,3727\,\text{Å}$ and $\lambda\,3726\,\text{Å}/\lambda\,3729\,\text{Å}$.

(28) [S II] $\lambda\,4072\,\text{Å}/\lambda\,6724\,\text{Å}$ and $\lambda\,6717\,\text{Å}/\lambda\,6731\,\text{Å}$.

(29) Electron density peak at 1.0 W and 10″ S of θ^1 Oric C.

(30) Pronik's method.

(31) 9900 K at the centre, falling to 8300 K 4′ away from centre, constant further away.

(32) Simultaneous solution of the equations expressing the ratios [O II] $[I(\lambda\,3729\,\text{Å})/I(\lambda\,3726\,\text{Å})]$ and [O II] $[I(\lambda\,7325\,\text{Å})/I(\lambda\,3726\,\text{Å} + 3729\,\text{Å})]$ which strongly depend on N_e and T_e, respectively. $N_e = 3.9 \times 10^3\,\text{cm}^{-3}$.

(33) Temperature estimates depend on the adopted $N(\text{N})/N(\text{H})$; see Appendix, Section II.3, Figure II.3.3; $T_e \simeq 7000$ K for $N(\text{N})/N(\text{H}) = 4 \times 10^{-4}$, $T_e \simeq 7500$ K for $N(\text{N})/N(\text{H}) = 3 \times 10^{-4}$, $T_e \simeq 8500$ K for $N(\text{N})/N(\text{H}) = 2 \times 10^{-4}$, and $T_e \simeq 9500$ K for $N(\text{N})/N(\text{H}) = 1.5 \times 10^{-4}$; temperature slowly varying over the nebula.

TABLE 2.2.IX

Electron temperature of M42 derived from radio recombination lines
and radio continuum measurements

No.	Method	Electron temperature T_e(K)	Remarks	References
1	H56α	$8\,400^{+2200}_{-1400}$	(22)	Sorochenko *et al.* (1969)
2	H56α	$7\,750 \pm 650$	(3)	Sorochenko and Berulis (1970)
3	H56α	$7\,670 \pm 500$		Berulis *et al.* (1975)
4	H65α	$6\,900^{+3400}_{-1700}$	(22)	Churchwell *et al.* (1970)
5	H66α	$7\,800 \pm 800$		Waltman and Johnson (1973)
6	H66α	$8\,200 \pm 200$	(2)	Pauls and Wilson (1977)
7	H66α	$9\,100 \pm 500$	(23), (3)	Chaisson and Dopita (1977)
8	H66α	$8\,200 \pm 300$	(4)	Wilson *et al.* (1979)
9	H66α	$7\,700 \pm 500$	(5)	Wilson *et al.* (1979)
10	H76α, He76α	$10\,400$	(6), (13)	
11	H76α, He76α	$13\,200 \pm 1\,600$	(7), (13)	
12	H76α, He76α	$12\,900 \pm 1\,400$	(8), (13)	Perrenod *et al.* (1977)
13	H76α	$9\,000 \pm 700$	(9), (13)	
14	H76α	$13\,000 \pm 2\,000$	(10), (13)	
15	H76α	$9\,500 \pm 2\,800$	(11), (13)	
16	H76α	$7\,800$	(3)	Pankonin *et al.* (1979)
17	H76α	$8\,500$	(12)	Pankonin *et al.* (1979)
18	H85α	$8\,710^{+450}_{-400}$	(22)	Churchwell and Mezger (1970)
19	H85α	$7\,890 \pm 200$	(22)	Gordon (1970)
20	H85α	$9\,000 \pm 500$	(3)	Doherty *et al.* (1972)
21	H86α	$8\,470 \pm 380$		Lichten *et al.* (1979)
22	H94α	$6\,820^{+370}_{-330}$		Mezger and Ellis (1968)
23	H94α	$6\,610 \pm 400$	(3)	Gordon and Meeks (1968)
24	H95β	$9\,500 \pm 500$	(14), (13)	Perrenod *et al.* (1977)
25	H95β	$7\,100$		Pankonin *et al.* (1979)
26	H104α	$7\,820^{+3300}_{-1720}$	(22)	Gudnov and Sorochenko (1968)
27	H104α	$7\,610^{+1280}_{-940}$	(22)	Dravskikh and Dravskikh (1967)
28	H106β	$9\,430^{+400}_{-360}$	(22)	Gordon (1970)
29	H108β	$7\,490 \pm 380$		Lichten *et al.* (1979)
30	H109α	$6\,400^{+680}_{-1500}$	(3)	Mezger and Höglund (1967)
31	H109α	$7\,480^{+570}_{-490}$	(22)	Palmer (1968)
32	H109α	$6\,690^{+360}_{-320}$		Mezger and Ellis (1968)
33	H109α	$6\,700$		Schraml and Mezger (1969)
34	H109α	$7\,240^{+500}_{-430}$	(22)	Churchwell and Mezger (1970)
35	H109α	$7\,000 \pm 800$		Reifenstein *et al.* (1970)
36	H109α	$7\,500 \pm 800$		Wilson *et al.* (1970)
37	H109α	$7\,500 \pm 800$		Mezger *et al.* (1970)
38	H109α	$7\,348$		Shaver and Goss (1970)
39	H109α	$8\,050$	(15)	Churchwell *et al.* (1978)
40	H110α	$6\,990 \pm 130$	(22)	Davies (1970)
41	H121γ	$11\,670^{+3200}_{-2000}$	(22)	Gordon (1970)
42	H126α	$6\,980$		McGee and Gardner (1968)
43	H126α + H127β	$6\,380^{+980}_{-740}$		Mezger and Ellis (1968)
44	H133δ	$9\,740^{+920}_{-760}$	(22)	Gordon (1970)
45	H134α	$7\,660$	(22)	Zuckerman and Palmer (1970)
46	H137β	$10\,120^{+880}_{-740}$	(22)	Palmer (1968)
47	H137β	$9\,220^{+1220}_{-950}$	(22)	Churchwell and Mezger (1970)
48	H137β	$11\,000 \pm 1\,900$		Mezger *et al.* (1970)

(continued)

Table 2.2.IX (continued)

No.	Method	Electron temperature $T_e(\text{K})$	Remarks	References
49	H137β	9 240	(15)	Churchwell *et al.* (1978)
50	H138β	9 150$^{+410}_{-370}$	(22)	Davies (1970)
51	H148δ	8 170$^{+2310}_{-1440}$	(22)	Gordon (1970)
52	H149α	6 480$^{+300}_{-280}$	(22)	
53	H150α	6 460$^{+320}_{-290}$	(22)	
54	H151α	6 670$^{+350}_{-320}$	(22)	
55	H152α	6 060$^{+300}_{-270}$	(22)	
56	H153α	6 160$^{+320}_{-290}$	(22)	Menon and Payne (1969)
57	H154α	6 430$^{+370}_{-330}$	(22)	
58	H155α	7 280$^{+510}_{-450}$	(22)	
59	H156α	7 790$^{+620}_{-530}$	(22)	
60	H156α	8 270$^{+1010}_{-800}$	(22)	Williams (1967)
61	H157α	6 140$^{+380}_{-340}$	(22)	Menon and Payne (1969)
62	H157γ	9 730$^{+1960}_{-1370}$	(22)	Churchwell and Mezger (1970)
63	H158α	6 460$\pm$500		Dieter (1967)
64	H156+H158α	7 130$^{+430}_{-380}$		Mezger and Ellis (1968)
65	H158α	6 890$^{+520}_{-440}$	(22)	Menon and Payne (1969)
66	H158β	13 830$^{+2240}_{-1660}$	(22)	Gardner and McGee (1967)
67	H158γ	9 130$^{+890}_{-740}$	(22)	Davies (1970)
68	H161α	6 290$^{+480}_{-410}$	(22)	
69	H162α	7 080$^{+650}_{-540}$	(22)	
70	H163α	6 480$^{+550}_{-470}$	(22)	
71	H164α	7 970$^{+930}_{-740}$	(22)	Menon and Payne (1969)
72	H165α	6 610$^{+630}_{-520}$	(22)	
73	H166α	8 650$^{+1200}_{-940}$	(22)	
74	H166α	8 380$^{+1200}_{-920}$	(22)	DeBoer *et al.* (1968)
75	H167α	8 160$^{+1100}_{-860}$	(22)	
76	H168α	7 900$^{+1080}_{-830}$	(22)	
77	H170α	7 450$^{+1030}_{-790}$	(22)	Menon and Payne (1969)
78	H171α	7 710$^{+1160}_{-870}$	(22)	
79	H172α	7 270$^{+1050}_{-800}$	(22)	
80	H172δ	10 350$^{+3100}_{-1890}$	(22)	Churchwell and Mezger (1970)
81	H173δ	10 260$^{+970}_{-800}$	(22)	Davies (1970)
82	H176α	7 010$^{+1150}_{-850}$	(22)	
83	H177α	6 470$^{+990}_{-740}$	(22)	Menon and Payne (1969)
84	H178α	12 700$^{+5500}_{-2900}$	(22)	
85	H172δ	10 350$^{+3100}_{-1890}$	(22)	Churchwell and Mezger (1970)
86	H173δ	10 260$^{+970}_{-800}$	(22)	Davies (1970)
87	H176α	7 010$^{+1150}_{-850}$	(22)	
88	H177α	6 470$^{+990}_{-740}$	(22)	Menon and Payne (1969)
89	H178α	12 700$^{+5500}_{-2900}$	(22)	
90	H186ε	10 430$^{+1620}_{-1200}$	(22)	Davies (1970)
91	H197β	21 700$^{+3000}_{-1940}$	(22)	Williams (1967)
92	H225γ	12 300$^{+2300}_{-1610}$	(22)	Williams (1967)
93	average of Hnα lines	10 000$^{+500}_{-500}$	(16)	Hjellming and Gordon (1971)
94	radio spectrum	3 000$^{+2000}_{-1000}$		Terzian *et al.* (1968)
95	brightness temperature	11 750$\pm$1 000	(17)	Pariiskii (1961b)
96	brightness temperature	7 600$\pm$800	(18)	Mills and Shaver (1968)

(continued)

Table 2.2.IX (continued)

No.	Method	Electron temperature $T_e(\mathrm{K})$	Remarks	Reference
97	brightness temperature	$7\,600 \pm 800$	(19)	Gordon (1969)
98	radio continuum	$7\,600 \pm 800$	(20)	Mills (1967)
99	radio continuum	$7\,000$	(21)	Shaver (1970)
100	radio continuum	$8\,550$	(20)	Shaver and Goss (1970)

Remarks to Table 2.2.IX

(1) Local Thermodynamic Equilibrium (LTE) solution unless otherwise stated; T_e values refer to the continuum peak unless otherwise stated.

(2) Value at $\alpha = 5^h32^m48^s$, $\delta = -5°25'17''$ (continuum peak near θ^1 Ori C); angular resolution: $40''$. T_e has a fairly constant value of 8200 ± 300 K within $2'$ from this position; for various T_e values within this area see the original paper.

(3) For values over the nebula see the original paper.

(4) T_e based on model $= 8500$ K.

(5) $\alpha = 5^h32^m50^s$, $\delta = -5°26'49''$ ($1.'6$ S of continuum peak).

(6) Method utilizing the line widths; position at the continuum peak ($\alpha = 5^h32^m50^s$, $\delta = -5°24'55''$).

(7) As in (6) but $3'$ S of continuum peak.

(8) As in (6) but $3'$ N of continuum peak.

(9) Position as in (6); method utilizing T_L/T_C of H76α line.

(10) Position as in (7); method as in (9).

(11) Position as in (8); method as in (9).

(12) Continuum peak position; non-LTE solution; T_e is constant (8000–8500 K) within a radius of $3'$ from the Trapezium.

(13) These observations show that the gas kinetic temperature increases by 3000 K at the nebular periphery.

(14) Method utilizing T_L/T_C of H95β line, value at the continuum peak (position as in (6)).

(15) T_e based on a non-LTE model: 11 170 K.

(16) Non-LTE solution; obtained by combining data from high order lines ($100 < n < 180$).

(17) At $\nu_1 = 0.4$ GHz, $\nu_2 = 3.2$ GHz.

(18) At $\nu_1 = 0.408$ GHz, $\nu_2 = 10.5$ GHz.

(19) At $\nu_1 = 0.408$ GHz, $\nu_2 = 15.5$ GHz.

(20) At $\nu = 408$ MHz.

(21) At $\nu = 85$ MHz.

(22) Recalculated by Hjellming and Gordon (1971) according to the formula:

$$T_e(\mathrm{LTE}) = (\nu_L/\mathrm{GHz})^{1.826}\left[\left\{\left(\frac{\Delta\nu_L}{\mathrm{kHz}}\ \frac{T_L}{T_C}\right)\left[1 + \frac{N(\mathrm{He^+})}{N(\mathrm{H^+})}\right]\right\}^{-1} \times \right.$$
$$\left. \times \ [(m-n)\mathrm{f}_{mn}/n]\right]^{0.87},$$

where ν_L is the frequency at the centre of a radio recombination line of hydrogen originating from the transition $m \to n$ with $m > n$; T_C is the continuum antenna temperature; T_L is the line antenna temperature; $\Delta\nu_L$ is the line width at half intensity; f_{mn} is the absorption oscillator strength; and $\mathrm{f}_{mn}/n = 0.194, 0.0271, 0.00841, 0.00365,$ and 0.00191 for $m - n = 1, 2, 3, 4,$ and 5, respectively. $N(\mathrm{He^+})/N(\mathrm{H^+})$ is taken to be 0.1.

(23) $\alpha = 5^h32^m49^s$, $\delta = -5°25'20''$.

TABLE 2.2.X

Electron temperature of M43 derived from radio recombination lines
and radio continuum measurements

No.	Method[1]	Electron temperature T_e(K)	Remarks	References
1	H76α	7 400	(2)	Pankonin et al. (1979)
2	H76α	8 500	(3)	Pankonin et al. (1979)
3	H86α	8 600 $\pm$ 1 100		Lichten et al. (1979)
4	H94α	5 630 $\pm$ 280		Gordon and Meeks (1968)
5	H108β	6 510 $\pm$ 1 100		Lichten et al. (1979)
6	H109α	6 366		Shaver and Goss (1970)
7	H109α	6 300 $\pm$ 1 000		Wilson et al. (1970)
8	H109α	5 800		Schraml and Mezger (1969)
9	H109α	6 460	(4)	Churchwell et al. (1978)
10	H137β	6 740	(4)	Churchwell et al. (1978)
11	radio continuum	4 400	(5)	Shaver and Goss (1970)
12	radio continuum	6 700	(6)	Thum et al. (1978)

Remarks to Table 2.2.X:

(1) LTE assumption, unless otherwise stated; T_e values refer to the continuum peak unless otherwise stated.
(2) For various values over the object see the original paper.
(3) Non-LTE solution.
(4) A non-LTE solution gives T_e = 7160 K.
(5) At ν = 408 MHz.
(6) At ν = 10.7 GHz.

TABLE 2.2.XI

Electron temperature of M42 as a function of distance from the Trapezium

Method	Angular resolution	Behaviour of temperature (versus distance from the Trapezium)	Remarks	References
a. Temperature fairly constant ($\sim$9000 $\pm$ 1000 K)				
Hα/[N II] ratio	3"	Stable up to 13'	(10)	Baudel (1970)
H85α line to continuum ratio	3'.2	Stable up to 5'	(1), (2)	Doherty et al. (1972)
Hα, [N II] and Hβ, [O III] profiles	40"	Stable up to 4'		Dopita et al. (1973)
[N II] ratio	14" $\times$ 60"	$\sim$9600K at a position $\sim$14' W of Trapezium	(3)	Dufour and Mathis (1975)
[N II] ratio, [O III] ratio	$\sim$5" $\times$ 80"	Stable up to 4'–5'		Peimbert and Torres Peimbert (1977)
H66α line to continuum ratio	40"	Stable up to 2'	(1), (11)	Pauls and Wilson (1977)
[N II] ratio, [O III] ratio	8" $\times$ 200"	Stable up to 4'–5'		Hawley (1978)
[N II] ratio, [O III] ratio	8" $\times$ 15"	Stable up to 7'		McCall (1979)

(continued)

Table 2.2.XI (continued)

Method	Angular resolution	Behaviour of temperature (versus distance from the Trapezium)	Remarks	References
		b. Temperature rising at the periphery		
H109α line to continuum ratio	6′	Stable within the 3′ core ($\sim 6800 \pm 200$ K), rising rather sharply to 8000 K at 6′ from the centre and subsequently decreasing to 5000 K at a distance 9′.	(1), (5)	Mezger and Ellis (1968)
Hα/[N II], Hβ/[O III] ratios	1′	Peak value (around 10^4 K) within the 3′ core, rising again to the same value towards the edge (13′).	(6)	Dopita (1973)
[O III] ratio	34″ × 42″	9200 K at the core, rising to 14 500 K at a distance of 6′.	(7)	Bohuski *et al.* (1974)
H76α, He76α profiles and H76α line to continuum ratio	2′	$\sim$9500 K at the core, rising by as much as 3000 K at 3′.	(1), (8), (9)	Perrenod *et al.* (1977)

Remarks to Table 2.2.XI:

(1) Calculated on the assumption of LTE.
(2) A trend towards higher temperature to the SW (up to 13 000 K) has however been detected.
(3) Bright filament 14′ W of the Trapezium.
(4) Theoretically predicted as a result of 'radiation hardening' (Hjellming, 1966; Rubin, 1968; the outer parts of an H II region are reached only by the most energetic photons which consequently produce more energetic electrons, i.e. electrons characterized by higher T_e than the ones produced in the core).
(5) Relatively old result probably influenced by the low angular resolution.
(6) The temperature between the central and peripheral peaks is stable and slightly more than 9000 K. Note however that the overall variation is no more than 1000 K.
(7) McCall (1979) has argued that contamination of the [O III] λ 4363 Å line with the Hg I λ 4358 Å can probably account for the observed temperature gradient.
(8) Perrenod *et al.* (1977) suggested that such a sharp gradient (750 K per arc min) may be the result of a combination of radiation hardening and a decrease of $N(O)/N(H)$ at the periphery by a factor of 4 (oxygen probably tied in grains); such a decrease, however, was not detected by Peimbert and Torres-Peimbert (1977), Kaler *et al.* (1979), and McCall (1979).
(9) McCall (1979) has argued that the difficulty of removing the C76α emission from the He76α profile, and the solution obtained by assuming LTE conditions and plane parallelism may account for such a sharp temperature gradient.
(10) Temperature peaks within the 4′ core and subsequently remains stable (till 13′). The difference between the peak value and that of the stable section is 1000 K; the actual value of temperature depends on the abundance of nitrogen (see note (33) of Table 2.2.VIII).
(11) Mean temperature 8200 $\pm$ 300 K.

TABLE 2.2.XII

Helium abundance of M42 derived from optical emission lines

No.	He lines used	$N(\text{He}^+)/N(\text{H}^+)$	$N(\text{He}^{++})/N(\text{H}^+)$	Remarks	References
1	He 4026 Å	0.16			
2	He 4471 Å	0.16			Wyse (1942)
3	He 5876 Å	0.27			
4	He 5876 Å	0.37			Johnson (1953)
5	He 4026 Å	0.16			
6	He 4471 Å	0.19			Mathis (1957)
7	He 5876 Å	0.18			
8	He 4026 Å	0.15			
9	He 4471 Å	0.12			Aller and Liller (1959)
10	He 5876 Å	0.28			
11	He 10830 Å	0.24			
12	He 5876 Å	0.117			Mathis (1962)
13	He 4471 Å	0.094			O'Dell et al. (1964)
14	He 5876 Å	0.110			
15	He 5876 Å	0.095			
16	He 5876 Å	0.088			Peimbert and Costero (1969)
17	He 5876 Å	0.072			
18	He 4686 Å		1×10^{-4}–1.4×10^{-4}		Peimbert and Goldsmith (1972)
19	He 6678 Å	0.110		(4)	Dufour and Mathis (1975)
20		0.090		(5)	Peimbert and Torres-Peimbert (1977)
21		0.082		(1)	
22		0.125		(2)	Hawley (1978)
23		0.082		(3)	
24	He 4686 Å		$< 3 \times 10^{-4}$	(6)	Lester et al. (1979)
25	He 4471 Å	0.076 ± 0.004		(7)	
26	He 4471 Å	0.074 ± 0.016		(8)	Kaler et al. (1979)
27	He 4471 Å	0.030 ± 0.016		(9)	

Remarks to Table 2.2.XII:

(1) Position 1: $(+0'', -120'')$ from Bond 628.
(2) Position 2: $(+0'', +120'')$ from Bond 628.
(3) Position 3: $(-59'', +300'')$ from Bond 628.
(4) Bright filament 14′ W of Trapezium.
(5) For values over the entire nebula see the original paper.
(6) Position at the electron density peak 1^s W and $1''$ S of θ^1 Ori C.
(7) 34″ S of θ^1 Ori C.
(8) 205″ S of θ^1 Ori C.
(9) 365″ N of θ^1 Ori C; region in the dark lane between NGC 1976 (M42) and NGC 1982 (M43).

TABLE 2.2.XIII

Helium abundance of M42 derived from radio recombination lines

No.	Lines used	$N(He^+)/N(H^+)$	Remarks	References
1	H66α, He66α	0.10 ± 0.03	(1)	Chaisson and Dopita (1977)
2	H85α, He85α	0.089 ± 0.004		Churchwell and Mezger (1970)
3	H85α, He85α	0.085 ± 0.010	(2)	Doherty et al. (1972)
4	H85α, He85α	0.084 ± 0.005		Churchwell et al. (1974)
5	H85α, He85α	0.080 ± 0.005	(3)	Balick et al. (1974)
6	H86α, He86α	0.072 ± 0.003		Lichten et al. (1979)
7	H94α, He94α	0.11 ± 0.01		Gordon and Meeks (1967)
8	H109α, He109α	0.077 ± 0.010		Mezger and Palmer (1968)
9	H109α, He109α	0.077 ± 0.010		Schraml and Mezger (1969)
10	H109α, He109α	0.083 ± 0.004		Palmer et al. (1969)
11	H109α, He109α	$0.078^{+0.016}_{-0.008}$		Reifenstein et al. (1970)
12	H109α, He109α	0.083 ± 0.016		Mezger et al. (1970)
13	H109α, He109α	0.080 ± 0.010		Mezger et al. (1970)
14	H109α, He109α	0.081 ± 0.02		Churchwell and Mezger (1970)
15	H109α, He109α	0.087 ± 0.019		Churchwell et al. (1974)
16	H109α, He109α	0.081 ± 0.015		Churchwell et al. (1974)
17	H110α, He110α	0.116 ± 0.010		Davies (1971)
18	H134α, He134α	0.087 ± 0.010		Zuckerman and Palmer (1970)
19	H138β, He138β	0.078 ± 0.024		Davies (1971)
20	Mean value from various lines	0.086 ± 0.004	(4)	Churchwell et al. (1974)
21	Mean value from various lines	0.084 ± 0.010		Churchwell et al. (1978)

Remarks to Table 2.2.XIII:
(1) Average value over six positions; may be slightly overestimated owing to the presence of a weak C66α feature.
(2) At the 2.8 cm continuum peak; for various positions over the nebula see the original paper; a contour map of the variations of $N(He^+)/N(H^+)$ is also available; $N(He^+)/N(H^+)$ is lower towards the NE than in the centre of the nebula.
(3) Value at the peak: 0.07 ± 0.01 at the 30% level and 0.06 ± 0.015 at the 10% contour (edge).
(4) $N(He^{++})/N(H^+) < 0.003$.

TABLE 2.2.XIV

Helium abundance of M43 derived from radio recombination lines

No.	Lines used	$N(He^+)/N(H^+)$	Remarks	References
1	H85α, He85α	≤ 0.120		Balick et al. (1974)
2	H86α, He86α	< 0.023		Lichten et al. (1979)
3	H91α, He91α	≤ 0.019		Thum et al. (1978)
4	H109α, He109α	0.047 ± 0.020		Churchwell et al. (1974)
5	H109α, He109α	≤ 0.010		Churchwell et al. (1977)

TABLE 2.2.XV

Ionic abundances of the Orion Nebula (M42)

Position	$\frac{N(C^+)}{N(H^+)}$	$\frac{N(C^{++})}{N(H^+)}$	$\frac{N(N^+)}{N(H^+)}$	$\frac{N(O^+)}{N(H^+)}$	$\frac{N(O^{++})}{N(H^+)}$	$\frac{N(Ne^+)}{N(H^+)}$	$\frac{N(Ne^{++})}{N(H^+)}$
M42 (continuum peak)	$(1.3\pm0.8)\times10^{-2}$						
M42 (bar, near θ^2 Ori A)			0.152×10^{-4}	1.14×10^{-4}	1.70×10^{-4}		
M42 (brightest position)							
M42 (bright filament 14' W of Trapezium)			0.33×10^{-4}		0.55×10^{-4}		
M42 (45" N of θ^1 Ori C)			0.166×10^{-4}	1.78×10^{-4}	3.63×10^{-4}		0.631×10^{-4}
M42 (25" S of θ^1 Ori C)			0.141×10^{-4}	1.05×10^{-4}	3.80×10^{-4}		0.708×10^{-4}
M42							
M42 (position 1; see remarks)			0.12×10^{-4}	2.8×10^{-4}	1.5×10^{-4}		0.40×10^{-4}
M42 (position 2; see remarks)			0.14×10^{-4}	3.0×10^{-4}	1.5×10^{-4}		0.33×10^{-4}
M42 (position 3; see remarks)			0.27×10^{-4}	2.8×10^{-4}	0.92×10^{-4}		0.17×10^{-4}
M42 (innermost area)			$(0.16\pm0.06)\times10^{-4}$	$(1.44\pm0.28)\times10^{-4}$	$(2.24\pm0.08)\times10^{-4}$		
M42 (outermost area)			$(0.20\pm0.11)\times10^{-4}$	$(1.51\pm0.37)\times10^{-4}$	$(0.77\pm0.11)\times10^{-4}$		
M42					$(5.0\pm0.9)\times10^{-4}$		
M42 (continuum peak)					1.1×10^{-4}		
M42 (continuum peak)					0.4×10^{-4}		
M42 (electron density peak)				$(1.6\pm0.18)\times10^{-4}$	$(2.4\pm0.15)\times10^{-4}$	$(0.26\pm0.02)\times10^{-4}$	$(0.55\pm0.02)\times10^{-4}$
M42 (34" S of θ^1 Ori C)			$(0.16\pm0.01)\times10^{-4}$	$(0.5\pm0.1)\times10^{-4}$	$(2.0\pm0.15)\times10^{-4}$		
M42 (205" S of θ^1 Ori C)			$(0.24\pm0.04)\times10^{-4}$	$(1.5\pm0.3)\times10^{-4}$	$(0.8\pm0.1)\times10^{-4}$		
M42 (365" N of θ^1 Ori C)			$(0.21^{+0.05}_{-0.04})\times10^{-4}$	$(1.6^{+1.3}_{-0.6})\times10^{-4}$	$(1.0^{+}_{-0.3})\times10^{-4}$		
M42 (bright core near θ^1 Ori C)	0.54×10^{-4}	1.8×10^{-4}		2.7×10^{-4}			
M42 (bar, near θ^2 Ori A)	0.74×10^{-4}	1.8×10^{-4}		2.7×10^{-4}			
M42 (mean value)							
M42 (45" N of θ^1 Ori C)	0.38×10^{-4}	1.86×10^{-4}					

Table 2.2.XV (continued)

No.	$\dfrac{N(S^+)}{N(H^+)}$	$\dfrac{N(S^{++})}{N(H^+)}$	$\dfrac{N(S^{+++})}{N(H^+)}$	$\dfrac{N(Cl^{++})}{N(H^+)}$	$\dfrac{N(Ar^{++})}{N(H^+)}$	$\dfrac{N(Fe^+)}{N(H^+)}$	$\dfrac{N(Fe^{++})}{N(H^+)}$	$\dfrac{N(Mg^+)}{N(H^+)}$	Remarks	References
1									(1), (4)	Doherty et al. (1972)
2									(1)	Simpson (1973)
3						3.11×10^{-7}	4.26×10^{-7}		(18)	Olthof and Pottasch (1975)
4	0.29×10^{-4}								(19)	Dufour and Mathis (1975)
5	0.004×10^{-4}	0.095×10^{-4}		0.0008×10^{-4}	0.027×10^{-4}				(1), (14)	Peimbert and Torres-Peimbert (1977)
6	0.0027×10^{-4}	0.071×10^{-4}			0.030×10^{-4}				(1), (14)	Peimbert and Torres-Peimbert (1977)
7		0.12×10^{-4}								Olthof et al. (1978)
8	0.005×10^{-4}	0.095×10^{-4}							(2)	} Hawley (1978)
9	0.005×10^{-4}	0.090×10^{-4}								
10	0.014×10^{-4}	0.086×10^{-4}								
11									(3)	} McCall (1979)
12										
13										
14									(20)	Melnick et al. (1979)
15									(21)	Storey et al. (1979)
16	0.0025×10^{-4}	$(0.12\pm0.006)\times10^{-4}$	$(9.6\pm0.02)\times10^{-6}$						(7)	Storey et al. (1979)
17										Lester et al. (1979)
18										} Kaler et al. (1979)
19										
20								$\le1.4\times10^{-7}$	(8)	} Perinotto and Patriarchi (1980)
21								$\le2.1\times10^{-7}$	(9)	
22						4.24×10^{-8}			(22)	Cosmovici et al. (1980)
23									(1)	Torres-Peimbert et al. (1980)

TABLE 2.2.XVI
Chemical composition of the Orion Nebula (M42)

No. Position	$\dfrac{N(\mathrm{He})}{N(\mathrm{H})}$	$\dfrac{N(\mathrm{C})}{N(\mathrm{H})}$	$\dfrac{N(\mathrm{N})}{N(\mathrm{H})}$	$\dfrac{N(\mathrm{O})}{N(\mathrm{H})}$	$\dfrac{N(\mathrm{Ne})}{N(\mathrm{H})}$
1 M42 (mean value)	0.109	5.13×10^{-4}	4.27×10^{-5}	6.17×10^{-4}	7.24×10^{-5}
2 M42 (θ^1 Ori C)				3.05×10^{-4}	
3 M42	0.064–0.081				
4 M42 (mean value)	0.101 ± 0.010				
5 M42 (mean value)			$(3.4 \pm 0.4) \times 10^{-5}$	$(2.2 \pm 0.3) \times 10^{-4}$	
6 M42 (bar, near θ^2 Ori A)			3.8×10^{-5}	2.84×10^{-4}	
7 M42 (mean value)	0.104–0.132				
8 M42			4.17×10^{-5}	5.01×10^{-4}	8.51×10^{-5}
9 M42 (mean value)	0.101		4.68×10^{-5}	7.76×10^{-4}	14.45×10^{-5}
10 M42 (mean value)	0.140				
11 M42 (brightest position)					
12 M42 (mean value)	0.100	3.31×10^{-4}	5.75×10^{-5}	5.62×10^{-4}	7.94×10^{-5}
13 M42					
14 M42 (mean value)	0.113		2.46×10^{-5}	4.17×10^{-4}	9.4×10^{-5}
15 M42 (innermost area)			$(3.98 \pm 0.92) \times 10^{-5}$	$(3.68 \pm 0.36) \times 10^{-4}$	
16 M42 (outermost area)			$(3.09 \pm 1.17) \times 10^{-5}$	$(2.28 \pm 0.48) \times 10^{-4}$	
17 M42 (electron density peak)				3.98×10^{-4}	8.13×10^{-5}
18 M42 (mean value)				$\sim 2.5 \times 10^{-4}$	
19 M42 (mean value)		$(2.52^{+1.46}_{-0.94}) \times 10^{-4}$			
20 M42 (mean value)		3.16×10^{-4}	5.01×10^{-5}	3.98×10^{-4}	
21 M42 (mean value)					
22 Solar Photosphere	0.0589	3.55×10^{-4}	8.51×10^{-5}	5.89×10^{-4}	9.33×10^{-5}
23 Average Planetary Nebula	0.170		12.58×10^{-5}	7.94×10^{-4}	7.94×10^{-5}
24 Interstellar Matter	0.100	2.3×10^{-4}	1.3×10^{-4}	6.0×10^{-4}	2.0×10^{-4}

Table 2.2.XVI (continued)

No.	$\frac{N(S)}{N(H)}$	$\frac{N(Cl)}{N(H)}$	$\frac{N(Ar)}{N(H)}$	$\frac{N(Fe)}{N(H)}$	Remarks	References
1	3.16×10^{-5}					Peimbert and Costero (1969)
2					(5)	Isobe (1970)
3					(13)	Robbins et al. (1971)
4					(12)	Batchelor and Brocklehurst (1972)
5						Dopita (1973)
6						Simpson (1973)
7						Dopita et al. (1974)
8						Dufour (1974)
9						Peimbert and Torres-Peimbert (1974)
10						Thomsen (1975)
11				1.16×10^{-6}	(18)	Olthof and Pottasch (1975)
12	2.57×10^{-5}	1.41×10^{-7}	5.01×10^{-6}		(14)	Peimbert and Torres-Peimbert (1977)
13	1.4×10^{-5}					Olthof et al. (1978)
14	1.4×10^{-5}					Hawley (1978)
15					(1), (3)	⎫ McCall (1979)
16	2.19×10^{-5}				(1), (3)	⎭
17					(7)	Lester et al. (1979)
18						Kaler et al. (1979)
19					(11)	Perinotto and Patriarchi (1980)
20						Torres-Peimbert et al. (1980)
21		5.94×10^{-8}		8.94×10^{-7}	(16), (23), (17), (22)	Cosmovici et al. (1980)
22	1.62×10^{-5}	(1.7×10^{-7})	(3.55×10^{-6})	6.61×10^{-6}	(6)	Lambert (1968), Lambert and Warner (1968), Bertsch et al. (1972), Lang (1978)
23	7.94×10^{-5}	7.94×10^{-6}	10×10^{-6}			Aller and Czyzak (1968)
24	2.1×10^{-5}		7.1×10^{-6}	3.2×10^{-5}	(24)	Unsöld (1972)

TABLE 2.2.XVII

Chemical composition of the Orion Nebula (M42)

(logarithmic scale; see remark (15))

Position	$\frac{N(\text{He})}{N(\text{H})}$	$\frac{N(\text{C})}{N(\text{H})}$	$\frac{N(\text{N})}{N(\text{H})}$	$\frac{N(\text{O})}{N(\text{H})}$	$\frac{N(\text{Ne})}{N(\text{H})}$	$\frac{N(\text{S})}{N(\text{H})}$	$\frac{N(\text{Cl})}{N(\text{H})}$	$\frac{N(\text{Ar})}{N(\text{H})}$	$\frac{N(\text{Fe})}{N(\text{H})}$	Remarks	References
M42 (mean value)	11.04	8.71	7.63	8.79	7.86	7.50					Peimbert and Costero (1969)
M42 (θ^1 Ori C)	10.8 – 10.9			8.48						(5)	Isobe (1970)
M42 (mean value)	11.004±0.041									(13)	Robbins et al. (1971)
M42 (mean value)			7.53±0.05	8.34±0.05						(12)	Batchelor and Brocklehurst (1972)
M42 (bar, near θ^2 Ori A)			7.58	8.45							Dopita (1973)
M42 (mean value)	11.02 – 11.12										Simpson (1973)
M42	11.00		7.62	8.70	7.93						Dopita et al. (1974)
M42 (mean value)	11.15		7.67	8.89	8.16						Dufour (1974)
M42 (mean value)	11.00		7.76	8.75							Peimbert and Torres-Peimbert (1974)
M42 (brightest position)											Thomsen (1975)
M42 (mean value)					7.90	7.41	5.15	6.7	6.06	(18)	Olthof and Pottasch (1975)
M42		8.52				7.15				(14)	Peimbert and Torres-Peimbert (1977)
M42 (mean value)			7.39	8.62	7.97	7.15					Olthof et al. (1978)
M42 (mean value)											Hawley (1978)
M42 (innermost area)			7.60±0.09	8.57±0.04						(1), (3)	McCall (1979)
M42 (outermost area)			7.49±0.14	8.36±0.08						(1), (3)	McCall (1979)
M42 (electron density peak)				8.60	7.91	7.34				(7)	Lester et al. (1979)
M42		8.4±0.2		8.40						(11)	Kaler et al. (1979)
M42 (mean value)											Perinotto and Patriarchi (1980)
M42 (mean value)		8.5	7.7	8.6							Torres-Peimbert et al. (1980)
M42 (mean value)							4.77		5.95	(16), (23), (17), (22)	Cosmovici et al. (1980)
Solar Photosphere	10.77	8.55	7.93	8.77	7.97	7.21	(5.23)	(6.55)	6.82	(6)	Lambert (1968), Lambert and Warner (1968), Bertsch et al. (1972), Lang (1978)
Average Planetary Nebula	11.23		8.10	8.90	7.90	7.90	6.90	7.00			Aller and Czyzak (1968)
Interstellar Matter	11.00	8.36	8.11	8.78	8.30	7.32	6.85		7.50	(24)	Unsöld (1972)

Remarks to Tables 2.2.XV, XVI, and XVII:

1) For values over the entire nebula see the original paper.

2) Position 1: $\Delta\alpha = +0''$, $\Delta\delta = -120'$ from Bond 628.
   Position 2: $\Delta\alpha = +0''$, $\Delta\delta = +120'$ from Bond 628.
   Position 3: $\Delta\alpha = -59''$, $\Delta\delta = +300''$ from Bond 628.

3) Position of innermost area 30″ W of θ^1 Ori C, position of outermost area 60′ E of Brun 405 (7.3 from θ^1 Ori C).

4) This is however from carbon, which does not originate in the same volume as the hydrogen line.

5) For $T_e = 8000$ K; for the distribution of $N(O)/N(H)$ versus distance and T_e see the original paper; $N(O)/N(H)$, 20′ away from θ^1 Ori C is 2.69×10^{-4} (8.43 in logarithmic scale).

6) Values in parenthesis are cosmic abundances.

7) 1″ W and 10″ S of θ^1 Ori C.

8) 34″ W and 3″ S of θ^1 Ori C in the area of the maximum optical surface brightness.

9) 39″ N of θ^2 Ori A (across the bar).

0) Very close to the solar value (recently published photospheric value 8.67, Lambert (1978); coronal value 8.50, Withbroe (1978)).

1) Average of three values: $(2.5 \pm 0.2) \times 10^{-4}$ at 34″ S of θ^1 Ori C, $(2.3 \pm 0.3) \times 10^{-4}$ at 205″ S of θ^1 Ori C and $(2.6^{+1.2}_{-0.7}) \times 10^{-4}$ at 365″ N of θ^1 Ori C. The corresponding logarithmic abundances are: $8.40^{+0.03}_{-0.03}$, $8.36^{+0.18}_{-0.06}$, $8.41^{+0.18}_{-0.13}$. No substantial depletion of O was found.

2) Radius of singly ionized helium sphere $\simeq 3'$; obtained from optical observation and a variable density model.

3) Neutral hydrogen was not taken into account in the calculation; result partly based on earlier not very accurate photographic observations.

4) For mean square temperature fluctuations $t^2 = 0.035$.

5) $\log N(X) = \log N(H) + \log [N(X)/N(H)]$ which for $\log N(H) = 12$ becomes $\log N(X) = 12 + \log [N(X)/N(H)]$; values of $N(X)/N(H)$ are taken from Table 2.2.XVI.

6) An appreciable amount of Cl should be contained in molecules like HCl^+ and HCl.

7) Fe is 35 times lower than in the Sun; missing Fe probably tied in grains.

8) Estimated for $T_e = 9000$ K; the respective values for $T_e = 7000$ K are:
   $\dfrac{N(Fe^+)}{N(H^+)} = 9.87 \times 10^{-7}$, $\dfrac{N(Fe^{++})}{N(H^+)} = 1.25 \times 10^{-6}$ and $\dfrac{N(Fe)}{N(H)} = 3.49 \times 10^{-6}$ or 6.54 in logarithmic scale.

9) The $N(N^+)/N(H^+)$ is in the range 0.26×10^{-4}–0.39×10^{-4} and the $N(S^+)/N(H^+)$ is in the range 0.23×10^{-4}–0.35×10^{-4}.

0) Radio data.

1) Far infrared data.

2) Estimated for $T_e = 9000$ K; the respective values for $T_e = 7000$ K are
   $\dfrac{N(Fe^+)}{N(H^+)} = 2.74 \times 10^{-7}$ and $\dfrac{N(Fe)}{N(H)} = 2.77 \times 10^{-6}$ or 6.44 in logarithmic scale.

3) $N(Cl)/N(H)$ estimated for $T_e = 9000$ K; the respective value for $T_e = 7500$ K is 8.01×10^{-8} or 4.90 in logarithmic scale.

4) Average values; cited in Mezger, (1972) as private communication.

TABLE 2.2.XVIII
Radial velocities of M42 derived from optical emission lines

No.	Line	Beam	E= emission A= absorption	$V_{LSR}^{(1)}$ (km s⁻¹)	Remarks	References
1	Hα		E	−0.1	(12)	Courtès (1960)
2	Hα	1ʺ;1ʺ6	E	−3.1±0.5	(12), (2)	Foukal (1969)
3	Hα	1ʺ3	E	−0.9±0.6	(12), (2)	Chaisson and Dopita (1977)
4	Hα	2′	E	−1.2	(12), (10), (2)	Balick et al. (1980)
5	H°		E	+2.0	(7), (8)	Kaler (1967a)
6	H°		E	−3.6±0.3	(16), (17)	Fehrenbach (1977)
7	He I		A	~−13	(26), (27)	Wilson (1939), Adams (1944), Wilson et al. (1959)
8	He I	1ʺ22	A	−15.4	(18), (19)	Vaughan (1968)
9	He I	1ʺ22	E	+14	(18), (19), (20), (24)	Vaughan (1968)
10	He I	1ʺ3	E	−1.5±1.0	(3), (2)	Chaisson and Dopita (1977)
11	He I		E	−3±0.5	(16), (17)	Fehrenbach (1977)
12	He I	2′	E	−0.8	(9), (10), (2)	Balick et al. (1980)
13	He I	45ʺ	E	−4 to +4	(21), (23), (24)	Münch and Taylor (1980)
14	He I	45ʺ	E	+11 to +15	(22), (23), (24)	Münch and Taylor (1980)
15	He		E	+3.3	(7), (8)	Kaler (1967a)
16	C I	40ʺ	E	+7.9	(6)	Hippelein and Münch (1978)
17	C II		E	−3±1.4	(16), (17)	Fehrenbach (1977)
18	N II		E	−3.5±2.0	(16), (17)	Fehrenbach (1977)
19	[N II]	2′	E	+4.2	(13), (10), (2)	Balick et al. (1980)
20	O I	50ʺ	E	+9.9±1	(4), (15)	Münch and Taylor (1974)
21	[O I]	50ʺ	E	+9.9±1	(5), (15)	Münch and Taylor (1974)
22	[O I]	2′	E	+10.4	(5), (10), (2)	Balick et al. (1980)
23	O II		E	−2.9±0.4	(16), (17)	Fehrenbach (1977)
24	[O II]		E	−4.1±2	(16), (17)	Fehrenbach (1977)
25	[O III]	1ʺ3	E	−0.1	(11), (25)	Wilson et al. (1959)
26	[O III]		E	−1.9±0.2	(16), (17)	Fehrenbach (1977)
27	[O III]	2′	E	+2.0	(11), (10), (2)	Balick et al. (1980)
28	O⁺⁺		E	−0.8	(7), (8)	Kaler (1967a)
29	[Ne III]		E	−4.9±1.5	(16), (17)	Fehrenbach (1977)
30	Ne⁺⁺		E	−4.2	(7), (8)	Kaler (1967a)
31	[S II]		E	+3.6	(16), (17)	Fehrenbach (1977)
32	[S II]	2′	E	+4	(14), (10), (2)	Balick et al. (1980)
33	S⁺⁺⁺		E	−1.4	(7), (8)	Kaler (1967a)
34	[Fe II]		E	+6.2±1	(16), (17)	Fehrenbach (1977)
35	Fe⁺		E	+8.8	(7), (8)	Kaler (1967a)
36	Fe⁺⁺		E	+9.4	(7), (8)	Kaler (1967a)
37	[Fe III]		E	−3±1	(16), (17)	Fehrenbach (1977)
38	Si II		E	−1.3±2	(16), (17)	Fehrenbach (1977)
39	[Ni II]		E	+6.7	(16), (17)	Fehrenbach (1977)

Remarks to Table 2.2.XVIII:

(1) $V_{LSR} = (V_{HeI} - 18.1)$ km s⁻¹.

(2) For variations of V_{LSR} over the nebula see the original paper.

(3) He I λ 6678 Å.

(4) O I λ 8446 Å, permitted line.

(5) [O I] λ 6300 Å.

(6) C I λ 9850 Å.

(7) From the study of a plate of a 20^h exposure at the coudé spectrograph of the $100''$ reflector at a dispersion of 20 Å mm^{-1}; many lines grouped according to ion.

(8) The H$^+$, He$^+$, Fe$^+$, O^{++}, Ne^{++}, and S^{+++} lines near θ^1 Ori C have radial velocities which show a dependence on the ionization potential of the emitting ions (Kaler, 1967a). This can be seen from the following table:

Emitting ion	Ionization potential (eV)	V_{LSR} (km s^{-1})	$V_{Heliocentric}$ (km s^{-1})
Fe$^+$	7.9	+8.8	+26.9
H$^+$	13.6	+2.0	+20.1
Fe^{++}	16.7	+9.4	+27.5
He$^+$	24.6	+3.3	+21.4
S^{+++}	35.0	−1.4	+16.7
O^{++}	35.1	−0.8	+17.3
Ne^{++}	41.1	−4.2	+13.9

(9) He I, λ 5876 Å.

(10) At θ^1 Ori C.

(11) [O III] λ 5007 Å.

(12) Hα λ 6563 Å.

(13) [N II] λ 6584 Å.

(14) [S II] λ 6717 Å.

(15) $45''$ W and $20''$ N of θ^1 Ori C.

(16) From a plate of a 6^h19^m exposure, at the coudé spectrograph of the 1.52 m Haute-Provence telescope, at a dispersion of 20 Å mm^{-1}; many lines grouped according to ion.

(17) Position on the 'bar' near θ^2 Ori A.

(18) He I λ 10830 Å; triplet with components λ 10829.081 Å, 10830.250 Å, and 10830.341 Å, respectively.

(19) $0\overset{''}{.}8$ S from θ^1 Ori C.

(20) V_{LSR} estimated from profiles observed at six positions over the nebula is in the range of +6 to +17 km s^{-1}.

(21) He I λ 10829 Å; one of the components of He I λ 10830 Å; see note (18).

(22) He I λ 10830 Å; components 10830.250 Å and 10830.341 Å; see note (18).

(23) Range of V_{LSR} estimated from the observation of 13 emission profiles over the nebula.

(24) Observed shift in V_{LSR} with respect to the V_{LSR} obtained from other lines is due to the complicated radiative transfer characterizing the line (helium both emitting and absorbing along the line of sight).

(25) Mean V_{LSR} velocity computed from the data of Wilson *et al.* (1959) for a position $0\overset{''}{.}8$ S of θ^1 Ori C.

(26) He I λ 3889 Å.

(27) Mean velocity of the main absorption feature seen against the Trapezium stars; analytical values including velocities observed against other prominent stars of the area are:

Stars	V_{LSR} of He I absorption line (km s^{-1})
θ^1 Ori A	−13
θ^1 Ori B	−12
θ^1 Ori C	−14, −49
θ^1 Ori D	−14, +6
θ^2 Ori A	−15.5
θ^2 Ori B	−6
HD 37062	−13.5
HD 37061	+0.5

(28) V_{LSR} estimated from profiles observed at six positions over the nebula is in the range of −6.5 to −18.1 km s^{-1}.

TABLE 2.2.XIX

Radial velocities of M42 derived from radio recombination and
neutral hydrogen lines

No.	Line	Beam	E = emission A = absorption	$V_{LSR}^{(1)}$ (km s^{-1})	Remarks	References
			a. Hydrogen lines			
1	H39α	1ʺ2	E	-2 ± 2	(26)	Gottlieb et al. (1978)
2	H41α		E	0 ± 4		Lovas et al. (1976)
3	H42α		E	-2 ± 2		Lovas et al. (1976)
4	H66α	1ʺ3	E	-1.3 ± 0.4	(4),(5)	Chaisson and Dopita (1977)
5	H66α	40ʺ	E	-2.1 ± 0.5	(2)	Pauls and Wilson (1977)
6	H66α	43ʺ	E	-1.4 ± 0.6	(3)	Wilson et al. (1979)
7	H76α	2′	E	-1.8	(27), (5)	Gull and Balick (1974)
8	H76α	2′	E	-1.8 ± 0.1	(6)	Perrenod et al. (1977)
9	H76α	0ʺ9	E	-1.3 ± 0.1	(7), (5)	Pankonin et al. (1979)
10	H83β	43ʺ	E	$+0.4\pm0.6$	(8)	Wilson et al. (1979)
11	H85α	3ʺ2	E	-2.9 ± 0.2	(5)	Doherty et al. (1972)
12	H85α	2ʺ8	E	-2.8		Balick et al. (1974)
13	H85α	3′	E	-2.5 ± 0.5		Churchwell et al. (1974)
14	H85α	2ʺ8	E	-3.49 ± 0.11	(17)	Jaffe and Pankonin (1978)
15	H86α	3ʺ5	E	-2.42 ± 0.04	(9)	Lichten et al. (1979)
16	H92α	4′	E	-3.0 ± 0.1	(9)	Chaisson (1974)
17	H94α	4ʺ2	E	-3.0 ± 0.8		Gordon and Meeks (1968)
18	H95β	2′	E	-1.9 ± 0.1	(6)	Perrenod et al. (1977)
19	H95β	0ʺ9	E	-1.7 ± 0.3	(7), (10)	Pankonin et al. (1979)
20	H104α		E	-2		Gudnov and Sorochenko (1969)
21	H107β	3ʺ5	E	-2.0 ± 0.1	(9)	Lichten et al. (1979)
22	H109α	6ʺ4	E	-2.0 ± 1.2		Mezger and Höglund (1967)
23	H109α	2′	E	-1.9		Schraml and Mezger (1969)
24	H109α	6ʺ5	E	-2.7 ± 0.5		Reifenstein et al. (1970)
25	H109α	4′	E	-2.8 ± 0.8		Wilson et al. (1970)
26	H109α	6′	E	-3.8 ± 0.5		Churchwell et al. (1974)
27	H109α	2ʺ6	E	-2.8 ± 0.1	(11), (5)	Jaffe and Pankonin (1978)
28	H109α	2ʺ6	E	-3.4 ± 0.1		Churchwell et al. (1978)
29	H126α +H127α		E	-7	(23), (24)	McGee and Gardner (1968)
30	H137β	2ʺ6	E	-3.4 ± 0.1		Churchwell et al. (1978)
31	H143α	13′	E	-5.7 ± 0.8	(21)	Lockman and Brown (1975)
32	H158α		E	-4.9		Lilley et al. (1966)
33	H158α	31ʺ4 × 35ʺ6	E	-4.9		Dieter (1967)
34	H158α(H II)		E	-7.0 ± 0.3	(12)	Chaisson and Lada (1974)
35	H158α(H I)		E	$+6.6\pm0.6$	(13)	Chaisson and Lada (1974)
36	H166α		E	0		Palmer and Zuckerman (1966)
37	H166α		E	0		McGee and Gardner (1967)
38	H166α	14′ × 19′	E	-4.2 ± 1.3		Pedlar and Davies (1972)
39	H166α(H II)	13′ × 13′	E	-7.5 ± 0.3	(14), (12)	Pedlar and Hart (1974)
40	H166α(H I)	13′ × 13′	E	$+5.5\pm1.5$	(14), (13)	Pedlar and Hart (1974)
41	H166α(H II)		E	-8.9 ± 0.4	(12)	Chaisson and Lada (1974)
42	H166α(H I)		E	$+6.5\pm0.3$	(13)	Chaisson and Lada (1974)
43	H183α	30′	E	-6.1 ± 2		Zuckerman and Ball (1974)
44	H192α	16′ × 22′	E	-18.9 ± 2.5		Pedlar and Davies (1972)

(continued)

Table 2.2.XIX (continued)

No.	Line	Beam	E=emission A=absorption	$V_{LSR}^{(1)}$ (km s^{-1})	Remarks	References
45	H197α	35′	E	−4.6±1.6	(21)	Lockman and Brown (1975)
46	H198α	36′	E	−6.8±3		Zuckerman and Ball (1974)
47	H210α +H211α	43′	E	−3±4		Zuckerman and Ball (1974)
48	H220α	31′×31′	E	−7.2±4.4		Pedlar and Davies (1972)

b. Helium lines

No.	Line	Beam	E=emission A=absorption	$V_{LSR}^{(1)}$ (km s^{-1})	Remarks	References
49	He66α	1′.3	E	−1.0±1.5	(4),(5)	Chaisson and Dopita (1977)
50	He76α	2′	E	−2.2±0.3	(6)	Perrenod *et al.* (1977)
51	He85α	3′	E	−3.1±1.0		Churchwell *et al.* (1974)
52	He85α	2′.8	E	−2.7		Balick *et al.* (1974)
53	He85α	2′.8	E	−2.0±0.9	(17)	Jaffe and Pankonin (1978)
54	He86α	3′.5	E	−2.6±0.3	(9)	Lichten *et al.* (1979)
55	He109α	6′	E	−4.9±1.0		Reifenstein *et al.* (1970), Churchwell *et al.* (1974)
56	He109α	2′.6	E	−2.6±0.5	(11), (5)	Jaffe and Pankonin (1978)
57	He109α	2′.6	E	−3.7±0.2		Churchwell *et al.* (1978)
58	He111α	6′	E	−5.1±0.6	(15)	Boughton (1978)
59	He126α	10′	E	−6.6±1.5	(15)	Boughton (1978)
60	He137β	2′.6	E	−3.8±1.3		Churchwell *et al.* (1978)
61	He137β	2′.6	E	−4.5±2.9	(11), (5)	Jaffe and Pankonin (1978)
62	He143β	13′	E	+5.0±3.0	(21)	Lockman and Brown (1975)
63	He158α		E	−9.4±2.5		Chaisson and Lada (1974)
64	He166α	21′	E	−7±5		Zuckerman and Ball (1974)
65	He166α		E	−7.1±2.9		Chaisson and Lada (1974)
66	He166α	13′	E	−11.7±3.0	(14)	Pedlar and Hart (1974)

c. Carbon lines

No.	Line	Beam	E=emission A=absorption	$V_{LSR}^{(1)}$ (km s^{-1})	Remarks	References
67	C75α	1′.3	E	+9.3±0.3	(16), (5)	Kuiper and Evans (1978)
68	C76α	2′	E	+9.8±0.5	(6)	Perrenod *et al.* (1977)
69	C85α	≳3′	E	+10.0		Gordon and Churchwell (1970)
70	C85α	3′.2	E	+8.0±1.4		Doherty *et al.* (1972)
71	C85α	2′.8	E	+9.7±0.8	(5)	Balick *et al.* (1974)
72	C85α		E	+9.28±0.60		Ahmad (1976)
73	C85α	2′.8	E	+8.0±0.3	(17)	Jaffe and Pankonin (1978)
74	C86α	3′.5	E	+10.8±0.3	(9)	Lichten *et al.* (1979)
75	C92α	4′	E	+9.8±0.2	(9)	Chaisson (1974)
76	C92α	4′	E	+9.7±1.2		Chaisson (1974a)
77	C94α	4′.2	E	+9.2		Chaisson (1972,1973)
78	C107β		E	+9.22±0.67		Ahmad (1976)
79	C109α	6′	E	+7.9		Palmer *et al.* (1967)
80	C109α	6′	E	+11.3±1.0		Churchwell (1970)
81	C109α	2′.6	E	+9.9±0.1		Churchwell *et al.* (1978)
82	C109α	2′.6	E	+10.7±0.6	(11), (5), (20)	Jaffe and Pankonin (1978)
83	C110α		E	+9.7±0.5		Balick *et al.* (1974)
84	C111α	6′	E	+9.2±0.4	(15),(22)	Boughton (1978)
85	C126α	10′	E	+8.9±0.6	(15),(22)	Boughton (1978)
86	C134α	11′	E	+8.2±1.2		Zuckerman and Palmer (1970)
87	C137β	2′.6	E	+9.3±0.4		Churchwell *et al.* (1978)

(continued)

Table 2.2.XIX (continued)

No.	Line	Beam	E= emission A= absorption	$V_{LSR}^{(1)}$ (km s^{-1})	Remarks	References
88	C137β	2.6	E	~ +8.1	(11), (5)	Jaffe and Pankonin (1978)
89	C137β	2.6	E	+7.9±0.4	(17), (5)	Jaffe and Pankonin (1978)
90	C143α	13'	E	+10.0±3.0	(21)	Lockman and Brown (1975)
91	C157α	18'	E	+11		Churchwell and Edrich (1970)
92	C158α		E	+6.7±0.3		Chaisson and Lada (1974)
93	C166α	21'	E	+7.0±0.5		Chaisson and Lada (1974)
94	C166α	13'	E	+6.1±0.6	(14)	Pedlar and Hart (1974)
95	C166α	21'	E	+7.0±1.0		Zuckerman and Ball (1974)
96	C183α	30'	E	+7.2±2.0		Zuckerman and Ball (1974)
97	C197α	35'	E	+9.0±3.0	(21)	Lockman and Brown (1975)
98	C198α	36'	E	+3.4±2.0		Zuckerman and Ball (1974)
99	C210α +C211α	43'	E	+6.0±5.0		Zuckerman and Ball (1974)
100	C220α	31'	E	+5.0		Pedlar and Davies (1972)

d. Heavy element lines[18]

No.	Line	Beam	E= emission A= absorption	$V_{LSR}^{(1)}$ (km s^{-1})	Remarks	References
101	Z85α		E	+2.5	(19)	Ahmad (1976)
102	Z92α	4'	E			Chaisson (1974α)
103	Z107β		E	+0.5	(19)	Ahmad (1976)
104	Z109α	2.6	E	−1.0±0.3	(17), (19)	Jaffe and Pankonin (1978)
105	Z109α	2.6	E	+9.3±0.8		Churchwell *et al.* (1978)
106	Z158α		E	−4.6	(19)	Chaisson and Lada (1974)
107	Z166α		E	−3.2	(19)	Chaisson and Lada (1974)
108	Z166α	13'	E	−0.6±0.5	(19)	Pedlar and Hart (1974)

e. Neutral hydrogen (H I, 21 cm) lines

No.	Line	Beam	E= emission A= absorption	$V_{LSR}^{(1)}$ (km s^{-1})	Remarks	References
109	H I		A	+4.5, +2.5		Clark (1965)
110	H I		A	+7.4		Vershuur (1969)
111	H I		A	+4.0		Menon (1970)
112	H I		A	+4.8		Menon (1970)
113	H I		A	+3.5		Gordon (1970)
114	H I		A	6±1		Radhakrishnan *et al.* (1972)
115	H I		A	5.51±0.04		Radhakrishnan *et al.* (1972)
116	H I	2'	A	−2.6, +5.0	(21), (25)	Lockhart and Goss (1978)
117	H I	2'×20'	A	~ +4	(28), (29)	Alferova *et al.* (1979)

Remarks to Table 2.2.XIX:

(1) $V_{LSR} = (V_{Hel} - 18.1)$ km s^{-1}; the quoted velocities were measured at the position of the continuum peak which is near the θ^1 Ori C star ($\alpha = 5^h32^m49^s$, $\delta = -5°25'16''$). The exact position is often quoted. The measured average velocity of the four Trapezium stars (θ^1 Ori A, B, C, D) is $V_{LSR} = +11$ km s^{-1} (Johnson, 1965).

(2) $\alpha = 5^h32^m48^s$, $\delta = -5°25'17''$; for variations of V_{LSR} within 2' of this position see the original paper; angular resolution of observations 40''.

(3) $\alpha = 5^h32^m47^s$, $\delta = -5°25'23''$; for a position 1.6 S: $V_{LSR} = -2.2 \pm 0.2$ km s^{-1}.

(4) $\alpha = 5^h32^m49^s$, $\delta = -5°25'20''$.

(5) For V_{LSR} variations over the nebula see the original paper.

(6) $\alpha = 5^h32^m50^s$, $\delta = -5°24'55''$.

(7) Position: $\Delta\alpha = -22''$, $\Delta\delta = 0''$ from θ^1 Ori C.

(8) $\alpha = 5^h32^m47^s$, $\delta = -5°25'23''$.

(9) $\alpha = 5^h32^m51^s$, $\delta = -5°25'39''$.

(10) For V_{LSR} variations around θ^1 Ori C see the original paper.

(11) Position: $\Delta\alpha = +20''$, $\Delta\delta = +10''$ from θ^1 Oric C.

(12) Originating from the H II region.

(13) Narrow line originating from a cool H I cloud lying along the line of sight.

(14) $\alpha = 5^h32^m49^s$, $\delta = -5°24'57''.6$.

(15) $\alpha = 5^h32^m50^s$, $\delta = -5°25'10''.8$.

(16) $\alpha = 5^h32^m48^s.6$, $\delta = -5°25'03''$.

(17) Position: $\Delta\alpha = +215'$, $\Delta\delta = +160'$ from θ^1 Ori C.

(18) Not identified; Chaisson (1974a) has attributed the observed blended feature to the elements ^{24}Mg, ^{28}Si, ^{32}S, and ^{56}Fe.

(19) Carbon velocity scale.

(20) The C109α data indicate the existence of three distinct regions of ionized carbon characterized by radial velocities 6, 8, 5, and 11 km s^{-1}. The 6 and 11 km s^{-1} regions appear to be localized near the optical dark bay. The 8.5 km s^{-1} component peaks to the NE of the radio continuum peak (see also comments in the legend to the Figure 4.1.3).

(21) $\alpha = 5^h32^m50^s.2$, $\delta = -5°25'12''$.

(22) Based on the analysis of the carbon recombination lines, Boughton (1978) suggested the existence of two distinct carbon clouds:
(1) A background C II region which is characterized by the emission of lines with $n \lesssim 135$.
(2) A foreground C II region which is characterized by the emission of lines with $n \gtrsim 190$.
The physical parameters of the two C II clouds are:

C II cloud	V_{LSR} (km s^{-1})	ΔV (km s^{-1})	Size of core	N_e (core) (cm^{-3})	N (total) (cm^{-3})	T_e (K)
Background	+9.5	4.5	3'	10^2	10^5	100–1000
Foreground	+4.5	8		0.2–10	4×10^2–2×10^4	20–1000

(23) $\alpha = 5^h32^m50^s$, $\delta = -5°25'48''$.

(24) Mean value.

(25) The existence of two H I clouds can be inferred; from a Gaussian analysis of the line profiles of the $-3(-2.6)$ km s^{-1} component, these may be physically associated with the H II region (the V_{LSR} of hydrogen recombination lines are of the same order). These H I concentrations are quite small (~ 0.5 pc) and contain $\lesssim 0.5 M_\odot$ of H I. Absorbing H I clouds lying along the line of sight (inferred from the $+5$ km s^{-1} component) have typical sizes of 1 pc, and H I masses of $13 M_\odot$. Typical H I densities are in the range of 100–200 cm^{-3}.

(26) $\alpha = 5^h32^m47^s$, $\delta = -5°24'20''$; KL Nebula.

(27) At θ^1 Ori C.

(28) Mean value.

(29) The nebula is probably surrounded by a rotating expanding envelope of H I gas with a mass of $100 M_\odot$.

Remarks to Table 2.2.XX:

(1) $V_{LSR} = (V_{Hel} - 18.1)$ km s^{-1}; the quoted velocities were measured at the continuum peak. The exact position is often stated.

(2) Position at $\Delta\alpha = 180''$, $\Delta\delta = 450''$ from θ^1 Ori C (continuum peak).

(3) For V_{LSR} variations over the object see the original paper.

(4) $\alpha = 5^h33^m4^s$, $\delta = -5°18'3''$.

(5) $\alpha = 5^h33^m3^s.8$, $\delta = -5°17'50''$.

(6) $\alpha = 5^h33^m3^s$, $\delta = -5°18'18''$.

(7) Two Gaussians fitted; alternative values $+6\pm0.3$, $+7.4\pm0.5$, if two Voigt profiles are fitted instead.

TABLE 2.2.XX*

Radial velocities of M43 derived from radio recombination and
neutral hydrogen lines

No.	Line	Beam	E = emission A = absorption	$V_{LSR}^{(1)}$ (km s^{-1})	Remarks	References
1	H76α	0''9	E	+5.2±0.4	(2), (3)	Pankonin et al. (1979)
2	H86α	3''5	E	+8.15±0.05	(4)	Lichten et al. (1979)
3	H91α		E	+7.3±0.1		Thum et al. (1978)
4	H108β	3''5	E	+7.8±0.2	(4)	Lichten et al. (1979)
5	H109α	2'	E	+8.4	(5)	Schraml and Mezger (1969)
6	H109α	4'	E	+7.7±1.1	(6)	Wilson et al. (1970)
7	H109α	2''6	E	+7.5±0.1		Churchwell et al. (1978)
8	H109α	2''6	E	+6.2±0.1, +9.2±0.4 }(7)		Jaffe and Pankonin (1978)
9	H137β	2''6	E	+7.2±0.4		Churchwell et al. (1978)
10	H162n		E	+4.6±1.1	(4)	Lichten et al. (1979)
11	C109α	2''6	E	+9.1±0.2		Jaffe and Pankonin (1978)
12	H I(21 cm)	2'	A	+6.6		Lockhart and Goss (1978)

*For remarks to Table 2.2.XX see page 77.

Remarks to Table 2.2.XXI:

(1) High velocity components with respect to the bulk of the ionized matter (in km s^{-1}).
(2) Large volumes (though small masses) of ionized material with radial velocities between -69 and $+119$ km s^{-1} with respect to the mean radial velocity of the nebula.
(3) Ionized condensation 43.5'' from θ^1 Orionis A.
(4) Non Gaussian wings.
(5) [O III] λ 5007 Å.
(6) [O III] λ 4959 Å.
(7) Possibly Hα.
(8) H7, λ 3970 Å.
(9) Hδ, λ 4101 Å.
(10) [O II], λ 3727 Å.
(11) [O II], λ 3729 Å.
(12) [O III], λ 4363 Å.
(13) Average most probable value from many measurements.
(14) The intensity of the -60 km s^{-1} feature varies across the nebula, ranging from 0.0005 to 0.005 relative to Hα.
(15) 15'' S and 0''5 W of θ^1 Ori C.
(16) 1' SE of θ^1 Ori C.
(17) Spectrographic observations of a 4' × 4' area centred on the Trapezium, made with the coudé spectrograph of the 200'' reflector at a dispersion of 4.5 Å mm^{-1} and a multislit of 31 slits separated by 1 mm ($\sim$1''3); see also Figures 2.2.2 and 2.2.3.
(18) From three interferograms taken in the [N II] λ 6584 Å line. Seeing < 2''.
(19) He I λ 3889 Å.
(20) Absorption feature against θ^1 Ori C.

TABLE 2.2.XXI*

Turbulence, line splitting and high velocity features in the Orion Nebula (M42, M43)

No.	Line	Beam	E=emission A=absorption	Turbulence $\langle Vt^2 \rangle^{1/2}$ (km s⁻¹)	Order of line splitting (km s⁻¹)	High velocity features[1] (km s⁻¹)	Remarks	References
					a. Optical emission lines (M42)			
1	[O II], [O III]	1".3	E	5-9	25		(17)	Wilson et al. (1959)
2	He I		A			−45	(19), (20)	Wilson et al. (1959)
3	Hα		E			−69 to +119	(2)	Shcheglov (1968a,b)
4	[O II]		E			−62.2	(10), (3)	
5	[O II]		E			−104, −59.4	(11), (3)	Lee (1969)
6	H7		E			−109.7 − 71.9	(8), (3)	
7	Hδ		E			−57.4	(9), (3)	
8	[O III]		E			−36.4?	(12), (3)	
9	Hβ		E			±100	(4)	Grachev (1970)
10	[O III]	7"−21"	E	7.6			(5), (13)	Smith and Weedman (1970)
11	[O III]		E	11		−50±5	(5)	Meaburn (1971)
12	[O III]		E			−40±5	(6)	Meaburn (1971)
13	Hβ, [O III]		E	4-8				Dopita et al. (1973)
14	Hα, [N II]		E	4.6-7.4				Dopita et al. (1973)
15	[N II]		E		25			Deharveng (1973)
16	Hα	73", 27"	E			−60	(16), (14)	Traub et al. (1974)
17	Hα	73", 27"	E			−100	(7), (16)	Traub et al. (1974)
18	Hα, He I	1.3	E	11.2				Chaisson and Dopita (1977)
19	Bα		E			−53±9	(15)	Smith et al. (1979)
					b. Radio recombination lines (M42)			
20	H66α	1.3	E	10.8				Chaisson and Dopita (1977)
21	H94α	4.2	E	17±0.9				Gordon and Meeks (1968)
22	H109α	6'	E	16.8				Mezger and Höglund (1967)
23	H109α	2'	E	17.2				Schraml and Mezger (1969)
24	H109α	6.5	E	17.8±1.0				Reifenstein et al. (1970)
25	H109α	4'	E	18.3±1.0				Wilson et al. (1970)
					c. Radio recombination lines (M43)			
26	H94α	4.2	E	13.4±0.8				Gordon and Meeks (1968)
27	H109α	2'	E	13.3				Schraml and Mezger (1969)
28	H109α	4'	E	14.6±2.3				Wilson et al. (1970)

*For remarks to Table 2.2.XXI see page 78.

TABLE 2.3.I

Technical informantion related to the photographs of the ionized stellar-size condensations (ISC) and ionized knots (IK) shown in Chapter 2.3

No.	Figure	Line	Telescope			Auxilliary device
1	2.3.1	Hα+[N II] λλ6563, 6584 Å	3.9 m (150 inch)	Anglo-Australian;	f:3.3 prime focus	
2	2.3.2a	Hα λ6563 Å	1.06 m (42 inch)	Pic-du-Midi;	f:15 Cassegrain focus	Lallemand electrono-graphic camera
3	2.3.2b	[O III] λ5007 Å	1.06 m (42 inch)	Pic-du-Midi;	f:15 Cassegrain focus	
4	2.3.2c	[N II] λ6584 Å	1.06 m (42 inch)	Pic-du-Midi;	f:15 Cassegrain focus	
5	2.3.5	[S II] λ6731 Å	1.90 m (74 inch)	Sutherland, South Africa;	f:18 Cassegrain focus	image tube
6	2.3.6	[N II] λ6584 Å	1.90 m (74 inch)	Sutherland, South Africa;	f:18 Cassegrain focus	
7	2.3.7	Hα+[N II] λλ6563, 6584 Å	3 m (120 inch)	Hale;	f:3.67 prime focus	
8	2.3.13	Hα+[N II] λλ6563, 6584 Å	2.54 m (100 inch)	Mt Wilson;	f:5 Newtonian focus	

No.	Type	Filter Peak wavelength $\lambda(\text{Å})$	Filter Full width $\Delta\lambda(\text{Å})$	Photo-cathode	Plate	Exposure time (min)	Print P = positive N = negative	Remarks	References
1	Schott RG630	6650	700		Kodak 098–04	5	P	(1)	Malin (1979)
2	Interference filter	6562	10	S20	Ilford G5	3	N		Laques and Vidal (1979)
3	Interference filter	5008	8.5	S11	Ilford G5	1	N	(2)	
4	Interference filter	6584	10	S20	Ilford G5	4	N		
5	Interference filter	6731	9	S25	Kodak IIα–D		N	(3)	Cantó et al. (1980)
6	Interference filter	6584	10	S25	Kodak IIα–D		P		
7	Schott RG2	6563	~300		Kodak III–E		N		Münch and Wilson (1962)
8	Schott RG630	6650	700		Kodak 103α–E	5	N	(4)	Münch (unpublished)

Remarks to Table 2.3.1:

(1) Hypersensitized emulsion; photograph printed through an unsharp mask.

(2) Photocathode 20 mm in diameter; field of view ~4'.

(3) Westinghouse WL – 30677 S25 image tube with phosphor output; IIaD hypersensitized emulsion.

(4) Filter transmitting $\lambda > 6300$ Å; effective bandwidth $\Delta\lambda \simeq 300$ Å; resolution of the plate set by the seeing conditions ~0″.5.

TABLE 2.3.II

Physical parameters of the ionized stellar-size condensations (ISC) and
ionized knots (IK) associated with the core of M42

| No. | Figure | Source | Nomenclature | | Angular size (arc sec) | Linear size for a distance of 500 pc (pc) | Electron density N_e (cm^{-3}) |
			Münch (1977), Taylor and Münch (1978)	Cantó et al. (1980)			
1	2.3.2a, b	ISC1-6			<1.5	<0.004	1.5×10^5–4.1×10^5
2	2.3.5	IK1	M42 HH1	M42 HH1	~9	~0.022	~4.8×10^3
3	2.3.5 2.3.6	IK2		M42 HH2	~11	~0.027	~1.0×10^4
4	2.3.7	IK3	Cloudlet A	M42 HH3	~10	~0.025	~7.6×10^3
5	2.3.7	IK4	Cloudlet B	M42 HH4	~6	~0.015	~1.3×10^4
6	2.3.7	IK5	Cloudlet C		~5	~0.012	—

Remarks to Table 2.3.II:

(1) $V_{Hel} = (V_{LSR} + 18.1)$ km s^{-1}.

(2) Probably as small as 0."2–0."4 with densities as high as 2.2×10^6–3.5×10^6 cm^{-3}; ISC1 coincides (within 2") with an infrared source (IRS 4, see Figure 3.1.2 later in the text); this is however believed to be coincidental (Laques and Vidal, 1979), since IRS 4 is embedded in the Orion molecular cloud (OMC 1; see Chapters 3 and 4).

(3) The electron densities of IK 1, 2, 3, and 4 were estimated by Cantó et al. (1980) from measurements of the [S II] (λ 6717 Å/λ 6731 Å) ratio. These densities characterize the velocity components originating within the objects and are much higher than those from the surrounding matter (between 1.7×10^3 cm^{-3} and 2.9×10^3 cm^{-3}). The electron density of IK 1 refers to an area of 2."6 × 2."6 whereas that of the rest (IK 2, 3, 4) refers to an area of 5."5 × 5."5.

(4) Tamura (1976) based on the assumption that the intensity fluctuations of the Hβ λ 4861 Å, [O III] (λ 4363 Å, λ 5007 Å, λ 4959 Å), and He I λ 4471 Å lines over the core of M42 are due to the existence of condensations, has worked out a statistical solution yielding 400 condensations of ~0.001 pc size each.

Table 2.3.II (continued)

Kinematics

Line	Heliocentric radial velocity V_{Hel} (km s^{-1})	Radial velocity with respect to the nebular material V (km s^{-1})	Remarks	References
			(2), (4)	Lacques and Vidal (1979)
[O I]	−266, −26	−240		Münch (1977)
[S II]	−266, −26	−240	(3)	Münch (1977)
[S II]	−254, −76, +17	−280, −102, −9		Cantó *et al.* (1980)
[S II]	−30, +26	−56	(3)	Cantó *et al.* (1980)
Hα	−48, +14	−62		Taylor and Münch (1978)
[N II]	−48, +14	−62	(3)	Taylor and Münch (1978)
He I λ10830 Å	−55, +19	−74		Taylor and Münch (1978)
[S II]	−46, +18	−64		Cantó *et al.* (1980)
Hα	−15, +15	−30		Taylor and Münch (1978)
[N II]	−25, −5, +15	−40, −20	(3)	Taylor and Münch (1978)
[S II]	−16, +20	−36		Cantó *et al.* (1980)
Hα	0, +9, +17	−17, −26		
[N II]	0, +14	−14,		Taylor and Münch (1978)
[O II]	+10, +17	−7		

TABLE 2.4.I

Evidence supporting the physical association of the dust and ionized gas in the Orion Nebula (M42, M43)

No.	Figure	Evidence	Remarks	References
1		The ultraviolet extinction of the θ Ori stars originates in the nebula	(1)	Schiffer and Mathis (1974)
2 {		Scattered continuum measurements are in agreement with those of light directly received from θ^1 Ori C (the main exciting star of the nebula)	(2)	Simpson (1973)
3 {		The far ultraviolet mapping of Orion resembles the visible continuum structure. The UV spectrum of the nebula is similar to that of the main stars (reflection spectrum)	(9), (10)	Carruthers and Opal (1977) Bohlin and Stecker (1975) Perinotto and Patriarchi (1980)
4 {		The He I λ 4868 line is seen in absorption in both the nebular continuum and the continuum of the main stars (θ^1, θ^2) of Orion.	(3)	Peimbert and Goldsmith (1972)
5	2.4.3	The extinction of light is closely correlated to the emission measure	(4)	{ Münch and Persson (1971) { Martin and Gull (1976) Barbieri et al. (1976)
6 {	2.4.4 2.4.5	{ The linear polarization measurements of the nebular light show centrosymmetric { patterns around the main ionizing stars (θ^1, θ^2 Ori in M42, NU Ori in M43)	(5)	Pallister et al. (1977) Khallese et al. (1980)
7 {	2.4.6	The structure of the infrared continuum maps is similar to that of the radio continuum maps (obtained with comparable angular resolution)	(6), (7)	Lemke et al. (1974) Lemke and Harris (1979)
8 {	2.4.7 2.4.8	The 'bar' in M42 exhibits the same structure in the infrared and radio continuum and in the optical line emission.	(6), (8)	Werner et al. (1976) Becklin et al. (1976)

Remarks to Table 2.4.1:

(1) Concluded from arguments based on a UV absorption feature at 2200 Å (Bless and Savages, 1972) which is probably due to graphite grains (Gilra, 1972).

(2) At least in the range of 4100–5800 Å; the light originating from the Trapezium stars is more reddened than that from the nebular region; this is probably due to circumstellar dust around the stars (Ney *et al.*, 1973; see also Table 3.1.IV). Münch and Persson (1971) report, however, no differences in the interstellar reddening law for the light of θ^1 Ori C and the visible nebular light.

(3) Indicating that the bulk of the continuum emission in this wavelength range is due to scattered light inside Orion.

(4) The variations of extinction are proportional to the variations of the Balmer line intensities (Münch and Persson, 1971) and the radio continuum emission (Martin and Gull, 1976), indicating that the gas and dust are mixed; see also Figure 2.4.3 where the minimum absorption (W of Trapezium) occurs at the brightest part of Orion.

(5) Indicating that the light emanating from θ^1, θ^2 and NU Orionis is scattered by the dust lying within the nebular materials of M42 and M43, respectively; the polarization pattern in the area between M42 and M43 strengthens the view that the dust lane is physically associated with the nebular material and does not represent a foreground feature. The origin of the polarization patterns is attributed to Mie scattering (grains of finite size); see Figures 2.4.4 and 2.4.5.

(6) The infrared radiation is due to the re-emission by dust of the UV radiation originally emitted by the hot ionizing stars. The dust is believed to absorb Lyman continuum directly from the stars and Lα photons which are due to the recombination of hydrogen and are resonantly trapped within the H II region (see Appendix, Section III.2).

(7) Compare for example the IR map of Figure 2.4.6 made at 20μ with an angular resolution of 1′ with the radio continuum maps of Figure 2.1.16 made with the comparable angular resolution of 2′. The bright compact source $\sim 1'$ NW of Trapezium (KL Nebula) lies within the massive molecular cloud which is on the backside of M42 and should not be considered. The rest of the contours are concentrated around the Trapezium stars and show a similar distribution with those of the radio maps, indicating that the dust and ionized gas are well mixed.

(8) See for example the IR map of Figure 2.4.7, made at 100μ with an angular resolution of 1′ or more distinctly the correlation of Figure 2.4.8 showing the distribution of 10μ radiation of the bar made with a 4″.5 angular resolution and the distribution of the radio continuum radiation taken from the high angular resolution (7″ × 20″) map of Figure 2.1.19. The broader shape of the bar at 10μ in the area around θ^2 Ori A is probably due to a circumstellar dust shell around this star. The distribution of the infrared radiation indicates the existence of hot dust at or near the bar which is most probably an ionization front (H I–H II interface) seen edge on.

(9) The brightness distribution in the range of 1000–2000 Å is probably due to starlight scattered by dust grains which are characterized by high albedo in the far ultraviolet.

(10) The spectrum of the nebula in the range of 1000–1900 Å shows a continuum, 'reflecting' that of the central stars.

TABLE 2.4.II

Gas to dust ratio in M42

Region	Gas to dust ratio[1] $(N_H/N_d)\,\sigma_\lambda$ $(10^{20}\,cm^{-2})$	Remarks	Reference
General Interstellar Medium	20		O'Dell *et al.* (1966), Wickramasinghe (1967)
M42 (inner part)	144	(2), (3), (4)	O'Dell *et al.* (1966)
M42 (outer part)	5	(3)	O'Dell *et al.* (1966)
M42 (mean)	4		Martin and Gull (1976)
M42 (near θ^1 Ori C)	17–47	(5)	Perinotto and Patriarchi (1980)
M42 (bar, near θ^2 Ori A)	19–59	(6)	Perinotto and Patriarchi (1980)

Remarks to Table 2.4.II:

(1) N_H is the hydrogen ion density, N_d is the density of the grains and σ_λ is the effective scattering cross section per unit volume presented by the grains of size distribution $N(\alpha)$. $N_d\sigma_\lambda$ is given by the relation:

$$N_d \cdot \sigma_\lambda = \int_0^\infty Q\left(\frac{\alpha}{\lambda}\right)\pi\alpha^2 N(\alpha)d,$$

where α is the radius of the grains of size distribution $N(\alpha)$ and $Q(\alpha/\lambda)$ is the scattering efficiency of the grains.

(2) At a distance of 1' ($\sim$0.15 pc) from the Trapezium.

(3) The dust is defficient in the central region by a factor of 20.

(4) The estimate is probably influenced by the adoption of an unrealistic spherical model (Martin and Gull, 1976).

(5) 34'' W and 3'' S of θ^1 Ori C.

(6) 39'' N of θ^2 Ori A (across the 'bar').

TABLE 2.4.III

Grain characteristics derived from infrared studies of M42

Region	Probable grain composition	Temperature of grains (K)	Remarks	Reference
Trapezium Nebula	silicate	250 (max.)	(1)	Gehrz *et al.* (1975)
Trapezium Nebula	silicate	160	(2)	Forrest *et al.* (1976)
Trapezium Nebula	silicate	140	(3)	Beichman (1978)
Bar		75	(4), (5)	Werner *et al.* (1976), Becklin *et al.* (1976)

Remarks to Table 2.4.III:

(1) Derived from the energy distribution of the Trapezium Nebula in the range of 8–13μ; T_{max} = 250 K on θ^1 Ori D, T = 160–180 K within 10''–15'' of θ^1 Ori D; no further decrease of T was observed for distances as far away as 50''.

(2) Derived from a broad emission feature peaking at 18.5μ; mixture with cooler grains is also possible.

(3) Single grain population with long wavelength characteristics similar to silicates.

(4) Average 50 to 100μ colour temperature; estimated from the observed ratio of average 50μ to average 100μ surface brightnesses (grey emission was assumed).

(5) The energy distribution of the 'bar' indicates the presence of grains of different temperatures; a bimodal particle size distribution is envisaged with temperatures of 60 K and 300 K and particle diameters of 0.1μ and 0.01μ, respectively: the smaller size particles must be more abundant by a factor of 10–100.

TABLE 2.4.IV

Probable constituents of dust associated with an H II region

(according to Panagia (1977))

Composition	Grain size	$M_d/M_g^{(1)}$	Absorption opacity k	Remarks
Frozen molecules?	0.1μ	5.5×10^{-3}	$k(\text{Ly-C}) = 9 \times 10^{-22} \; n_H$	(2), (3), (7)
Graphite	0.01μ	2×10^{-3}	$k(\text{He}) = 1.5 \times 10^{-21} \; n_H$	(4), (5), (7)
Silicate	0.15μ	1.3×10^{-3}	$k(\text{Ly-C}) \sim 8 \times 10^{-23} \; n_H$	(3), (6), (7)
Water ice	0.2μ	$\sim 4 \times 10^{-4}$		

Remarks to Table 2.4.IV:

(1) Dust to gas ratio by mass.

(2) This component dominates the UV absorption (500–1000 Å) and the far infrared (30–300μ) and submillimetter (300–1000μ) emission; absorptivity $Q \sim (2\pi a/\lambda)$ for $\lambda < 100\mu$ (see Appendix, Section III.2).

(3) Opacity in the Lyman-continuum ($\sim$912 Å).

(4) Responsible for the absorption in the He-ionizing continuum.

(5) Opacity at 500 Å.

(6) Opacity at optical wavelengths $\simeq 1.3 \times 10^{-22} \; n_H$.

(7) n_H is the number density of hydrogen atoms.

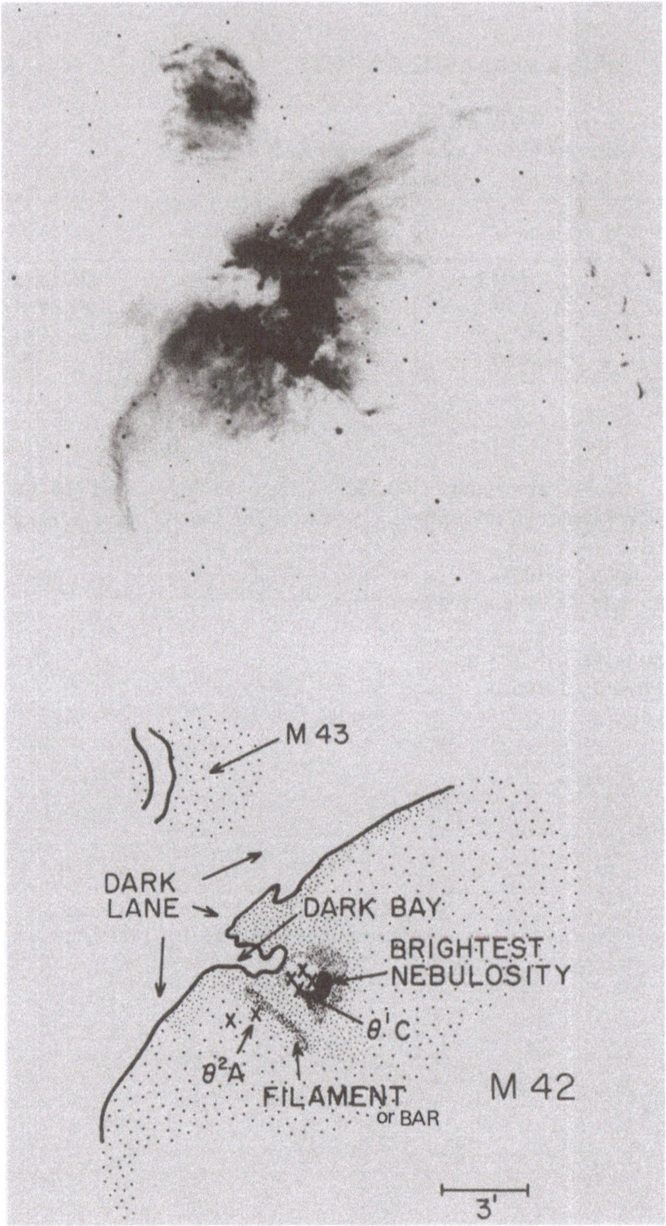

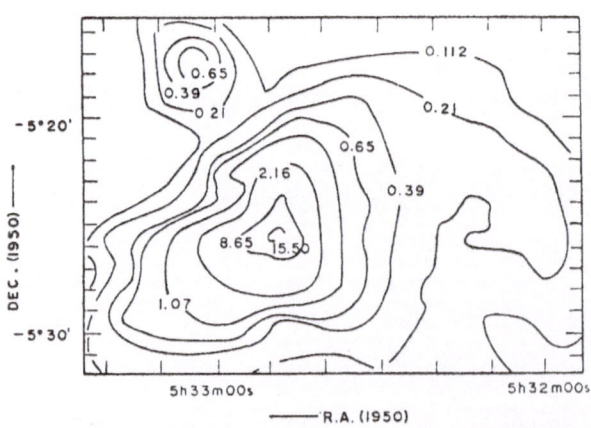

Fig. 2.1.1. Schematic representation of the most important optical features of the Orion nebulae (M42, M43) as appear in the light of many emission lines. A narrow band [S II] $\lambda\lambda$ 6717, 6731 Å photograph of Orion (M42, M43), taken by T. R. Gull at the Kitt Peak National Observatory is also shown for comparison (after Balick *et al.*, 1974).

Fig. 2.1.2. Hα λ 6563 Å contour map of Orion (M42 + M43) (absolutely calibrated). The units are expressed in 1×10^{-2} ergs cm^{-2} s^{-1} sr^{-1}. The resolution of the map is 1.'5 (after Schmitter, 1971).

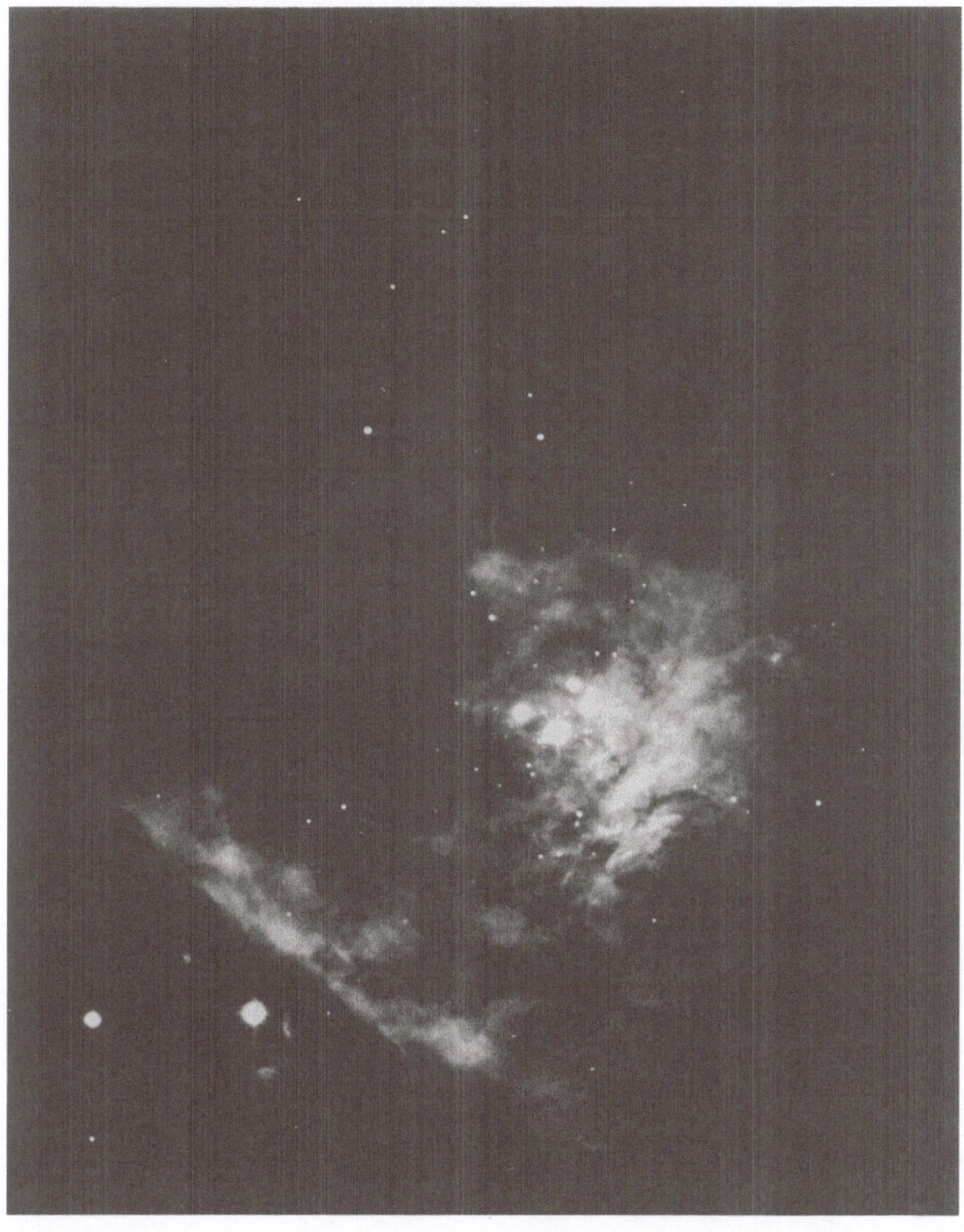

Fig. 2.1.3. Broad band Hα + [N II] photograph of the core of Orion (M42). For technical information concerning this photograph see Table 2.1.IV (after Münch and Wilson, 1962).

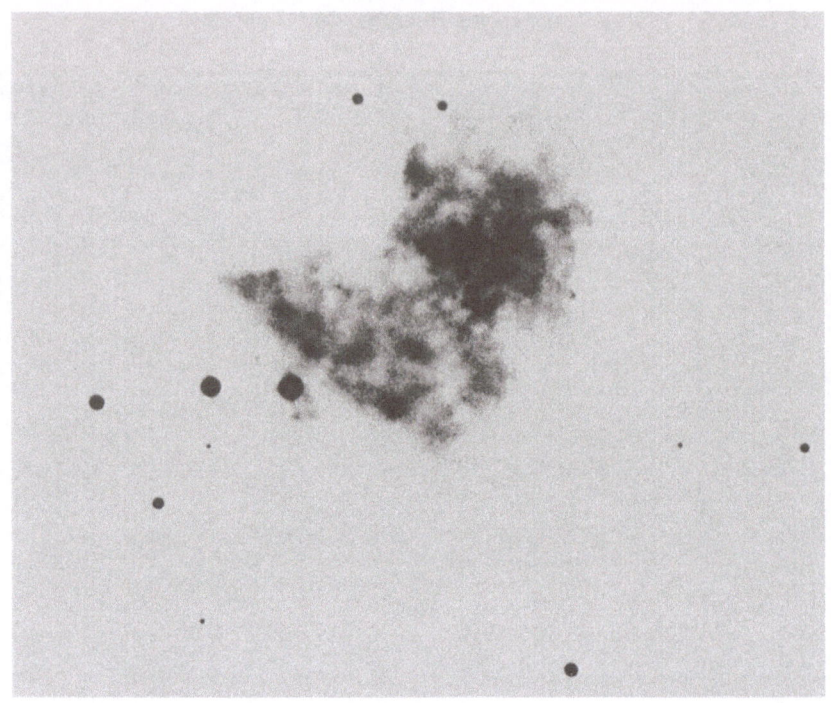

Fig. 2.1.4.   Hβ λ 4861 Å photograph of the core of M42. For technical information concerning this photograph see Table 2.1.IV (after Wurm and Rosino, 1959).

Fig. 2.1.5.   [N II] λ 6584 Å photograph of the core of M42. For technical information concerning this photograph see Table 2.1.IV (after Elliott and Meaburn, 1974).

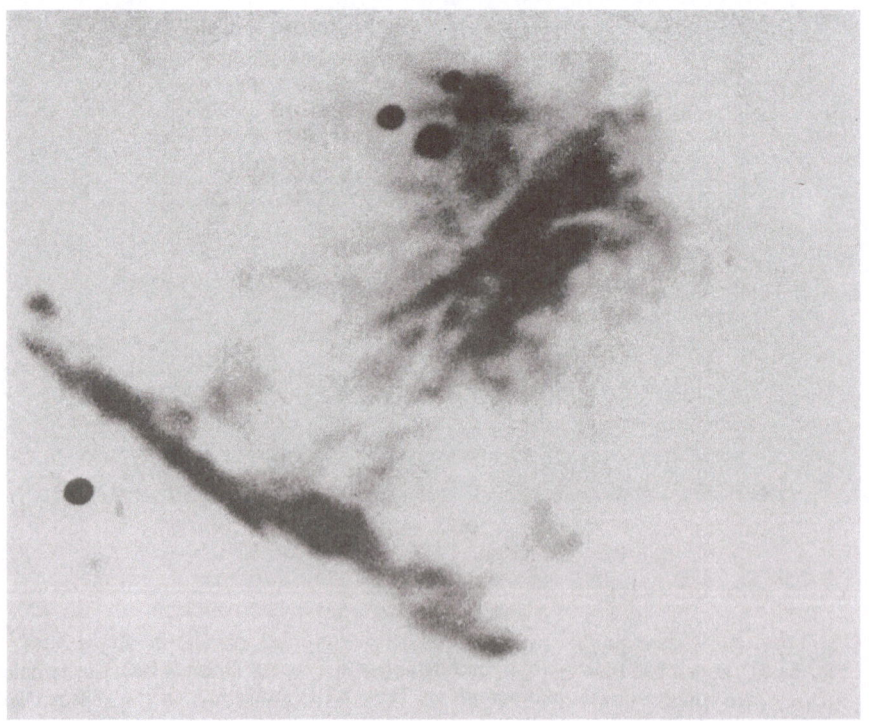

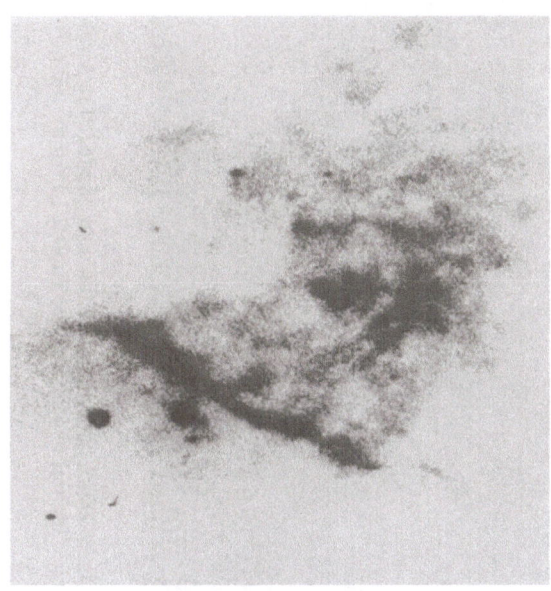

Fig. 2.1.6. [O II] $\lambda\lambda$ 3726, 3729 Å photograph of the core of M42. For technical information concerning this photograph see Table 2.1.IV (after Courtès and Viton, 1967).

Fig. 2.1.7. [O III] λ 5007 Å photograph of the core of M42. For technical information concerning this photograph see Table 2.1.IV (after Elliott and Meaburn, 1974).

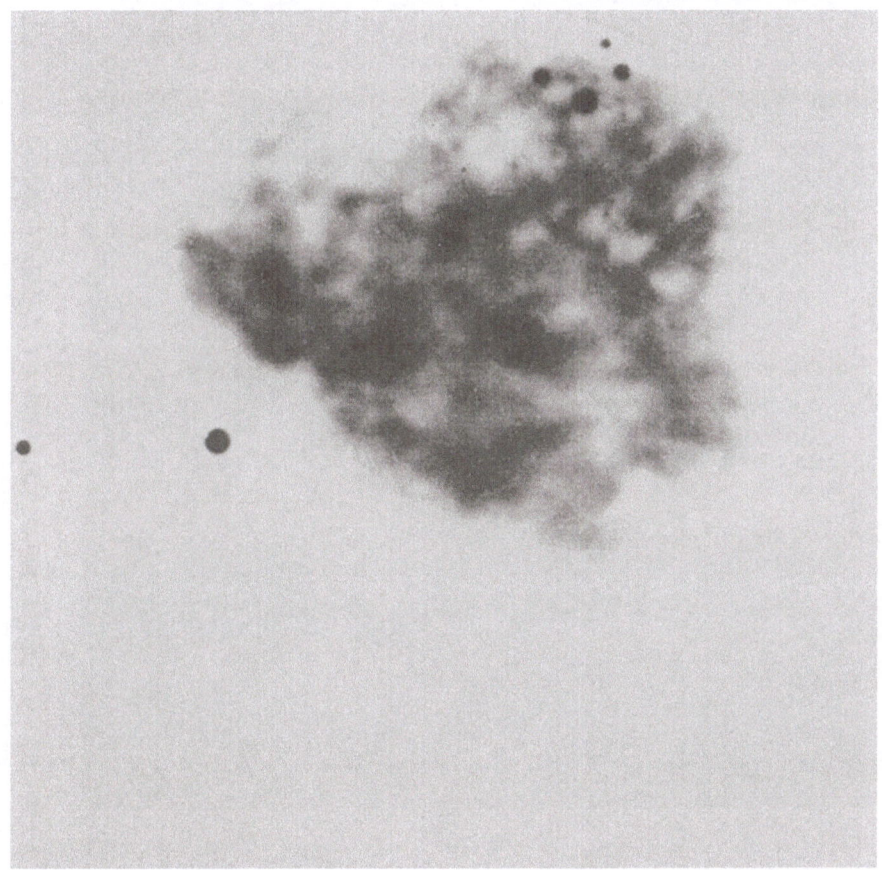

Fig. 2.1.9. [O I] λ 6300 Å photograph of the core of M42. For technical information concerning this photograph see Table 2.1.IV (after Münch and Taylor, 1974).

Fig. 2.1.8. O I λ 8446 Å photograph of the core of M42. For technical information concerning this photograph see Table 2.1.IV (after Münch and Taylor, 1974).

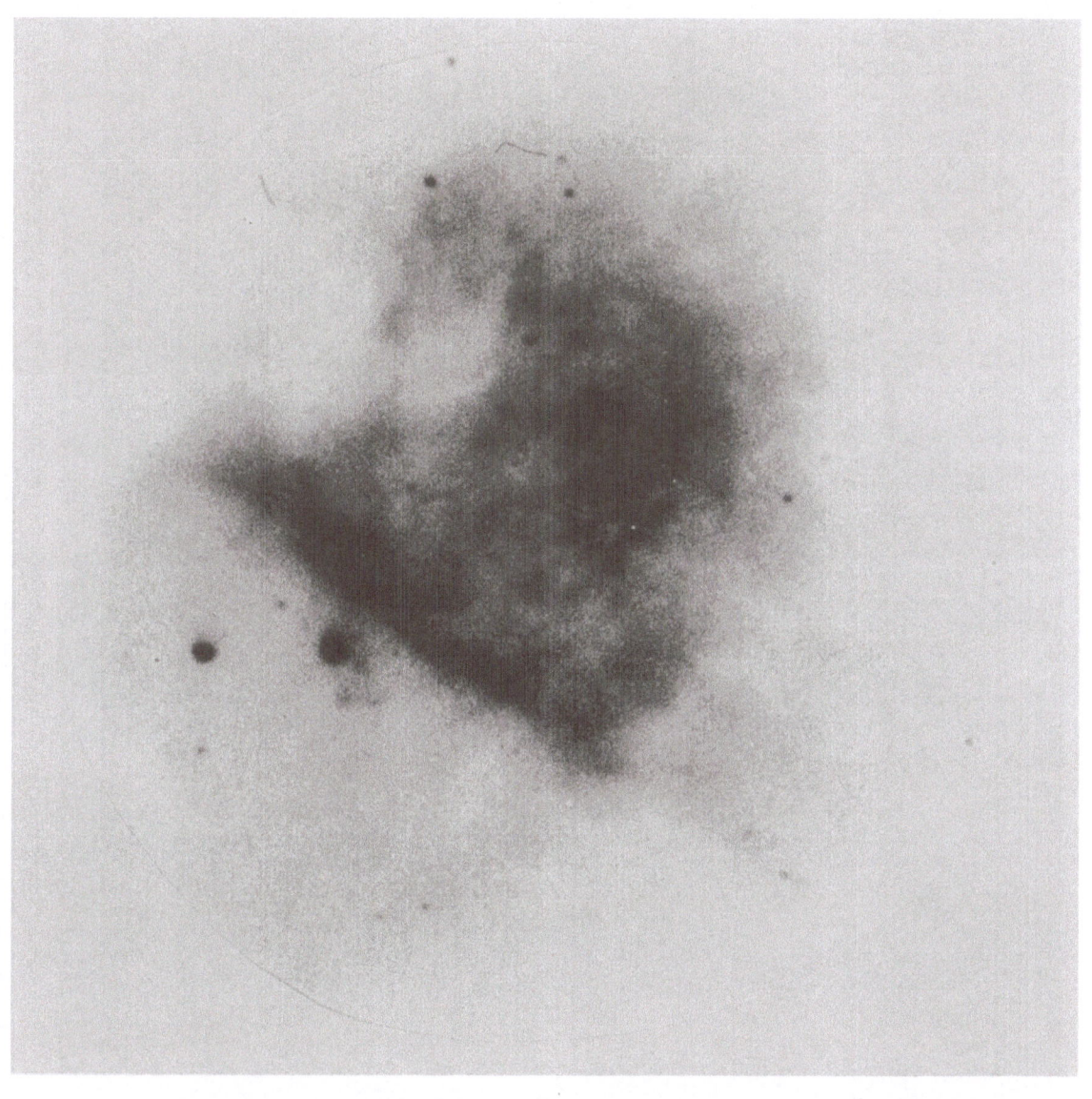

Fig. 2.1.10.   [S III] λ 9069 Å photograph of the core of M42. For technical information concerning this photograph see Table 2.1.IV (after Andrillat and Duchesne, 1974).

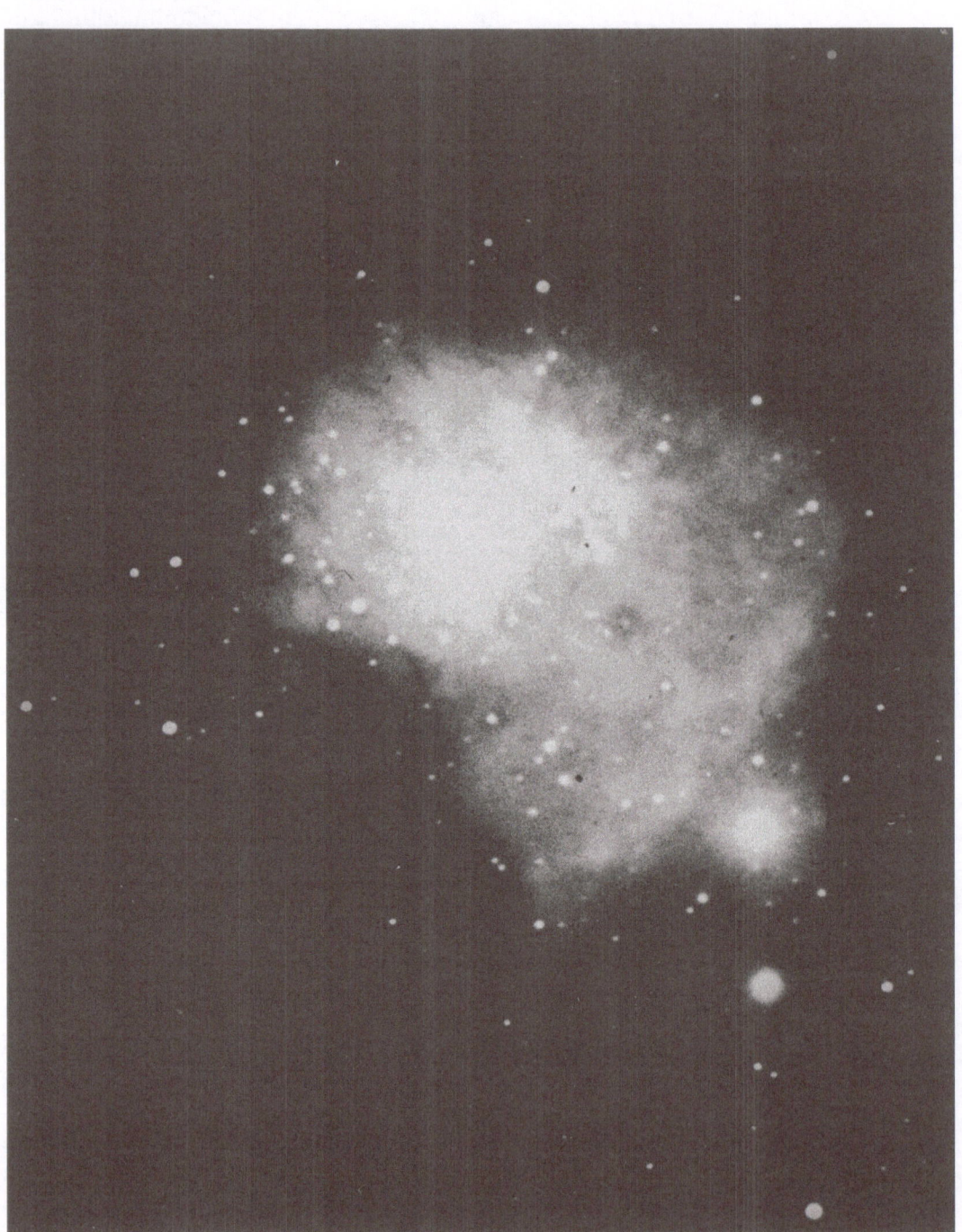

Fig. 2.1.11.   He I λ 10830 Å photograph of the core of M42. For technical information concerning this photograph see Table 2.1.IV. (Courtesy of Prof. G. Münch.)

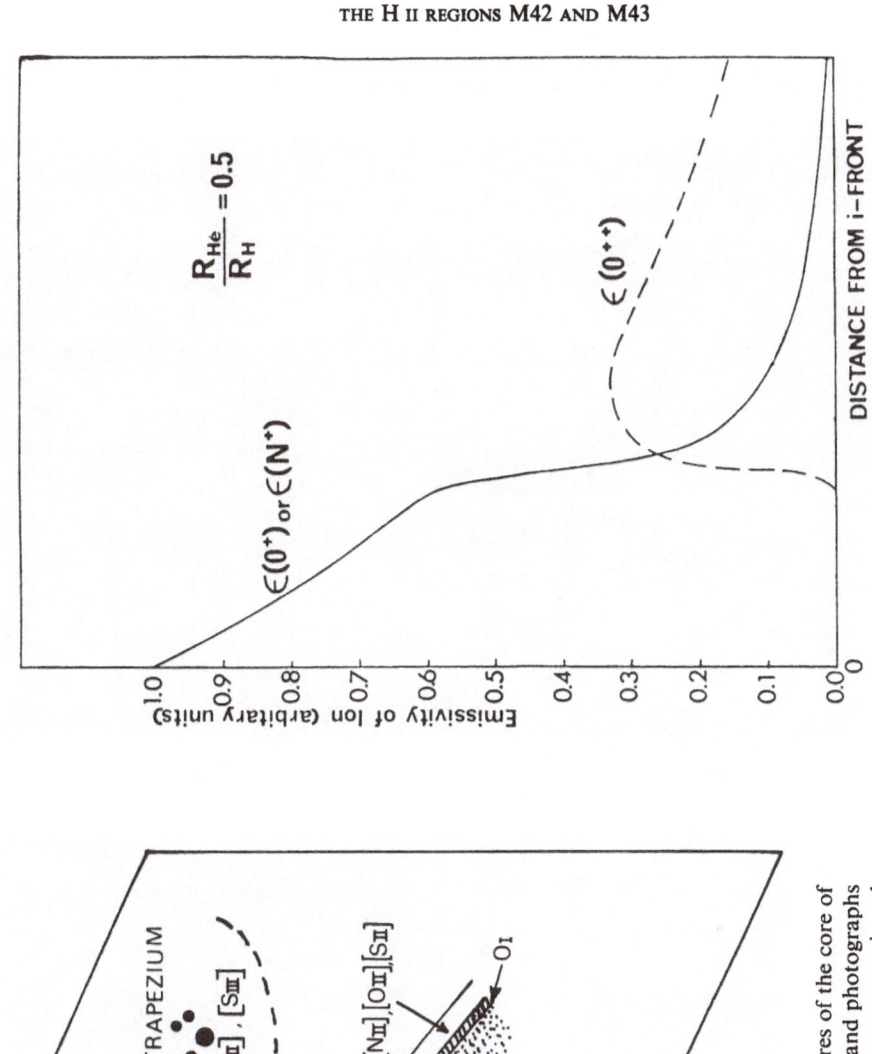

Fig. 2.1.12. Schematic representation of the main features of the core of M42, derived from a comparative study of the narrow band photographs of Orion shown in Figures 2.1.1. to 2.1.10 The phenomena are viewed tangentially by the observer. The ionization front is encroaching into a neutral/molecular layer which is probably a foreground folding originating from the large molecular cloud lying behind the Trapezium stars. It should be noted that the formation of the bar is also influenced by the ionizing radiation of the star θ^2 Ori A which lies on the outer side of the neutral folding. The effects of the θ^2 Ori A radiation were omitted in the sketch for simplicity.

Fig. 2.1.13. The variations of the volume emissivities of the [O II], [N II], and [O III] lines as a function of the distance from the bar (ionization front) (after Dopita et al., 1974).

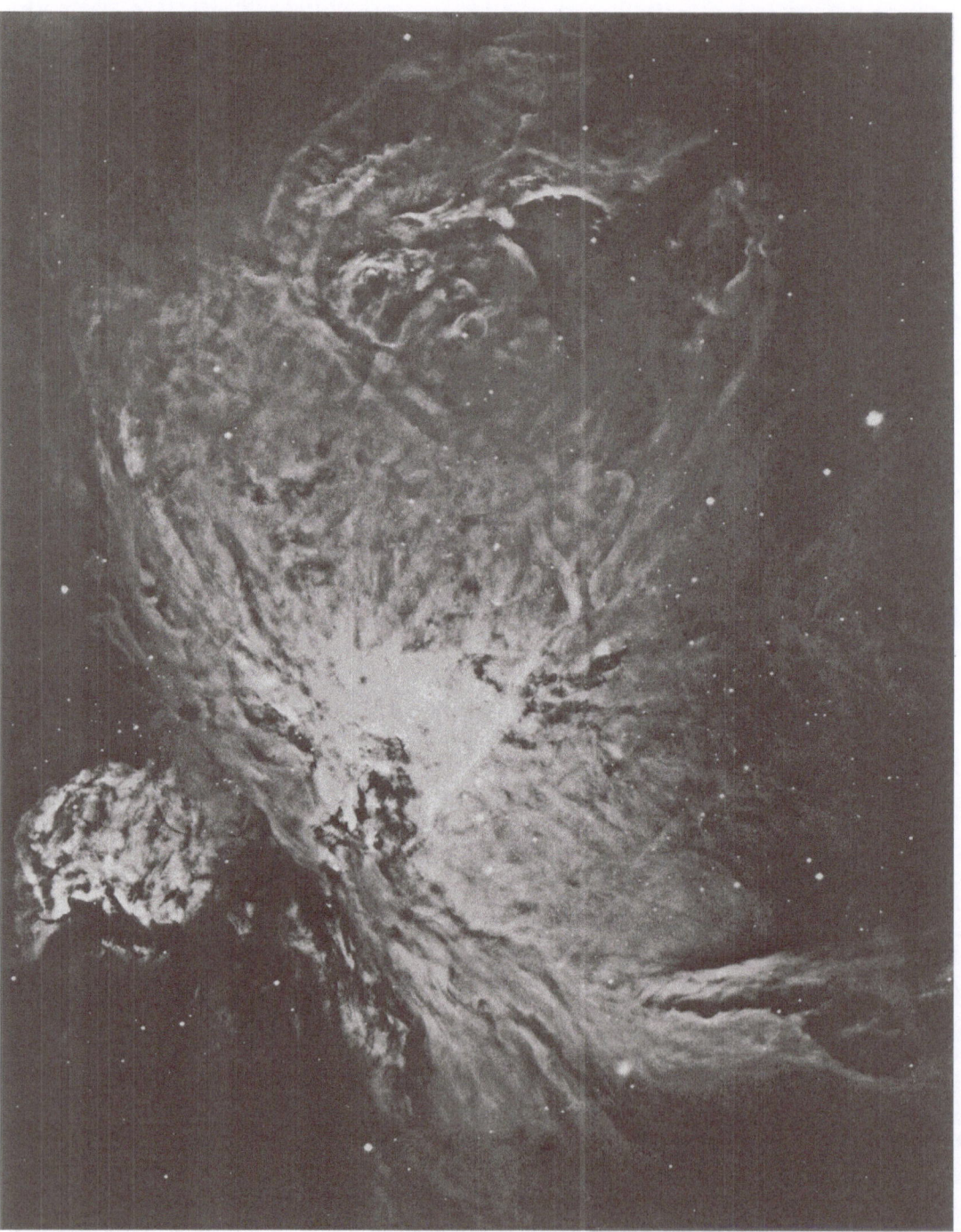

Fig. 2.1.14. Broad band Hα + [N II] photograph of M42 and M43 taken with the 3.9 m Anglo-Australian-Telescope through a Schott RG 630 filter (bandpass 6300–7000 Å). The exposure time was 5ᵐ; the print was made by Dr. D. F. Malin through an unsharp mask, to demonstrate the complicated filamentary structure surrounding the core of M42 (after Malin 1979)

Fig. 2.1.15. Broad band Hα + [N II] photograph of the Orion Region. M43 is the nebulosity to north of the Great Nebula (M42). This photographe was taken by K. Birkle with an image tube cam attached to the Cassegrain focus of the 1.2 m telescope at Calar Alto. The band width of the filter $\Delta\lambda = 150$ Å. The exposure time was 15^m (after Thum *et al.*, 1978).

Fig.   2.1.16a.

Fig. 2.1.16a, b, c.   Radio continuum contour maps of Orion A (M42 + M43) at 408 MHz, 1
MHz, and 15 000 MHz made with an angular resolution of 3′, 2′, and 2′, respectively. The cont∣
of the 15 000 MHz map (c) are in units of 1 K antenna temperature or 2 K brightness tempera∣
(after Shaver and Goss, 1970; Lockhart and Goss, 1978; and Schraml and Mezger, 1969, respectiv∣
The 408 MHz contour map (a) is superimposed on a photograph of Orion.

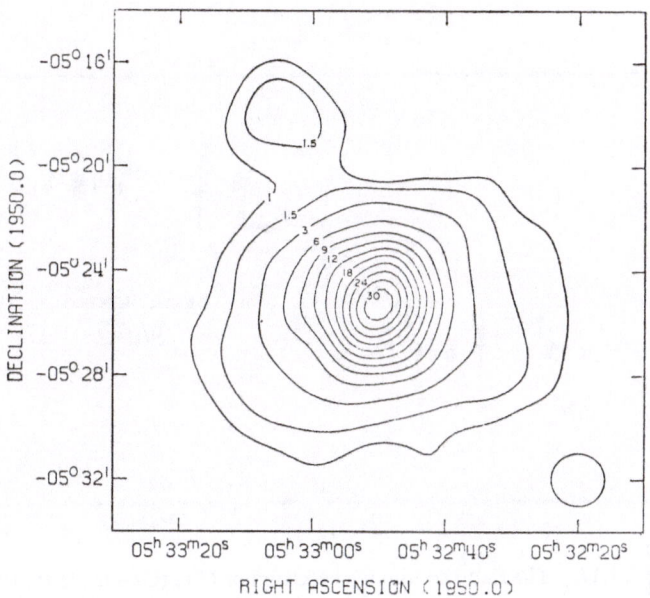

Fig. 2.1.16b.

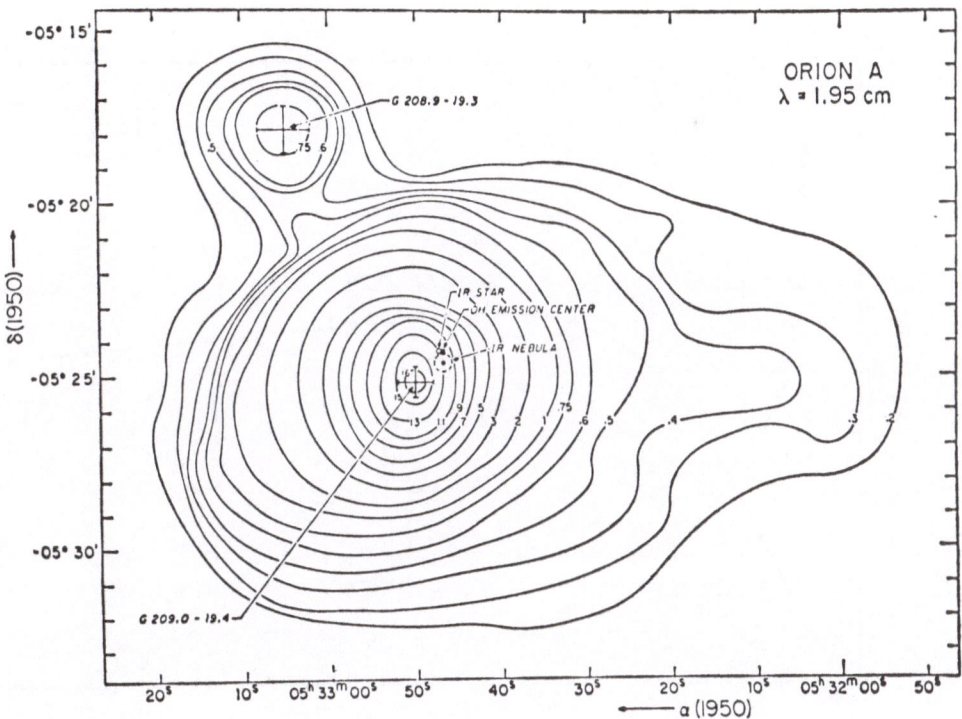

Fig. 2.1.16c.

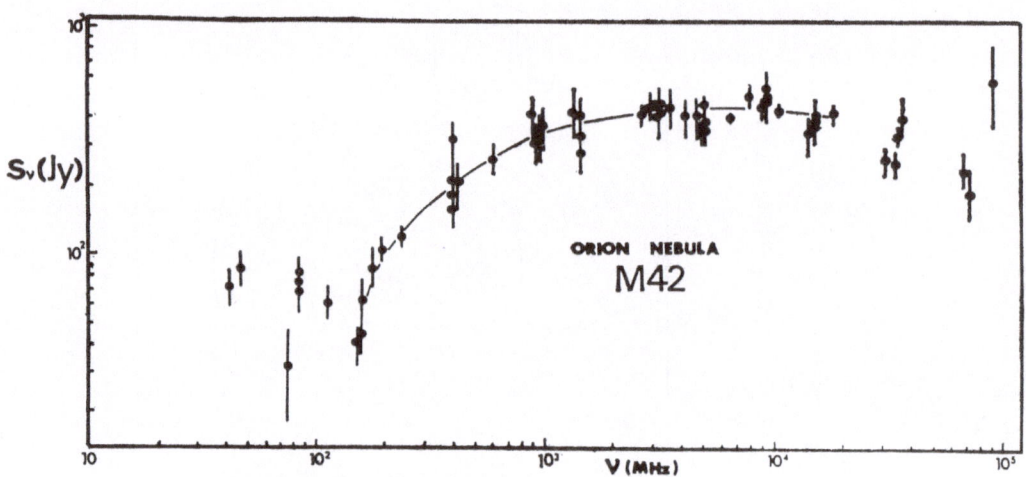

Fig. 2.1.17.   The radio continuum spectrum of M42 (Goudis, 1975, 1977).

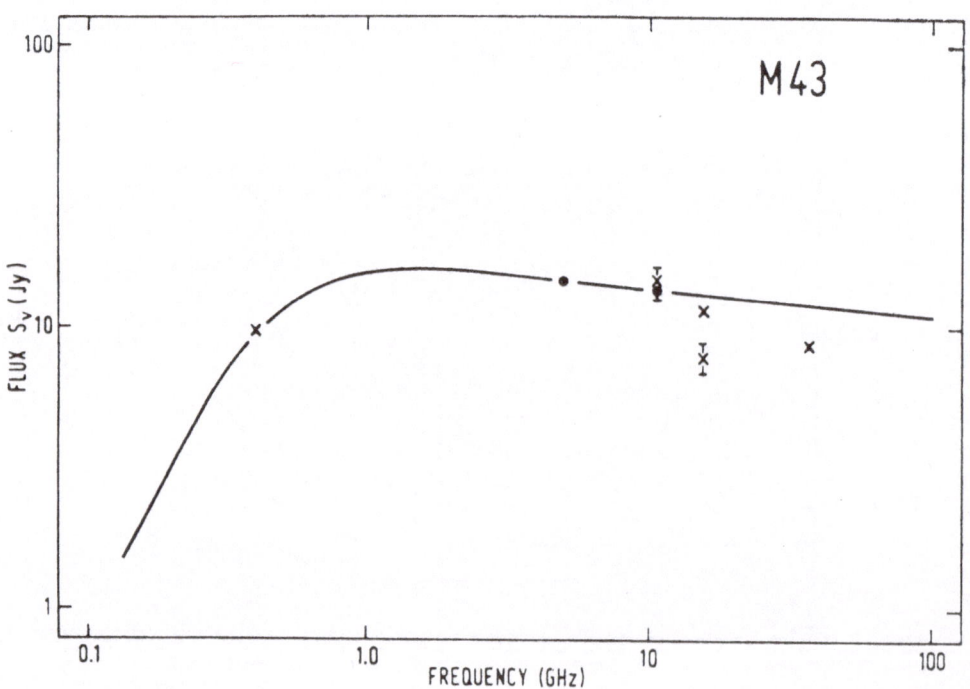

Fig. 2.1.18.   The radio continuum spectrum of M43 (after Thum *et al.*, 1978).

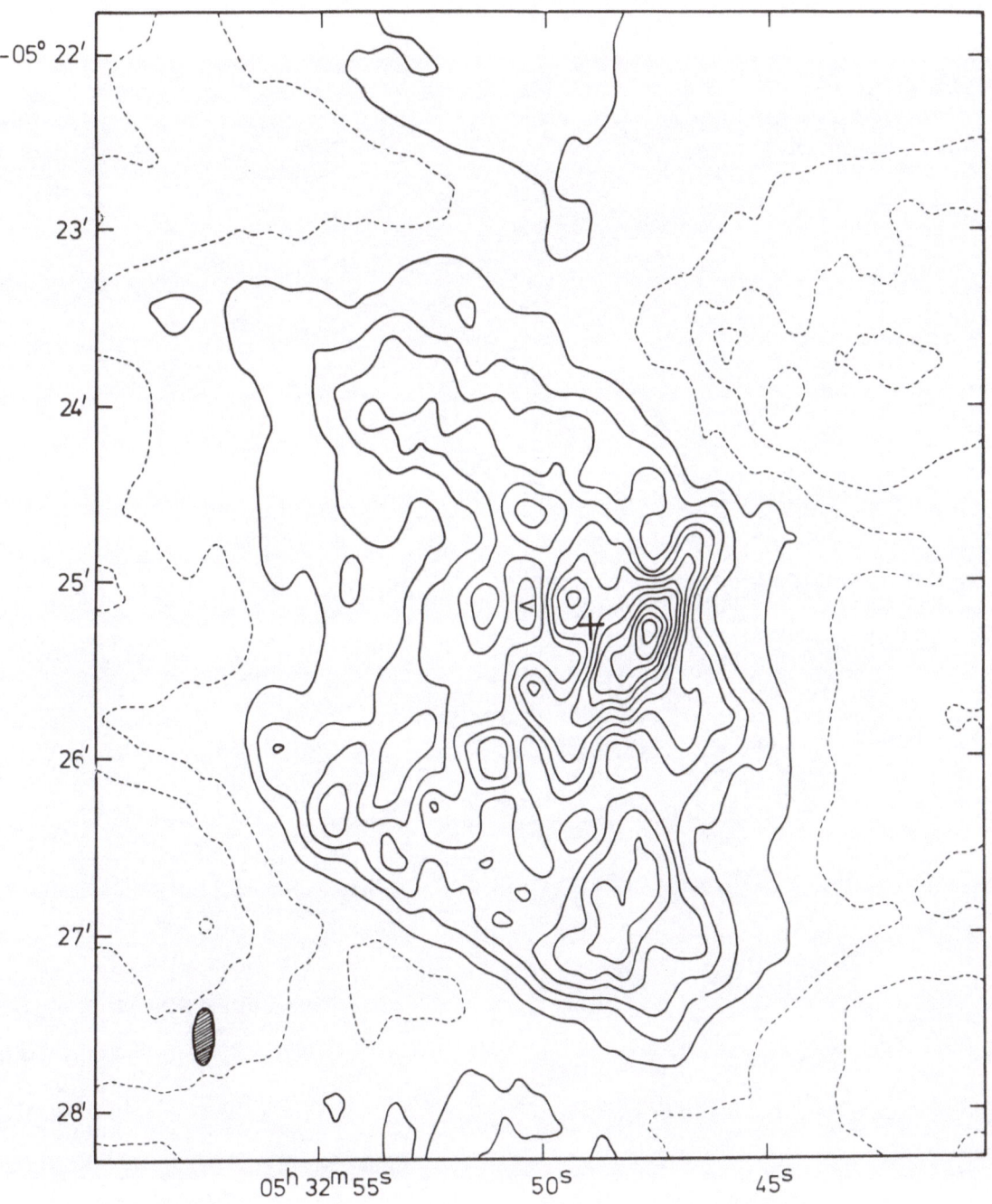

Fig. 2.1.19.   Radio continuum contour map of the core of M42 made at 5000 MHz with an angular resolution of 7″ × 20″. The contour interval is 58 K. The cross indicates the position of the exciting star θ^1 Orionis C (after Martin and Gull, 1976).

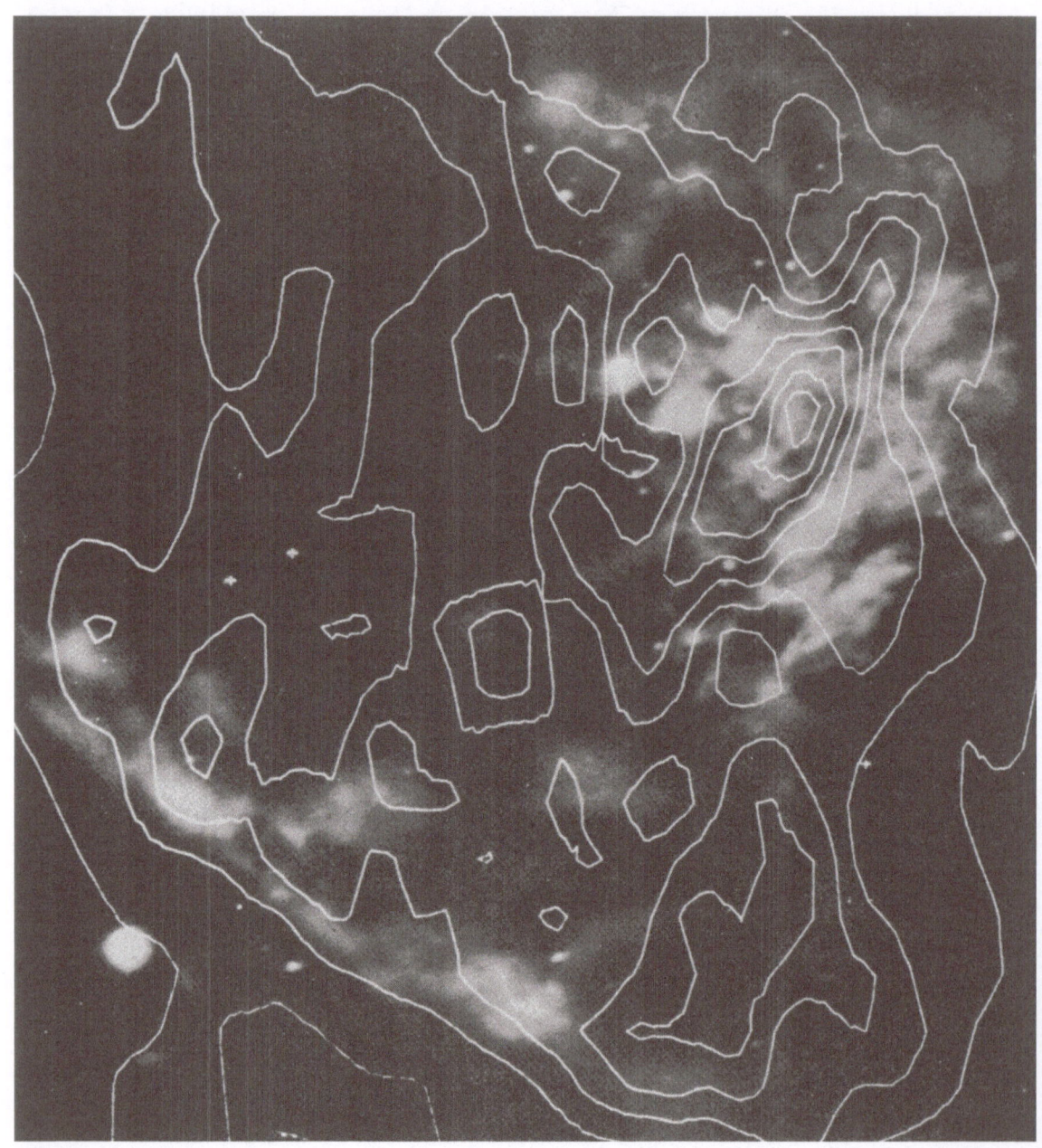

Fig. 2.1.20.   Superposition of the radio contour map of Figure 2.1.19. on the optical photograph of Figure 2.1.3. The contour interval is 77 K. The crosses were used for alignement purposes (after Martin and Gull, 1976).

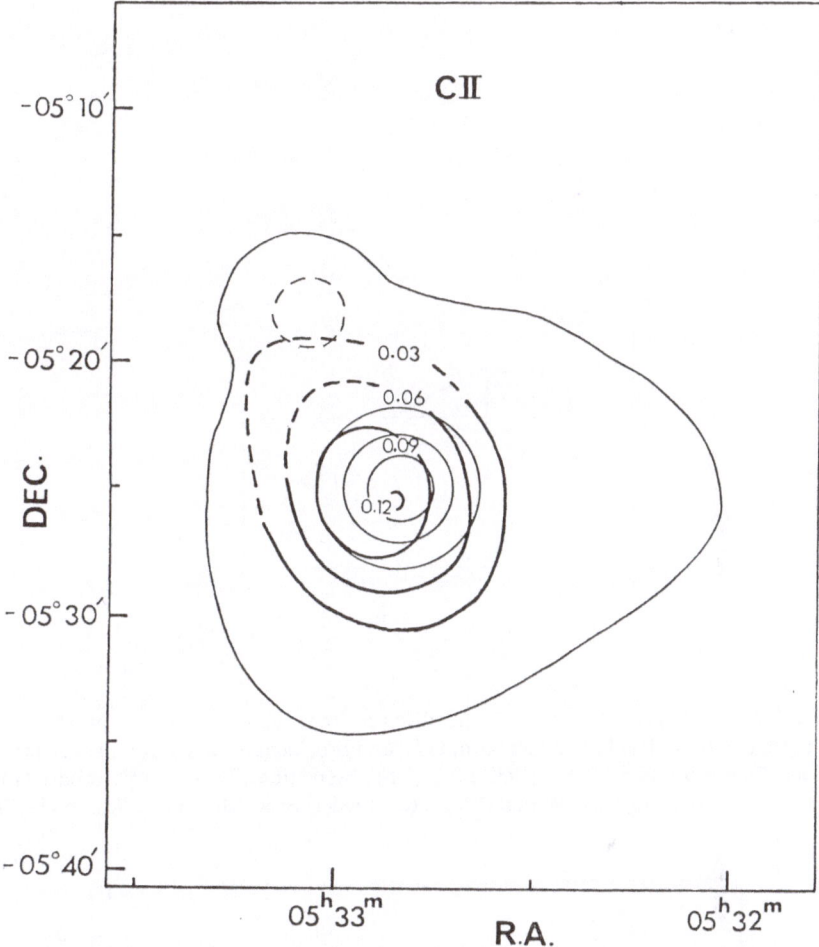

Fig. 2.1.21. Contour map of the C II distribution over the Orion Nebula derived from observations of the C85α line made with 3′ angular resolution. The C85α map is superimposed on a 10 GHz radio continuum contour map of M42 and M43 made with similar (2.′8) angular resolution by MacLeod and Doherty (1968) (light contours). The 75%, 50%, and 25% contours of the peak level are to be compared with the corresponding contours of the ionized hydrogen (after Balick *et al.*, 1974a). Compare the elliptical distribution of C II with the similar ridge-like shape of the core of the molecular cloud lying behind the H II region (e.g. Figures 3.3.5, 3.3.7, and 3.3.8 later in the text). The C recombination lines are believed to originate in a zone or shell of partially ionized matter located in the interface between the H II region and the neutral/molecular complex lying behind. This is further supported (1) by the observed radial velocity of the carbon lines which is similar to that of the molecular lines ($\simeq 9$ km s^{-1}) and distinctly different from that of the ionized region ($\simeq -5$ km s^{-1}) and (2) by the narrow width of the C lines implying a cold place of origin ($\simeq 100$–1000 K). According to this view, the carbon map represents the outer surface of the near edge of the molecular cloud associated with the H II region (M42).

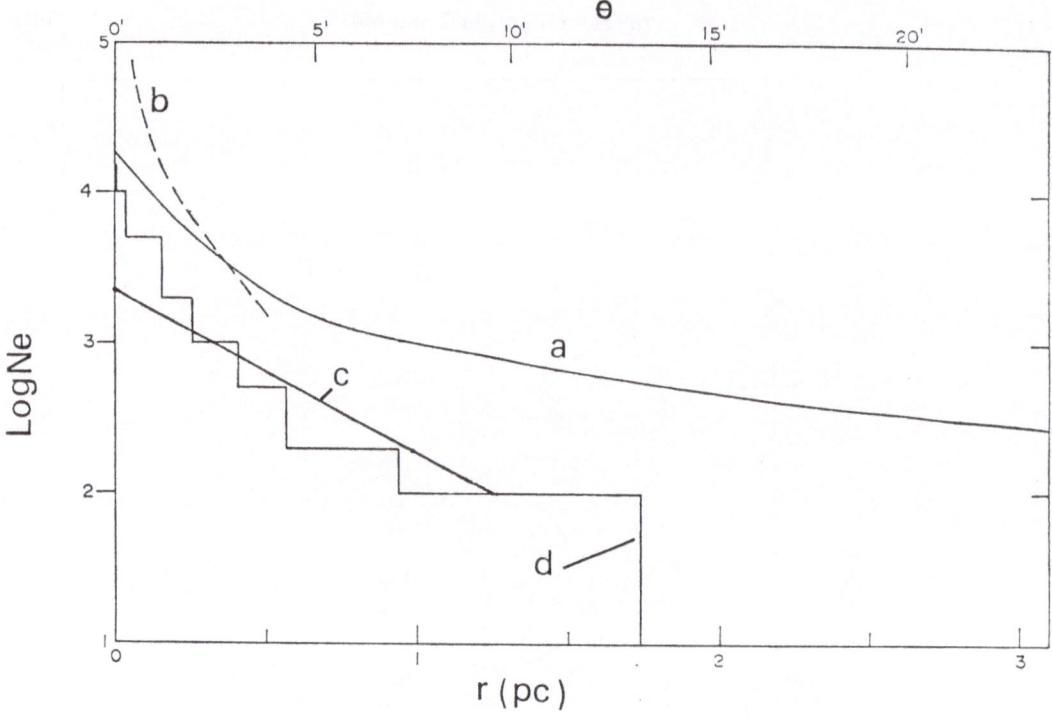

Fig. 2.2.1.   Density distribution versus distance from the density centre of M42 ($\simeq 30''$ W of the Trapezium) (after Danks and Meaburn, 1971, and Brocklehurst and Seaton, 1972); (a); [O II] observations, Osterbrock and Flather (1959); (b): [S II] observations, Danks and Meaburn (1971); (c): radio observations, Menon (1961); (d): model, Brocklehurst and Seaton (1972).

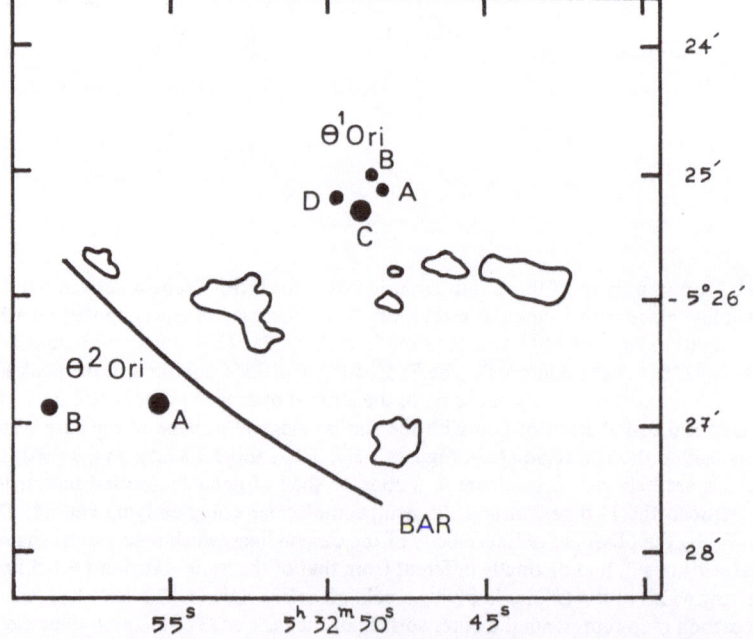

Fig. 2.2.2.   Areas in M42 for which the [O III] line is split (after Dopita *et al.*, 1974, from observations made by Wilson *et al.*, 1959). The sizes of the areas involved are in the range of 0.02 to 0.07 pc.

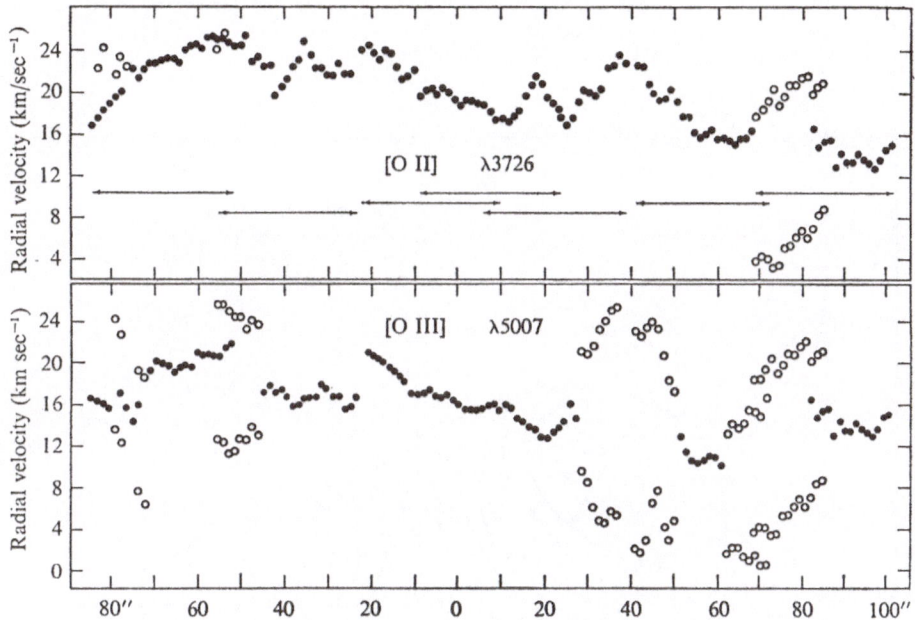

Fig. 2.2.3.   The splitting of the [O III] and [O II] lines along an E–W line 25″ S of θ^1 Ori C in the core of M42 (after Münch, 1958). Solid circles indicate points at which lines are measured as single; empty circles indicate points at which lines are measured as double.

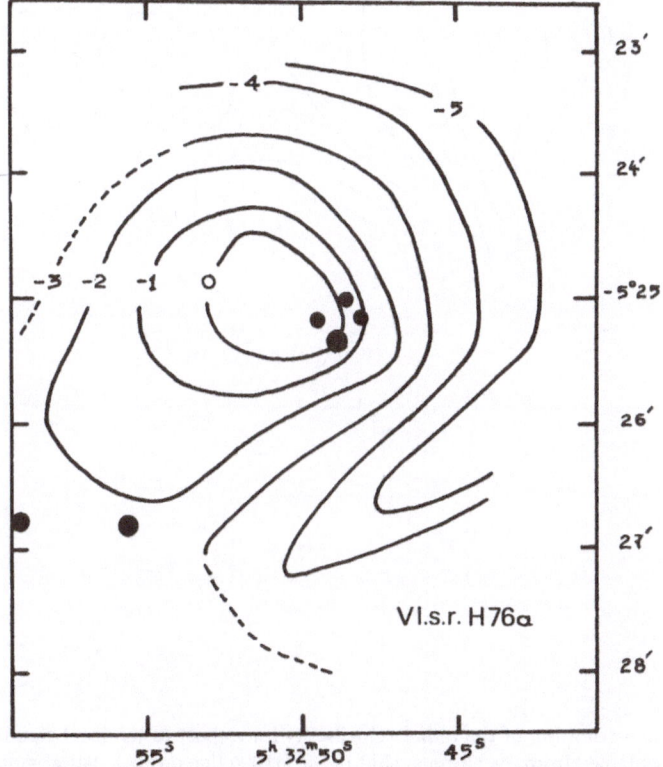

Fig. 2.2.4.   The distribution of the radial velocities (with respect to the local standard of rest) over the core of M42, obtained from the Doppler shift of the H76α line profiles. The angular resolution of the observations was 1′ (after Pankonin et al., 1979).

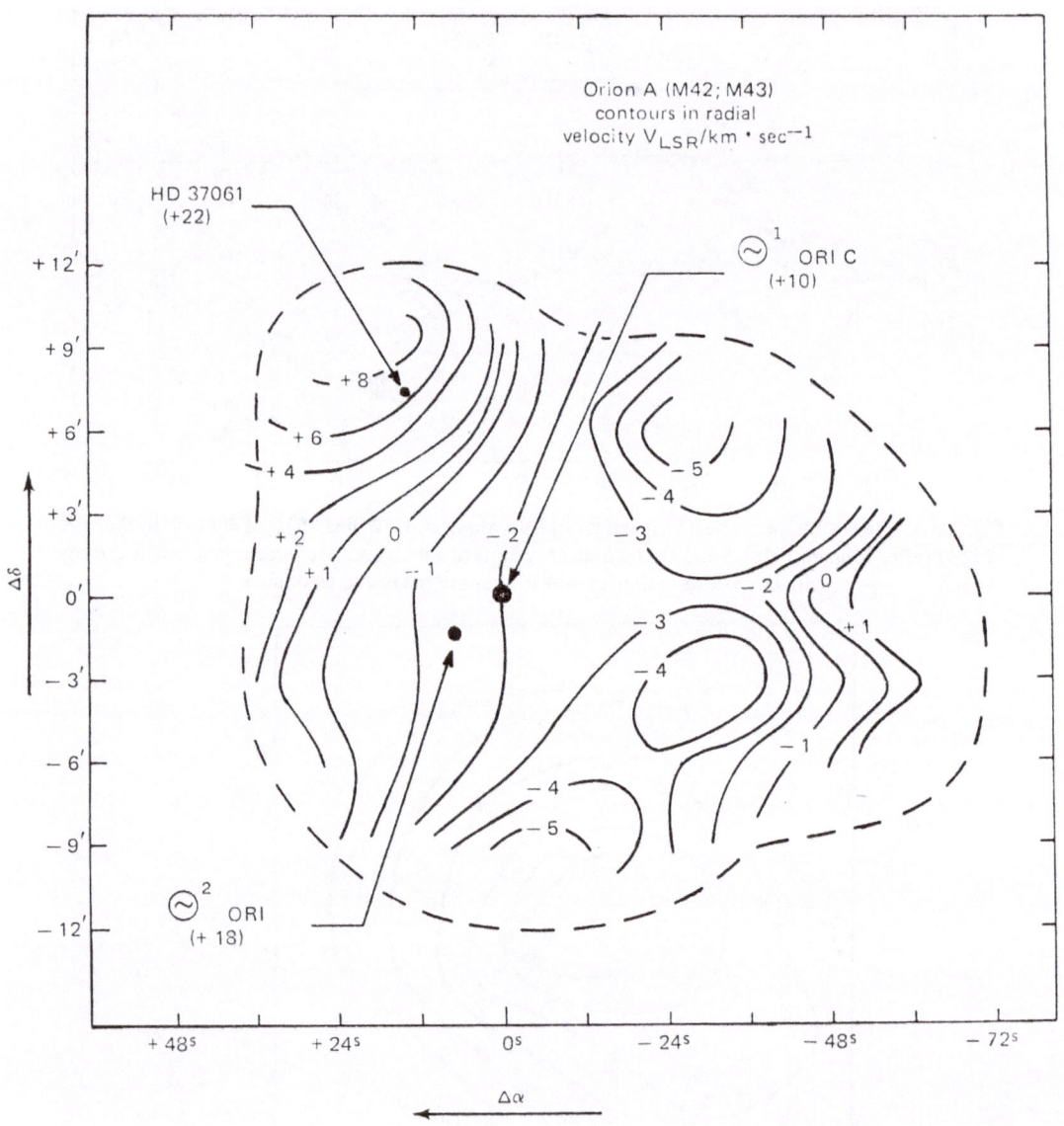

Fig. 2.2.5.   The distribution of the radial velocities (with respect to the local standard of rest) over M42 and M43, obtained from the Doppler shift of the H109α line profiles. The angular resolution was 6′. The dashed line is the outer contour from a 5 GHz continuum map of the area (after Mezger and Ellis, 1968).

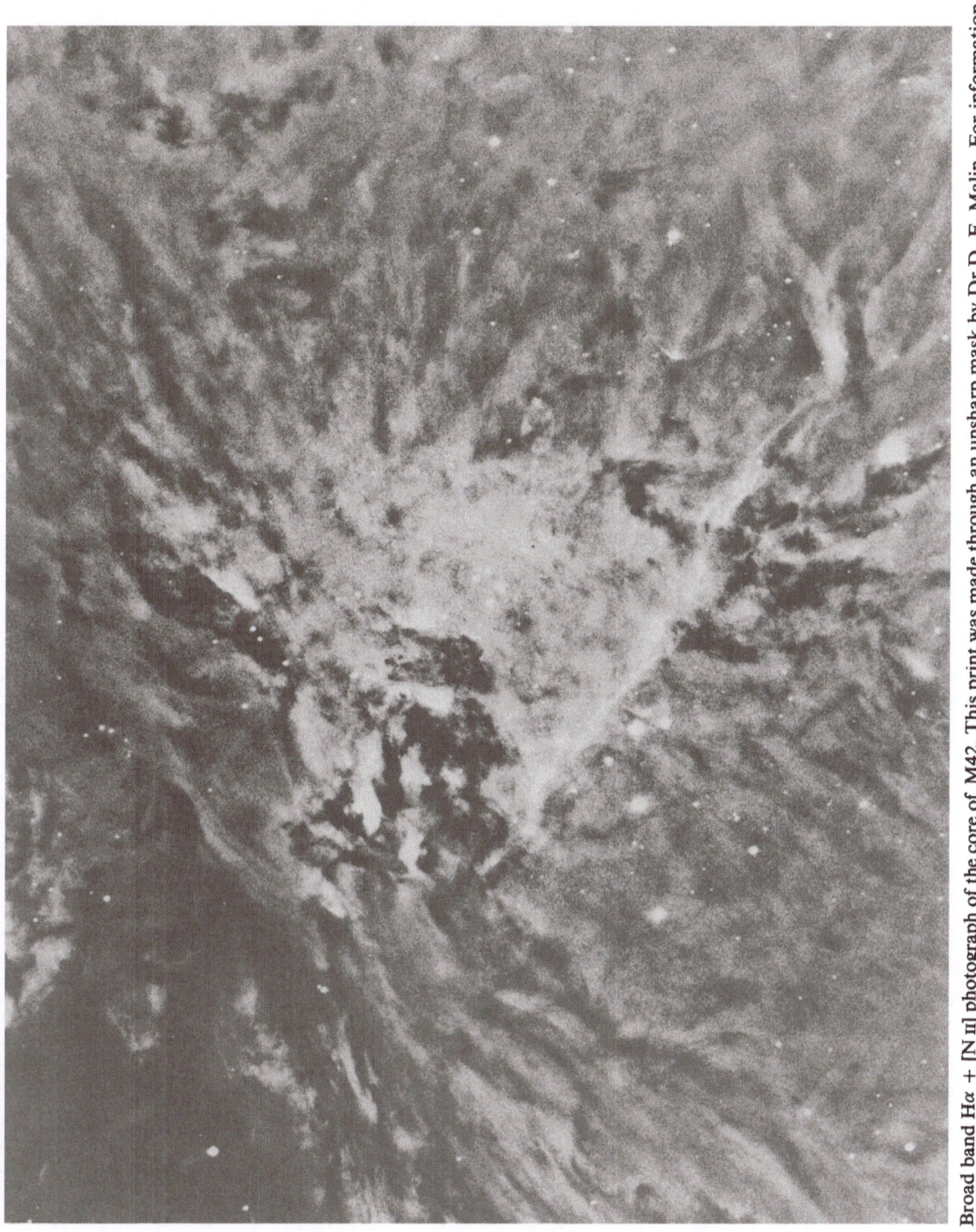

Fig. 2.3.1.   Broad band Hα + [N II] photograph of the core of M42. This print was made through an unsharp mask by Dr D. F. Malin. For information concerning this photograph see Table 2.3.I. (after Malin, 1979).

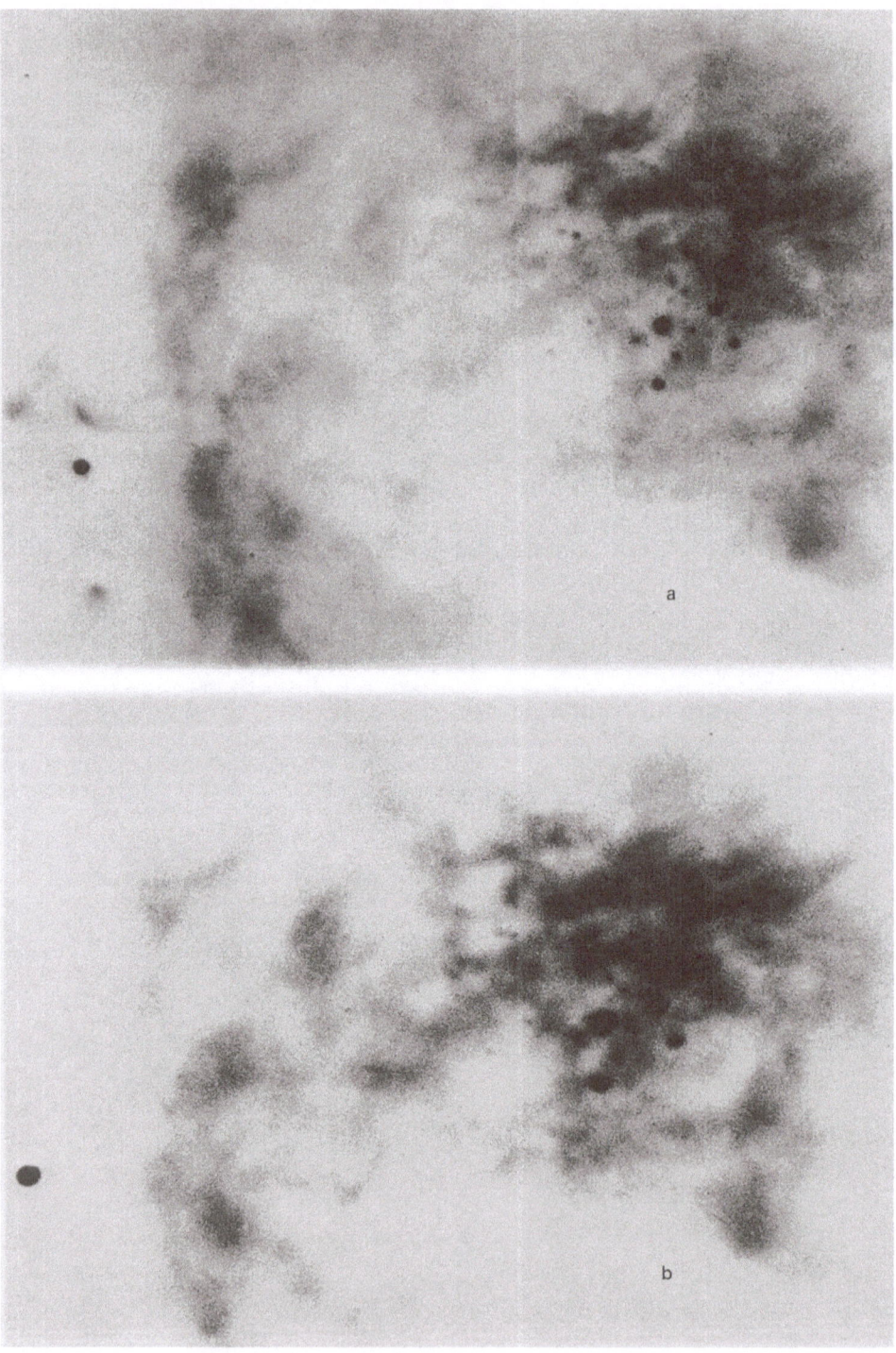

Figs.   2.3.2a–b.

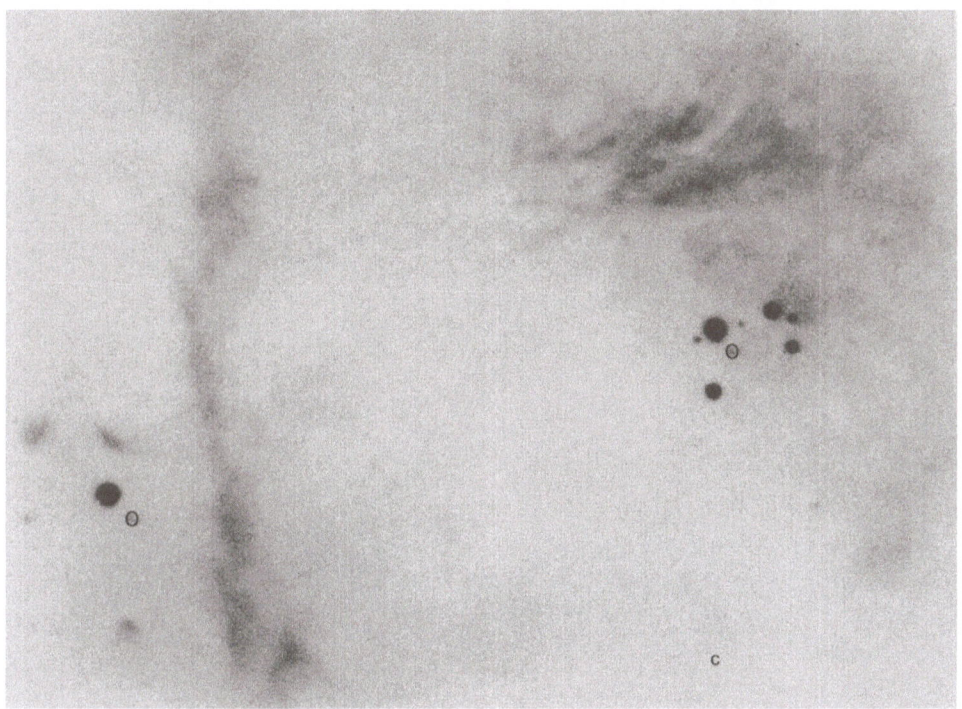

Fig. 2.3.2c.

Fig. 2.3.2a–c.   (a) Hα λ 6563 Å photograph of the core of M42. (b) [O III] 5007 Å photograph of the core of M42. (c) [N II] 6584 Å photograph of the core of M42. The two little circles enclose images due to filter reflection. Note the absence of the condensations which are apparent in the Hα and [O III] photographs. For information concerning the above photographs see Table 2.3.I (after Laques and Vidal, 1979).

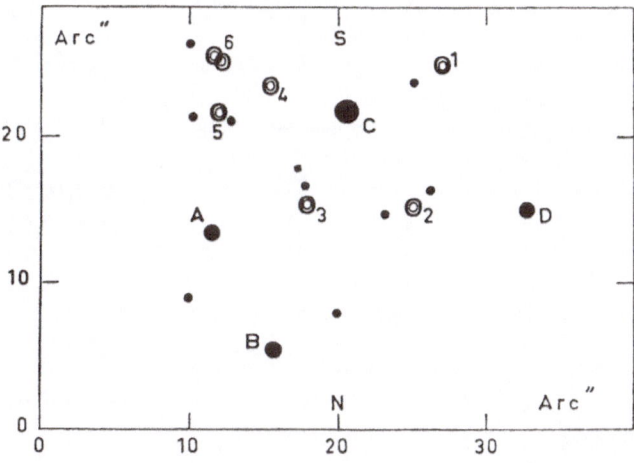

Fig. 2.3.3.   Identification chart of the six Ionized Stellarlike Condensations (ISC 1 to 6) with respect to the Trapezium (A, B, C, D) and other stars in the area. ISC 1 to 6 are marked with numbered open circles. Stars are designated as black points (after Laques and Vidal, 1979).

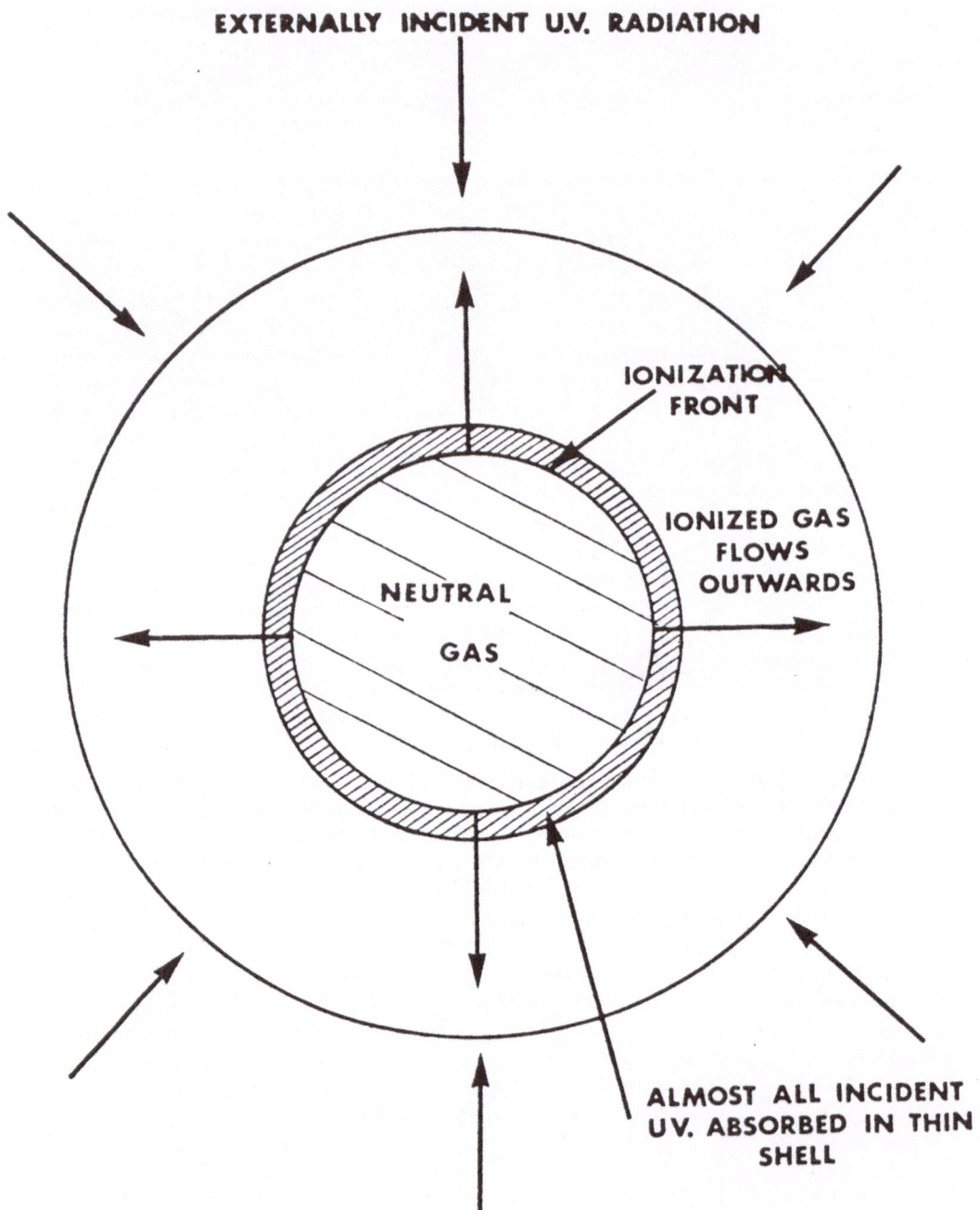

Fig. 2.3.4.   Schematic model of a spherically symmetric neutral globule having an ionization front on its surface. The existence of these small (< 0.01 pc) condensations within the H II regions (named also PIGs: Partially Ionized Globules) was first suggested by Dyson (1968) (after Dopita *et al.*, 1974).

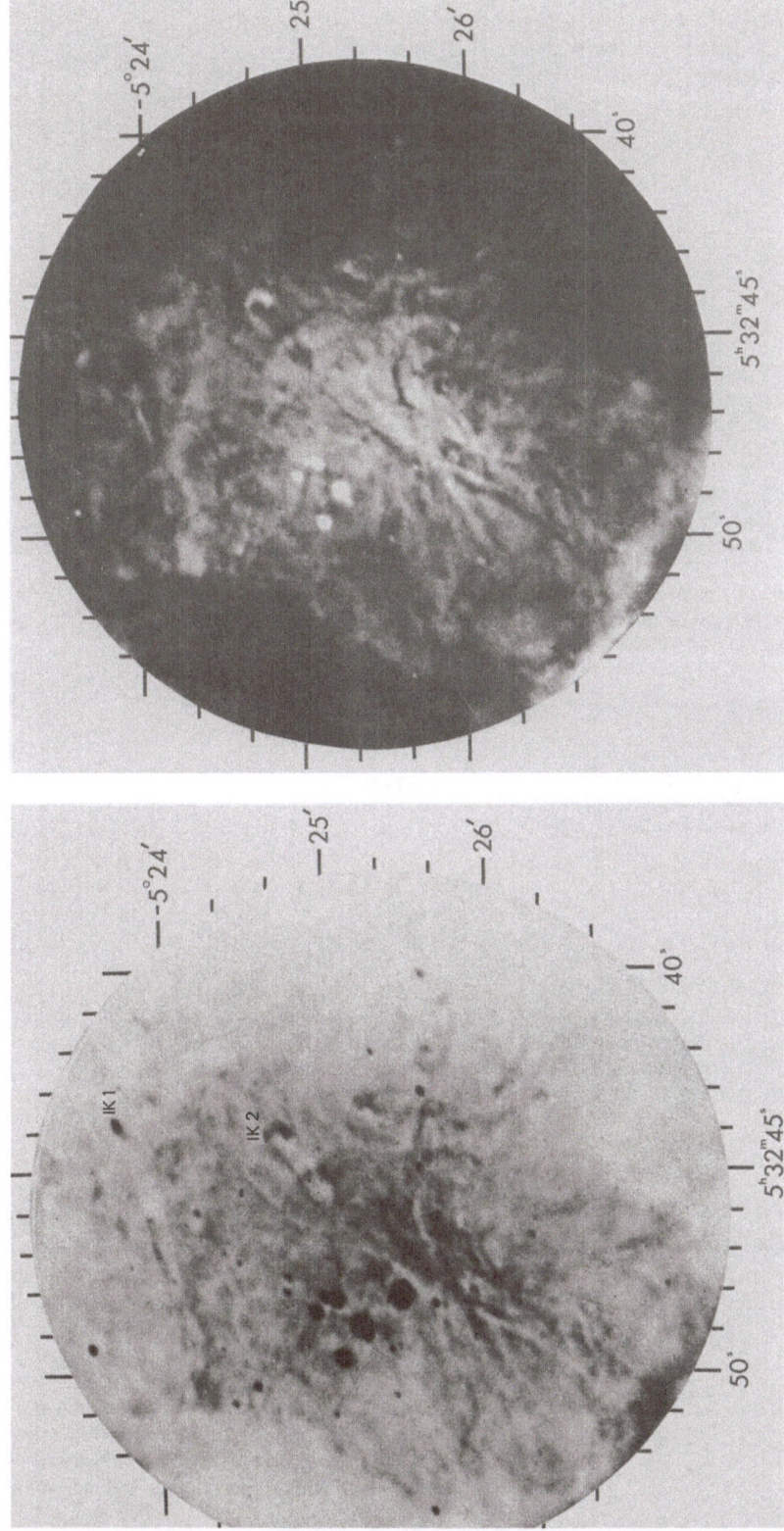

Fig. 2.3.5. [S II] 6731 Å photograph of the core of M42. For information concerning this photograph see Table 2.3.I (after Cantó et al., 1980). Two ionized knots (IK 1, IK 2) can be distinctly seen.

Fig. 2.3.6. [N II] 6584 Å photograph of the core of M42. For information concerning this photograph see Table 2.3.I Note that IK 1 has virtually disappeared (after Cantó et al., 1980).

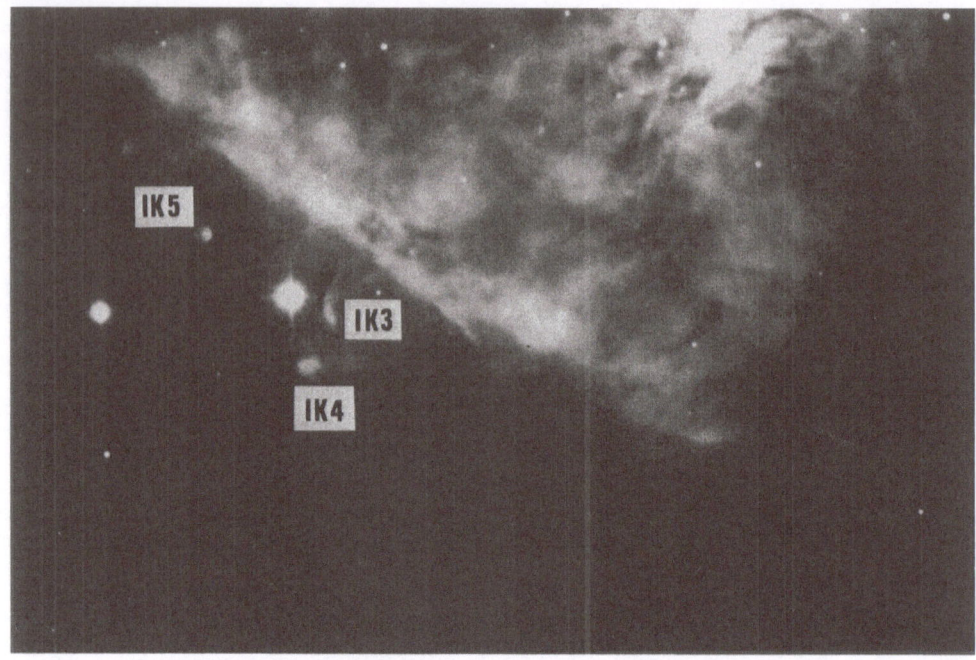

Fig. 2.3.7. Part of Figure 2.1.3 showing the ionized knots IK 3, 4, and 5. (After Taylor and Münch, 1978). For information concerning this photograph see Table 2.3.I.

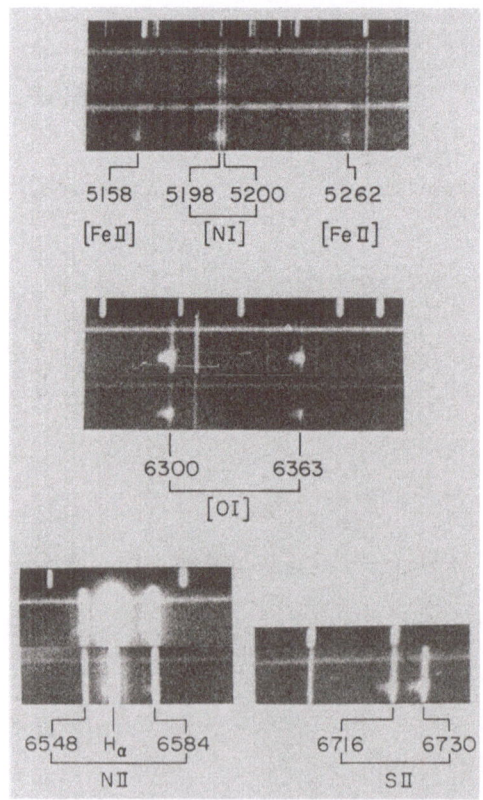

Fig. 2.3.8. Spectra of IK 1 with average dispersion 30 Å mm⁻¹. The spectrum of the star π 1782 appears near the top of each strip (after Münch, 1977). Note the greatly enhanced [O I], [N I], [S II], and [Fe II] lines with respect to those in the background nebula, which characterize the behaviour of a typical Herbig Haro object (Herbig, 1951, 1969; Haro, 1950, 1952; Böhm, 1977). IK 1 is also referred to as M42 HH1. Note also the splitting of the O I and S II lines implying the presence of a component moving with the supersonic velocity of -240 km s⁻¹ with respect to the nebular material.

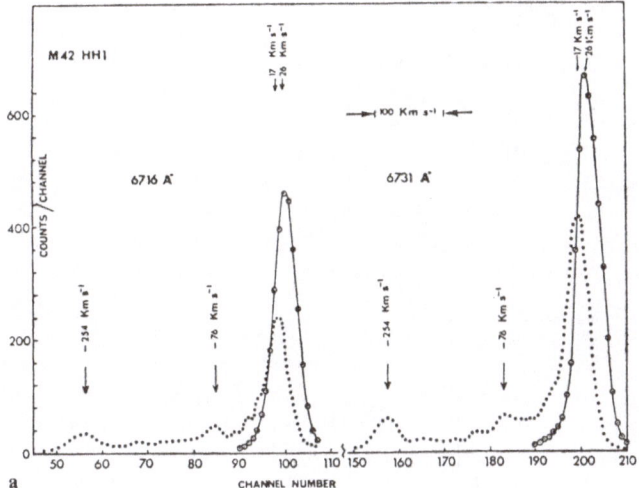

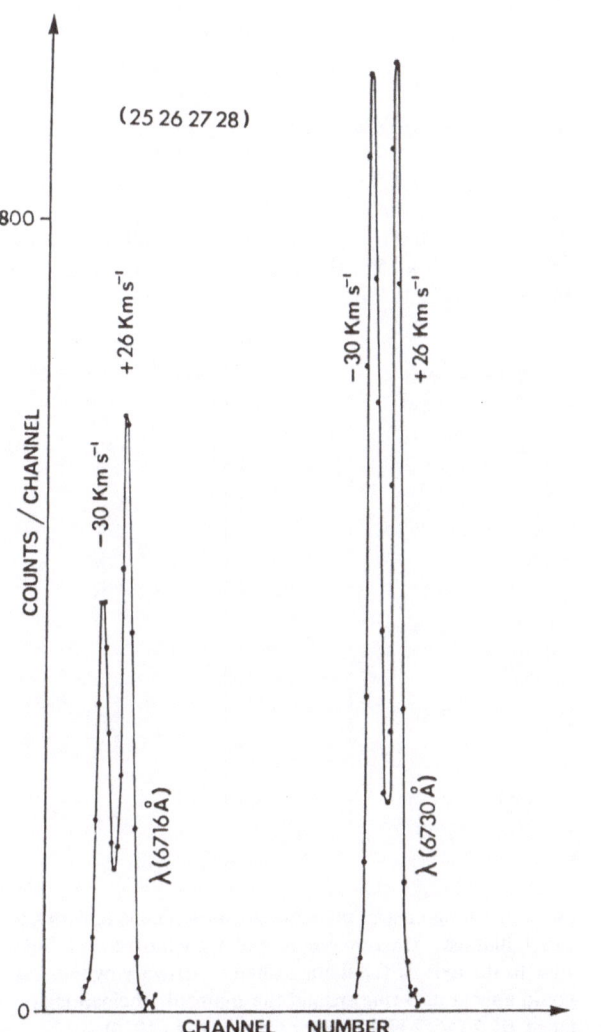

Fig. 2.3.9. [S II] ($\lambda\lambda$ 6731 Å and 6716 Å) profiles from IK 1 (M42 HH1) observed with the RGO Spectrograph at 5 Å mm^{-1} combined with the $f/7.9$ focus of the 3.9 m Anglo–Australian Telescope. The image photon counting system (IPCS) (Boksenberg and Burgess, 1973) was the detector. The slit width was 50 microns which produced a measured instrumental halfwidth of 23 km s^{-1} at 6730 Å. The IPCS was used in the full two-dimensional mode with 60 data taking windows (each $\equiv$ 1''29 long) placed over a long slit of total length 77''5. The profiles shown here (dotted lines) are the ones obtained in two data taking windows over the object, after subtracting the contribution of the background nebulosity obtained in two nearby data taking windows (dots within circles). The heliocentric radial velocities of the most important features (-254 km s^{-1}, -76 km s^{-1}, $+17$ km s^{-1}) are marked. The positive component ($+26$ km s^{-1}) originates in the adjacent nebular material of M42. The heliocentric radial velocity components which originate in IK 1 (with respect to the adjacent nebular material) are then -280 km s^{-1}, -102 km s^{-1} and -9 km s^{-1} (after Cantó et al., 1980).

Fig. 2.3.10. [S II] (λ 6731 Å and λ 6716 Å) profiles from IK 2 (M42 HH2) obtained in four data taking windows; see Figure 2.3.9 for details (after Cantó et al., 1980). The heliocentric radial velocity of the two main features (-30 km s^{-1}, $+26$ km s^{-1}) are marked. The positive velocity feature ($+26$ km s^{-1}) appears to originate within the general nebular material of M42. This makes the heliocentric radial velocity of the object with respect to the nebular material -56 km s^{-1}.

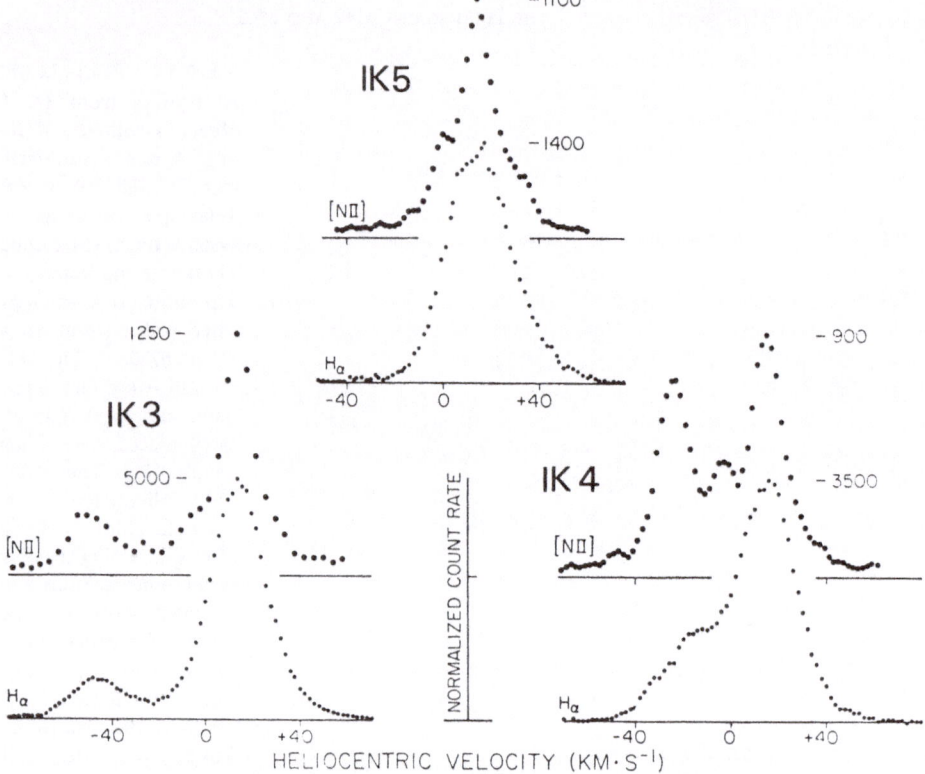

Fig. 2.3.11. Hα λ 6563 Å and [N II] λ 6584 Å profiles from IK 3, 4, and 5 obtained with the Fabry-Perot spectrometer of the Palomar 1.5 m reflector, with a 15″ field stop; see also Table 2.3.II (after Taylor and Münch, 1978).

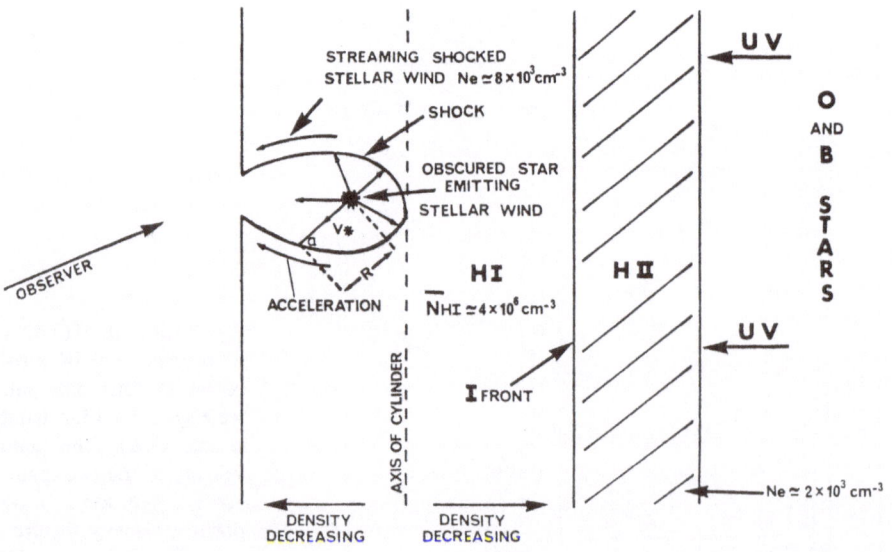

Fig. 2.3.12. A model of an HH object formed in a long neutral filament. A star which is emitting a stellar wind is located near the centre of one such filament. The ram pressure of the wind makes a hole into the filament which projects perpendicular to the axis. If the filament had a perfectly cylindrical cross-section then the wind-driven cavity would appear as a ring around the filament. Incidentically, such a configuration resembles that of IK 2 (M42 HH2) (after Cantó et al., 1980).

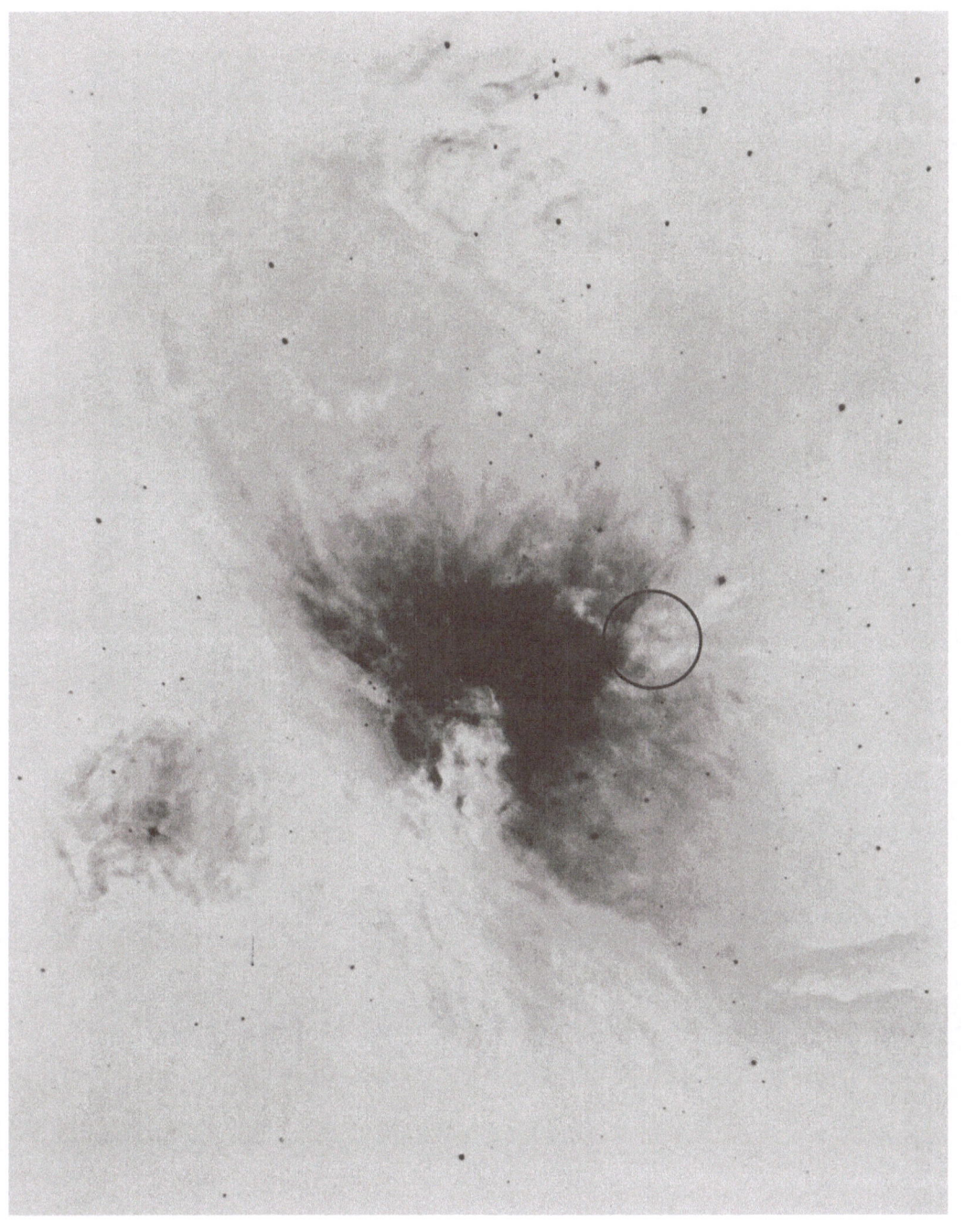

Fig. 2.3.13.   A red photograph of M42, printed through an unsharp mask (courtesy of Prof. G. Münch); for details concerning this photograph see Table 2.3.I. Note the 'cartwheel' filamentary structure within the circle.

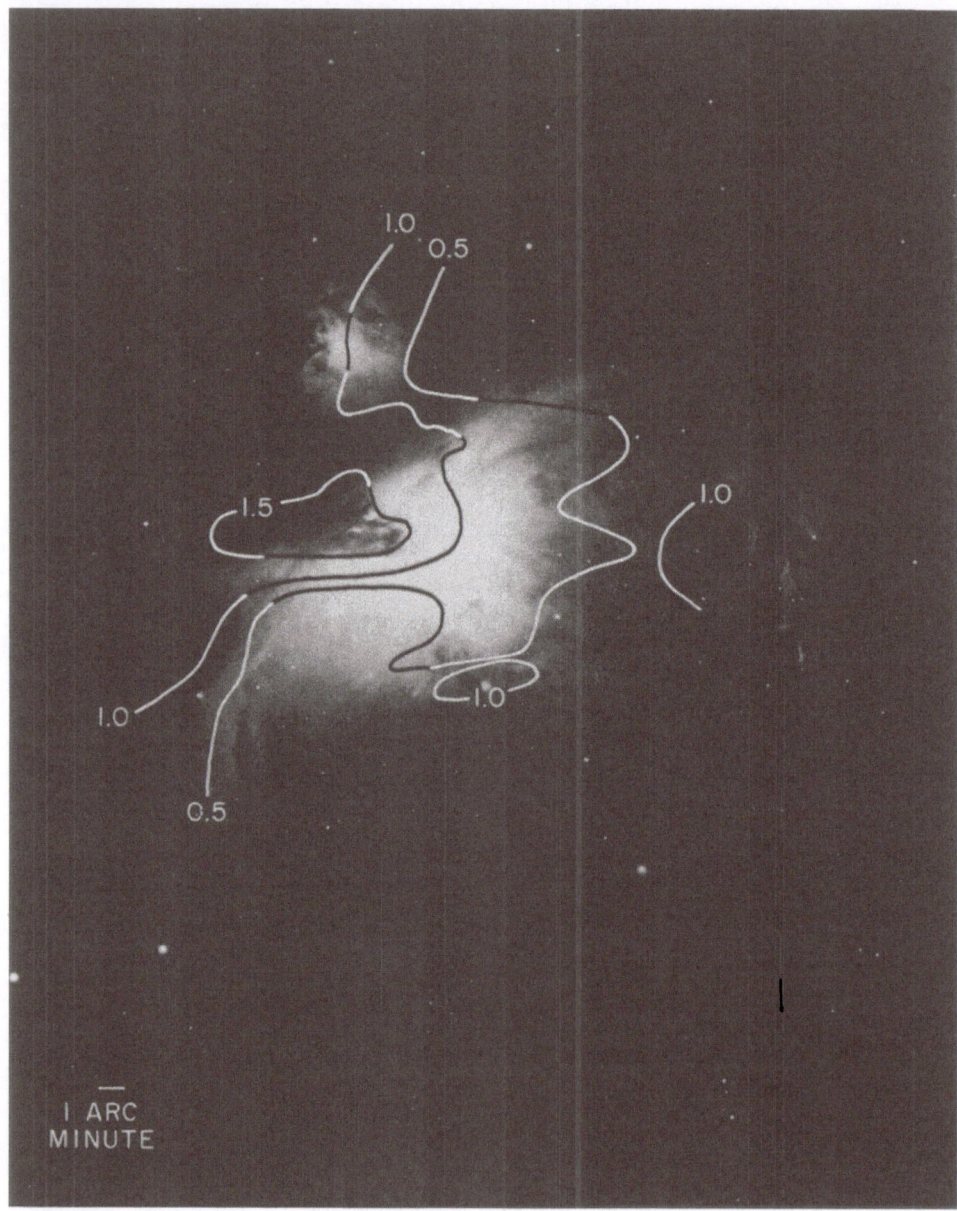

Fig. 2.4.1.   Distribution of A_v across the Orion Nebula (M42 + M43). The angular resolution is $1.'7$ (after Chaisson and Dopita, 1977). Note that the maximum A_v ($\simeq 1.^m5$) occurs in the 'bay' region where the H I optical depth is also maximum (Figure 2.4.9 later in the text). This is an indication that the dust is physically associated with the H I feature. The bay may be a neutral feature originating in the massive H I/Molecular Cloud which lies behind M42 and enveloping part of the ionized gas.

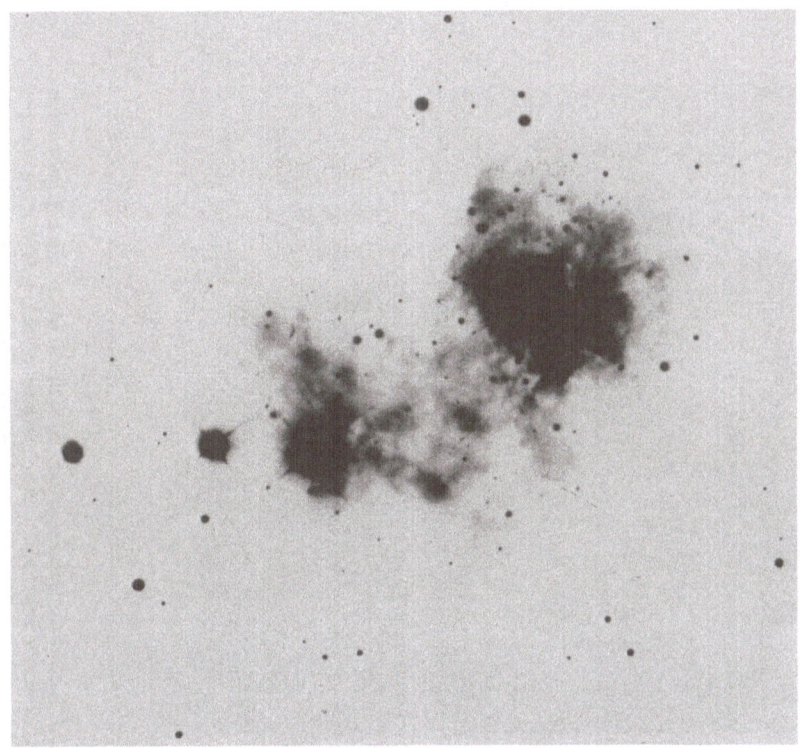

Fig. 2.4.2.   A continuum photograph of M42 taken through a 200 Å wide dye filter centred on 5200 Å (after Wurm and Rosino, 1965). Note the different appearance of the nebula from that of the line photographs in Figures 2.1.4 to 2.1.9 indicating the dominant role of the dust scattered component with respect to the atomic component of the continuum.

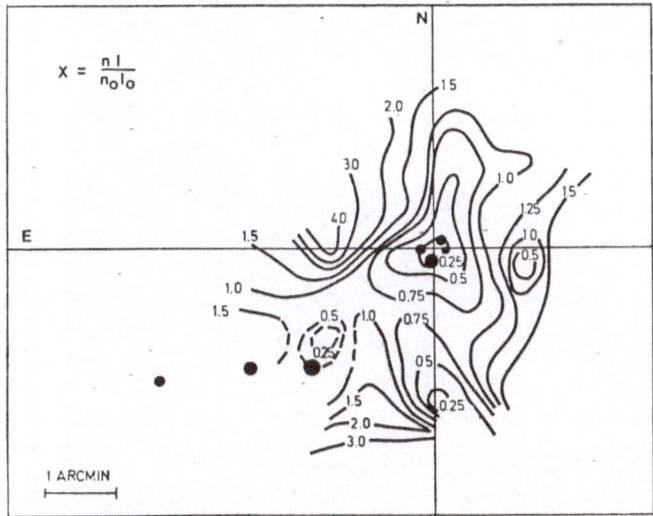

Fig. 2.4.3.   A contour map of the 'absorbing matter' of the Orion nebula made with an angular resolution of 38″. The map represents the distribution of the dust column density ratio $r = nl/n_0l_0$ where n, l are the column density and the path along the line of sight, and n_0, l_0 are the corresponding parameters in the direction of the Trapezium. The map was produced by comparing $P\alpha - 12$ ($\lambda = 8750$ Å) intensities with the $H\alpha$ ($\lambda = 6563$ Å) intensities observed by Münch and Persson (1971) (after Barbieri et al., 1976). Note the 'hole' in the dust around the Trapezium. Note also that the absorption ridge to the NW of Trapezium coincides with the brightest part of the nebula, implying that the gas and dust are well mixed. An absorption maximum also occurs around the 'bar' area just north of θ^2 Ori A.

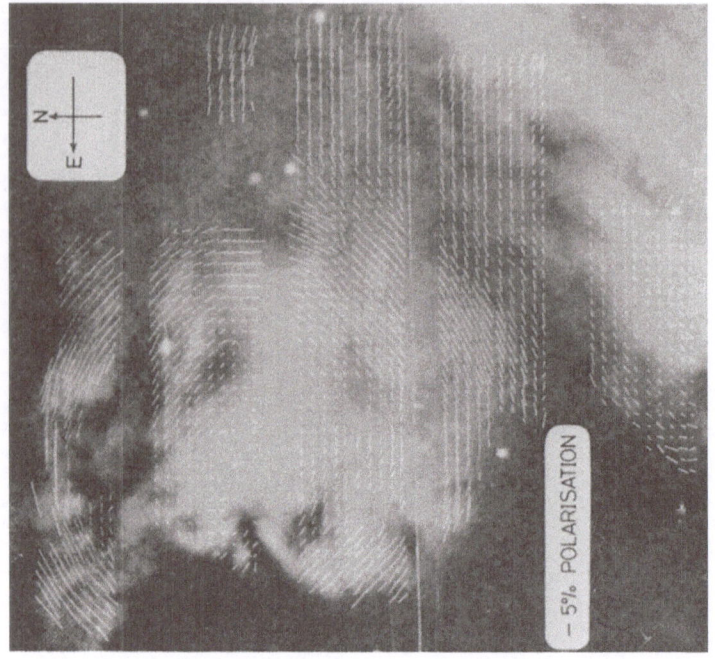

Fig. 2.4.5. Linear optical polarization map of M43 (after Khallesse *et al.*, 1980). Note the centrosymmetric pattern of the polarization vectors around the star NU Orions confirming its role as exciting star of M43. Note also the abrupt change of the polarization orientation in the densest part of the dust lane which indicates a physical association of the lane with M42 and M43.

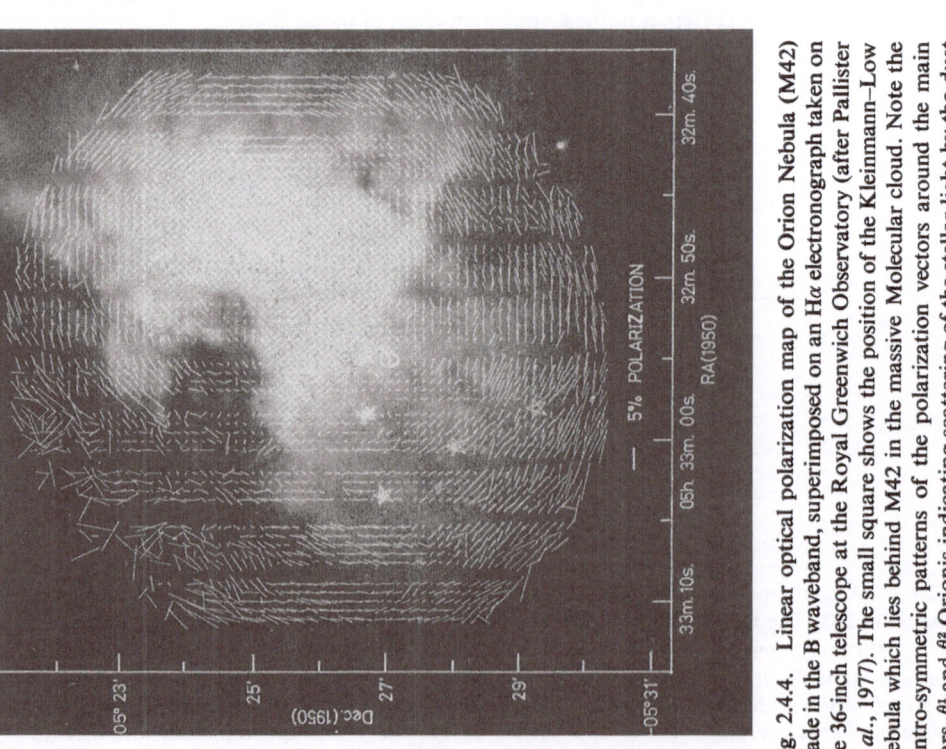

Fig. 2.4.4. Linear optical polarization map of the Orion Nebula (M42) made in the B waveband, superimposed on an Hα electronograph taken on the 36-inch telescope at the Royal Greenwich Observatory (after Pallister *et al.*, 1977). The small square shows the position of the Kleinmann–Low Nebula which lies behind M42 in the massive Molecular cloud. Note the centro-symmetric patterns of the polarization vectors around the main stars θ¹ and θ² Orionis indicating scattering of the stellar light by the dust contained in the surrounding medium.

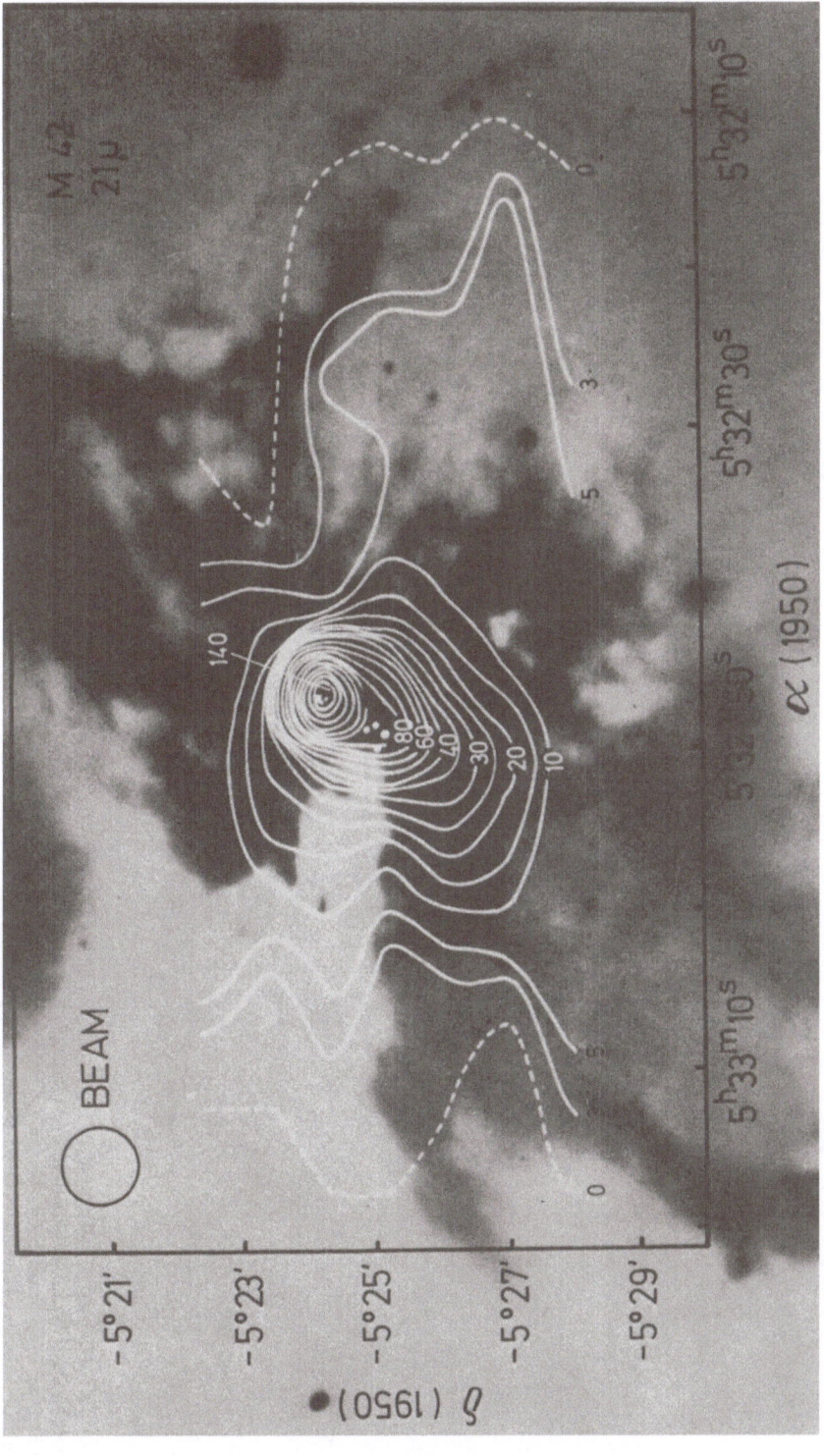

Fig. 2.4.6. Infrared map of M42 made at 21μ with an angular resolution of 1′ (after Lemke *et al.*, 1974). The bright compact source ∼1′ NW of the Trapezium is the Kleinmann–Low Nebula which lies in the massive Molecular Cloud on the backside of M42. The rest of the contours are concentrated around the Trapezium and show a similar distribution with the contours of the radio continuum emission (Figure 2.1.16 made with comparable angular resolution) indicating that the dust and gas are well mixed. The IR radiation is due to the thermal reradiation of the UV stellar radiation absorbed by dust.

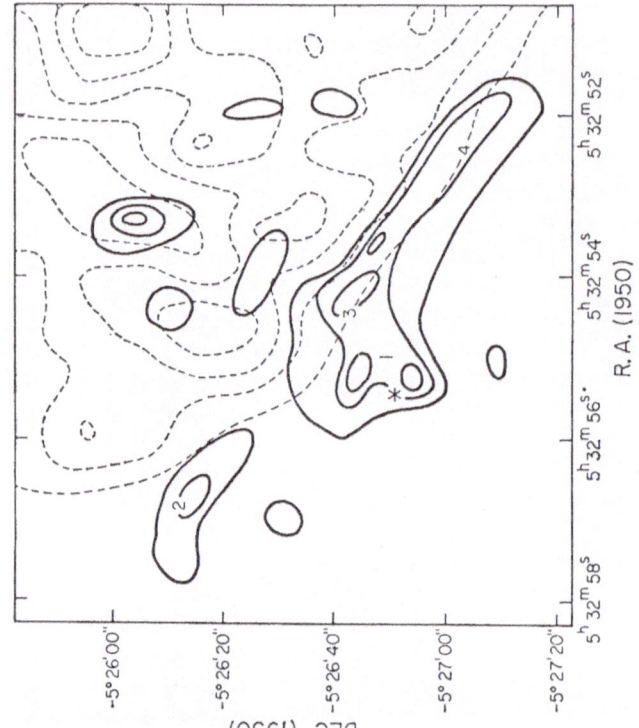

Fig. 2.4.8. The distribution of the 10μ radiation in the vicinity of θ^2 Orionis A (bar) made with an angular resolution of $4''.5$ (solid curve) and superposed on the corresponding part of the radio continuum map of Gull and Martin (1976, Figure 2.1.19) made with a $7'' \times 20''$ angular resolution (after Becklin et al., 1976). Note the coincidence of the IR with the radio 'bar' indicating the physical association of dust and gas. The infrared radiation around θ^2 Ori A probably originates in a circumstellar dust envelope surrounding the star.

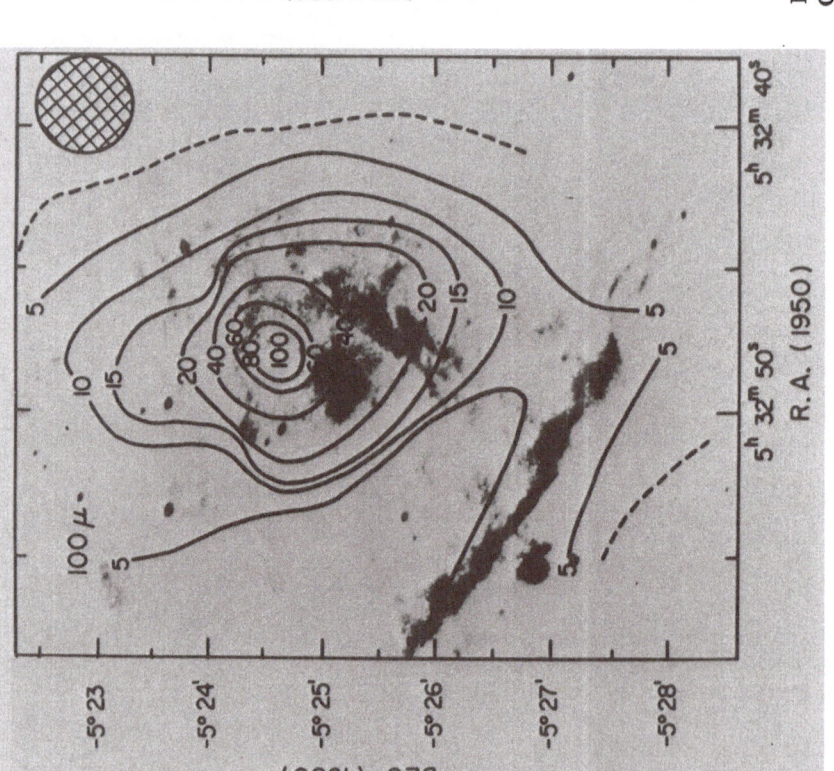

Fig. 2.4.7. Infrared map of M42 made at 100μ with an angular resolution of $1'$, superimposed on a [O I] λ 6300 Å photograph taken by T. R. Gull at the Kitt Peak National Observatory (after Werner et al., 1976). Note the spatial coincidence of the infrared ridge with the optical bar indicating that the gas and dust are physically associated.

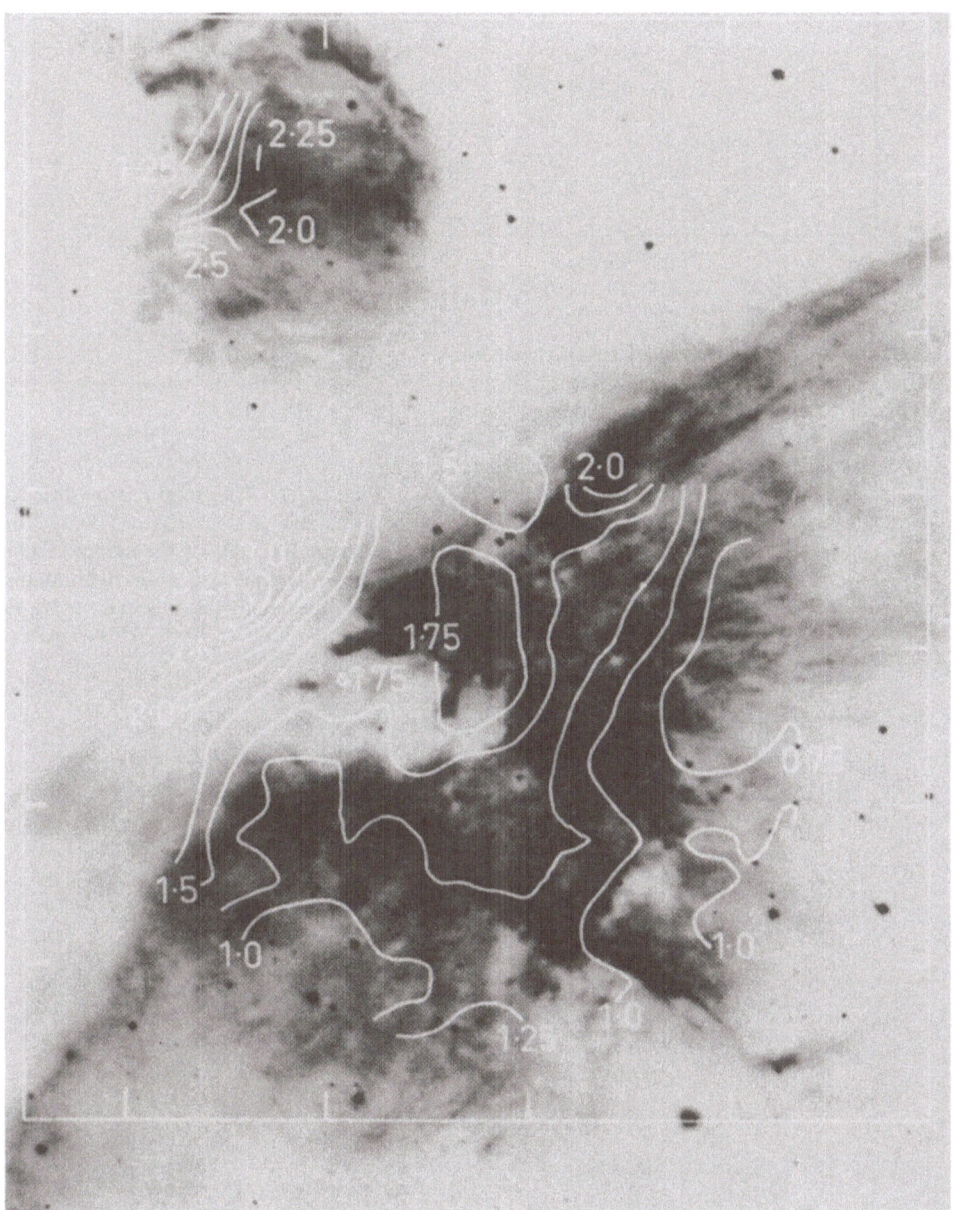

Fig. 2.4.9.   Distribution of the integrated H I optical depth made with a 2′ angular resolution and superimposed on an [S II] λλ 6717, 6731 Å photograph of Orion (M42 + M43) taken by T. R. Gull (after Lockhart and Goss, 1978). Note the similarity with the A_v contour map of Figure 2.4.1 made with a comparable angular resolution (1.̇7) and the comment made in the caption to that figure concerning the physical association of dust and neutral hydrogen in the bay area.

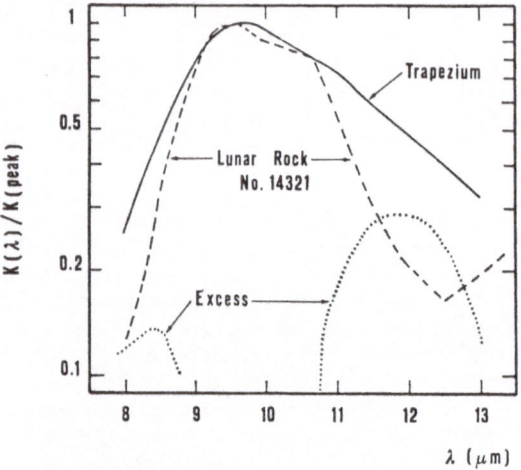

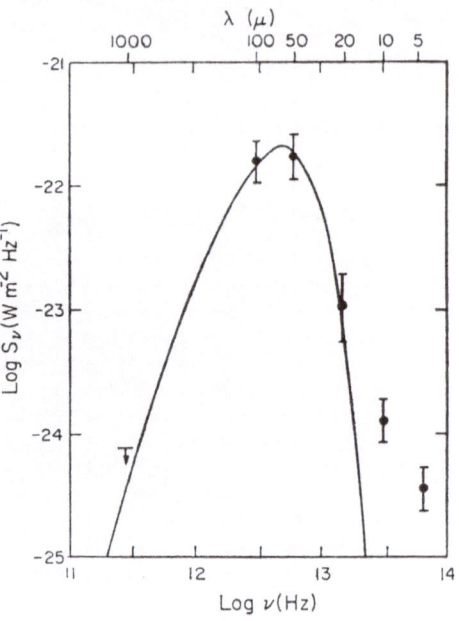

Fig. 2.4.10. A comparison of the 10μ emission band observed in the Trapezium Region (Forrest *et al.*, 1975; solid line) with that computed for the lunar silicate No. 14321 (Bussoletti and Zambetta, 1976; dashed line). The opacities are normalized to the peak values (after Panagia, 1977). Note the excesses around 8.5μ and 11.8μ (dotted lines).

Fig. 2.4.11. The total energy distribution of the 'bar' in the range of $5-1000\mu$ for the entire solid angle of the source. The solid curve represents the energy distribution of a 60 K blackbody (after Becklin *et al.*, 1976).

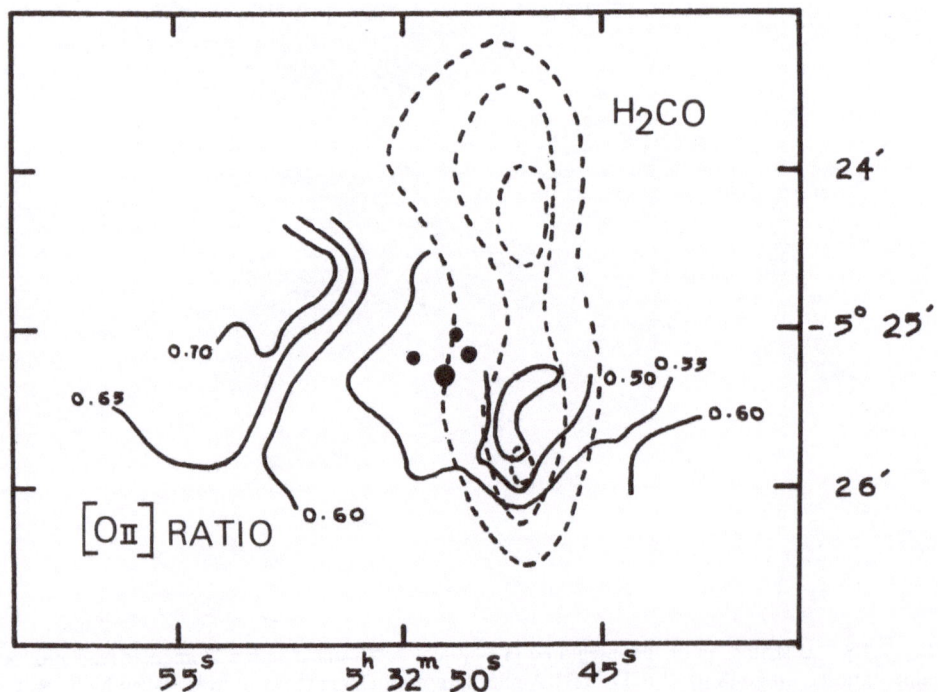

Fig. 2.4.12. The brightness distribution of the H_2CO molecular emission (Thaddeus *et al.*, 1971) superimposed on the distribution of the intensity ratio of the [O II] lines (λ 3729 Å/λ 3726 Å) (after Elliott and Meaburn, 1973). Low values of this ratio indicate high values of local electron density. Note the coincidence of the densest peak of M42 with one peak of the molecular ridge.

THE ORION COMPLEX (M42/OMC 1, M43/OMC 2)

3.1. Infrared Structure

The Orion Nebula and the associated region have been extensively studied over the whole range of the infrared part of the spectrum (near infrared, $1-3\mu$; middle infrared, $3-30\mu$; far infrared, $30-500\mu$; and submillimeter range, $500\mu-1$ mm). A compilation of the available maps produced from observations of the infrared continuum emission is presented in Table 3.1.I. The infrared continuum is thought to be thermal re-emission of the stellar UV radiation absorbed by the dust contained within the ionized gas (M42) and the associated neutral/molecular complex. Some comments concerning the physics of the infrared continuum emission are included in the Appendix, Section A.III.2.

Two extended sources dominate the middle infrared ($3-30\mu$) continuum emission, the Trapezium Nebula (or Ney–Allen Nebula discovered by Ney and Allen, 1969) and the Kleinmann–Low Nebula (discovered by Kleinmann and Low, 1967). The structure of the central $2'$ of the Orion Complex is shown in Figure 3.1.1 in a sequence of contour maps made at 8.6μ, 10.1μ, 11.2μ, 12.3μ, and 13.1μ with high angular resolution ($9''$). The Trapezium Nebula is physically associated with M42 and has a similar overall structure with its corresponding optical and radio counterparts. The Kleinmann–Low Nebula (KL), lying $\sim 1'$ NW of Trapezium and having a diameter of the order of $30''$, does not coincide with any obvious optical or radio feature.

Embedded within the KL nebula are five infrared compact sources the most prominent of which is the so called Becklin–Neugebauer Object (BN) discovered by Becklin and Neugebaurer (1967). These sources, first studied by Rieke *et al.* (1973), are probably of stellar nature (protostars?) and are thought to be members of a cluster (IR-cluster). Their coordinates are given in Table 3.1.II.

At longer wavelengths the KL Nebula becomes more prominent relative to the Trapezium Nebula. This can be seen in the sequence of contour maps shown in Figures 3.1.2 to 3.1.5. In the contour map of Figure 3.1.2, made at 20μ with a $5''$ angular resolution, both nebulae are shown. Also shown is the position of the BN object (especially prominent at $\lambda < 10\mu$) and the association of KL with the H_2O and OH masers observed in the Orion area. In the map of Figure 3.1.3, made at 33μ with a $10''$ angular resolution, the KL nebula is the dominant source, although the Trapezium Nebula is still discernible.

In Figure 3.1.4 a map of a more extended area ($8' \times 3'$) is shown, made at 1 mm with an angular resolution of $1'$. The structure of the map and the correlation with the distribution of the 2 cm H_2CO molecular emission, from a contour map of the same angular resolution, clearly demonstrates the intrinsic relation of the 1 mm emission with the molecular gas of the Orion Molecular Complex (OMC 1; see Section 3.3). The peak of the 1 mm map ($\sim 40''$) coincides with the KL Nebula, the members of the IR cluster and the peak of the molecular cloud, a result which strongly indicates that the KL Nebula/IR cluster lie within the molecular cloud. This is also supported by the absence of any 1 mm feature associated with the ionized gas (M42), the position of which is also marked in the map. A synopsis of the main evidence supporting the generic association of the KL Nebula/IR cluster with the molecular cloud OMC 1 is presented in Table 3.1.III. In addition, the intrinsic association of the KL Nebula with the IR cluster is supported not only by the positional proximity of the sources but also by energetic considerations. The IR cluster is the only source capable of powering the infrared emission of the KL Nebula (the Trapezium cluster of stars being inadequate, Werner *et al.* (1976)). In Figure 3.1.5 a 91μ contour map with a $2\rlap{.}''2$ angular resolution of an extended area ($15' \times 10'$) in Orion is shown. The map is characteristic of the far-infrared emission, the distribution of which (peaking near KL) supports the conclusion reached above. However, an extended diffuse component ($6' \times 8'$) is also present and this may be powered by the stars of the Trapezium cluster. The map also covers the M43 nebula. An extended infrared component is discernible around NU Orionis, the exciting star of M43.

Apart from the extended sources mentioned above, there are also smaller infrared sources ($\sim 25''$ diameter) which have been detected around certain stars and are probably due to the emission from circumstellar shells surrounding them. The most prominent of these discovered by Ney *et al.* (1973), are presented in Table 3.1.IV.

Numerous infrared sources have also been discovered in the area $\sim 6'$ to the north of M43 which is associated with the second Orion Molecular Cloud (OMC 2, $\sim 4'$ N, $1'$ W of M43; see Figure 3.3.6 in Section 3.3). These sources, discovered by Gatley *et al.* (1974), are shown in the 2.2μ continuum map of Figure 3.1.6, made with an angular resolution of $7\rlap{.}''5$. Many of the sources have been identified with visible stars, but five of them (IRS 1, 2, 3, 4, 5/OMC 2) seem to be members of an infrared cluster of very young objects (probably similar to the IR cluster associated with OMC 1).

The emission from dust and molecular gas is not contained in the two molecular clouds OMC 1, OMC 2 but is probably extended along the central ridge ($\sim 10' \times 40'$) of the huge dust cloud L1641 (see Chapter 1). The Orion Molecular Clouds OMC 1 and OMC 2, and probably a third cloud $16'$ to the north of OMC 2 (OMC 3), may be the results of a fragmentation process which occured in L1641. The extent of the dust and molecular gas emission originating from the central part of L1641 is shown in Figure 3.1.7 in which three contour maps, an infrared map made at 400μ with $3'$ resolution, a ^{13}CO molecular map at 2.6 mm with $2\rlap{.}'6$ resolution and a radio continuum map at 5 GHz with $4'$ resolution, are presented together for comparison.

3.2. Physical Parameters Derived from Infrared Observations

The physical parameters characterizing the dust associated with the Orion Complex are presented in Tables 3.2.I to 3.2.IX. Comments on the estimation of the various parameters are included in the Appendix, Section A.III.2. Tables 3.2.I to 3.2.IV contain information concerning the infrared fluxes, energy distribution, dust temperature and infrared luminosity of the Trapezium Nebula, the area of Orion containing both the Trapezium and the Kleinmann–Low (KL) Nebula, the KL Nebula and the IR cluster in OMC 1, respectively. It should be noted that the information included in Table 3.2.II, although obtained from observations made with low angular resolution (large beams containing both the Trapezium and the KL Nebula), reflects the properties of the KL Nebula which is the dominant contributor in the far infrared.

Some information from infrared studies concerning the properties of the dust within M42 has already been presented in the previous Section 2.4 in connection with optical and radio studies of the ionized gas. Information relating to the energy spectrum of the infrared sources of the Orion Complex is presented here in Figures 3.2.1 to 3.2.4. The 10μ 'silicate' feature, present in emission in the spectrum of the Trapezium Nebula (see also Section 2.4) and in absorption in the spectrum of KL is shown in Figure 3.2.1. The Trapezium emission feature is thought to be the result of circumstellar shells of silicate dust, whereas the KL absorption feature is thought to originate from the extended dust cloud surrounding the embedded IR cluster. The $10–1000\mu$ energy distribution of the 2' core of KL is shown in Figure 3.2.2, fitted with curves representing a uniform distribution of dust with dust temperature $T_d \sim 70$ K and a clumply distribution with $T_d \sim 100$ K. The energy distribution in the 1.4 core of KL in the range $30–125\mu$ is shown in Figure 3.2.3, fitted with two curves indicating that T_d is in the range 70–95 K. The $1–1000\mu$ energy spectrum of the 1' core of KL in comparison with that of BN is shown in Figure 3.2.4 fitted with a curve estimated from a model by Scoville and Kwan (1976). Note the predominance of BN for $\lambda < 10\mu$.

The dust temperature distribution over the Trapezium and KL Nebulae is shown in Figure 3.2.5. The dust temperature in the KL is $\sim 80–85$ K (see also Table 3.2.III) and is similar to the molecular gas temperature which is derived from various molecules present in the core of KL (see Table 3.3.III). This is a further indication that the gas and dust in the area are well mixed and belong to the same complex (see also Table 3.1.III).

It should be noted that the dust in the ionized gas (M42) must be related to the dust in the molecular cloud (OMC 1), since the H II region was born out of this cloud and has evolved in physical association with it. It is possible that some of the dust observed in the ionized region was present in that part of the molecular cloud which became ionized after the formation of the Trapezium cluster of stars. Since the harsh conditions prevailing in the H II region are expected to cause the destruction of grains within 10^3–10^5 yr, it is possible that the molecular cloud has served as a 'reservoir' of dust replenishing these losses (Isobe, 1977).

Information concerning the polarization of the infrared continuum emission originating in the extended KL Nebula ($\sim 30''$) and the starlike ($< 2''$) BN object is presented in Table 3.2.V. The emission from the BN object exhibits high linear polarization around the 3μ and 10μ absorption features. These features are attributed to the presence of water ice and silicate grains. To explain this phenomenon two mechanisms have been proposed: (a) The 'Davis–Greenstein' mechanism attributes the observed linear polarization of anisotropic extinction by elongated (needle like) dust particles aligned under the influence of a magnetic field (Dyck and Beichman, 1974). Such an interpretation however has to assume dust temperatures $T_d < 6$ K which is not realistic. (b) The alternative 'Elsässer–Staude' mechanism explains the phenomenon by scattering dust grains and electrons in non-spherical circumstellar shells (Elsässer and Staude, 1979). The predicted energy distribution and linear polarization of BN, derived from a model based on silicate/ice particles with a size distribution $n(a) \propto a^{-3.5}$, are in good agreement with the observations. This is shown in Figure 3.2.6. Configurations used in explaining the phenomenon are shown in Figure 3.2.7. Some details concerning this mechanism are included in the caption to the figure. Another view, proposed by Knacke and Capps (1979), circumvents the 'temperature problem' of the 'Davis–Greenstein' mechanism by invoking the presence of a cold cloud of aligned absorbing grains in front of the BN and KL sources. The advantage of this model with respect to the 'Elsässer–Staude' mechanism is that it deals with the recently discovered highly polarized 20μ feature (Table 3.2.V).

Information concerning the dust component of OMC 2 is presented in Table 3.2.VI. The ~ 2–20μ energy distribution of the $7''.5$ core of the infrared sources IRS 1, 2, 3, 4, 5/OMC 2 is shown in Figure 3.2.8 while the 20–1000μ energy distribution of the $3'$ core of both OMC 1 and OMC 2 is shown in Figure 3.2.9.

More information concerning the dust in the OMC 1, OMC 2, and OMC 3 clouds is presented in Table 3.2.VII which includes estimates of the mass of dust and gas (M_d, M_g), optical depth (τ), column densities of dust and gas (N_D, N_{H_2}) and visual extinction (A_v), as derived from observations made at different wavelengths with different beams. Comments concerning the methods used for estimating these parameters are made in the Appendix, Section A.III.2.

The information presented up to this point has been obtained from observations of the infrared continuum emission which is characteristic of the dust contained in the Orion Complex. However, atomic infrared emission originating in the ionized gas has been both predicted (Petrosian, 1972; Simpson, 1975) and detected in Orion (M42) and the BN object.

Information concerning the IR lines detected in Orion (M42) and relevant parameters derived from them is presented in Table 3.2.VIII (near infrared lines observed in M42 have been included in Table 2.1.III together with the optical lines).

Similar information derived from high resolution spectroscopic studies of the BN object is presented in Table 3.2.IX. The detection of various hydrogen recombination lines (Bα, Bγ, Pfβ, Pfγ) in the spectrum of the BN source has helped to clarify the nature of the object which was a subject of considerable debate. Both the earlier and more recent views concerning the nature of the object are presented in Table 3.2.X and the accompanying commentary. The detection of the hydrogen recombination lines indicates the presence of a compact H II region around it and supports the 'early type star' hypothesis in the 'highly reddened evolved star' versus 'protostar-early type star' debate.

3.3. Structure and Physical Parameters of the Molecular Complex (OMC 1, OMC 2)

Two molecular clouds, OMC 1 and OMC 2 are associated with the H II regions M42 and M43. These are parts of a larger molecular complex covering a 10′ × 30′ area (1.5 × 5 pc for a distance of 0.5 kpc). The structure of this dense ($n_{H_2} \simeq 5 \times 10^4$ cm^{-3}) molecular complex, the mass of which is $7 \times 10^3\ M_\odot$, is shown in Figures 3.3.1 (H$_2$CO emission) and 3.3.2 (^{13}CO emission), respectively. Both OMC 1, OMC 2, and possibly a third feature to the North of OMC 2 (OMC 3) appear to be the result of the fragmentation of this molecular complex. This complex is in turn part of the massive ($\sim 1 \times 10^5\ M_\odot$) extended diffuse dust cloud L1641 (Lynds, 1962) from which CO emission has been detected over an area of 18 square degrees (Southern Complex; Figures 1.1.8, 1.1.9, and 1.1.10 in Chapter 1; Kutner *et al.* (1977)).

Evidence for rotation of the L1641/Southern Molecular Complex has been already discussed in Chapter 1. Furthermore, there is evidence for rotation of both the 10′ × 30′ OMC 1/OMC 2 molecular complex and the central part of OMC 1 which exhibits a ridge-like shape. There is also evidence for contraction of the Orion cluster of stars which may imply contraction of the Orion Molecular Cloud. The observed velocity gradients and their interpretation in terms of rotation or contraction of the whole or parts of the molecular complex are presented in Table 3.3.I and Figure 3.3.3.

The structure of the OMC 1/OMC 2 molecular complex, derived from the mapping of the CO emission over a 1° × 1° area around the Orion Nebula, is shown in Figure 3.3.4. This map shows the N–S extension of the molecular complex containing OMC 1 and OMC 2, and in addition a previously undetected condensation 24′E of the main peak. The structures of the CO emission in OMC 1 and OMC 2 are shown in Figures 3.3.5 and 3.3.6, respectively. The ridge like shape of the central part of OMC 1, the peak of which coincides with the KL Nebula, can be clearly seen in Figure 3.3.5(b) and the subsequent Figures 3.3.7 and 3.3.8 where the distribution of various molecules are shown. Information relating to the available contour maps of the emission of the numerous molecules detected in OMC 1 and OMC 2 is presented in Table 3.3.II. It should be noted that the extent of the emission from the different molecules is not similar since it depends on their excitation requirements. This means that molecules requiring high H$_2$ densities for excitation ($n_{H_2} \geqslant 10^4$ cm^{-3}) as for example CS ($J = 1$–0) are found near the dense core of the cloud (within 2′ $\simeq$ 0.3 pc) whereas low excitation molecules as for example CO ($J = 1$–0) can be traced far away (several degrees or tens of pc). This can also be seen in Table 3.3.VI.

Information concerning the kinetic temperature of the central part of the molecular clouds OMC 1 and OMC 2, inferred from observations of the optically thick ^{12}CO lines, is given in Tables 3.3.III and 3.3.IV, repsectively.

Information concerning the molecular abundance in the central part of OMC 1 and OMC 2 is presented in Table 3.3.V. (The column density of C^0 which originates in the interface between the H II region and the molecular cloud is also included for comparison).

Estimates of the H_2 density and mass of OMC 1, OMC 2 and their various components are presented in Table 3.3.VI. Some details of the structure of OMC 1 are also included in the accompanying commentary.

An outline of the procedures employed to infer or estimate the various physical parameters mentioned above is included in the Appendix, Section A.IV.1.

Radial velocities and line widths for OMC 1 and OMC 2 are presented in Tables 3.3.VII, 3.3.VIII, 3.3.IX, and 3.3.X. Note that the vast majority of molecular lines are seen in emission. This supports the view that the molecular cloud (OMC 1) lies behind the ionized gas (M42). Otherwise optical molecules such as CN, CH and CH$^+$ would be observed in absorption in front of the Trapezium stars (Kutner *et al.*, 1971). No such molecular optical lines have been detected (Adams, 1949).

The core of OMC 1 (KL Nebula) is characterized by three kinematic components. The main component which characterizes all molecules found in the 'ridge' of OMC 1 exhibits a narrow width of 4 km s^{-1}, and a radial velocity ~ 8.5 km s^{-1} which is the same as that of the surrounding molecular cloud. This is also called the 'spike' and probably originates from the interaction of the molecular cloud with the infrared cluster embedded in KL. The second component shows a broad velocity feature or 'plateau' with a width of $\gtrsim 30$ km s^{-1} and is localized to a region $\leqslant 1'$ in diameter, approximately centred on KL.

The 'plateau' feature is more prominent in various inorganic molecules (e.g. H_2S, SO, SO_2, and SiO; see Table 3.3.IX) which are more abundant in the high velocity core than in the extended molecular cloud. This is demonstrated in Figure 3.3.9 where the spectra of two lines with similar excitation requirements are presented. However only one of them (SiO) shows the broad 'plateau' feature which is characteristic of the high velocity core region, the other (CS) exhibits only the narrow 'spike' feature which characterizes every molecular emission originating from the extended cloud. The observed difference in the two spectra is attributed to abrupt variations of chemical abundances between the two regions (Scoville, 1980).

The spatial coincidence of the high velocity 'plateau' component with the KL Nebula is thought to indicate a generic relation between the two. Suggestions have been made attributing the broad profiles to collapsing clouds leading to the creation of protostars, or to expansion due to energetic events (explosions, stellar winds) originating in the stars belonging to the IR cluster which is embedded in KL. Studies along these lines have been made by Kwan and Scoville (1976) and Scoville (1980), who interprete the observed high velocity wings in terms of a differentially expanding envelope, possibly caused by an explosive event within the IR cluster. Scoville (1980) estimates the total kinetic energy present in the 37" core as $\simeq 4 \times 10^{47}$ ergs (15 $M_\odot$ moving at 50 km s^{-1}) which amounts to 0.1% of the energy of a supernova and suggests that the BN object should be considered as the prime candidate for such an energetic event. The main difficulty with such an interpretation seems to be the requirement of a shock front moving with a velocity of 50 km s^{-1} which should have caused dissociation of the observed vibrationally excited H$_2$. Alternative interpretations due to turbulence or rotation have also been suggested (e.g. Clark et al., 1979).

A third low velocity (~ 3 km s^{-1}) kinematic component has been recently detected in the profiles of a few molecular lines, predominantly in NH$_3$. Such a profile originating in the core of KL and exhibiting all three kinematic components is shown in Figure 3.3.10. A schematic interpretation of the broad 'plateau' feature based on the assumption of expanding envelopes around the stars of the IR cluster embedded in KL is shown in Figure 3.3.11. The low velocity feature termed 'hot core' component, because of its high kinetic temperature ($\gtrsim 220$ K), is probably associated with IRc 2 (Morris et al., 1980).

Typical estimates of the physical parameters of the 'plateau' and 'hot core' components are presented in Table 3.3.XI. The association of these features with the individual sources of the IR cluster in KL will be discussed in Section 3.4 together with the numerous maser sources found in the core of KL/OMC1.

Finally the detection of an ultrahigh density ($n_H > 10^{10}$ cm^{-3}) compact region (~ 1 AU) near BN in the core of KL should be noted. This was inferred from observations of CO band emission made by Scoville et al. (1979).

3.4. Maser Sources

Four types of strong masers have been observed in the core of OMC 1 (KL Nebula). These are: hydroxyl (OH), water vapour (H$_2$O), SiO and methanol (CH$_3$OH) masers. In addition, a water vapour (H$_2$O) maser source has been observed in the core of the OMC 2 cloud. The general characteristics of masers associated with H II regions are shown in Table 3.4.I. Spectra of all four masers present in Orion and exhibiting the typical numerous narrow velocity features are shown in Figure 3.4.1; a non maser line is also shown for comparison.

The spectrum of the OH (1665 MHz) masering source in KL is shown in Figure 3.4.2. The dominant feature occurs at ~ 7.8 km s^{-1} and appears as a Zeeman doublet at 7.1 km s^{-1} in right-hand circular polarisation (RC) and at 8.6 km s^{-1} in left-hand circular polarisation (LC). This line probably originates from one source (Norris et al., 1980). The other features appear to be contained in two clusters one at ~ 3.5 km s^{-1} and the other at ~ 20 km s^{-1}. The dominant feature and the features in the range 2–11 km s^{-1} appear to coincide spatially with IRc 4/OMC 1, whereas the features in the range 17–23 km s^{-1} appear to be associated with IRc 2 (IRS 3)/OMC 1. Their spatial distribution is shown in Figure 3.4.3.

The composite H$_2$O spectrum in KL is shown in Figure 3.4.4. This is characterized by strong low velocity features in the range -10 to $+30$ km s^{-1}, and weak high velocity features in the range -90 to $+80$ km s^{-1}. The time variations of the low velocity features are shown in Figure 3.4.5. The low velocity features are grouped in nine or more clusters, the sizes of which are 2″. The high velocity features appear to be dispersed over a 60″ × 60″ area centred on KL. Their distribution is shown in Figure 3.4.6 together with that of the various infrared sources in KL and the 21μ emission from KL. It appears that the H$_2$O cluster designated 'source A' coincides spatially with IRc 4/OMC 1 and the OH maser spot at 7.1/8.6 km s^{-1} and probably also with the OH cluster (in the range 2–11 km s^{-1}) (Genzel and Downes, 1977). The H$_2$O spectrum of 'Source A' purged of lines which do not originate in the source is shown in Figure 3.4.7 in comparison with the corresponding OH and SiO spectra. The H$_2$O spectrum is characterized by an inner group of lines at 3.4 and 7.6 km s^{-1} and an outer group of lines at -5.5 km s^{-1} and $+16$ km s^{-1}. Both groups are symmetrical with a centre of symmetry at 5.5 km s^{-1}. The outer lines are present in all three spectra and are separated by a further 2.5 km s^{-1} in the spectra of H$_2$O and SiO. The inner group is however absent from the SiO spectrum. On the basis of the shape of these spectra Genzel and Downes (1977) have proposed that the H$_2$O maser emission originates in the dust envelope (cocoon) of a massive star ($\geqslant 10\ M_\odot$) the core of which has already reached the main sequence. Their model is shown in Figure 3.4.8. Recent observations (Genzel et al., 1979), however, have shown that the 'shell features' of H$_2$O, SiO and probably OH (the -6 km s^{-1} and $\sim +18$ km s^{-1} features which are suggestive of a shell structure) originate in fact in the IRc 2 (IRS 3)/OMC 1 source with which they are spatially associated (see Table 3.4.VI). The 'shell features' of the H$_2$O spectrum are shown in Figure 3.4.9 (shaded parts) together with the corresponding SiO features. Further studies by Genzel et al. (1980) have shown that the 'shell source' is also associated with a broad SiO feature (thermal or weak maser emission) originating in an expanding circumstellar envelope. They conclude that the IRc 2 (IRS 3)/OMC 1 shell source is probably an evolved giant or supergiant embedded in the core of OMC 1. Norris et al. (1980) have, however, associated the low velocity (~ 3 km s^{-1}) NH$_3$ 'hot core' feature (see Section 3.3) with IRc 2 (IRS 3)/OMC 1 and suggested that this object is a very young protostar.

Recent observations of the proper motions of the Orion/KL H$_2$O maser features (a few milliarcsec per year) have been interpreted by Genzel et al. (1980) in terms of two outflows consisting of compact cloudlets of masing gas: (a) a low 18 km s^{-1} outflow and (b) a high (30–100 km s^{-1}) outflow. The centroid of both flows is the same and coincides within 5″ with the position of IRc 2. The 18 km s^{-1} flow is most probably physically associated with IRc 2 and is thought to be powered by the loss of rotational, magnetic or gravitational energy of a massive star in the last stages of its formation; the high velocity flow, however, may be associated with the BN object.

Another infrared source which is spatially associated with H_2O masers is IRS 2/ OMC 1. A similar spatial coincidence has also been found for IRS 4/OMC 2 which is probably a young lower mass (4–10 $M_\odot$) B star (Genzel and Downes, 1979).

The infrared sources IRc 2 (IRS 3)/OMC 1, IRc 4/OMC 1, and IRS 2/OMC 1 are also associated with broad H_2 line emission (Nadeau and Geballe, 1979). This, in combination with the high velocity maser emission and the broad plateau emission observed in various molecular lines (Section 3.3; Tables 3.3.IX, 3.3.XI) is suggestive of mass loss from the stars of the KL/IR cluster (Downes and Genzel, 1980). The correlation of the positions of the H_2O masers, the infrared sources of KL and the H_2 emission at 2μ in KL in Orion is shown in Figure 3.4.10.

The CH_3OH maser emission is relatively weak and does not show any spatial coincidence with the H_2O, OH, and SiO masers which are associated with circum-stellar envelopes. The CH_3OH sources, investigated by Matsakis et al. (1980) exhibit velocity features in the range 7–10 km s^{-1} which are similar to the velocities of the various molecules in OMC 1 (Table 3.3.VII). The distribution of the CH_3OH sources, which appear to be dense concentrations within OMC 1, is shown in Figure 3.4.11.

The absolute positions of the OH, H_2O, SiO, and CH_3OH masers detected in OMC 1 and OMC 2 are shown in Table 3.4.II. Information concerning the strong circular polarization of the OH masers and the strong linear polarization of the H_2O masers in OMC 1 (attributed to the presence of a magnetic field) are presented in Table 3.4.III. Typical sizes of the individual maser features (maser spots) and of the masering regions in which they are clustered are shown in Tables 3.4.IV and 3.4.V, respectively. Finally the positions of the IR and maser sources present in OMC 1 and OMC 2, which are probably physically associated, and relevant information concerning their probable nature are presented in Table 3.4.VI. Further information and clarification of various points are included in the commentaries accompanying the tables.

3.5. Magnetic Field in the Orion Complex

Estimates of the strength of the magnetic field present in the Orion Complex have been obtained from the study of certain lines exhibiting Zeeman splitting (H and OH lines) and from infrared linear polarization measurements of the KL Nebula in the core of OMC 1. The uncertainties in the estimated field strength are large (several orders of magnitude). This was pointed out by Zuckerman and Palmer (1975) who discussed in detail the difficulties associated with each method.

The difficulties in the first method are mainly due to distorted Zeeman patterns of the OH lines which may cause detectable lines to be produced only where the Zeeman splitting is equal to the line widths. The difficulties in the second method are due to the assumed conditions in the Orion Molecular Cloud ($T_{gas} \simeq 70$ K, $T_{dust} \simeq 7$ K instead of $T_{gas} \simeq T_{dust} \simeq 70$–80 K), needed to explain the infrared linear polarisa-tion measurements in terms of the 'Davis–Greenstein' mechanism (alignement of grains by paramagnetic relaxation; see also Section 3.2). Furthermore, different methods estimate the magnetic strength of different areas of the complex, which are not characteristic of the whole.

The estimates of the Orion magnetic field according to various workers are presented in Table 3.5.I and briefly discussed in the accompanying commentary. It should be noted that the magnetic field is not only confined to the Orion Complex but is also extended over the surrounding area (Barnard's Loop) as shown by Appenzeller (1974) (see also Chapter 1 and Figure 1.2.3a).

TABLE 3.1.I

Infrared continuum contour maps of the Orion Complex

No.	Figure	Beam (arcsec)	Wavelength λ (μ)	Remarks	References
1	3.1.7	180	400	(5)	Smith *et al.* (1979)
2	3.1.5	132	91	(1)	Harper (1974)
3		100	700–2000	(3)	Harvey *et al.* (1974)
4		96	400	(3)	Soifer and Hudson (1974)
5		60	21	(3)	Lemke *et al.* (1974)
6		60	69	(3)	Fazio *et al.* (1974)
7		60	20	(3)	Werner *et al.* (1976)
8		60	50	(3)	Werner *et al.* (1976)
9	2.4.7	60	100	(3)	Werner *et al.* (1976)
10	3.1.4	60	1000	(6)	Westbrook *et al.* (1976)
11		56	350	(3)	Gezari *et al.* (1974)
12		26	20	(3)	Ney and Allen (1969)
13		25	34		Low *et al.* (1973)
14		13	11.6	(3)	Ney and Allen (1969)
15	3.1.3	10.2 × 10.5	33	(3)	Beichman *et al.* (1978)
16		9.2	7.6		
17		9.2	8.6		
18	3.1.1	9.2	10.1	(3)	Gehrz *et al.* (1975)
19		9.2	11.2		
20		9.2	12.2		
21		9.2	13.1		
22		7.5	2	(4)	Becklin *et al.* (1976)
23		5.5	10.5	(2)	Rieke *et al.* (1973)
24	3.1.2	5	20	(3)	Wynn-Williams and Becklin (1974)
25		5	21	(4), (2)	Rieke *et al.* (1973)
26	2.4.8	4.5	10	(6), (4)	Becklin *et al.* (1976)
27		3.5	5	(4), (2)	Rieke *et al.* (1973)
28		3	10.5	(4), (2)	Rieke *et al.* (1973)

Remarks to Table 3.1.I:
(1) Map of an extended area covering the whole M42 Region (Trapezium + KL Nebula) and M43.
(2) KL Nebula.
(3) Trapezium + KL Nebula.
(4) Bar (near θ^2 Ori A).
(5) Distribution of IR continuum emission over a 40′ × 25′ region containing the molecular clouds OMC 1 (near the Orion Nebula ≡ M42), OMC 2 (12′ N of OMC 1), and OMC 3 (16′ N of OMC 2).
(6) OMC 1.

TABLE 3.1.II

Positions of the members of the IR cluster/OMC 1*

No.	Infrared sources	Coordinates (1950) α	δ
1	IRc 1 ≡ BN object	$5^h32^m46\overset{s}{.}7$	$-5°24'17''$
2	IRc 2 ≡ IRS 3	5 32 47.0	−5 24 24
3	IRc 3	5 32 46.7	−5 24 25
4	IRc 4	5 32 46.8	−5 24 29
5	IRc 5	5 32 46.9	−5 24 33

* According to Rieke *et al.* (1973); another infrared source, IRS 2, detected by Wynn-Williams and Becklin (1974), lies to the north of the cluster ($\alpha = 5^h32^m46\overset{s}{.}3$, $\delta = -5°23'55''$) and is not associated with it.

TABLE 3.1.III

Evidence supporting the physical association of the KL nebula
with the molecular cloud OMC 1

No.	Evidence	Remarks
1	The KL nebula is not detectable at optical and radio wavelengths	(1)
2	The KL nebula spatially coincides with the core of the molecular cloud OMC 1	(2)
3	The dust temperature in KL is roughly equal to the temperature of the molecular gas(~ 80 K)	(3)
4	The KL nebula spatially coincides with the IR cluster of stars and the numerous OH and H_2O masers	(4)

Remarks to Table 3.1.III:

(1) Compare the position of KL ($\alpha = 5^h32^m47^s$, $\delta = -5°24'21''$) with the photographs or radio continuum maps of Orion presented in Chapter 2; no corresponding optical or radio feature is disscernible.

(2) See Figure 3.1.4.

(3) See Tables 3.2.III and 3.3.III.

(4) See Figures 3.1.2 and 3.1.4; see also Tables 3.1.II and 3.4.VI.

TABLE 3.1.IV

Small infrared shells ($\sim 25''$) surrounding visible stars in Orion (M42, M43)

No.	Infrared source	Star surrounded by the source	Remarks
1	IR 1	θ^2 Ori A	(1)
2	IR 2	θ^2 Ori B	(1)
3	IR 3	θ^1 Ori	(1)
4	IR 4	V361 Ori	(2)
5	IR 5	LP Ori	(3)
6	IR 6	NU Ori	(1)

Remarks to Table 3.1.IV:

(1) For parameters of visible star see Tables 2.2.III and 2.2.IV.

(2) Spectral type of visible star: B5V, $m_v = 8.29$ (Penston, 1972).

(3) Spectral type of visible star: B1.5Vp, $m_v = 8.60$ (Penston, 1972).

TABLE 3.2.I

Physical parameters of the Trapezium (Ney–Allen) Nebula derived from infrared continuum measurements

No.	Airborne (A) or Ground (G) observations	Bandwidth (μ)	λ_{eff} (μ)	Beam	Beam separation	Flux (Jy) 1 Jy = 10^{-26} W m^{-2} Hz^{-1}	Flux (10^{-17} W cm^{-2} μ^{-1})	Dust temperature T_d (K)	Luminosity L ($L_\odot$)	Remarks	References
1	G		11.6	26″	51″	9.38×10^2	204	} 220		} (7), (8)	Ney and Allen (1969)
			11.6	13″	51″	3.27×10^2	71				
2	G	28–40	350	1′	1.8	$(1.82 \pm 0.34) \times 10^3$					Harper et al. (1972)
3	G		34	25″	45″	1×10^3					Low et al. (1973)
4	A		69	1′ × 1.5	5.3	$4^{+3}_{-1} \times 10^5$					Fazio et al. (1974)
5	G		21	1′	1.3	1.4×10^5		102 ± 9	3.9×10^5	(1), (2)	Lemke et al. (1974)
6	G		8.6	9″2	80″	6.4×10^2	261	} ~200		} (3), (4), (5)	} Gehrz et al. (1975)
			10.1	9″2	80″	1.74×10^3	512				
			11.2	9″2	80″	2.28×10^3	544				
			12.2	9″2	80″	2.4×10^3	491				
			13.1	9″2	80″	1.9×10^3	339				
7	A	16–40		25″	90″			150–200		(6)	Forrest et al. (1976)

Remarks to Table 3.2.I:

(1) The quoted temperature is the colour temperature estimated from the ratio of the fluxes in the 21μ and 350μ bands and characterize a 2′ × 4′ area with the Trapezium in the lower part. The temperature was estimated on the assumption that the emissivity of the grains Q_{IR} does not show any dependence on the wavelength (λ). For $Q_{IR} \propto \lambda^{-1}$ the colour temperature is lower (~ 70 K).

(2) The quoted luminosity was estimated from the total infrared flux of the 2′ × 4′ area (see note (1) above) which is 5.1×10^{-12} W cm^{-2} for $T = 102$ K. The IR luminosity of the Trapezium Nebula is only 1.15 times higher than the UV luminosity of the Trapezium stars.

(3) Trapezium Nebula (Ney Allen I); a small source W of the Trapezium (Ney–Allen II) is not included.

(4) Integrated fluxes deduced from the isophote maps shown in Figure 3.1.1 the values of fluxes in units of 10^{-17} W cm^{-2} μ^{-1} are 261 (8.6μ), 512 (10.1μ), 544 (11.2μ), 491 (12.2μ), and 339 (13.1μ), respectively.

(5) The 8–13μ spectrum of the Trapezium Nebula (Ney–Allen I) is characterized by an emission feature at 10μ indicating the presence of silicate grains; the maximum grain temperature is estimated ~ 250 K and occurs around θ^1 Ori D; the temperature decreases to 160–180 K within 10″–15″ of θ^1 D; no further decrease of temperature is observed as far as 50″ away from θ^1 Orionis D.

(6) The 16–40μ spectrum of the Trapezium Nebula exhibits a broad emission feature indicating the presence of silicate grains.

(7) The cited temperature is the colour temperature estimated from the ratio of the fluxes in the 11.6μ and 20μ bands. The brightness temperature of the nebula observed in a 26″ beam is 98 K at 11.6μ and 70 K at 20μ.

(8) Integrated flux density.

TABLE 3.2.II

Physical parameters of M42/OMC 1 (Trapezium Nebula, KL Nebula etc.) derived from infrared continuum measurements

No.	Airborne (A) or Ground (G) observations	Bandwidth (μ)	λ_{eff}(12) (μ)	Beam	Beam separation	Flux (Jy) 1 Jy = 10^{-26} W m^{-2} Hz^{-1}	Flux (10^{-12} W cm^{-2})	Dust temperature T_d (K)	Luminosity L ($L_\odot$)	Remarks	References
1	G	45–70	900			$(4.5\pm1.6)\times10^4$				(6), (7)	Park et al. (1970)
2	A			8'.4±1'			1.86	68	1.6×10^5	(5)	Harper and Low (1971)
3	A	50–300		8'±1'	9'		0.9±0.1	100^{+20}_{-12}	1.6×10^5	(1), (2)	Low and Auman (1971)
		10–300		8'±1'	9'		2.2	75			
4	A	40–350	350	4'.5	7'.5		1.06				Furniss et al. (1972)
5	A			0'.9	1'.8	$(3.19\pm1.24)\times10^4$				(8)	Harper et al. (1972), Ade et al. (1974)
6	A	40–350		4'.5	7'.5		1.35		1.05×10^5	(24)	Emerson et al. (1973)
		40–350		4'.5	7'.5		2.04		1.6×10^5	(23)	
7	A	55–200	1400	7'	12'		1.2	70		(30)	Erickson et al. (1973)
8	G			65"±6'		$(5.84\pm1.17)\times10^2$		30–70	$2.4^{+1.5}_{-0.6}\times10^5$	(3), (4), (9)	Ade et al. (1974)
9	A		69	1'×1'.5	5'.3	5.5×10^5					Fazio et al. (1974)
10	G		21	1'	1'.3	1.5×10^5				(10)	Lemke et al. (1974)
11	G		350	3'.5	10'	$(2\pm0.4)\times10^4$ (4.38×10^4)				(11)	Gezari et al. (1974), Ade et al. (1974)
12	A	56–500	99	5'		3×10^5	2.5	85		(13), (14)	Harper (1974)
		30–500	72	5'		3.3×10^5					
		45–500	91	5'		3.1×10^5					
		30–45	42	5'		3.5×10^5					
		45–80	59	5'		4.2×10^5					
		65–110	78	5'		4.1×10^5					
		125–500	183	5'		1.4×10^5					
		45–750	91	8'.4		3.9×10^5					
		45–330	91	5'		3.1×10^5					
		30–300		5'							
		45–200	91	2'.2		7.3×10^5				(15), (16)	
13	A	20–40		4'.7	15'		1.8	~100		(17)	Houck et al. (1974)
14	G	388–444	400	1'.6	5'.9	2.5×10^4		~35		(18),(19), (20)	Soifer and Hudson (1974)
15	A	80–125	100	5'		3.3×10^5		60–100		(31), (36)	Ward and Harwit (1974)
16	A	80–135	100	7'.4	8'	4.2×10^5		90		(32)	Brandshaft et al. (1975)
17	A	75–120		5'	15"			75		(33)	Ward (1975)

Table 3.2.II (continued)

No.	Airborne (A) or Ground (G) observations	Bandwidth (μ)	λeff[12] (μ)	Beam	Beam separation	Flux (Jy) 1 Jy=10^{-26} W m^{-2} Hz^{-1}	Flux (10^{-12} W cm^{-2})	Dust temperature T_d (K)	Luminosity L ($L_\odot$)	Remarks	References
18	A	42–115		4′ × 5′				100		(34)	Ward *et al.* (1976)
19	G	700–1500	1000	1′	4′	1.2×10^3			1.5×10^5	(25)	Westbrook *et al.* (1976)
20	A	40–130						55–85		(22), (21)	Werner *et al.* (1976)
21	A	100–500		2:1	8′			85		(28)	Pipher *et al.* (1978)
22	A		39	3:5	13′	$(3 \pm 0.15) \times 10^5$					
			56	3:5	13′	$(3.9 \pm 0.13) \times 10^5$		60	2.5×10^5	(29), (35)	Thronson *et al.* (1978)
			73	3:5	13′	$(4.0 \pm 0.25) \times 10^5$					
			140	3:5	13′	$(4.1 \pm 0.12) \times 10^5$					
			300	9′	13′	$(5.7 \pm 0.2) \times 10^4$					
23	G	300–1000	400	3′	5′	$(6.7 \pm 0.7) \times 10^3$		45		(27)	Smith *et al.* (1979)
			340	3′	5′	$(9.2 \pm 0.9) \times 10^3$					
	G	700–1500	400	3′	5′	2.1×10^4				(26)	

Remarks to Table 3.2.II:

(1) The emission is dominated by the cool wings of the KL Nebula.

(2) The spectral distribution in the range 10–100μ is dominated by a 75 K blackbody. Beyond 100μ the spectrum falls off much more steeply.

(3) Beam characterized by a broad error pattern of 6′.

(4) The flux is the sum of the flux from KL (1′; 84 ± 17 Jy) and the flux of the extended region (~3′) on which it is superimposed (500 ± 100 Jy).

(5) The quoted temperature is the colour temperature derived from the ratio of fluxes in the 45–750μ band and the 60–750μ band. A similar value (105 $^{+12}_{-8}$) is derived from the respective ratio in the 44.5 and 79μ bands.

(6) The spectral distribution in the range of $10^{10.5}$–$10^{13.5}$ Hz is fitted with a 68 K blackbody of 1′ angular diameter.

(7) Ade *et al.* (1974) have argued that the observed flux represents the 350μ flux due to the atmospheric conditions prevailing at the time of the observations and the altitude at which the observations were made (Pic du Midi Observatory, altitude: 2900 m).

(8) Results by Harper *et al.* (1972) corrected by Ade *et al.* (1974).

(9) The energy distribution in the range of 10μ–10 cm is also deduced; the results are interpreted by assuming an optically thin thermal model consisting of an 1′ source (KL Nebula) at T = 70 K, set on a more extended region, 3′ in diameter, at T = 30 K.

(10) The quoted flux is the sum of the fluxes quoted for the KL and the Trapezium Nebulae.

(11) The flux in parenthesis is the corrected flux for partial resolution of the extended source by the beam (corrected by Ade *et al.*, 1974).

(12) The wavelength in the second column is given by the relation:

$$\lambda_{eff} = \int_0^\infty B(\lambda, T_c)G(\lambda)\,\lambda\,d\lambda \Big/ \int_0^\infty B(\lambda, T_c)G(\lambda)\,d\lambda,$$

where $B(\lambda, T_c)$ is the spectral emittance of a blackbody of colour temperature T_c and $G(\lambda)$ is the instrumental response.

(13) The quoted temperature is the temperature of a blackbody curve, which was fitted to match the energy distribution in the 30–300μ range.

(14) Observations made at $\alpha = 5^h32^m50^s \pm 4^s$, $\delta = -5°25' \pm 1'$.

Remarks to Table 3.2.II (continued)

(15) The quoted flux density is the total flux density deduced from the 91μ map and includes an estimated contribution from the area under contours not closed in the map (Figure 3.1.5); the flux from the mapped area is 6.3×10^5 Jy.

(16) The total flux originating in the IRc4(M43) is 4.5×10^4 Jy. The infrared luminosity of this source is $\sim 0.29 \times 10^5 \, L_\odot$.

(17) The cited temperature is from the 20–40μ spectrum measured with the $4\overset{'}{.}7$ beam; similar value is obtained from a model in which the dust within the $2'$ molecular cloud has a clumpy distribution; this represents the best fit of the energy spectrum in the 30–1000μ range. The clumps have effective grain temperatures of $T = 100$ K and are optically thick for $\lambda < 400\mu$. The total dust mass M_d is less than $20M_\odot$. An alternative model based on the assumption of a $2'$ uniform molecular cloud gives grain temperature $T = 68$ K for a cloud becoming optically thick for $\lambda < 180\mu$. These models predominantly characterize the behaviour of the KL Nebula.

(18) Size of mapped area $4' \times 4'$.

(19) Colour temperature obtained from the ratio of the fluxes in the 91μ and 350μ bands; indicates that the extended component is cooler than the compact core of the nebula.

(20) Wavelength dependence of the grain emissivities: $Q_{\mathrm{IR}} \propto \lambda^{-2}$.

(21) The quoted luminosity refers to a $5' \times 5'$ area.

(22) The quoted temperature is the colour temperature estimated from the ratio of the fluxes in the 50μ and 100μ bands; it varies over the nebula within the quoted range; $T_{\mathrm{KL}} \simeq 85$ K; $T_{\mathrm{BAR}} \simeq 75$ K, T (extended component in the north) $\simeq 55$ K.

(23) Total flux density.

(24) The cited flux density and luminosity refer to the peak signals.

(25) Total 1 mm flux density of a $8' \times 3'$ area of the molecular cloud centred on the infrared cluster.

(26) Total flux density of the $30' \times 6'$ lane containing OMC 1 and a ridge to the north (Figure 3.1.7); this lane of dust comprises the dense core of the L1641 dark/molecular cloud.

(27) Mass of dust within the $3'$ beam centred on KL ($\alpha = 5^{\mathrm{h}}32^{\mathrm{m}}47^{\mathrm{s}}$, $\delta = -5°24'30''$), $M_g = 4.4 M_\odot$; optical depth at 400μ averaged over the beam, $\tau(400\mu) = 0.015$, dust column density averaged over the beam $N_D = 4.1 \times 10^{-3}$ g cm^{-2}.

(28) The cited temperature is the temperature of a diluted blackbody curve fitted to match the data of the energy spectrum in the 10–1000μ range.

(29) The energy spectrum in the 10–1000μ range is fitted with a 60 K blackbody curve.

(30) The best fit of the energy spectrum in the 55–200μ range is provided by a 70 K blackbody curve.

(31) The best fit of the energy distribution in the range 80–125μ is provided by a model in which the dust within the molecular cloud has a clumpy distribution. The clumps have effective grain temperatures of $T = 100$ K and are optically thick for $\lambda < 400\mu$. The total dust mass is $< 20 M_\odot$.

(32) The best fit of the energy distribution in the range 80–135μ is provided by a 90 K diluted blackbody curve.

(33) The best fit of the energy distribution in the range 75–120μ is provided by a 75 K blackbody curve.

(34) The best fit of the energy distribution in the range 42–115μ is provided by a diluted 100 K blackbody curve.

(35) The cited luminosity is the total infrared luminosity into a $3\overset{'}{.}5$ beam.

(36) The flux density at 100μ is 10^{-14} W cm^{-2} μ^{-1}.

TABLE 3.2.III

Physical parameters of the Kleinmann–Low (KL) Nebula derived from infrared continuum measurements

No.	Airborne (A) or Ground (G) observations	Bandwidth (μ)	λ_{eff} (μ)	Beam	Beam separation	Flux (Jy) 1 Jy = 10^{-26} W m^{-2} Hz^{-1}	Flux (10^{-17} W cm^{-2} μ^{-1})	Dust temperature T_d (K)	Luminosity L ($L_\odot$)	Remarks	References
1	G		22	30″		$>8 \times 10^3$	>500	~ 70	$>1 \times 10^5$	(24), (25)	Kleinmann and Low (1967)
2	G		11.6	26″	51″	9.38×10^2	204			(26)	Ney and Allen (1969)
3	G		1000			10^2					Low (1971)
4	G		350	1′	1.8	$(4.7 \pm 0.6) \times 10^3$				(5)	Harper et al. (1972)
5	G	8–13.5	10.5	2″ 3″; 5″.5		5.1×10^2	138			(6)	Rieke et al. (1973)
			21	5″		5.2×10^3					
6	G	28–40	34	25″	45″	2.9×10^4					Low et al. (1973)
7	G	800–2000	1400	65″ ±6′		$(8.4 \pm 1.7) \times 10$		70		(1), (2)	Ade et al. (1974)
8	A		69	1′ × 1.5	5.3	1.5×10^5		70	7×10^4	(3)	Fazio et al. (1974)
9	G		21	1′	1.3	1.1×10^4				(4)	Lemke et al. (1974)
10	A	45–200	91	2.2	8′	1.3×10^5			1.3×10^5	(15)	Harper (1974)
11	G	388–444	400	1.6	5.9	$\sim 6 \times 10^3$				(10)	Soifer and Hudson (1974)
12	G	700–2000	1000	1.6	2.25	$(3 \pm 1) \times 10^3$				(7)	Harvey et al. (1974)
13	G	700–2000	1000	1′	2′	$(2.4 \pm 0.75) \times 10^2$					Werner et al. (1974)
14	G		350	56″	4′	$(8.8 \pm 1.6) \times 10^3$				(18)	Gezari et al. (1974)
15	G		8.6	9″.2	80″	5×10^2	204			(8), (9)	Gehrz et al. (1975)
			10.1	9″.2	80″	3.5×10^2	103				
			11.2	9″.2	80″	6.6×10^2	156				
			12.2	9″.2	80″	1.8×10^2	363				
			13.1	9″.2	80″	3.4×10^3	590				
16	A	16–40	25	25″	90″			~ 85		(10)	Forrest et al. (1976)
17	A	16–25	17	17″	25″			~ 80		(11)	Forrest and Soifer (1976)
18	G	10–1000	1′	1′				70	7×10^4	(12)	Werner et al. (1976)

(continued)

Table 3.2.III (continued)

No.	Airbone (A) or Ground (G) observations	Bandwidth	λ_{eff} (μ)	Beam	Beam separation	Flux (Jy) 1 Jy = 10^{-26} W m^{-2} Hz^{-1}	Flux (10^{-17} W cm^{-2} μ^{-1})	Dust temperature T_d (K)	Luminosity L ($L_\odot$)	Remarks	References
19	A		20	1'	2'	1.7×10^4					*(continued)*
			50	1'	2'	1.1×10^5		85		(13), (14)	Werner et al. (1976)
			100	1'	2'	9×10^4					
20	G	700–1500	1000	1'	4'	$(2.15 \pm 0.13) \times 10^2$				(19)	Westbrook et al. (1976)
21	A	29–125	54	1'.4	3'.1	2×10^5		70–95	1.1×10^5	(16), (17)	Erickson et al. (1977)
22	A	4.5–8	6	30"	2'	$\sim 3 \times 10^2$	250			(22)	Russell et al. (1977)
23	A		39	50"	2'.4	1.3×10^5					
			56	50"	2'.4	1.4×10^5		70	1×10^5	(20), (23)	Thronson et al. (1978)
			73	50"	2'.4	1.2×10^5					
			140	50"	2'.4	4.1×10^5					
			390	1'.25	2'.8	4.4×10^3		60		(21)	

Remarks to Table 3.2.III:

(1) Beam characterized by a broad error pattern of ~6'.

(2) Temperature of KL obtained from a model introduced to interpret the Orion energy spectrum in the 10μ–10 cm range (see note (9) in the commentary on Table 3.2.II).

(3) The quoted temperature is the temperature of a blackbody 35" in diameter fitted to match the energy distribution of KL in the range of 20–300μ. The 69μ size of KL is 45" ± 15".

(4) The flux is corrected for background emission; the uncorrected value is 1.5×10^4 Jy.

(5) Flux 1' N of KL: $(1.38 \pm 0.44) \times 10^3$ Jy, flux 1' S of KL: $(1.64 \pm 0.62) \times 10^3$ Jy.

(6) The quoted flux is the total flux of the Nebula; it includes the fluxes of the five compact infrared sources (IRc 1 = BN, IRc 2, IRc 3, IRc 4, IRc 5); the total flux of KL without them is 1.8×10^2 Jy at 10.5μ and 4.15×10^3 Jy at 21μ.

(7) Flux at the centre of the KL source (central 1'.6 of the nebula).

(8) Integrated flux densities deduced from the isophotes of the maps shown in Figure 3.1.1.

(9) The 8–13μ spectrum of the KL Nebula is characterized by a 10μ absorption feature indicating the presence of silicate grains (see Figure 3.2.1).

(10) The 16–40μ spectrum exhibits a continuum with colour temperature ~85 K.

(11) Brightness temperature; the contribution from the BN source is not significant ($< 25\%$ of the flux at 21μ); the temperature is nearly equal to the peak 2.6 mm CO brightness temperature (75 ± 7 K by List et al., 1974, 1' beam); the gas and dust are probably well mixed and $T_{gas} \simeq T_{dust} \simeq$ 70–80 K (Table 3.1.III).

Remarks to Table 3.2.III (continued)

(12) Energy spectrum observed with a 1′ beam centred on the IR cluster containing the BN object, the KL Nebula etc. The best fit to the data is provided by a 70 K blackbody filling the 1′ beam, for which the emissivity is 1 at 20μ and decreases as λ^{-1} for $\lambda > 20\mu$. $L_{KL} > 1.2 \times 10^5 L_\odot$.

(13) The flux densities quoted are the peak flux densities into the 1′ beam.

(14) The colour temperature is estimated from the ratio of the 50μ and 100μ surface brightnesses.

(15) Total flux density derived from the 91μ map.

(16) The wavelength cited in the second column (54μ) is the wavelength at the peak of the energy distribution in the 29–125μ range; the cited flux density is the flux density measured in the 1.4 beam at this wavelength; the luminosity refers to the 29–125μ range which includes most of the radiation from the KL nebula.

(17) The temperature is in the range of 70–95 K; it is estimated by assuming a simple model for the emission; a good fit to the data is provided by adopting thermal emission by dust at $T = 71$ K and absorption cross section proportional to frequency.

(18) The 350μ optical depth at the peak of the source is 0.1 for $T = 100$ K and 0.2 for $T = 60$ K.

(19) Column density of dust $N_D = 2 \times 10^{-2}$ g cm^{-2}, expected visual extinction $A_v = 700$ mag, column density of gas $N_{H_2} = 7 \times 10^{23}$ molecules cm^{-2}; the parameters were measured along the line of sight through the peak.

(20) A 70 K blackbody curve is the best fit to the energy distribution in the range of 10–100μ (deduced from high angular resolution (50″) observations).

(21) The energy spectrum in the range of 10–100μ (deduced from low angular resolution data) is fitted with a 60 K blackbody curve.

(22) The cited luminosity is the total infrared luminosity into a 50″ beam.

(23) A composite spectrum in the 2–25μ range is also deduced.

(24) The cited luminosity is the estimated total luminosity.

(25) Position of KL: $\alpha = 5^h32^m47^s$, $\delta = -5°24'30''$; angular diameter $\simeq 30''$; distance 500 pc; linear diameter: 2.3×10^{17} cm; mass $M = 10^2$–10^3 $M_\odot$; lifetime: $> 2 \times 10^3$ yr.

(26) Integrated flux density.

TABLE 3.2.IV

Physical parameters of the members of the IR cluster/OMC 1 (IRc 1, IRc 2, IRc 3, IRc 4, IRc 5) derived from infrared continuum measurements

No.	Airborne (A) or Ground (G) observations	Band-width (μ)	λ_{eff} (μ)	Beam	Flux (Jy) $1\,Jy = 10^{-26}$ W m^{-2}Hz$^{-1}\mu^{-1}$	Flux $(10^{-17}$ W cm$^{-2}\mu^{-1})$	Flux (W m^{-2})	Dust temperature T_d (K)	Luminosity L $(L_\odot)$	Remarks	References
					a. IRc 1 ($\equiv$BN object)/OMC 1						
1	G		22	13″	$<1.6\times10^3$	<10		~610	1.4×10^3	(2),(5)	Kleinmann and Low (1967)
2	G		1.65 2.2 3.4 10	6′ 6′ 6′ 6′	0.13 ± 0.03 5.48 ± 0.32 44 ± 2.7 113 ± 10	1.4 ± 0.3 34 ± 2 132 ± 8 34 ± 3		~700		(4)	Becklin and Neugebauer (1967)
3	G	1.6–10	1.62 2.2 3.4 5.0		0.15 7.11 59.3 20.9	1.7 44.1 154 251	13×10^{-15}	600	1×10^3		Low et al. (1970)
					b. IRc 2 ($\equiv$IRS 3)/OMC 1						
4	G	4.5–5.5 8–13.5 17–25	5 10.5 21	3″.5, 7″ 2″,3″, 5″.5 5″	1.7×10^2 2.6×10^2 4.2×10^2			440		(3)	Rieke et al. (1973)
5	G		20	$<5″$	4×10^2			530	1.5×10^3	(1), (2)	Becklin et al. (1973)
					c. IRc 3/OMC 1						
6	G	4.5–5.5 8–13.5 17–25	5 10.5 21	3″.5, 7″ 2″, 3″, 5″.5 5″	10 30 110			335		(6)	Rieke et al. (1973)
7	G	4.5–5.5 8–13.5 17–25	5 10.5 21	3″.5, 7″ 2″, 3″, 5″.5 5″	3 15 170			265		(6)	Rieke et al. (1973)

Table 3.2.IV (continued)

No.	Airborne (A) or Ground (G) observations	Band width (μ)	λ_{eff} (μ)	Beam	Flux (Jy) $1\,\text{Jy}=10^{-26}$ W m^{-2} Hz^{-1} cm^{-2} μ^{-1} Flux 10^{-17} W	Flux (W m^{-2})	Dust temperature T_d (K)	Luminosity L ($L_\odot$)	Remarks	References
					d. IRc 4/OMC 1					
8	G	4.5–5.5 8–13.5 17–25	5 10.5 21	3″.5, 7″ 2″, 3″, 5″.5 5″	<1.5 23 250		<235		(6)	Rieke *et al.* (1973)
					e. IRc 5/OMC 1					
9	G	4.5–5.5 8–13.5 17–25	5 10.5 21	3″.5, 7″ 2″, 3″, 5″.5 5″	<1.5 <10 110		<255		(6)	Rieke *et al.* (1973)

Remarks to Table 3.2.IV:

(1) A 530 K blackbody curve is the best fit to the energy distribution in the range of 1.65–20μ.
(2) The cited luminosity is the total luminosity of the BN object.
(3) The quoted temperature is the colour temperature in the 5–21μ band.
(4) The cited temperature is from a 700 K blackbody curve fitted to the 1.65–10μ energy distribution of the source.
(5) The cited temperature is from a 610 K blackbody curve fitted to the 1.65–22μ energy distribution of the source.
(6) The temperature quoted is the colour temperature in the 5–21μ band.

TABLE 3.2.V

Polarization measurements of the KL nebula and associated objects (BN, IRS 2)

No.	Source	Airborne(A) or Ground(G) observations	λ_{eff} (μ)	Beam	Beam separation	Linear polarization P (%)	Position angle θ (°)	Remarks	References
					a. Linear polarization measurements				
1	KL nebula	G	11.1	8."8	56"–70"	~3–15	~80–110	(2)	Dyck and Beichman (1974)
2	KL nebula	A	85	6'	8'	2.5±2.5	80	(1), (9)	Dennison et al. (1977)
3	KL nebula	A	>28	1'	2.5	~2±2		(9)	Gull et al. (1978)
4	KL nebula	G	11.1	10."8		2–5	~70–100	(11), (2)	Knacke and Capps (1979)
5	KL nebula	G	19.6	10."8		2–5	~100–120		Serkowski and Rieke (1973)
6	BN object	G	3.4			9.1±0.4	116		
7		G	3.6	12"		7.4±0.6	117		
8		G	4.9	12"		4.5±0.8	63		
9		G	8.3	12"		2.6±0.7	177		
10	BN object	G	8.51	12"		3.2±1.3	95	(3), (7)	Dyck et al. (1973)
11		G	9.15	12"		8.9±1.8	96		
12		G	9.95	12"		14.1±1.4	120		
13		G	10.7	12"		8.4±2.7	112		
14		G	11.1	12"		5.4±0.9	162		
15	BN object	G	2.2	~10"		16±1	122±2	(3), (6)	Loer et al. (1973)
16	BN object	G	3.4	~10"		7±1	110±4		
17	BN object	G	1.6	~10"		25.5±3.1	117	(3)	Breger and Hardrop (1973)
18	BN object	G	2.2	~10"		14.2±1.3	115		
19		G	3.4	8."8	56"–70"	10.3±1.2	117±3	(5), (10)	Dyck and Beichman (1974)
20		G	4.9	8."8	56"–70"	1.3±1.7	78±5.2		
21	BN object	G	8.4	8."8	56"–70"	2.6±0.4	112±4		
22		G	11.1	8."8	56"–70"	10.0±0.9	97±3		
23		G	12.6	8."8	56"–70"	7.6±0.5	107±2		
24		G	1.6	10"	15"	38±2	117	(3), (4)	Dyck and Capps (1978)
25	BN object	G	2.2	10"	15"	17.1±0.2	113		
26		G	3.8	10"	15"	5.9±0.3	102		
27	BN object	G	11.1	10."8		8.4±0.5	116±2	(11)	Knacke and Capps (1979)
28	BN object	G	19.6	10."8		6.8±0.7	115±3		
29	BN object	G	2.2	10"	20"	18.4±0.6	115		Lonsdale et al. (1980)
30	IRS 2	G	2.2	10"	20"	10.8±1.0	136		

(see Figure 3.1.2)

Table 3.2.V (continued)

No. Source	Airborne (A) or Ground (G) observations	λ_{eff} (μ)	Beam	Beam separation	Circular polarization q (+ = right − = left) (%)	Ellipticity $e = q/p$ (%)	Remarks	References
				b. Circular polarization measurements				
1 BN object	G	3.45			+0.86±0.15	9		Serkowski and Rieke (1973)
2 BN object	G	2.2	10″	20″	+1.56±0.18	8±1	} (8)	} Lonsdale *et al.* (1980)
3 IRS 2	G	2.2	10″	20″	+0.66±0.15	6±1		
(see Figure 3.1.2)								

Remarks to Table 3.2.V:

(1) The whole M42 was included within the 6′ beam; the emission at 85μ originates predominantly in the KL Nebula.

(2) For detail polarization measurements over the KL Nebula see the original paper.

(3) $\alpha = 5^h32^m46\overset{s}{.}7 \pm 0\overset{s}{.}1$, $\delta = -5°24'17'' \pm 1''$.

(4) The IR size of the object is $< 0\overset{''}{.}7$; the diameter is $\le 5 \times 10^{15}$ cm for a distance of 500 pc.

(5) The magnitude of linear polarization of KL is correlated with the silicate absorption feature at 10μ; the polarization is probably caused by preferential extinction of aligned dust grains which do not radiate significantly at the wavelength of the observations ($\sim 10\mu$). The alignement may be produced by a magnetic field of 7 mG by the Davis–Greenstein mechanism.

(6) The polarization is attributed to the scattering of radiation by dust grains; see also note (7).

(7) The polarization at 10μ is attributed to a cloud of cool, absorbing, elongated, aligned dust particles around or in front of BN, a model consistent with the idea that BN is a heavily reddened star; an alternative explanation is scattering of radiation in an assymetric circumstellar shell.

(8) The high ellipticities (ratio of circular to linear polarization) is attributed to polarization produced by aligned grains with a twist in the alignement ($< 60°$).

(9) Low polarization is compatible with observations at shorter wavelengths only if the grains exhibit limited alignement.

(10) The strong linear polarization of the BN object was recently interpreted as due to scattering dust grains and electrons in non spherical circumstellar shells (Elsässer and Staude, 1978; see Figures 3.2.6 and 3.2.7).

(11) The polarization is probably caused by a cold cloud of aligned absorbing grains, lying in front of the BN and KL objects.

TABLE 3.2.VI

Physical parameters of the molecular clouds OMC 2 and OMC 3 derived from infrared continuum measurements

No.	Airborne (A) or Ground (G) observations	Bandwidth (μ)	λ_{eff} (μ)	Beam	Beam separation	Flux (Jy) $1\,Jy=10^{-26}$ $W\,m^{-2}\,Hz^{-1}$	Dust temperature T_d (K)	Luminosity L ($L_\odot$)	Remarks	References
						a. OMC 2				
1	G	1.6–2.0		5″–1′				$<5\times10^2$	(6)	Gatley et al. (1974)
2	G		400	1′.6	4′.1	365 ± 140			(1)	Hudson and Soifer (1976)
3	G	700–1500	1000	1′	4′	9 ± 2				Westbrook et al. (1976)
4	A		42	50″	2′.4	140 ± 38	30	6×10^2	(2), (4)	Thronson et al. (1978)
			61	50″	2′.4	660 ± 45				
			105	50″	2′.4	1700 ± 32				
			145	50″	2′.4	1600 ± 21				
			390	1′.25	2′.8	370 ± 29	30		(2)	
			42	3′.5	13′	<3000		3.5×10^3	(2), (5)	
			61	3′.5	13′	3400 ± 700	30			
			105	3′.5	13′	9500 ± 590				
			145	3′.5	13′	7200 ± 450				
			327	9′.0	13′	4800 ± 270	30		(2)	
						b. IRS 3/OMC 2				
5	G	1.6–20		5″–1′			~500			Gatley et al. (1974)
6	A		42	18″×28″	1′.3	<28		$(1.10\pm0.33)\times10^2$	(3)	Thronson et al. (1978)
			61	18″×28″	1′.3	<56				
						c. IRS 4/OMC 2				
7	A		42	18″×28″	1′.3	300 ± 12		$(1.3\pm0.4)\times10^2$	(3)	Thronson et al. (1978)
			61	18″×28″	1′.3	570 ± 25				
						d. OMC 3				
8	A		61	3′.5	13′	4500 ± 1500	30	3×10^3	(5)	Thronson et al. (1978)
			105	3′.5	13′	8300 ± 840	30			
			327	9′	13′	<490	30			

Remarks to Table 3.2.VI:

(1) Column density of dust $N_D = 1 \times 10^{-3}$ g cm^{-2}, expected visual extinction $A_v \simeq 30$ mag., column density of gas $N_{H_2} = 3 \times 10^{22}$ molecules cm^{-2}; the parameters were measured along the line of sight through the peak.

(2) A 30 K blackbody curve fits the energy spectrum in the range of 10–1000μ.

(3) $\lambda < 60\mu$ luminosities.

(4) The cited luminosity is the total infrared luminosity into a 50" beam.

(5) The cited luminosity is the total infrared luminosity into a 3.5 beam.

(6) Luminosity for $\lambda < 20\mu$.

Remarks to Table 3.2.VII:

(1) The parameters were measured along a line of sight through the peak; peak positions:

OMC 1: $\alpha = 5^h32^m47^s$, $\delta = -5°24'30''$;

OMC 2: $\alpha = 5^h32^m59^s$, $\delta = -5°12'10''$.

(2) Extended component: 6'–8' in diameter.

(3) The values relate to the H II region.

(4) The dust mass, M_d, was estimated on the assumption that the emitting grains consist of a lunar rock core of radius $a = 0.1\mu$ with a water ice mantle 0.01μ thick, and are characterized by density $\rho = 2$ g cm^{-3} and an infrared emission efficiency Q_{IR} as given by Aannestad (1975) (see Appendix, Section III.2, Relation III.2.10). A constant grain temperature T_d was also assumed.

(5) The dust mass M_d was estimated on the assumption of a grain radius $a = 0.1\mu$, a grain density $\rho = 1$ g cm^{-3}, an infrared emission efficiency $Q_{IR} = 0.005$ and constant grain temperature T_d (see Appendix, Section III.2).

(6) $M_{dust}/M_{gas} = 0.01$ is an adopted value.

(7) Reference beam probably set in another source; in that case, the reported values of M_{dust}, M_{gas}, τ_{IR}, and N_{H} are lower limits.

TABLE 3.2.VII*

Physical parameters of the dust and gas in M42/OMC 1, M43/OMC 2 and OMC 3

No.	Source	λ (μ)	Beam	$\dfrac{M_{dust}}{M_{gas}}$	M_{dust} (M')	M_{gas} ($M_\odot$)	Optical depth $\tau_{IR}(\lambda)$	H_2 column density, N_{H_2} (molecules cm^{-2})	Dust column density N_D ($g\ cm^{-2}$)	Extinction A_v (mag.)	Remarks	References
							a. M42/OMC 1 Complex					
1	M42	45–750	8'.4	0.011	0.074						(3)	Harper and Low (1971)
2	M42	92	2'.2	0.04	0.28						(2), (3), (5)	Harper (1974)
3	KL/OMC 1	92	2'.2		0.110						(5)	Harper (1974)
4	OMC 1	1000	1'					7×10^{23}	2×10^{-2}	700		Westbrook et al. (1976)
5	OMC 1	80	0'.83	0.01	0.6	60	0.4	3×10^{23}			(4), (6)	Thronson et al. (1978)
		80	3'.5	0.01	6	600	0.2	2×10^{23}				
		300	9'	0.01	20	2000	1×10^{-2}	1×10^{23}				
		390	1'.25	0.01	3	300	4×10^{-2}	8×10^{23}				
6	OMC 1	400	3'			4.4	0.015		4.1×10^{-3}			Smith et al. (1979)
							b. M43/OMC 2 Complex					
1	M43	92	2'.2	0.03	0.026						(3), (5)	Harper (1974)
2	OMC 2	1000	1'					3×10^{22}	1×10^{-3}	30		Westbrook et al. (1976)
3	OMC 2	80	0'.83	0.01	0.15	15	0.1	8×10^{22}			(4), (6)	Thronson et al. (1978)
		80	3'.5	0.01	1.5	150	0.05	4×10^{22}				
		327	9'	0.01	9	900	0.03	4×10^{22}				
		390	1'.25	0.01	0.8	80	0.01	20×10^{22}				
4	OMC 2	400	3'			0.9	0.0031		8.3×10^{-4}			Smith et al. (1979)
							c. OMC 3					
1	OMC 3	80	3'.5	0.01	1.5	150	0.05	4×10^{22}			(4), (6)	Thronson et al. (1978)
		327	9'	0.01	$\leqslant 1$	$\leqslant 100$	$\leqslant 4 \times 10^{-4}$	$\leqslant 1 \times 10^{22}$			(4), (6), (7)	Thronson et al. (1978)

*For remarks to Table 3.2.VII see page 147.

TABLE 3.2.VIII

Infrared atomic lines detected in M42

No.	Line	Airborne (A) or Ground (G) observations	Beam	Beam separation	Flux (W cm⁻²)	Remarks	References
1	[Ar III] $\lambda 8.99\mu$	G	10″	26″	1.77×10^{-17}	(6)	Lester et al. (1979)
2	[S IV] $\lambda 10.50\mu$	G	50″		$(2.4 \pm 0.8) \times 10^{-16}$	(3)	Anderegg et al. (1976)
3	[S IV] $\lambda 10.51\mu$	G	10″	100″	$(2.44 \pm 0.73) \times 10^{-17}$	(6)	Lester et al. (1979)
4	[Ne II] $\lambda 12.78\mu$	G	10″	26″	$(8.1 \pm 2.4) \times 10^{-18}$	(6)	Lester et al. (1979)
5	[S III] $\lambda 18.70\mu$	A	55″	10′	$(1.6 \pm 0.2) \times 10^{-16}$	(3), (8)	Baluteau et al. (1976)
6	[S III] $\lambda 18.68\mu$	G	50″		$<2.1 \times 10^{-16}$	(3)	Anderegg et al. (1976)
7	[S III] $\lambda 18.71\mu$	A	2″7	8′1	$(23.8 \pm 0.8) \times 10^{-16}$	(12), (7)	
8	[S III] $\lambda 18.71\mu$	A	2″7	8′1	$(20.1 \pm 1.2) \times 10^{-16}$	(13), (10)	McCarthy et al. (1979)
9	[S III] $\lambda 18.71\mu$	A	2″7	8′1	$(14.1 \pm 0.6) \times 10^{-16}$	(14), (11)	
10	[O III] $\lambda 51.80\mu$	A	4′×6′	16′	$(70 \pm 8) \times 10^{-16}$	(1)	Melnick et al. (1979a)
11	[O I] $\lambda 63.00\mu$	A	4′×6′	16′	$(80 \pm 40) \times 10^{-16}$	(2)	Melnick et al. (1979b)
12	[O I] $\lambda 63.20\mu$	A	1′3	2′–6′	1.3×10^{-16}	(3), (15)	Storey et al. (1979)
13	[O III] $\lambda 88.16\mu$	A	5′×5′	15′	$<20 \times 10^{-16}$	(9)	Ward (1975)
14	[O III] $\lambda 88.35\mu$	A	90″	10′	$(3 \pm 0.6) \times 10^{-16}$	(3), (8)	Baluteau et al. (1976)
15	[O III] $\lambda 88.35\mu$	A	1′3	2′–6′	1.2×10^{-16}	(3), (4)	Storey et al. (1979)

Remarks to Table 3.2.VIII:

(1) The electron density derived from the [O III] $51.8\mu/88.35\mu$ ratio is $N_e = 7080^{+7045}_{-3960}$ cm⁻³.

(2) This line strength represents the 0.3% of the energy radiated at all wavelengths.

(3) At $a = 5^h32^m49^s.6$, $\delta = -5°25'16''$ (Trapezium).

(4) Predicted intensity: 1.6×10^{-16} W cm⁻².

(5) Predicted intensity: 1.9×10^{-16} W cm⁻².

(6) At the electron density peak position $1.'0$ W and $10''$ S of θ^1 Ori C.

(7) Predicted line intensity: 29.3×10^{-16} W cm⁻².

(8) Sky chopping was not employed; instead a reference sky position was measured (10′ away from the source).

(9) Predicted value (for a 5′ beam): 3.9×10^{-15} W cm⁻² (Petrosian, 1970) and 1.4×10^{-15} W cm⁻² (Simpson, 1975).

(10) Predicted line intensity: 20.4×10^{-16} W cm⁻².

(11) Predicted line intensity: 16.7×10^{-16} W cm⁻².

(12) Trapezium.

(13) KL Nebula.

(14) θ^2 Ori.

TABLE 3.2.IX

High resolution infrared spectroscopy of the BN object. Hydrogen recombination lines detected in BN and physical parameters of the object

No.	Line	Transition $m \to n$	λ [1] (μ)	Angular resolution	Beam separation [2]	Line flux (W cm^{-2})	Continuum flux (W cm^{-2} μ^{-1})	Radial velocity V_{LSR} (km s^{-1})	Angular size [3] (arc sec)	Density [3] (cm^{-3})	Mass [3] ($M_\odot$)	Remarks	References
1	Bα	5-4	4.0512	6"×12"	20"	$(1.59 \pm 0.13) \times 10^{-18}$			<0.7	$>3 \times 10^5$	$<10^{-4}$		Grasdalen (1976)
2	Bα	5-4	4.0512	11"	12"	$(1.80 \pm 0.15) \times 10^{-18}$						(4)	Joyce et al. (1978)
3	Bα	5-4	4.0512	11"	180"	1.7×10^{-18}			<12				
4	Bγ	7-4	2.1655	11"	12"	$(1.5 \pm 0.7) \times 10^{-19}$							
5	Bα	5-4	4.0512	2".8		2×10^{-18}	1.01×10^{-14}	21.5 ± 0.2				(5), (6)	Hall et al. (1978)
6	Bγ	7-4	2.1655	2".8		1.9×10^{-19}	1.10×10^{-5}	21.2 ± 0.2	~3			(7), (8)	
7	Pfβ	7-5	4.6525	2".8		—	—	17 ± 5					
8	Pfγ	8-5	3.7395	2".8		2.6×10^{-19}	8.7×10^{-15}	—					
9	Bα	5-4	4.0512	11"	120"	$(2.5 \pm 0.6) \times 10^{-18}$	2.5×10^{-15}		<0.3			(9), (10)	Smith et al. (1979)
10	Pfβ	7-5	4.6525	11"	120"	$(2.9 \pm 0.7) \times 10^{-18}$	2.4×10^{-15}					(11)	
11	Pfβ	7-5	4.6525	11"	120"	$<0.7 \times 10^{-18}$							

Remarks to Table 3.2.IX:

(1) Wiese et al. (1966).

(2) Chopper throw.

(3) Inferred parameter of the emitting region.

(4) Bα line centre to continuum flux ratio $\simeq 0.08$.

(5) Bα line centre to continuum flux ratio $\simeq 0.09$.

(6) The observed Bα profile is characterized by weak high velocity wings (± 100 km s^{-1}) similar to the ones found in early type stars with stellar winds.

(7) CO lines (^{12}CO, $J = 1-0$; ^{13}CO, $J = 1-0$) in absorption have also been observed; mean velocity of 13 ^{12}CO lines, $V_{LSR} = -17.7$ km s^{-1}; ^{13}CO line appears doubled; mean velocity of the ^{13}CO blue component $V_{LSR} = -16.1 \pm 1.1$ km s^{-1}, of the red component $V_{LSR} = -9.9 \pm 0.5$ km s^{-1}. The rotational temperature of the two ^{13}CO components is of the order of 85 ± 15 K respectively; this is an indication that the CO absorption originates in the molecular cloud in which BN is embedded and not in the hot circumstellar shell around BN.

(8) The continuum fluxes are from Gillett et al. (1975).

(9) The line flux at a point located 15" S and 0".5 W of θ^1 Orionis is $(4.5 \pm 0.7) \times 10^{-18}$ W cm^{-2} (observations made in February, 1978).

(10) Continuum fluxes are from Gillett and Forrest (1973).

(11) The Pfβ line flux $(2.9 \pm 0.7) \times 10^{-18}$ W cm^{-2} observed in December, 1976, is anomalously high in comparison with the flux of the same line observed in February, 1978 ($< 0.7 \times 10^{-18}$ W cm^{-2}). This may be due to a masing mechanism caused by a hydrogen laser contained in the BN object or in the dense small region which envelopes it; rapid variations of H recombination lines have already been observed in the small dense H II regions surrounding Be stars (Doazan, 1976).

TABLE 3.2.X

Nature of the BN object

No.	Possible nature of the BN object	Visual extinction A_v (mag.)	Remarks	References
1	Protostar			Becklin and Neugebauer (1967) Ney and Allen (1969) Low et al. (1970) Becklin et al. (1973)
2	M supergiant	~35		Becklin and Neugebauer (1967)
3	F supergiant	~80	(1)	Penston et al. (1971)
4	Proto–HII region	~24	(2)	Grasdalen (1976)
5	Pre-main sequence star surrounded by a circumstellar shell	~50	(3)	Grasdalen (1976)
6	Early B star surrounded by a compact H II region and an expanding dust envelope		(4)	Hall et al. (1978)

Remarks to Table 3.2.X:

(1) With $L \sim 10^5 L_\odot$.

(2) Ionized by the UV continuum of a star approaching or recently having reached the main sequence. $T_{eff} \simeq 1.7 \times 10^4$–$2.4 \times 10^4$ K.

(3) T_{eff} is in the range 3×10^3–8×10^3 K; this is based on the assumption that the observed infrared radiation is characteristic of the stellar photosphere; if however part of the observed radiation originates in the circumstellar shell, the T_{eff} can rise up to the values predicted for the Proto-H II region case.

(4) The Bα profile exhibits weak high velocity wings ($\pm$ 100 km s^{-1}); this is a characteristic feature of early type stars with stellar winds.

TABLE 3.3.I

Velocity gradients observed in the Orion Molecular Complex (OMC 1/OMC 2)

No.	Region	Size (longer dimension for a distance of 500 pc) (pc)	Velocity gradient (km s^{-1} pc^{-1})	Probable cause	Remarks	References
1	L1641 (Southern Complex)	~70	0.135	Rotation	(1)	Kutner et al. (1977)
2	OMC 1/OMC 2 (N–S strip)	~10	0.600	Rotation		Linke and Wannier (1974)
3	OMC 1 (ridge)	~0.6		Rotation	(2), (4)	Liszt et al. (1974)
4	Orion cluster of stars	~6.5	4.600	Contraction	(3)	Fallon et al. (1977)

Remarks to Table 3.3.I:

(1) Apparent axis of the cloud at $b^{II} = -19.4°$ (Figure 1.1.10). Angular frequency of rotation 4.5×10^{-15} s^{-1} (or period of 4×10^7 yr). The denser fragments of the collapsing cloud are expected to have a higher velocity gradient; this is corroborated by the findings of Linke and Wannier (1974) (note (2) in this table).

(2) Rotating about an E–W axis running through KL with velocity of rotation ~1 km s^{-1}.

(3) Contraction of the cluster up to a distance of $0°4$ (3.5 pc) from the Trapezium stars with velocity directly proportional to the distance from the cloud centre (Trapezium). Velocity of contraction at the edge of the cluster (3.5 pc from Trapezium) $\simeq$ 16 km s^{-1}. The collapse of the cluster, whose mass is less than the critical Jeans mass, can be understood if both cluster and molecular cloud are within each other and contract together.

(4) Ho and Barrett (1978), based on the study of NH$_3$ emission, have argued that OMC 1 consists of two separately rotating clouds moving at 9.6 km s^{-1} and 7.6 km s^{-1}, respectively.

TABLE 3.3.II
Molecular contour maps of OMC 1, OMC 2 and OMC 1/OMC 2

No.	Molecule	Transition	Frequency ν (GHz)	λ	HPBW (arc sec)	References
			a. OMC 1			
1	H_2	$v=1 \to 0$, $S(1)$		~2.1μ	5	Beckwith *et al.* (1978)
2	H_2	$v=1 \to 0$, $S(1)$		~2.1μ	13	
3	$^{12}C^{16}O$	$J=1-0$	115.2712	~2.6 mm	60	Liszt *et al.* (1974)
4	$^{13}C^{16}O$	$J=1-0$	110.2014	~2.7 mm	60	
5	$^{12}C^{16}O$	$J=1-0$	115.2712	~2.6 mm	120	Loren (1979)
6	$^{13}C^{16}O$	$J=1-0$	110.2014	~2.7 mm	120	
7	$^{12}C^{14}N$	$N=1-0$	113.4910	~2.6 mm	65	Turner and Gammon (1975)
8	$^{12}C^{14}N$	$N=1-0$	113.4910	~2.6 mm	63	Turner and Thaddeus (1978)
9	$^{12}C^{32}S$	$J=3-2$	146.9692	~2 mm	60	Liszt *et al.* (1974)
10	$^{12}C^{32}S$	$J=3-2$	146.9692	~2 mm	108	Liszt and Linke (1975)
11	$^{12}C^{32}S$	$J=2-1$	97.9810	~3.1 mm	150	Liszt and Linke (1975)
12	$H^{12}C^{14}N$	$J=1-0$	88.6319	~3.4 mm	77	Buhl (1972), Clark *et al.* (1974)
13	$H^{12}C^{14}N$	$J=1-0$	88.6319	~3.4 mm	75	Turner and Thaddeus (1978)
14	HCO^+		89.1886	~3.4 mm	75	Turner and Thaddeus (1978)
15	C_2H	$N=1-0$	87.3487	~3.4 mm	75	Tucker and Kutner (1978)
16	N_2H^+		93.1736	~3.2 mm	71	Turner and Thaddeus (1978)
17	$H_2^{12}C^{16}O$	$2_{12}-1_{11}$	140.8395	~2 mm	61	Thaddeus *et al.* (1971)
18	$H_2^{12}C^{16}O$	$2_{11}-2_{12}$	14.4887	~2 cm	72	Harvey *et al.* (1974)
19	$H_2^{12}C^{16}O$	$2_{11}-2_{12}$	14.4887	~2 cm	72	Evans *et al.* (1975)
20	$H_2^{12}C^{16}O$	$2_{11}-2_{12}$	14.4887	~2 cm	126	Scoville and Wannier (1979)
			b. OMC 2			
21	$^{12}C^{16}O$	$J=1-0$	115.2712	~2.6 mm	120	Gatley *et al.* (1974)
22	$H^{12}C^{14}N$	$J=1-0$	88.6319	~3.4 mm	78	Morris *et al.* (1974)
			c. OMC 1/OMC 2			
23	$^{12}C^{16}O$	$J=1-0$	115.2712	~2.6 mm	192	Gillespie and White (1980)
24	$^{13}C^{16}O$	$J=1-0$	110.2014	~2.7 mm	156	Kutner *et al.* (1976)
25	$H_2^{12}C^{16}O$	$2_{12}-1_{11}$	140.8395	~2 mm	120	Kutner *et al.* (1976)

TABLE 3.3.III
Kinetic temperature of the molecular gas in OMC 1 inferred from ^{12}CO measurements[1],[2]

No.	Kinetic temperature T_k (K)	Remarks	References
1	76	(4)	Phillips *et al.* (1973)
2	75 ± 7	(3)	Liszt *et al.* (1974)
3	~80	(4)	Goldsmith *et al.* (1975)
4	90	(4)	Plambeck and Williams (1979)
5	70	(4)	Phillips *et al.* (1979)

Remarks to Table 3.3.III:

(1) Kinetic temperature of the 'ridge' or 'spike' component of OMC 1.

(2) Kinetic temperature inferred from NH_3 observations: $T_{exc} \sim T_k = 56.9 \pm 3.5$ K (Sweitzer *et al.*, 1979; Wilson *et al.*, 1979); kinetic temperature inferred from H_2O (183 GHz) observations: $T_k \sim 80$ K (Waters *et al.*, 1980).

(3) From ^{12}CO, $J = 1-0$ observations.

(4) From ^{12}CO, $J = 2-1$ observations.

TABLE 3.3.IV

Kinetic temperature of the molecular gas in OMC 2 inferred from ^{12}CO measurements

No.	Kinetic temperature T_k (K)	Remarks	References
1	53	(1)	Gatley et al. (1974)
2	40	(2)	Phillips et al. (1979)
3	~40	(2)	Goldsmith et al. (1975)
4	55	(2)	Plambeck and Williams (1979)

Remarks to Table 3.3.IV:

(1) From ^{12}CO, $J = 1–0$ observations.
(2) From ^{12}CO, $J = 2–1$ observations.

TABLE 3.3.V

Molecular abundances* in the central part of OMC 1 and OMC 2

No.	Molecule	Column density N (molecules cm^{-2})	Remarks	References
		a. OMC 1		
1	H_2	2×10^{23}		Thaddeus et al. (1971)
2	H_2	7×10^{23}	(1)	Westbrook et al. (1976)
3	H_2	$\gtrsim 1 \times 10^{23}$		Phillips et al. (1977)
4	H_2	$\sim 10^{19}$		Beckwith et al. (1978)
5	H_2	$1 \times 10^{23}–8 \times 10^{23}$	(1)	Thronson et al. (1978)
6	H_2	$\sim 6 \times 10^{19}$	(2)	Beck et al. (1979)
7	H_2	$\sim 3 \times 10^{20}$	(13)	Beckwith et al. (1979)
8	OH	3×10^{14}		Turner (1974)
9	CO	1.5×10^{19}	(16)	Wilson et al. (1974)
10	CO	2.1×10^{19}	(16)	Liszt et al. (1974)
11	CO	$\gtrsim 6 \times 10^{18}$	(16)	Phillips et al. (1977)
12	CN	1.1×10^{15}		Jefferts et al. (1970)
13	CN	$\sim 2 \times 10^{14}$	(3)	Turner and Gammon (1977)
14	CN	1.6×10^{14}		Turner and Thaddeus (1977)
15	CS	1.8×10^{14}		Turner et al. (1973)
16	CS	2.7×10^{14}		Linke and Goldsmith (1980)
17	SiO	6×10^{13}		Dickinson (1972)
18	SiS	$< 3 \times 10^{14}$		Gottlieb et al. (1978)
19	SO	1×10^{15}		Gottlieb and Ball (1973)
20	SO	1×10^{16}		Gottlieb et al. (1978)
21	H_2O	$10^{17}–10^{18}$	(14)	Waters et al. (1980)
22	$H_2^{18}O$	$\sim 5 \times 10^{18}$	(14)	Phillips et al. (1978)
23	HDO	2×10^{15}		Turner et al. (1975)
24	HDO	$2 \times 10^{15}–2 \times 10^{16}$		Waters et al. (1980)
25	H_2S	9×10^{14}		Thaddeus et al. (1972)
26	HCN	$\sim 3 \times 10^{15}$	(8)	Snyder and Buhl (1973)
27	HCN	2×10^4		Gottlieb et al. (1975)
28	N_2H^+	10^{13}		Turner and Thaddeus (1977)
29	C_2H	$\sim 8 \times 10^{14}$	(12)	Tucker and Kutner (1978)
30	SO_2	$\sim 2 \times 10^{16}$	(4)	Snyder et al. (1975)
31	NH_3	1.3×10^{16}		Schwartz et al. (1977)
32	NH_3	1.2×10^{16}	(11)	Wilson et al. (1979)
33	H_2CO (ortho)	3×10^{14}	(10)	Thaddeus et al. (1971)
34	H_2CO (ortho)	2.2×10^{14}		Evans et al. (1975)

(continued)

Table 3.3V (continued)

No.	Molecule	Column density N (molecules cm^{-2})	Remarks	References
		a. OMC 1		
35	H$_2$CO (ortho)	2×10^{14}	(5)	Kutner *et al.* (1976)
36	H$_2$CO (para)	3×10^{13}	(9)	Kaifu *et al.* (1975)
37	CH$_4$	$\sim 10^{22}$		Fox and Jennings (1979)
38	HC$_3$N	1.8×10^{14}	(7)	Morris *et al.* (1977)
39	CH$_3$OH	5×10^{16}		Barrett *et al.* (1971)
40	CH$_3$OH	2.5×10^{16}	(6)	Jennings and Fox (1978)
41	CH$_3$OH	2.4×10^{15}		Gottlieb *et al.* (1979)
42	CH$_3$CN	1×10^{14}		Johnson *et al.* (1977)
43	CH$_3$CH$_2$N	1.8×10^{14}		Johnson *et al* (1977)
44	C$^\circ$	$>7.5 \times 10^{17}$		Phillips *et al.* (1980)
		b. OMC 2		
45	H$_2$	3×10^{22}		Westbrook *et al.* (1976)
46	H$_2$	3×10^{22}–2×10^{23}		Thronson *et al.* (1978)
47	CO	$\sim 10^{19}$	(15)	
48	CH$_3$OH	6×10^{13}		Gottlieb *et al.* (1979)
49	HCN	1.6×10^{13}		Turner and Thaddeus (1977)
50	N$_2$H$^+$	5×10^2		Turner and Thaddeus (1977)
51	CS	4×10^{13}		Linke and Goldsmith (1980)
52	C	$>1.7 \times 10^{17}$		Phillips *et al.* (1980)

* Carbon abundances are also included for comparison.

Remarks to Table 3.3.V:
   (1) See Table 3.2.VII.
   (2) In the range 4×10^{19}–8×10^{19} cm^{-2}.
   (3) In the range 5.2×10^{13}–4.8×10^{14} cm^{-2}.
   (4) In the range 3×10^{15}–3.5×10^{16} cm^{-2}.
   (5) In the range 1.2×10^{14}–2.9×10^{14} cm^{-2}.
   (6) In the range 2×10^{16}–3×10^{16} cm^{-2}.
   (7) $(1.8 \pm 0.2) \times 10^{14}$ cm^{-2}.
   (8) In the range 8.6×10^{14}–5.2×10^{15} cm^{-2}.
   (9) In the range 2×10^{13}–5×10^{13} cm^{-2}.
   (10) In the range 6×10^{13}–1.8×10^{15} cm^{-2}.
   (11) ^{14}NH$_3$ (1,1): 1.4×10^{16} cm^{-2}; ^{14}NH$_3$ (2, 2): 1.1×10^{16}; ^{15}NH$_3$ (1, 1): $< 0.8 \times 10^{13}$ cm^{-2}; ^{15}NH$_3$ (2, 2) $\leqslant 3.1 \times 10^{13}$ cm^{-2}.
   (12) Column density greater than $(2$–$5) \times 10^{14}$ cm^{-2} and lesser than $(1$–$3) \times 10^{15}$ cm^{-2}.
   (13) From measurements of the $v = 1 \rightarrow 0$, $Q(3)/S(1)$ intensity ratio of molecular H$_2$.
   (14) For a 1.3 source at 460 pc.
   (15) Assuming $N_{H_2} = 10^{23}$ molecules cm^{-2} and $N_{H_2}/N_{Co} = 10^4$.
   (16) An isotopic abudance [^{12}CO]/[^{13}CO] = [^{12}C]/[^{13}C] = 40 derived by Wannier (1976) has been questioned by Kutner *et al.* (1976) who find 89 (consistent with the terrestial value).

TABLE 3.3.VI

H_2 densities and masses of the molecular clouds OMC 1 and OMC 2[1]

No.	Molecule observed	Region	H_2 density n_{H_2} (cm^{-3})	Mass ($M_\odot$)	Remarks	References
a. OMC 1						
1	CO		4×10^3		(2)	Plambeck and Williams (1979)
2	CO, CS	$4' \times 9'$ ridge	$>2 \times 10^2$	>200		Liszt et al. (1974)
3	CN	$<1'$ core $2'$ ridge outside the ridge	8×10^5 3×10^4 2×10^4			Turner and Gammon (1975)
4	CS	ridge	$>6 \times 10^4$			Liszt and Linke (1975)
5	CS	ridge	3×10^4			Gottlieb et al. (1975)
6	CS	$2' \times 11'$ ridge	2×10^5			Linke and Goldsmith (1980)
7	HCN	ridge	1×10^5			Gottlieb et al. (1975)
8	HCN	ridge	4.5×10^5			Padman et al. (1980)
9	HCO$^+$	core	$\sim 3 \times 10^5$			Huggins et al. (1979)
10	H_2CO	$3' \times 5'$ ridge	$\sim 2 \times 10^5$	~ 200		Thaddeus et al. (1971)
11	H_2CO	$3'$ core	$>7 \times 10^5$			Kutner et al. (1971)
12	H_2CO	$<1'$ core $2' \times 5'$ ridge $30'$ extended component	$\sim 10^7$ $\simeq 10^5$ $\sim 10^3$	60 500		Harvey et al. (1974), Evans et al. (1975)
13	HC$_3$N	$\leqslant 2'$ core	$\sim 10^6$			Morris et al. (1977)
14	CH$_3$OH	$<1'$ core	$\sim 2 \times 10^6$	>20		Barrett et al. (1971)
15	CH$_3$OH	$<1'$ core	10^7–10^8	~ 1000		Kutner et al. (1973)
16	1mm IR emission	$1'$ core	2×10^6	200	(3)	Westbrook et al. (1976)
17	CO	Fragment ($\sim 3'$) lying 24' E of KL	8×10^4	2000	(5)	Gillespie and White (1980)
b. OMC 2						
18	CO	$6'$ ridge	$\simeq 10^4$	1000		Gatley et al. (1974)
19	CO		8×10^3			Plambeck and Williams (1979)
20	CS		2×10^5			Linke and Goldsmith (1980)
21	NH$_3$		2×10^3			Morris et al. (1974)
22	H_2CO	peak	$>3 \times 10^4$			Kutner et al. (1976)
23	H_2CO	$3'$ core	$<5 \times 10^4$			Lucas et al. (1977)
24	1 mm IR emission	$1'$ core	7×10^4	10	(4)	Westbrook et al. (1976)

Remarks to Table 3.3.VI:

(1) Both OMC 1 and OMC 2 are parts of a dense molecular cloud ($n_{H_2} \simeq 5 \times 10^4$ cm^{-3}) covering a $10' \times 30'$ area (1.5×5 pc for a distance of 0.5 kpc) (Figure 3.3.1). The total mass of this complex is 7×10^3 $M_\odot$. This is part of the extended diffuse dust cloud Lynds 1641 (Lynds, 1962); the mass of the cloud, M_{H_2}, is $\simeq 1 \times 10^5$ $M_\odot$ (Lynds, 1962) (Figure 1.1.9). OMC 1 is a rather uniform density ($n_{H_2} \simeq 10^5$ cm^{-3}) disklike or barlike object centred on the KL Neb, (Figures 3.3.5, 3.3.7, and 3.3.8) and embedded in a lower density ($n_{H_2} \simeq 10^3$ cm^{-3}) cloud. Embedded in the disk is a still denser condensation ($n_{H_2} \simeq 10^7$ cm^{-3}). A similar structure was inferred by Balick et al. (1974). The $60 M_\odot$ core contains H_2O, OH, and CH$_3$OH molecules. The $500 M_\odot$ disk contains HCN, CS, H_2S, H_2CO. The mass of the disk is comparable to the mass of the Orion Nebula plus the Trapezium cluster ($\simeq 450 M_\odot$; Strand, 1958). The extended low density component ($\simeq 1°$; $M \simeq 10^4$ $M_\odot$) contains H, CO, H_2CO and more or less coincides with the Orion cluster of stars (Balick et al., 1974).

(2) Densities derived from CO observations are typically 10 times lower than the corresponding densities derived from observations of other molecules (CS, CN, HCN, H_2O); for possible explanations see Plambeck and Williams (1979).

(3) 1 mm emission. The peak ($\alpha = 5^h32^m47^s$, $\delta = -5°24'30''$) coincides with the IR cluster and the core of OMC 1.

(4) 1 mm emission. Peak at $\alpha = 5^h32^m59^s$, $\delta = -5°12'10''$.

(5) Peak at $\alpha = 5^h34^m22^s$, $\delta = -5°27'36''$. $V_{LSR} = 9$ km s^{-1}. Parameters derived by assuming a source dimension of 0.5 pc and intrinsic line width of 2.9 km s^{-1}. The source may be a protostellar region; alternatively, it can be caused by a shock front moving away from the Trapezium stars.

TABLE 3.3.VII
Radial velocities of OMC 1 derived from molecular lines

No.	Molecular line	Beam	E= emission A= absorption	$V_{LSR}^{(1)}$ (km s^{-1})	Remarks	References
1	H$_2$	23″ × 35″	E	+9.5±4	(2), (3), (4)	Joyce et al. (1978)
2	H$_2$	10″6 × 10″6	E	+9.3±1.5	(2), (4)	Ogden et al. (1978)
3	H$_2$	10″	E	+13	(2), (5)	Nadeau and Geballe (1979)
4	H$_2$	7″	E	~+7	(158), (159)	Beck et al. (1979)
5	H$_2$	23″ × 35″	E	+9.5±4	(207), (208), (209)	Scoville (1980)
6	OH		A	+5.7±0.3	(6)	Goss (1968)
7	OH		A	+7.1±1.4	(7)	Goss (1968)
8	OH		E	+6 to +24	(13), (144)	Goss (1968)
9	OH		E	−10 to +26	(8), (144)	Goss (1968)
10	OH	{	E	+4 to +4.5, +6.8 to +7.9, +17.6 to 23.7	(8), (9), (11), (144)	Raimond and Eliasson (1969)
11	OH	18′	A	+6.4±0.1	(6), (12)	Chaisson (1974)
12	OH		E	+6.93±0.04, +8.44±0.04	(8), (14)	Chaisson and Beichman (1975)
13	OH		E	+7.76±0.04, +7.64±0.04	(13), (14)	Chaisson and Beichman (1975)
14	OH		E	−7.0 to +22.6	(8), (9), (10), (11), (44)	Hansen et al. (1977)
15	CH		E	+10	(15), (16)	Rydbeck et al. (1973), (1974)
16	CH	6′4	E	+8.1	(54)	Whiteoak et al. (1978)
17	CO	70″	E	+9	(17)	Wilson et al. (1974)
18	CO	70″	E	+9	(20)	Wilson et al. (1974)
19	CO	1′	E	+9	(17)	Liszt et al. (1974)
20	CO	1′75	E	+9	(21), (18)	Phillips et al. (1973)
21	CO	1′75	E	+9	(21), (184)	Phillips et al. (1973)
22	CO	1′	E	+9	(18), (21)	Phillips et al. (1974)
23	CO	4′5	E	+9.5	(18), (23)	White et al. (1980)
24	CO	4′5	E	+10	(190), (23)	White et al. (1980)
25	CO	66″	E	~+9	(17), (16), (22)	Kwan and Scoville (1976)
26	CO	66″	E	~+9	(20), (16), (22)	Kwan and Scoville (1976)
27	CO	2′	E	+9	(17), (25)	Loren (1979)
28	CO	65″	E	~+9.3	(17), (22)	Zuckerman et al. (1976)
29	CO	64″	E	+9.0	(17)	Ulich and Haas (1976)
30	CO	64″	E	+9.0	(18)	Ulich and Haas (1976)
31	CO	64″	E	+9.0	(167)	Ulich and Haas (1976)
32	CO		E	+9.0	(17), (23), (22)	Wannier and Phillips (1977)
33	CO		E	+9.0	(18), (23), (22)	Wannier and Phillips (1977)
34	CO	2′	E	+9.3	(17)	Wannier et al. (1976)
35	CO	2′	E	+9.3	(167), (23)	Wannier et al. (1976)
36	CO	2′	E	+9.3	(20)	Wannier et al. (1976)
37	CO	36″±3″	E	~9	(19), (22), (149)	Phillips et al. (1977)
38	CO	2′1	E	+9.0	(17), (105)	Scoville and Wannier (1979)
39	CO	0′95±3′5	E	+9.0	(18), (163), (23), (22)	Plambeck and Williams (1979)
40	CO	3′2	E	+9.0	(17), (21)	Gillespie and White (1980)
41	CO	2′25	E	+9.0	(18), (23)	Goldsmith et al. (1975)
42	CO	2′25	E	+10.0	(184), (23)	Goldsmith et al. (1975)
43	CN		E	+9.0		Penzias et al. (1974)
44	CN	63″	E	+8.4	(24), (25), (26)	Turner and Thaddeus (1977)
45	CN	65″	E	+8.4	(24), (25)	Turner and Gammon (1975)
46	CN	65″	E	+8.4	(147), (25)	Turner and Gammon (1975)
47	CS	1′	E	+9.0	(29)	Liszt et al. (1974)
48	CS	1′1	E	+9.2	(27), (25)	Turner et al. (1973)
49	CS	1′1	E	+8.6	(28), (25)	Turner et al. (1973)

Table 3.3.VII (*continued*)

No.	Molecular line	Beam	E= emission A= absorption	$V_{LSR}^{(1)}$ (km s^{-1})	Remarks	References
50	CS	4.7	E	+9.0	(27), (23)	Liszt and Linke (1975)
51	CS	2.5	E	+8.9	(28), (23)	Liszt and Linke (1975)
52	CS	1.8	E	+9.0	(29), (23)	Liszt and Linke (1975)
53	CS	2.6	E	+8.5	(27), (23)	Linke and Goldsmith (1980)
54	CS	2.1	E	+8.2	(28), (23)	Linke and Goldsmith (1980)
55	SO	1.3	E	+8	(30), (16)	Kaifu *et al.* (1974)
56	SO		E	+9	(30), (16), (148)	Clark and Johnson (1974)
57	SO		E	~ +10	(31), (23), (22)	Wannier and Phillips (1977)
58	SO		E	+8.6±2	(32), (21)	Gottlieb and Ball (1973)
59	SO		E	+9.6±1.6	(33), (21)	Gottlieb and Ball (1973)
60	SO	1.0	E	$\begin{cases}+8.1\pm0.3\\+9.2\pm0.2\end{cases}$	(31), (23), (146)	
61	SO	1.1	E	$\begin{cases}+8.3\pm0.1\\+8.7\pm0.1\end{cases}$	(32), (23), (146)	
62	SO	1.9	E	$\begin{cases}+8.6\pm0.2\\+10\pm3\end{cases}$	(33), (23), (146)	Gottlieb *et al.* (1978)
63	SO	1.1	E	$\begin{cases}+7.0\pm0.7\\+9.8\pm0.7\end{cases}$	(34), (23), (146)	
64	SO	1.2	E	+7.8±0.6	(35), (23)	
65	SO	1.3	E	$\begin{cases}+7.7\pm0.2\\+9.1\pm0.2\end{cases}$	(30), (23), (146)	
66	SO	2'	E	+7.5	(168), (25)	Rydbeck *et al.* (1980)
67	SiO		E	+8	(36), (150)	Dickinson (1972)
68	SiO	160″	E	−7	(42), (178)	Snyder and Buhl (1975)
69	SiO	160″	E	−7	(38), (178)	Snyder and Buhl (1975)
70	SiO		E	+15.3	(39), (40), (11)	Davis *et al.* (1974)
71	SiO	70″	E	−7.1 to +17.3	(38), (16), (11), (41), (144)	Snyder and Buhl (1974)
72	SiO		E	−6.5, +17	(42), (40), (11)	Thaddeus *et al.* (1974)
73	SiO	160″	E	−6, +16	(43), (11)	Buhl *et al.* (1974)
74	SiO	1.3	E	~ +10	(36)	Buhl *et al.* (1975)
75	SiO		E	7.2±0.8	(36), (21), (26), (152)	Dickinson *et al.* (1976)
76	SiO		E	9.7±0.7	(37), (21), (153)	Dickinson *et al.* (1976)
77	SiO	78″	E	−8 to +16.4	(38), (166)	Ulich and Haas (1976)
78	SiO		E	10±3	(36), (16)	Lovas *et al.* (1976)
79	SiO		E	−6.7 to +17.8	(42), (11), (44), (144)	Moran *et al.* (1977)
80	SiO	78″	E	7.6±0.3	(36), (21), (26), (141)	Lada *et al.* (1978)
81	SiO		E	−7 to +19.5	(42), (178), (191)	Snyder *et al.* (1978)
82	SiO	24″	E	−5.5	(169), (170), (171)	Genzel *et al.* (1980)
83	H$_2$O		E	+0.95, +7.6, +(9.2−10.8)	(45)	Hills *et al.* (1972)
84	H$_2$O		E	−8 to +26	(45), (144), (145)	Sullivan (1973)
85	H$_2$O		E	+11.0	(45), (46)	Baudry *et al.* (1974)
86	H$_2$O		E	+9, +11	(45), (46)	Bologna *et al.* (1975)
87	H$_2$O		E	−50 to +25	(45), (47), (144)	Goss *et al.* (1976)
88	H$_2$O	40″ × 43″	E	+5.5, +18	(45), (46)	Genzel and Downes (1977)
89	H$_2$O		E	+0.8 to +17.4	(45), (48), (144)	Johnston *et al.* (1977)
90	H$_2$O		E	−6.7 to +28.2	(45), (49), (144)	Moran *et al.* (1977)
91	H$_2$O		E	+10	(45), (50)	Cesarsky *et al.* (1978)
92	H$_2$O		E	+7.6 to +26	(45), (51), (144)	Forster *et al.* (1978)
93	H$_2$O	40″	E	−6.1 to +31.3	(45), (52), (144)	Genzel *et al.* (1978)
94	H$_2$O	7.5	E	+8	(160), (21), (161)	Waters *et al.* (1980)
95	H$_2$O	1'	E	+9	(162), (21)	Phillips *et al.* (1978)
96	HDO	1.5	E	+7.8	(53)	Turner *et al.* (1975)

(*continued*)

Table 3.3.VII (continued)

No.	Molecular line	Beam	E= emission A= absorption	$V_{LSR}^{(1)}$ (km s^{-1})	Remarks	References
97	H$_2$S	0.75	E	+8.4	(154), (138)	Thaddeus *et al.* (1972)
98	HCN		E	+10		Snyder and Buhl (1971)
99	HCN		E	+8.56±0.34	(55)	Snyder and Buhl (1973)
100	HCN	1.3	E	+10	(57), (16)	Kaifu *et al.* (1974)
101	HCN	70″	E	+9, +14, +2	(55), (61)	Clark (1975)
102	HCN		E	+8.51±0.09	(55), (55a), (21), (26)	Gottlieb *et al.* (1975)
103	HCN		E	+8.58±0.04	(55), (55b), (21), (26)	Gottlieb *et al.* (1975)
104	HCN		E	+8.71±0.08	(55), (55c), (21), (26)	Gottlieb *et al.* (1975)
105	HCN	3.1	E	+8.6	(188),(185)	Baudry *et al.* (1980)
106	HCN		E	+9.46±0.06	(56), (56a), (21)	Gottlieb *et al.* (1975)
107	HCN		E	+9.75±0.08	(56), (56b), (21)	Gottlieb *et al.* (1975)
108	HCN		E	+9.60±0.20	(56), (56c), (21)	Gottlieb *et al.* (1975)
109	HCN	75″	E	+8.56	(55), (25), (26)	Turner and Thaddeus (1977)
110	HCN	78″	E	+9.4	(55)	Ulich and Haas (1976)
111	HCN	47″	E	~8−9	(58)	Huggins *et al.* (1979)
112	HCN	60″	E	+9.8	(176), (177)	Padman *et al.* (1980)
113	HNC	47″	E	~ +8−9	(59)	Huggins *et al.* (1979)
114	HNC	2.9	E	+8.5	(179), (62)	Snell and Wootten (1979)
115	HNC	3′	E	+6.8	(180), (62)	Snell and Wootten (1979)
116	HNC	3.9	E	+8.5	(179), (185)	Baudry *et al.* (1980)
117	DNC	1.7	E	+8.0	(181), (62)	Snell and Wootten (1977)
118	HCO$^+$	75″	E	+8.66	(63), (25), (26)	Turner and Thaddeus (1977)
119	HCO$^+$		E	+8.66	(156), (25), (157)	Huggins *et al.* (1979)
120	HCO$^+$	3.5	E	+9.5	(187), (185), (26)	Baudry *et al.* (1980)
121	N$_2$H$^+$	71″	E	+9.29	(64), (25), (26)	Turner and Thaddeus (1977)
122	C$_2$H	75″	E	+8.7	(155), (16), (26)	Tucker *et al.* (1974)
123	OCS		E	+7.5±0.5	(137)	Turner (1974)
124	OCS	69″	E	+7.4±0.2	(65), (21)	Lada *et al.* (1978)
125	CCH	3.5	E	+8.8	(189), (185)	Baudry *et al.* (1980)
126	SO$_2$	74″	E	+8.5	(66), (16), (67)	Snyder *et al.* (1975)
127	NH$_3$	2′×1.3	E	+7.7	(69), (25)	
128	NH$_3$	2′×1.3	E	+7.7	(70), (25)	
129	NH$_3$	2′×1.3	E	+9.0	(71), (25)	
130	NH$_3$	5.3	E	+8.3	(69), (25)	Morris *et al.* (1973)
131	NH$_3$	5.3	E	+9.5	(70), (25)	
132	NH$_3$	5.3	E	+6.0	(71), (25)	
133	NH$_3$	5.3	E	+5.2	(72), (25)	
134	NH$_3$	2.1	E	+7	(69), (68)	Schwartz *et al.* (1977)
135	NH$_3$	1.4	E	+8.0	(69), (25)	
136	NH$_3$	1.4	E	+8.5	(70), (25)	
137	NH$_3$	1.4	E	+8.5	(71), (25)	
138	NH$_3$	1.4	E	+8.0	(72), (25)	Barrett *et al.* (1977)
139	NH$_3$	1.4	E	+6.2	(73), (25)	
140	NH$_3$	1.4	E	+6.5	(74), (25)	
141	NH$_3$	1.4	E	+8.63±0.03	(70), (78), (25), (26)	
142	NH$_3$	1.4	E	+7.56±0.04, +9.47±0.13	(70), (79), (25)	Ho and Barrett (1978)
143	NH$_3$	1.4	E	+6.7	(80), (84)	
144	NH$_3$	1.4	E	+6.4	(81), (84)	
145	NH$_3$	1.4	E	+3.4	(82), (84)	Sweitzer *et al.* (1979)
146	NH$_3$	1.4	E	+3.0	(83), (84)	
147	NH$_3$	43″	E	+8.0	(69)	
148	NH$_3$	43″	E	+7.7	(70)	
149	NH$_3$	43″	E	+7.3	(71)	

Table 3.3.VII (continued)

No.	Molecular line	Beam	E= emission A= absorption	$V_{LSR}^{(1)}$ (km s⁻¹)	Remarks	References
150	NH₃	43″	E	+8.0±0.2	(75)	Wilson and Pauls (1979)
151	NH₃	43″	E	+7.0±0.3	(76)	
152	NH₃	43″	E	+6.7±0.3	(77)	
153	NH₃	1.4	E	+7.5, +9.7	(69), (25), (26)	
154	NH₃	1.4	E	+7.6, +9.5	(70), (25), (26)	
155	NH₃	1.4	E	+8.2	(71), (25)	Ho et al. (1979)
156	NH₃	1.4	E	+5.2±8.8	(72), (25)	
157	NH₃	1.4	E	+6.8	(74), (25)	
158	NH₃	40″	E	+7.2±0.5, +4.4±1.9, ~ +3.8	(80), (84), (198)	
159	NH₃	40″	E	+6.4±0.7	(81), (84), (199)	
160	NH₃	40″	E	+6.9±0.5	(82), (84), (198), (201)	
161	NH₃	40″	E	+5.5±0.9	(192), (84), (200)	
162	NH₃	40″	E	+5.9±1.0	(83), (84), (200)	Morris et al. (1980)
163	NH₃	40″	E	+5.6±0.7	(193), (84), (200)	
164	NH₃	40″	E	+5.4±0.6	(194), (84), (200)	
165	NH₃	40″	E	+4.7±1.2	(195), (84), (200)	
166	NH₃	40″	E	+5.4±1.0	(196), (84), (200)	
167	NH₃	40″	E	+5.4±1.2	(197), (84), (200)	
168	H₂CO	61″	E	+8.5	(172)	
169	H₂CO	61″	E	+8.5	(173)	Thaddeus et al. (1971)
170	H₂CO	61″	E	+5	(174)	
171	H₂CO	1.5	E	+10	(172)	Kutner et al. (1971)
172	H₂CO	1′	E	+8.3		Kutner (1972)
173	H₂CO	1′	E	+8.1	(94), (21), (26)	Harvey et al. (1974), Evans et al. (1975)
174	H₂CO	2.6	E	+8.7	(91), (21), (26)	Zuckerman et al. (1975)
175	H₂CO	2.4	E	+8.0	(92), (25), (26)	Kaifu et al. (1975)
176	H₂CO	1.2	E	+9.0	(93), (40)	Evans et al. (1979)
177	H₂CO	2.1	A	+8.5	(94), (105)	Scoville and Wannier (1979)
178	H₂CO	6.6	A	+9.0	(91), (142), (26)	Kutner and Thaddeus (1971)
179	H₂CO		A	+6.1	(91)	Zuckerman and Ball (1974)
180	H₂CO	4.2	A	+6.1	(91), (95)	Whiteoak and Gardner (197
181	H₂CO	2.6	A	+5.1	(91), (21), (26)	Zuckerman et al. (1975)
182	H₂CO	1.2	E	+8.6±0.1	(186)	Myers and Buxton (1980)
183	H₂CS		E	+9.0	(96), (23)	Liszt (1978)
184	CH₄		E	+8.5	(97), (16)	Fox and Jennings (1979)
185	HC₃N	70″	E	+6.4±2		Clark et al. (1976)
186	HC₃N	70″	E	+3.6±2		
187	HC₃N	2.8	E	+9.0	(98), (25), (26)	
188	HC₃N	2.0	E	+9.7	(99), (25), (26)	Morris et al. (1977)
189	HC₃N	1.9	E	+8.7	(100), (25), (26)	
190	HC₃N	1.8	E	+9.5	(101), (25), (26)	
191	CH₃OH	1.3	E	+7.8, +8.2, +7.3	(202), (203), (204), (102), (103)	Barrett et al. (1975)
192	CH₃OH	1.3	E	+8.0±0.5	(143)	Barrett et al. (1971)
193	CH₃OH	1′	E	+7.5	(139), (140)	Kutner et al. (1973)
194	CH₃OH	7.8	E	+8.3	(102), (21)	Chui et al. (1974)
195	CH₃OH	7.8	E	+8.3	(103), (21)	
196	CH₃OH	35″	E	+7.8	(102),(106),(26),(107)	Hills et al. (1975)
197	CH₃OH	35″	E	+7.8	(103),(106),(26),(107)	
198	CH₃OH		E	+8.3±0.1	(108), (21)	Gottlieb et al. (1979)
199	CH₃OH	87″	E	+8.5	(109), (16)	Jennings and Fox (1978)
200	CH₃OH	63″	E	+7.5±1.0	(108)	Hocking et al. (1979)

(continued)

Table 3.3.VII (continued)

No.	Molecular line	Beam	E= emission B= absorption	$V_{LSR}^{(1)}$ (kms $^{-1}$)	Remarks	References
201	CH₃OH		E	+8.0	(23), (210)	⎫
202	CH₃OH		E	+7.6	(23), (211)	⎪
203	CH₃OH		E	+7.8	(23), (212)	⎬ Matsakis *et al.* (1980)
204	CH₃OH		E	+7.0	(23), (213)	⎪
205	CH₃OH		E	+8.8	(23), (214)	⎪
206	CH₃OH		E	+6.4	(23), (215)	⎭
207	CH₃CN		E	+8.5±2	(110), (16)	⎫
208	CH₃CN		E	+8.5±2	(111), (16)	⎬ Lovas *et al.* (1976)
209	CH₃CN		E	+8.5±3	(112), (16)	⎪
210	CH₃CN		E	+9±3	(113), (16)	⎭
211	CH₃CHO		E	+5±3	(114), (16)	⎫ Lovas *et al.* (1976)
212	CH₃CCH		E	+8±1	(115), (16)	⎭
213	CH₃NH₂	1ʹ3	E	+5	(116), (16)	Kaifu *et al.* (1974)
214	CH₃NHD	2ʹ8	E	+10	(117), (68)	Fourikis *et al.* (1977)
215	CH₃OCH₃	72″	E	+8.65±1.6	(118), (16)	⎫
216	CH₃OCH₃	94″	E	+6±10	(119), (16)	⎪
217	CH₃OCH₃	94″	E	+15±10	(120), (16)	⎬ Snyder *et al.* (1974)
218	CH₃OCH₃	77″	E	+11.9±3.5	(121), (16)	⎪
219	CH₃OCH₃	77″	E	+15.6±3.5	(122), (16)	⎭
220	CH₃OCH₃		E	+6±3	(123), (16)	⎫
221	CH₃OCH₃		E	+5.8±4	(124), (16)	⎪
222	CH₃OCH₃		E	+5.7±3	(125), (16)	⎪
223	CH₃OCH₃		E	+4.3±4	(126), (16)	⎪
224	CH₃OCH₃		E	+7.3±4	(127), (16)	⎪
225	CH₃OCH₃		E	+6.5±3	(128), (16)	⎪
226	CH₃OCH₃		E	+5.8±4	(129), (16)	⎬ Clark *et al.* (1979)
227	CH₃OCH₃		E	+7.9±3	(130), (16)	⎪
228	CH₃OCH₃		E	+7.6±5	(131), (16)	⎪
229	CH₃OCH₃		E	+7.4±3	(132), (16)	⎪
230	CH₃OCH₃		E	+7.1±5	(133), (16)	⎪
231	CH₃OCH₃		E	+6.0±4	(134), (16)	⎪
232	CH₃OCH₃		E	+8.6±2	(135), (16)	⎭
233	CH₃CH₂CN		E	+4.5±1.0	(136)	Johnson *et al.* (1977)
234	X-OGEN	78″	E	+9.5	(164), (165)	Ulich and Haas (1976)
235	CI*	2ʹ5	E	+10.5	(104), (23)	Phillips *et al.* (1980)

* Carbon is included for comparison.

Remarks to Table 3.3.VII:

(1) $V_{LSR} = (V_{Hel} - 18.1)$ km s^{-1}. The reported V_{LSR} refers to the core of OMC 1/KL Nebula. The exact po is often stated.

(2) Near infrared molecular hydrogen (H₂) emission at $\lambda = 2.1\mu$. $S(1) = (v = 1 \rightarrow 0, J = 3 \rightarrow 1)$.

(3) H₂ emission is collisionally excited in a thin schock-heated layer inside the cool molecular cloud.

(4) $\alpha = 5^h32^m46^s$, $\delta = -5°24'5''$.

(5) V_{LSR} of the peak line intensity at the edges of the cloud. The H₂, S(1) line emission can be described as composed of a blue and a red component. The analysis of the line profiles shows that the emitting expanding from a region near the BN-KL infrared sources at a typical velocity of 40 km s^{-1} but as high a km s^{-1} with respect to the bulk of the molecular cloud.

(6) Ground state transition $^2\Pi_{3/2}$, $J = \frac{3}{2}$, $F = 2-2$ at 1667.358 MHz.

(7) Ground state transition $^2\Pi_{3/2}$, $J = \frac{3}{2}$, $F = 2-1$ at 1720.533 MHz.

(8) Ground state transition $^2\Pi_{3/2}$, $J = \frac{3}{2}$, $F = 1-1$ at 1665.401 MHz.

(9) For the V_{LSR} of the various OH maser components and their positions see Table 3.4.II.

(10) V_{LSR} features at -7.0, -5.7, $+3.8$, $+7.1$, $+7.9$, $+8.6$, $+11.1$, $+15.7$, $+17.8$, $+18.9$, $+19.9$, $+20.9$, and · km s^{-1}.

(11) Maser source.

(12) $\alpha = 5^h32^m51^s$, $\delta = -5°25'39''$.

(13) Ground state transition $^2\Pi_{3/2}$, $J = \frac{3}{2}$, $F = 1$–2 at 1612.231 MHz.

(14) σ^-, σ^+ values respectively; Zeeman pattern magnetic field of several miligauss.

(15) Average of three lines of the ground state Λ-doublet ($^2\Pi_{1/2}$, $J = \frac{1}{2}$); the lines are: (1) $F = 1$–0 at 334? MHz; (2) $F = 1$–1 at 3335.481 MHz; and (3) $F = 0$–1 at 3263.794 MHz.

(16) $\alpha = 5^h32^m47^s$, $\delta = -5°24'21''$ (KL Nebula).

(17) $^{12}C^{16}O$, $J = 1$–0 at 115 271.2 MHz.

(18) $^{12}C^{16}O$, $J = 2$–1 at 230 537.97 MHz.

(19) $^{12}C^{16}O$, $J = 3$–2.

(20) $^{13}C^{16}O$, $J = 1$–0 at 110 201.4 MHz.

(21) $\alpha = 5^h32^m46^s.8$, $\delta = -5°24'24''$ (KL Nebula).

(22) The profile consists of a 'plateau' component (broad, $\Delta V \simeq 50$ km s^{-1}) and a 'spike' component (narrow, Δ km s^{-1}, $V_{LSR} = +9$ km s^{-1}).

(23) $\alpha = 5^h32^m47^s$, $\delta = -5°24'30''$ (KL Nebula).

(24) CN, $K = 1$–0, $J = \frac{3}{2} - \frac{1}{2}$, at 113 491 MHz.

(25) $\alpha = 5^h32^m47^s$, $\delta = -5°24'20''$ (KL Nebula).

(26) For the distribution of V_{LSR} over the nebula see the original paper.

(27) $^{12}C^{32}S$, $J = 1$–0 at 48 991 MHz.

(28) $^{12}C^{32}S$, $J = 2$–1 at 97 981 MHz.

(29) $^{12}C^{32}S$, $J = 3$–2 at 146 969 MHz.

(30) ^{32}SO, 2_2–1_1 at 86 100MHz.

(31) ^{32}SO, 2_3–1_2 at 109 300 MHz.

(32) ^{32}SO, 3_2–2_1 at 99 300 MHz.

(33) ^{32}SO, 4_3–3_2 at 138 200 MHz.

(34) ^{32}SO, 4_5–4_4 at 100 000 MHz.

(35) ^{34}SO, 3_2–2_1 at 97 715 MHz.

(36) SiO, $v = 0$, $J = 2$–1 at 86 847 MHz.

(37) SiO, $v = 0$, $J = 3$–2 at 130 268.4 MHz.

(38) SiO, $v = 1$, $J = 2$–1 at 86 243.27 MHz.

(39) SiO, $v = 1$, $J = 3$–2 at 129 363.1 MHz.

(40) $\alpha = 5^h32^m47^s$, $\delta = -5°24'25''$ (KL Nebula).

(41) Velocity features at -7.1, -5.8, -3.0, -1, $+9$, $+15.3$, $+17.3$ km s^{-1}.

(42) Sio, $v = 1$, $J = 1$–0 at 43 122 MHz.

(43) Sio, $v = 2$, $J = 1$–0 at 42 820.51 MHz.

(44) Velocity features at -6.7, -5.5, $+14.5$, $+16.2$, $+17.8$ km s^{-1}.

(45) H_2O, 6_{16}–5_{23} at 22 235.080 MHz (maser emission).

(46) For further details see Table 3.4.II.

(47) Velocity features at -50, -22, -6, 0, $+3$, $+4$, $+9$, $+11$, $+15$, $+17$, $+25$ km s^{-1}.

(48) Velocity features at $+0.8$, $+2.2$, $+3.4$, $+4.5$, $+6.7$, $+8.1$, $+9.0$, $+9.8$, $+10.6$, $+14.6$, $+17.4$ km s^{-1}.

(49) Velocity features at -6.7, -4.0, $+0.5$, $+3.4$, $+5.1$, $+6.4$, $+7.7$, $+9.3$, $+10.9$, $+15.0$, $+17.2$, $+18.5$, $+22.0$, $+25.3$, $+28.2$ km s^{-1}.

(50) $\alpha = 5^h32^m58^s.0 \pm 0^s.7$, $\delta = -5°7'25'' \pm 10''$ (maser source).

(51) Velocity features at $+7.6$, $+10.8$, $+14.8$, $+16.8$, $+26$ km s^{-1}.

(52) Velocity features at -1.7, $+2.7$, $+3.2$, $+5.3$, $+7.6$, $+8.3$, $+8.9$, $+9.9$, $+10.3$, $+10.8$, $+11.9$, $+12.4$, $+24.8$, $+27.1$, $+28.1$ km s^{-1}.

(53) H_2O, 1_{10}–1_{11} at 80 578 MHz.

(54) CH, $^2\Pi_{1/2}$, $J = \frac{1}{2}$, $F = 0$–1 at 3264 MHz.

(55) $H^{12}C^{14}N$, $J = 1$–0 at 88 631.85 MHz.

(55a) $F = 1$–1.

(55b) $F = 2$–1.

(55c) $F = 0$–1.

(56) $H^{13}C^{14}N$, $J = 1$–0 at 86 340.05 MHz.

(56a) $F = 2$–1.

(56b) $F = 1$–1.

(56c) $F = 0$–1.

(57) $H^{12}C^{15}N$, $J = 1$–0 at 86 055.05 MHz.

(58) $H^{12}C^{14}N$, $J = 3$–2 at 265 886.4 MHz.

(59) HNC, $J = 3$–2 at 271 982 MHz.

(60) $D^{12}C^{14}N$, $J = 2$–1 at 144 827.86 MHz.
(61) The assymetric line profiles (three peaks with high velocity wings) are probably due to small radial mo in OMC 1.
(62) $\alpha = 5^h32^m46^s$, $\delta = -5°24'25''$ (KL Nebula).
(63) HCO^+, $J = 3$–2 at 89 188.545 MHz.
(64) N_2H^+ at 93 173.58 MHz ($F = 1$–2, $F = 1$–1, and $F = 1$–0).
(65) OCS, $J = 8$–7 at 97 300 MHz.
(66) SO_2, 8_{17}–8_{08} at 83 688.074 MHz.
(67) The narrow component or 'spike' ($V_{LSR} = 8.5$ km s^{-1} and $\Delta V = 6$ km s^{-1}) is due to SO_2 present i 'ridge' of OMC 1; the underlying broad SO_2 'plateau' component ($\Delta V \simeq 30$ km s^{-1}) is probably d emission originating in a circumstellar envelope associated with a late type star embedded in the KL N
(68) $\alpha = 5^h32^m48^s$, $\delta = -5°24'18''$.
(69) $^{14}NH_3$; J, $K = 1$, 1, at 23 694.48 MHz.
(70) $^{14}NH_3$; J, $K = 2$, 2, at 23 722.71 MHz.
(71) $^{14}NH_3$; J, $K = 3$, 3 at 23 870.11 MHz.
(72) $^{14}NH_3$; J, $K = 4$, 4 at 24 129.39 MHz.
(73) $^{14}NH_3$; J, $K = 5$, 5.
(74) $^{14}NH_3$; J, $K = 6$, 6 at 25 056.04 MHz.
(75) $^{15}NH_3$; J, $K = 1$, 1 at ~23 GHz.
(76) $^{15}NH_3$; J, $K = 2$, 2 at ~23 GHz.
(77) $^{15}NH_3$; J, $K = 3$, 3 at ~23 GHz.
(78) One Gaussian fit.
(79) Two Gaussians; fit based on the assumption of two separate NH_3 clouds in OMC 1.
(80) $^{14}NH_3$, J, $K = 2$, 1 at ~ 23 GHz.
(81) $^{14}NH_3$, J, $K = 3$, 2 at ~ 23 GHz.
(82) $^{14}NH_3$, J, $K = 4$, 3 at ~ 23 GHz.
(83) $^{14}NH_3$, J, $K = 5$, 4 at ~ 23 GHz.
(84) Mean value at $\alpha = 5^h32^m46^s.8$, $\delta = -5°24'26''$ (KL Nebula).
(85) V_{LSR} of broad feature $= +5.5$ km s^{-1}, $\Delta V = 46$ km s^{-1}.
(86) V_{LSR} of molecular ridge is $+10.0$ km s^{-1} for a position $\Delta\delta = +40''$ from the core, and $+7.9$ km s^{-1} position $\Delta\delta = -40''$ from the core.
(87) V_{LSR} of broad feature $= \sim 0$ km s^{-1}, $\Delta V = 50$ km s^{-1}.
(88) V_{LSR} of molecular ridge is $+10.0$ km s^{-1} for a position $\Delta\delta = +40''$ from the core, and $+8.3$ km s^{-1} position $\Delta\delta = -40''$ from the core.
(89) V_{LSR} of broad feature $= +3$ km s^{-1}, $\Delta V = 55$ km s^{-1}.
(90) V_{LSR} of molecular ridge is $+8.8$ km s^{-1} for a position $\Delta\delta = +40''$ from the core, and $+7.0$ km s^{-1} position $\Delta\delta = -40''$ from the core.
(91) H_2CO, 1_{11}–1_{10} at 4829.66 MHz.
(92) H_2CO, 1_{01}–0_{00} at ~ 73 GHz.
(93) H_2CO, 3_{12}–2_{11} at 225 700 MHz.
(94) H_2CO, 2_{12}–2_{11}.
(95) $\alpha = 5^h32^m50^s$, $\delta = -5°25'24''$.
(96) H_2CS, 3_{13}–2_{12}.
(97) Assumed V_{LSR} for various transitions.
(98) HC_3N, 10–9 at 90 979 MHz.
(99) HC_3N, 14–13 at 127 368 MHz.
(100) HC_3N, 15–14 at 136 464 MHz.
(101) HC_3N, 16–15 at 145.561 MHz.
(102) CH_3OH, 6_2–6_1 (E_1) at 25 018.123 MHz.
(103) CH_3OH, 7_2–7_1 (E_1) at 25 124.873 MHz.
(104) $C I$ ($^3P_1$–$^3P_0$); $J = 1$–0 at 492 162.3 MHz.
(105) $\alpha = 5^h32^m47^s$, $\delta = -5°25'39''$.
(106) $\alpha = 5^h32^m46^s.55 \pm 0^s.5$, $\delta = -5°24'30'' \pm 8''$.
(107) Velocity features at $+7.2$, $+7.8$, and $+8.2$ km s^{-1}, the emission is probably due to weak maser acti
(108) CH_3OH, $J = 2$–1 at ~ 96.7 GHz.
(109) Assumed V_{LSR} for transitions 11_1–$10_2(A^-)$ and 5_0–4_1 (E) at ~ 76 GHz.
(110) CH_3CN, 5_0–4_0 at ~ 91 987 MHz.
(111) CH_3CN, 5_2–4_2 at ~ 91 979 MHz.
(112) CH_3CN, 5_3–4_3 at ~ 91 970 MHz.

(113) CH_3CN, 5_4-4_4 at $\sim 91\ 959$ MHz.

(114) CH_3CHO, $2_{12}-1_{01}$ at $\sim 83\ 584$ MHz.

(115) CH_3CCH, 5_0-4_0 at $\sim 85\ 457$ MHz.

(116) CH_3NH_2, s, $4_{14}-4_{04}$ at $\sim 86\ 075$ MHz.

(117) CH_3NHD, $2_{02}(+)-1_{10}(-)$ at $10\ 310$ MHz.

(118) CH_3OCH_3, $6_{06}-5_{15}$, EE at $90\ 937.539$ MHz.

(119) CH_3OCH_3, $2_{11}-2_{02}$, $A_1E + EA_1$ at $31\ 105.26$ MHz.

(120) CH_3OCH_3, $2_{11}-2_{02}$, EE at $31\ 106.20$ MHz.

(121) CH_3OCH_3, $2_{20}-2_{11}$, EA_1 at $86\ 225.67$ MHz.

(122) CH_3OCH_3, $2_{20}-2_{11}$, EE at $86\ 226.728$ MHz.

(123) CH_3OCH_3, $3_{13}-2_{02}$; AA, EE, AE + EA at $82\ 650$ MHz.

(124) CH_3OCH_3, $4_{22}-4_{13}$, AA at $82\ 691$ MHz.

(125) CH_3OCH_3, $4_{22}-4_{13}$, EE at $82\ 689$ MHz.

(126) CH_3OCH_3, $4_{22}-4_{13}$, AE + EA at $82\ 687$ MHz.

(127) CH_3OCH_3, $3_{21}-3_{12}$, EE at $84\ 634$ MHz.

(128) CH_3OCH_3, $3_{21}-3_{12}$, AE + EA at $84\ 632$ MHz.

(129) CH_3OCH_3, $2_{20}-2_{11}$, AE at $86\ 224$ MHz.

(130) CH_3OCH_3, $4_{23}-4_{14}$, EE at $93\ 857$ MHz.

(131) CH_3OCH_3, $4_{23}-4_{14}$, AE + EA at $93\ 854$ MHz.

(132) CH_3OCH_3, $5_{24}-5_{15}$, AA at $96\ 853$ MHz.

(133) CH_3OCH_3, $5_{24}-5_{15}$, EE at $96\ 850$ MHz.

(134) CH_3OCH_3, $5_{24}-5_{15}$, AE + EA at $96\ 847$ MHz.

(135) CH_3OCH_3, $4_{14}-3_{03}$, AA, EE, AE + EA at $99\ 325$ MHz.

(136) Value obtained from 24 CH_3CH_2N emission lines.

(137) OCS, $J = 7-6$.

(138) H_2S, $1_{10}-1_{01}$ at $168\ 700$ MHz.

(139) CH_3OH, $J = 3-2$, $\Delta K = 0$ at ~ 145.1 MHz.

(140) $\alpha = 5^h32^m46^s.9$, $\delta = -5°24'26''$ (KL Nebula).

(141) Profile consists of a narrow 'spike' of emission ($V_{LSR} \simeq +7.6$ km s^{-1}) superiumposed on a broad mol line emission ('plateau'; $\Delta V = 21.5$ km s^{-1}).

(142) $\alpha = 5^h32^m48^s$, $\delta = -5°30'$.

(143) From observations of the $J = 4, 5, 6, 7$, and 8, $\Delta J = 0$, $K = 2 \rightarrow 1$ transitions (at ~ 25 GHz); positior $5^h32^m48^s$, $\delta = -5°24'20''$ (KL Nebula).

(144) The emission peaks occur within the quoted velocity range.

(145) Bright features at $+1, +3, +5, +6.5, +9$, and $+11$ km s^{-1}.

(146) The profiles are described in terms of two Gaussians: (1) a narrow line ('spike', top value) with $\Delta V \simeq$ s^{-1}; and (2) a broad line ('plateau', bottom value) with $\Delta V \simeq 30$ km s^{-1}.

(147) CN, $K = 1-0$, $J = \frac{1}{2} - \frac{1}{2}$ at $113\ 170$ MHz.

(148) The excessive width of the SO line profile ($\simeq 30$ km s^{-1}) is interpreted as due to the effect of Zeeman sp which yields an upper limit of the magnetic field of the central features in Orion of 6 G (possibly redu a factor of 2 or 3 if kinematics and other factors are responsible for the observed width) (see however S 3.5).

(149) The spectrum is the result of the convolution of the narrow line which originates in the extended molecular ($V_{LSR} \simeq 9$ km s^{-1}) and the broad line which originates in a small region ($< 30''$) centred on $\alpha = 5^h32^r$ $\delta = -5°24'19''$ (about 5'' N of KL). The high velocity gas (broad component $\Delta V \simeq 50$ km s^{-1}) pro originates near or in front of the molecular cloud, has a kinetic temperature $T_k \gtrsim 100$ K, a mass $M_{H_2} \gtrsim 1$ and a density $n_{H_2} \gtrsim 5 \times 10^5$ cm^{-3}.

(150) V_{LSR} of 'spike' around 8 km s^{-1} superimposed on a broader feature with $\Delta V \simeq 15$ km s^{-1}.

(151) Broad line ammonia probably originates in expanding shells enveloping the stars of the KL infrared c The narrow line at $+9$ km s^{-1} originates in the unresolved NH_3 'core' which is external to these env and is heated by the radiation emitted from the cluster. The 'core' region is in turn embedded in the density 'ridge' of the OMC 1; (Figure 3.3.11). The NH_3 'core' coincides within the observational error IRc 4 (Rieke et al., 1973) which consistitutes the core of KL (Wynn-Williams and Becklin, 1974) ai various OH, H_2O, SiO maser sources.

(152) Feature superimposed on a broader velocity component (plateau) with $\Delta V = 30$ km s^{-1}.

(153) Feature superimposed on a broader velocity component (plateau) with $\Delta V = 28$ km s^{-1}.

(154) $\alpha = 5^h32^m43^s$, $\delta = -5°24'11''$.

(155) C_2H; $N = 1 \rightarrow 0$ at $87\ 300$ MHz.

(156) HCO^+: $J = 3-2$ at $267\ 600$ MHz.

(157) The HCO$^+$, $J = 3\text{-}2$ profile shows a weak broad velocity emission feature ± 25 km s^{-1} from the line c The V_{LSR} of the line centre has been taken from Turner and Thaddeus (1977).

(158) Molecular hydrogen emission at 12.28μ ($v = 0 \rightarrow 0$, S(2) line).

(159) The line profiles at various positions are similar; their intrinsic widths are of the order of 55 km s^{-1}.

(160) H$_2$O; $3_{13}\text{-}2_{20}$ rotational transition at 183 GHz.

(161) The profile consists of a narrow 'spike' centred at ~ 9.5 km s^{-1} with $\Delta V = 4$ km s^{-1} and a broader 'pl: centred at ~ 8 km s^{-1}. The 'plateau' emission appears to originate in the same region as the plateau em of other molecules (size $< 1'$); the 'spike' originates in the Orion molecular ridge (size $\simeq 3' \times 7'$).

(162) H$_2^{18}$O; $3_{13}\text{-}2_{20}$ at 203 407.5 MHz.

(163) With a 3.5 error pattern surrounding the central lobe.

(164) Unknown; probably HCO$^+$ $v = 0$, $J = 1\text{-}0$ (Klemperer, 1970; Buhl and Snyder, 1970; Snyder et al., 1975

(165) At 89 188.5 MHz (3.36 mm).

(166) Features at -8.0, -6.9, $+13.3$, $+14.7$, $+16.4$ km s^{-1}.

(167) ^{12}C^{18}O, $v = 0$, $J = 1\text{-}0$ at 109 782.18 MHz.

(168) ^{32}SO, $1_0\text{-}0_1$ at 30 GHz.

(169) $\alpha = 5^h32^m47^s$, $\delta = -5°24'23''$.

(170) SiO, $v = 0$, $J = 1\text{-}0$ at 43 GHz.

(171) The narrow 'spike' at -5.5 is probably maser emission; this is superimposed on a broad flat-topped em feature ($\Delta V \simeq 35 \pm 5$ km s^{-1}) which is due either to optically thin thermal emission or to weak maser em originating in an expanding circumstellar envelope.

(172) H$_2$CO; $2_{12}\text{-}1_{11}$ at 140 839.3 MHz.

(173) H$_2$CO; $2_{11}\text{-}1_{10}$ at 150 500 MHz.

(174) H$_2$CO; $2_{02}\text{-}1_{01}$ at 145 600 MHz.

(175) $\alpha = 5^h32^m46^s.9$, $\delta = -5°23'54''$ (KL Nebula).

(176) HCN; $J = 4\text{-}3$ at 345 MHz.

(177) V_{LSR} of broad feature ('plateau' component): $+11.9$ km s^{-1}, $\Delta V = 18$ km s^{-1}.

(178) $\alpha = 5^h32^m47^s$, $\delta = -5°24'18''$ (Becklin's star).

(179) HNC, $J = 1\text{-}0$ at 90 663.59 MHz.

(180) HN^{13}C, $J = 1\text{-}0$ at 87 090.85 MHz.

(181) DNC, $J = 2\text{-}1$ at 152 609.77 MHz.

(182) Originates in the dense central ridge of OMC 1.

(183) Probably originates in the less massive neutral cloud lying in front of the H II region (Zuckerman and Ball, 1

(184) ^{13}CO; $J = 2\text{-}1$ at 220 398.7 MHz.

(185) $\alpha = 5^h32^m46^s.8$, $\delta = -5°24'32''$.

(186) H$_2$CO; $J_{k-1} - J_{k+1} = 3_{12} \rightarrow 3_{13}$ at 28 974.804 MHz.

(187) HCO$^+$, $J = 1\text{-}0$.

(188) HCN, $J = 1\text{-}0$, $F = 2\text{-}1$ at 88 631.85 MHz.

(189) CCH, $J = \frac{3}{2}\text{-}\frac{1}{2}$, $F = 2\text{-}1$.

(190) C^{18}O, $J = 2\text{-}1$ at 219 GHz.

(191) Maser emission; velocity features at -7, -5, $+1$, $+2$, $+5$, $+13.5$, $+17$, and $+19.5$ km s^{-1} with width: 1.5, 1, 1, 1, 1, 1.5, and 1 km s^{-1}, respectively; a weak maser emission pedestal at $V_{LSR} = -6.9 \pm 0.1$ and 5.1 ± 0.2 km s^{-1} has also been detected.

(192) NH$_3$; $J, K = 4, 2$.

(193) NH$_3$; $J, K = 6, 5$.

(194) NH$_3$; $J, K = 7, 6$.

(195) NH$_3$; $J, K = 8, 7$.

(196) NH$_3$; $J, K = 9, 8$.

(197) NH$_3$; $J, K = 10, 9$.

(198) The profile consists of: (1) a 'spike' component at 7.2 km s^{-1} with $\Delta V = 3.1$ km s^{-1} (at the same r velocity with that of the surrounding molecular cloud; (2) a 'plateau' component at 4.4 km s^{-1} with $\Delta V =$ km s^{-1}; and (3) a lower velocity feature at ~ 3.8 km s^{-1}.

(199) Peak V_{LRS} of 'spike' + 'lower velocity' components (note (198)); average $V_{LSR} \simeq +4.8 \pm 1.2$ km s^{-1}; ('plateau') $= 5.3 \pm 1.8$ km s^{-1}.

(200) The profile is dominated by the low velocity feature (see note (198)).

(201) Peak value; average $V_{LSR} = 5 \pm 1.5$ km s^{-1}; all components are present (see note (198)).

(202) CH$_3$OH; $J_{k'}\text{-}J_k = 4_2\text{-}4_1$ at 24 993.468 MHz.

(203) CH$_3$OH; $J_{k'}\text{-}J_k = 5_2\text{-}5_1$ at 24 959.080 MHz.

(204) Maser source; the cited velocity features are the strongest components.

(205) HC$_3$N, V_7, $J = 10\text{-}9$ at 91 202.6 MHz.

(206) HC_3N, V_7, $J = 10-9$ at 91 332.2 MHz.
(207) $\alpha = 5^h32^m46^s.1$, $\delta = -5°24'5''$.
(208) The close agreement of the observed velocity with that of the extended molecular cloud indicates that the H_2, S(1) line originates in the molecular cloud.
(209) True line width of the 'plateau' component < 30 km s^{-1}.
(210) CH_3OH; $J = 2$, $K = 2-1$ (E) at 24 934.382 MHz.
(211) CH_3OH; $J = 3$, $K = 2-1$ (E) at 24 928.70 MHz.
(212) CH_3OH; $J = 4$, $K = 2-1$ (E) at 24 933.468 MHz.
(213) CH_3OH; $J = 6$, $K = 2-1$ (E) at 25 018.12 MHz.
(214) CH_3OH; $J = 7$, $K = 2-1$ (E) at 25 124.87 MHz.
(215) CH_3OH; $J = 10$, $K = 2-1$ (E) at 25 878.18 MHz.

TABLE 3.3.VIII

Radial velocities of OMC 2 derived from molecular lines

No.	Molecular line	Beam	E= emission A= absorption	$V_{LSR}^{(1)}$ (km s^{-1})	Remarks	References
1	CO		E	+11	(18), (13)	Loren (1979)
2	CO		E	+11	(2), (3)	Plambeck and Williams (1979)
3	CO	2''.25	E	+10	(5), (2)	Goldsmith et al. (1975)
4	CO	2''.25	E	+12	(5), (17)	Goldsmith et al. (1975)
5	CH		E	+11	(9)	Rydbeck et al. (1975)
6	CS	2''.6	E	+11.3	(14), (15)	Linke and Goldsmith (1980)
7	CS	2''.1	E	+11.4	(14), (16)	Linke and Goldsmith (1980)
8	H_2O		E	+11.6	(12), (13)	Morris and Knapp (1976)
9	HCN	78''	E	$+11.14\pm0.10$	(4), (5)	Morris et al. (1974)
10	HCN		E	+10.9	(5), (6)	Turner and Thaddeus (1977)
11	HCO^+		E	+11.42	(5), (6)	Turner and Thaddeus (1977)
12	N_2H^+		E	+11.75	(5), (6)	Turner and Thaddeus (1977)
13	NH_3	78''	E	$\sim +10$	(7), (5)	Morris et al. (1974)
14	NH_3		E	+11.1	(7), (11), (6)	Ho et al. (1979)
15	NH_3		E	+11.2	(10),(11), (6)	Ho et al. (1979)
16	H_2CO		A	+10		Kutner and Thaddeus (1971)
17	CH_3OH		E	$+10.8\pm0.2$	(8), (5)	Gottlieb et al. (1979)
18	C I*	2''.5	E	+12	(19)	Phillips et al. (1980)

* Carbon is included for comparison

Remarks to Table 3.3.VIII:

(1) $V_{LSR} = (V_{Hel}-18.1)$ km s^{-1}. Values refer to the core of the Orion Molecular Cloud 2 (OMC 2) which lies 12' NE of the Trapezium. The exact position is often quoted.
(2) ^{12}CO, $J = 2-1$ at 230 537.97 MHz.
(3) $\alpha = 5^h32^m59^s.1$, $\delta = -5°12'10''$.
(4) HCN, $J = 1-0$.
(5) $\alpha = 5^h33^m00^s$, $\delta = -5°12'34''$.
(6) For the distribution of V_{LSR} over OMC 2 see the original paper.
(7) NH_3; $J, K = 1, 1$.
(8) CH_3OH; $J = 2-1$.
(9) Average of three lines ($F = 1-0, 1-1, 0-1$) of the ground state Λ-doublet ($^2\Pi_{1/2}$, $J = \frac{1}{2}$).
(10) NH_3, $J, K = 2, 2$.
(11) Core of OMC 2; $\Delta\alpha = 3'$ N, $\Delta\delta = 12'$ E of KL ($\alpha = 5^h32^m47^s$, $\delta = -5°24'20''$).
(12) H_2O; $6_{16}-5_{23}$ at 22 235.080 MHz.
(13) Water maser; see also Table 3.4.II.
(14) $\alpha = 5^h32^m59^s.1$, $\delta = -5°12'00''$.
(15) $^{12}C^{36}S$; $J = 1-0$ at 48 991 MHz.
(16) $^{12}C^{36}S$; $J = 2-1$ at 97 981 MHz.
(17) ^{13}CO; $J = 2-1$ at 220 398.7 MHz.
(18) ^{12}CO; $J = 1-0$ at 115 271.2 MHz.
(19) C I ($^3P_1 - ^3P_0$); $J = 1-0$ at 492 162.3 MHz.

TABLE 3.3.IX

Full width at half intensity, ΔV (km s⁻¹), of various lines detected in M42/OMC 1[a]

Lines originating in the H II region (M42)			Lines originating in the partially ionized medium (H I–H II interface layer)		Lines originating in the neutral/molecular complex (OMC 1)			
Forbidden lines	H–lines	He lines	C lines	Heavy element (Z) lines	H I (21 cm) line	Narrow lines from diatomic molecules[b] ('spike')	Narrow lines from polyatomic molecules[b] ('spike')	Broad molecular lines[b] ('plateau')
[N II]: 20[56]	$H\alpha$: 30[57]	He I: 24[56]	C75α: 3[10]	Z109α: 9[4]	H I: 4[20]	CH: 6[23]	H_2O: 4[54]	H_2: 18–58[22]
[O I]: 15[56]	$H\beta$: 30[57]	He85α: 20[3]	C85α: 6[11]	Z109α: 6[5]	H I: 2–6[21]	^{12}CO: 7[24]	HDO: 9[36]	H_2: 55[35]
[O III]: 18[56]	H39α: 28[1]	He85α: 18[4]	C85α: 4[3],[4]			^{13}CO: 6[24]	H_2S: 7[37]	CO: ~50[31]
[S II]: 22[56]	H76α: 25[2]	He109α: 20[4]	C92α: 4[12]			CN: 4[25]	HCN: 4[38]	SO: ~25[32]
	H85α: 27[3],[4]	He109α: 23[5]	C92α: 6[13]			CN: 5[26],[27]	HCN: 2[39]	SO: ~20[33]
	H109α: 30[4],[5]	He111α: 24[9]	C94α: 11[14]			CS: 5[28]	HCO^+: 4[40]	SO: 20–30[30]
	H126α/127α: 38[6]	He126α: 27[9]	C109α: 4[15]			CS: 4[29]	N_2H^+: 3[40]	SiO: 30[34]
	H166α: 39[7]	He166α: 10[7]	C109α: 5[5]			SO: 4–9[30]	C_2H: 4[41]	SiO: 22[35]
	H183α: 40[7]	He137β: 25[5]	C109α: 8[4]				NH_3: 3–6[42],[44]	H_2S: >25[37]
	H198α: 40[7]	He137β: 14[4]	C111α: 4[9]				NH_3: 4[43]	SO_2: ~30[50]
	H210α/211α: 13[7]		C126α: 6[9]				H_2CO: 5[45]	NH_3: ~50[44]
	H95β: 29[8]		C134α: 5[16]				H_2CO: 4[46]	H_2O: 15[54]
	H137β: 34[6]		C157α: 13[17]				H_2CO: 2[7]	
			C158α: 9[18]				CH_3OH: 2[47]	
			C166α: 9[18]				CH_3OH: 4[48]	
			C166α: 10[7]				CH_3OH: 1[49]	
			C183α: 5[7]					
			C198α: 16[19]					
			C210α/211α: 12[7]					
			C220α: 16[19]					

Remarks to Table 3.3.IX:

(a) V_{LSR} of H and He lines ~ −3 km s⁻¹, V_{LSR} of C and molecular lines ~ +9 km s⁻¹ (see Tables 2.2.XIX and 3.3.VII).

(b) The narrow line of 'spike' component ($V \sim$ +9 km s⁻¹, $\Delta V \sim$ 4 km s⁻¹) originates in the diffuse molecular cloud OMC 1 in and around the KL nebula ('ridge'); the broad line or 'plateau' component originates in the core of KL; the broad molecular emission indicates a differentially expanding envelope which is possibly the outcome of an explosive event (see also Section 3.3 and notes (5), (22), (67), (149), and (151) to Table 3.3.VII).

TABLE 3.3.X
Full width at half intensity (ΔV, km s^{-1})
of various lines detected in M43/OMC 2[a]

Line	ΔV (km s^{-1})
H76α	20[8]
H109α	22[4],[5]
H137β	23[5]
C109α	3[4]
CO	3[51]
CH	4[52]
HCN	2[53]
NH$_3$	1[53]

(a) V_{LSR} of H lines (M43) $\sim +7$ km s^{-1}
(Table 2.2.XX); V_{LSR} of molecular lines
(OMC 2) $\sim +11$ km s^{-1} (Table 3.3. VIII).

References to Table 3.3.IX and 3.3.X:

(1) Gottlieb *et al.* (1978).
(2) Perrenod *et al.* (1977).
(3) Balick *et al.* (1974).
(4) Jaffe and Pankonin (1978).
(5) Churchwell *et al.* (1978).
(6) McGee and Gardner (1968).
(7) Zuckerman and Ball (1974).
(8) Pankonin *et al.* (1979).
(9) Boughton (1978).
(10) Kuiper and Evans (1978).
(11) Gordon and Churchwell (1970).
(12) Chaisson (1974).
(13) Chaisson (1974).
(14) Chaisson (1972).
(15) Churchwell (1970).
(16) Zuckerman and Palmer (1970).
(17) Churchwell and Edrich (1970).
(18) Chaisson and Lada (1974).
(19) Pedlar and Davies (1972).
(20) Radakrishnan *et al.* (1972).
(21) Lockhart and Goss (1978).
(22) Nadeau and Geballe (1979).
(23) Rydbeck *et al.* (1973).
(24) Wilson *et al.* (1974).
(25) Penzias *et al.* (1974).
(26) Turner and Gammon (1975).
(27) Turner and Thaddeus (1977).
(28) Turner *et al.* (1973).
(29) Liszt and Linke (1975).
(30) Gottlieb *et al.* (1978).
(31) Kwan and Scoville (1976), Zuckerman *et al.* (1976), Wannier and Phillips (1977), Phillips *et al.* (1977).
(32) Clark and Johnson (1974).
(33) Gottlieb and Ball (1973).
(34) Dickinson *et al.* (1976).
(35) Lada *et al.* (1978).
(36) Turner *et al.* (1975).
(37) Thaddeus *et al.* (1972).
(38) Snyder and Buhl (1973).
(39) Gottlieb *et al.* (1975).
(40) Turner and Thaddeus (1977).
(41) Tucker *et al.* (1974).
(42) Morris *et al.* (1973).
(43) Ho and Barrett (1978).
(44) Wilson *et al.* (1979).
(45) Kutner *et al.* (1971).
(46) Whiteoak and Gardner (1974).
(47) Barrett *et al.* (1971).
(48) Kutner *et al.* (1973).
(49) Chui *et al.* (1974).
(50) Snyder *et al.* (1975).
(51) Gatley *et al.* (1974).
(52) Rydbeck *et al.* (1975).
(53) Morris *et al.* (1974).
(54) Waters *et al.* (1980).
(55) Beck *et al.* (1979).
(56) Balick *et al.* (1980).
(57) Wilson *et al.* (1959).

TABLE   3.3.XI

Physical parameters of the 'plateau' source ($<1'$ high velocity core of OMC 1)
and the 'hot core' component

Source	Molecule observed	Size (arc sec)	Temperature T_k(K)	H_2 density n_{H_2}(cm^{-3})	Mass $M(M_\odot)$	Probable association	References
Plateau	CO	37–52	~100	$>5 \times 10^5$	>10		Phillips *et al.* (1977)
Plateau	HCN	29	~100	2×10^6			Huggins *et al.* (1979)
Plateau	HCN	<30	~250	1×10^6			Padman *et al.* (1980)
Plateau	CO	37			~15	IRc 1(BN)	Scoville (1980)
Hot core	NH$_3$	~6	>220	$>5 \times 10^7$		IRc 2(IRS 3)	Morris *et al.* (1980)

TABLE   3.4.I

Characteristics of masers associated with H II regions (according to Moran, 1980)

Quantity	H_2O	OH	SiO	CH_3OH
Transitions observed	1	9	4	8
Number known	170	~100	1[1]	1[2]
Linewidth (km s^{-1})	0.5–2	0.1–1	1–2	0.5–2
T_k (K)[3]	100–1500	5–500	3500	150
Number of spectral features	1–200	1–50	~5	~10
Velocity range (km s^{-1})	1–400	1–30	25	4
Polarization (%)	linear (0–20)	linear (0–100) circular (0–100)	none	none known
Lifetime of feature[4] (s)	10^6–10^8	10^7–10^8	$>10^7$	$>10^7$
Spot size (cm)	10^{13}–10^{14}	10^{14}–10^{15}	10^{14}	~10^{16}
T_B (K)	10^{13}–10^{15}	10^{12}–10^{13}	10^9	10^3–10^4
Cluster size (cm)	10^{16}–10^{17}	10^{16}–10^{17}	$<10^{15}$	3×10^{17}
Power[5] (ergs s^{-1})	10^{25}–10^{33}	10^{25}–10^{30}	10^{29}	10^{28}

Remarks to Table 3.4.1:

(1) The only source known is in Orion A. The classification of the SiO maser is controversial.
(2) The only source known is in Orion A.
(3) Assuming no line narrowing or mass motions.
(4) There are some cases of shorter time scales.
(5) Assuming isotropic radiation. 10^{25} ergs s^{-1} is the sensitivity limit for present radio telescopes for masers at a distance of 300 pc.

Remarks to Table 3.4.II:

(1) Relative positions of the various velocity features can be found in the following papers:

OH   masers: {Raimond and Eliasson (1969)
{Norris *et al.* (1980)

H_2O masers: {Goss *et al.* (1976)
{Moran *et al.* (1977)
{Genzel *et al.* (1978)

SiO   masers: {Moran *et al.* (1977)
{Genzel *et al.* (1979)

CH$_3$OH masers: Hills *et al.* (1975).

(2) OH, $^2\Pi_{3/2}$, $J = \frac{3}{2}$, $F = 1$–1 at 1665.402 MHz.
(3) Maser clusters within 1″ of each other.
(4) H_2O, 6_{16}–5_{23} at 22 235.08 MHz.
(5) V_{LSR} centroid; 'source A' associated with KL Nebula.
(6) V_{LSR} centroid; 'source B' associated with KL Nebula.

TABLE 3.4.II

Positions of masers[1] in OMC 1 and OMC 2

No.	Maser	Velocity feature V_{LSR} (km s^{-1})	Coordinates (1950) α	δ	Remarks	References
			a. OMC 1			
1	OH	+6.8/+7.9	5^h 32^m 46^{s}8±0.1	−5° 24′ 25″±3		
2	OH	+4.0/+4.5	5 32 47.0±0.2	−5 24 19±3		Raimond and Eliasson
3	OH	+17.6 to +23.7	5 32 46.9±0.2	−5 24 18±3		(1969)
4	OH	+7.1/+8.6, +3.8	5 32 46.85±0.1	−5 24 29±5	(2), (3)	Hansen et al. (1977)
5	OH	+7.1/+8.6	5 32 46.76±0.04	−5 24 27±5	(2), (15)	Norris et al. (1980)
6	H$_2$O	+7.6	5 32 46.8±0.2	−5 24 26±3		
7	H$_2$O	+9.2/+10.8	5 32 47.8±0.2	−5 24 8±3	(4)	Hills et al. (1972)
8	H$_2$O	+0.95	5 32 47.7±0.2	−5 24 0±3		
9	H$_2$O	+11.0	5 32 47.76±0.16	−5 24 8±3	(4)	Baudry et al. (1974)
10	H$_2$O	+5.5	5 32 46.71±0.05	−5 24 28±1	(4), (5), (12)	Genzel and Downes (1
11	H$_2$O	+18.0	5 32 46.57±0.05	−5 24 31.5±1	(4), (6), (7)	
12	H$_2$O	+10	5 32 58.0±0.7	−5 07 25±10	(4)	Cesarsky et al. (1978)
13	H$_2$O	+7.6	5 32 46.59±0.02	−5 24 31.5±5		
14	H$_2$O	+10.8	5 32 47.58±0.02	−5 24 9.3±5		
15	H$_2$O	+14.8	5 32 46.92±0.02	−5 24 22.6±5	(4)	Forster et al. (1978)
16	H$_2$O	+16.8	5 32 47.00±0.02	−5 24 23.5±5		
17	H$_2$O	+26	5 32 46.33±0.02	−5 24 33.0±5		
18	H$_2$O		5 32 47.0	−5 24 23	(4), (8)	Genzel et al. (1979)
19	H$_2$O	+18 to +19.5	5 32 46.82±0.02	−5 24 30.6±0.5		Downes (1980)
20	H$_2$O	+6.5	5 32 46.37±0.02	−5 23 53.2±0.5		
21	H$_2$O	+69 to +70.5	5 32 46.33±0.02	−5 23 54.7±0.5	(16)	Downes (1980)
22	H$_2$O	+73.1	5 32 46.33±0.02	−5 23 54.5±0.5		
23	SiO	+16.8	5 32 46.9±0.8	−5 24 24±12	(9), (8)	Moran et al. (1977)
24	CH$_3$OH	+7.8	5 32 46.55±0.5	−5 24 30±8	(10), (11)	Hills et al. (1975)
			b. OMC 2			
25	H$_2$O	+12	5 32 59.9±4	−5 11 29±4	(4),(13),(14)	Genzel and Downes (

(7) 'Source B'; strong low velocity features clustered near the core of the KL Nebula; diameter < 2″ (≃ 10^{16} cr

(8) 'Shell' source with stable velocity features at ∼ −6 and +16 km s^{-1}; it is probably associated with the ci stellar envelope of IRc 2 (IRS 3)/OMC 1; see Table 3.4.VI.

(9) SiO, $v = 1$, $J = 1-0$ at 43 122.03 MHz.

(10) CH$_3$OH, $K = 2-1$, $J = 6$, and $J = 7$ at 25 018.123 MHz and 25 124.873 MHz, respectively; the emiss probably produced by weak maser action.

(11) CH$_3$OH in the $J = 4$, 5, 6, 7, and 8 transitions was first detected by Barrett et al. (1971). Chui et al. (1974) detected temporal variations in the $J = 6$ and 7 transitions and suggested the possibility of maser action. poral variations in the emission of CH$_3$OH have also been detected in the $J = 4$, 5, 6, and 7 transitic Barrett et al. (1975). More recently Matsakis et al. (1980) have detected ten CH$_3$OH components.

(12) 'Source A'; strong low velocity features clustered near the core of the KL Nebula; diameter < 2″ (≃10^{1} probably associated with IRc 4/OMC 1; see Table 3.4.VI.

(13) Group of lines with a 'triple' profile spread over 16 km s^{-1} (observations made in November, 1977).

(14) The H$_2$O position agrees with the position of IRS 4/OMC 2 (Gatley et al., 1974); the H$_2$O maser source prc originates in the shell of a young B star (mass ∼5$M_\odot$).

(15) The two dominant features (7.1 km s^{-1} RHC polarisation and 8.6 km s^{-1} LHC polarisation) probably con a Zeeman doublet emitted by one source. Their mean velocity (7.8 km s^{-1}) is close to that of the OI (∼8 km s^{-1}). The OH source is probably associated with the H$_2$O master 'source A' and the IRc 4/ON see Table 3.4.VI.

(16) Associated with the IRS 2/OMC 1; see Table 3.4.VI.

TABLE 3.4.III

Polarization of masers in Orion (OMC 1)

No.	Maser	Velocity feature V_{LSR} (km s^{-1})	Linear polariza-tion $P(\%)$	Position angle $\theta(°)$	Circular polarization ($+$ = right, $-$ = left) ($q\%$)	Remarks	References
1	OH	+7.1			+60		
2	OH	+8.6			−80	(1), (2)	Hansen *et al.* (1977)
3	OH	+3.8			+40		
4	H$_2$O	+11	17±1	42±5		(3),(4)	Bologna *et al.* (1975)
5	H$_2$O	+9	8±1	65±10			

Remarks to Table 3.4.III:

(1) The +8.6 and +7.1 components of the OH (1665) MHz maser are spatially close (0″06 apart) and circularly polarized in opposite senses; these are probably Zeeman components originating in a cloud moving with $V_{LSR} = +7.9$ km s^{-1} and having a longitudinal magnetic field of 2.5 mG. An alternative idea is that, since all three components have a diameter of ∼400 AU, they originate in the infalling envelope of a newly forming $4M_\odot$ star.

(2) Analytical information of the various OH (1665) features:

OH feature	V_{LSR} (km s^{-1})	Diameter (AU)	Flux density S (1977), Jy	Circular polarization (%)
1	− 7.0	>70	7	+30
2	− 5.7	>90	8	+20
3	+ 3.8	13 (core)	16	+40
4	+ 7.1	25	44	+60
5	+ 7.9	−	9	−20
6	+ 8.6	25	38	−80
7	+11.1	>30	8	−30
8	+15.7	>70	9	−20
9	+17.8	>70	12	+20
10	+18.9	>70	19	−20
11	+19.9	>70	22	0
12	+20.9	>70	24	+20
13	+22.6	>70	15	0

(3) See also Sullivan (1971) for strong linear polarization (up to 50%) of the features +1, +3, and +9 km s^{-1}; the features +1 and +3 were < 1% polarized at the time of their study; the degree of polarization and the intensity of the various features vary considerably with time.

(4) The cause of polarization is probably a large mangetic field (magnetic field of a protostellar cloud form2d by the the the collapse of interstellar gas).

TABLE 3.4.IV

Typical 'maser spot' sizes in Orion (KL Nebula)

No.	Maser and comparison sources	Angular diameter (arc sec)	Order of linear diameter for a distance of 500 pc		Remarks	References
			$(10^{13}$ cm$)$	(AU)		
1	H_2O	0.006	5	3	(1), (2)	Johnston *et al.* (1977)
2	OH	0.05	40	25	(3)	Hansen *et al.* (1977)
3	SiO	0.02	15	10	(4)	Genzel *et al.* (1979)
4	CH_3OH	7	5200	3500	(5)	Matsakis *et al.* (1980)
5	KL Nebula	30	22500	15000		
6	Planetary system		120	80		

Remarks to Table 3.4.IV:

(1) $> 0\overset{''}{.}004 – 0\overset{''}{.}008$.

(2) Apart from the H_2O features associated with IRc 2 which have an angular diameter in the range of $0\overset{''}{.}03$ to $0\overset{''}{.}13$ (i.e larger by a factor of ten than that of the other H_2O maser features; Genzel *et al.* (1979)).

(3) The size of the various features is in the range of $0\overset{''}{.}025$ to $> 0\overset{''}{.}2$.

(4) $0\overset{''}{.}02 \pm 0\overset{''}{.}01$.

(5) Ten components have been detected: six are unresolved in RA; the size of the rest are $6'' – 8''$.

TABLE 3.4.V

Sizes of masering regions in Orion (KL nebula)

No.	Maser	Velocity feature V_{LSR} (km s^{-1})	Size of masering region		Remarks	References
1	H_2O	-10 to $+30$	9 or more clusters, $2''$ each		(1)	Genzel and Downes (1977)
2	H_2O	-90 to $+80$	zone of $60''$ in diameter, centred at KL		(2)	Genzel *et al.* (1978)
3	OH	7.1/8.6	maser spot, $\sim 0\overset{''}{.}05$		(3)	
4	OH	$+2$ to $+11$	elongated cluster $\sim 1'' \times 3''$		(4)	Norris *et al.* (1980)
5	OH	$+17$ to $+23$	cluster $\sim 2'' \times 7''$		(5)	
6	SiO	$-6/+19$	'shell source' $\sim 0\overset{''}{.}05 \times 0\overset{''}{.}5$		(6)	Genzel *et al.* (1979)
7	CH_3OH	$+7$ to $+10$	10 clumps distributed over a $8'' \times 45''$ area around KL		(7)	Matsakis *et al.* (1980)

Remarks to Table 3.4.V:

(1) Strong, low velocity features concentrated in 'centres of activity' or clusters near the core of KL; the mean size of each cluster is $2''$; one of them, 'source A', is probably associated with IRc 4/OMC 1; another cluster the 'H_2O shell source' is probably associated with the SiO maser and IRc 2; see Table 3.4.VI. All these clusters are probably associated with circumstellar expanding shells enveloping certain young massive stars. The H_2O/SiO/IRc 2 may however be an evolved giant or supergiant.

(2) Weak high velocity features spread over an area of $60'' \times 60''$ centred on KL; they may be dense fragments blown out by intense stellar winds (emitted from stars which have reached the last stages of their formation).

(3) Maser spot associated with IRc 4/OMC 1; two Zeeman splitted components, spatially closed ($0\overset{''}{.}06$) apart), of a cloud with $V_{LSR} \simeq 7.8$ km s^{-1}.

(4) Cluster very near to the OH maser spot (note (3)); also associated with IRc 4/OMC 1; see Table 3.4.VI.

(5) Cluster probably associated with IRc 2(IRS 3)/OMC 1; see Table 3.4.VI.

(6) SiO 'shell source' probably associated with the H_2O 'shell source' and the infrared star IRc 2 (IRS 3)/OMC 1; see Table 3.4.VI.

(7) The CH_3OH masers do not coincide with the H_2O, OH, and SiO masers; they are denser concentrations embedded in OMC 1 and are excited by the emission of the infrared objects of the IR/OMC 1 cluster.

TABLE 3.4.VI

Association of IR and maser sources in the Orion Complex (OMC 1, OMC 2)

No.	IR/maser source	Velocity feature V_{LSR} (km s⁻¹)	Coordinates (1950) α	δ	Probable nature	References
			a. OMC 1			
1	IRc 2 (IRS 3)		5ʰ 32ᵐ 47ˢ.00	−5°24'24"	Evolved giant or supergiant[1]	Rieke et al. (1973), Wynn-Williams and Becklin (1974)
2	H₂O 'shell source'	−6, +16	5 32 47.00	−5 24 23		Genzel et al. (1979)
3	SiO 'shell source'	−6, +19	5 32 46.90	−5 24 24		Moran et al. (1979)
4	IRc 4		5 32 46.80	−5 24 29		Rieke et al. (1973)
5	H₂O 'source A'	3.4, 7.6	5 32 46.71	−5 24 28	Massive young star with circumstellar envelope ($M>10$ $M_\odot$)	Genzel and Downes (1977)
6	OH	7.1/8.6	5 32 46.76	−5 24 27		Norris et al. (1980)
7	OH	7.1/8.6	5 32 46.80	−5 24 25		Raimond and Eliasson (1969)
8	OH	7.1/8.6	5 32 46.85	−5 24 29		Hansen et al. (1977)
9	IRS 2		5 32 46.30	−5 23 55		Wynn-Williams and Becklin (1974)
10	H₂O	+6.5	5 32 46.37	−5 23 53	Young star(s) with circumstellar envelope	Downes and Genzel (1980)
11	H₂O	+69/+70.5	5 32 46.33	−5 23 55		Downes (1980)
12	H₂O	+73.1	5 32 46.33	−5 23 55		
			b. OMC 2			
13	IRS 4		5 32 59.9	−5 11 29	Young star ($M\sim M_\odot$; B5?)	Gatley et al. (1974)
14	H₂O	+12				Genzel and Downes (1979)

Remark to Table 3.4.VI:

(1) Source associated with a broad ($\Delta V \sim 35$ km s⁻¹) SiO ($v = 0$, $J = 1$–0) velocity feature which is probably thermal (Genzel et al., 1980); it is also associated with the low velocity (~ 3 km s⁻¹) 'hot core' component which appears in the NH₃ profiles (Morris et al., 1980; see also Figure 3.3.10); according to Morris et al. (1980) the object is a very young protostar.

TABLE 3.5.I

Magnetic field in the Orion (M42/OMC 1) Complex[1]

No.	Region	Density (cm^{-3})	Inferred from	Inferred strength of magnetic field B (mG)	Remarks	References
1	H I foreground cloud	n_H: 10^2–10^3	Zeeman splitting of H I (21 cm) line	0.06 ± 0.02	(2), (6)	Verschuur (1970)
2	M42 (H II region)	n_e: $10^{3.5}$–$10^{4.5}$	Zeeman splitting of H85α line	<0.21	(2)	Troland and Heiles (1977a)
3	M42 (H II region)	n_e: $10^{3.5}$–$10^{4.5}$	Zeeman splitting of H90 line	0.4 ± 0.1	(2)	Troland and Heiles (1977b)
4	KL/OMC 1 (core)	n_{H_2}: ~10^6	Zeeman splitting of SO (2_2–2_1) line	~6000	(4)	Clark and Johnson (1974)
5	KL/OMC 1 (core)	n_{H_2}: ~10^6	Zeeman splitting of SO (2_3–2_2) line	<1000	(5)	Brown et al. (1980)
6	KL/OMC 1 (core)	n_{H_2}: ~10^6	Infrared linear polarization measurements	10	(3)	Beichman and Chaisson (1974)
7	KL/OMC 1 (core)	n_{H_2}: ~10^6		7	(3)	Dyck and Beichman (1974)
8	KL/OMC 1 (core)	n_{H_2}: ~10^6		0.003	(3)	Gnedin and Mitrofanov (1975)
9	IRc 4/OMC 1	n_H: ~10^8	Zeeman splitting of OH (1612,1665) lines	3–5	(2), (7)	Chaisson and Beichman (1975)
10	IRc 4/OMC 1	n_H: ~10^8	Zeeman splitting of OH (1665) line	2.5	(2), (7)	Hansen et al. (1977)

Remarks to Table 3.5.I:

(1) Values uncertain by several orders of magnitude (Zuckerman and Palmer, 1975; Chaisson and Vrba, 1978).

(2) Longitudinal (along the line of sight) magnetic field.

(3) Transverse (perpendicular to the line of sight) magnetic field.

(4) Broad line width ('plateau') attributed to Zeeman effect; there are however several molecules such as HCN, SiO, and H$_2$O (see Table 3.3.IX) which exhibit broad profiles not caused by magnetic broadening and molecules such as CN, C$_2$H which show narrow profiles despite the existence of such a strong magnetic field. In addition the existence of such a magnetic field places severe dynamical constraints on the region in which the SO lines originate. These objections were raised by Zuckerman and Palmer (1975) and Chaisson and Vrba (1978). The SO broadening is now attributed to kinematic causes (see Section 3.4).

(5) Probably no more than a few milligauss.

(6) At $V_{LSR} = +7$ km s^{-1} the magnetic field $B = 0.050\pm0.015$ mG; at $V_{LSR} = +2$ km s^{-1}, $B = 0.070\pm0.020$ mG. The n_H densities of the H I cloud are 680 cm^{-3} and 350 cm^{-3}, respectively (Clark, 1965).

(7) Inferred from two Zeeman components (at 7.1 km s^{-1} and 8.6 km s^{-1}) originating in a cloud having a 7.8 km s^{-1} velocity.

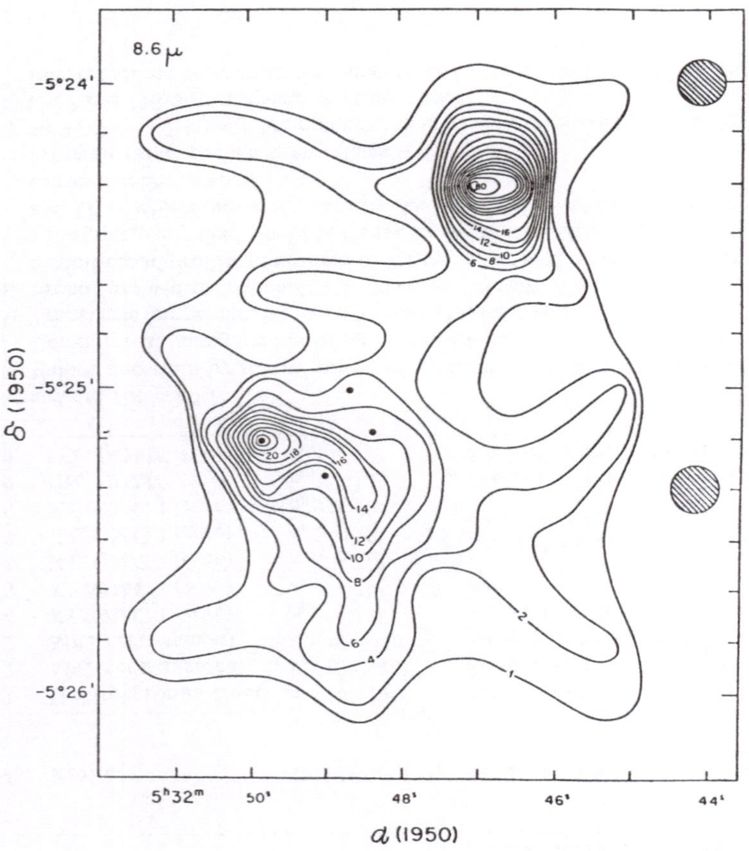

Fig.   3.1.1.b.

Fig.   3.1.1.a.

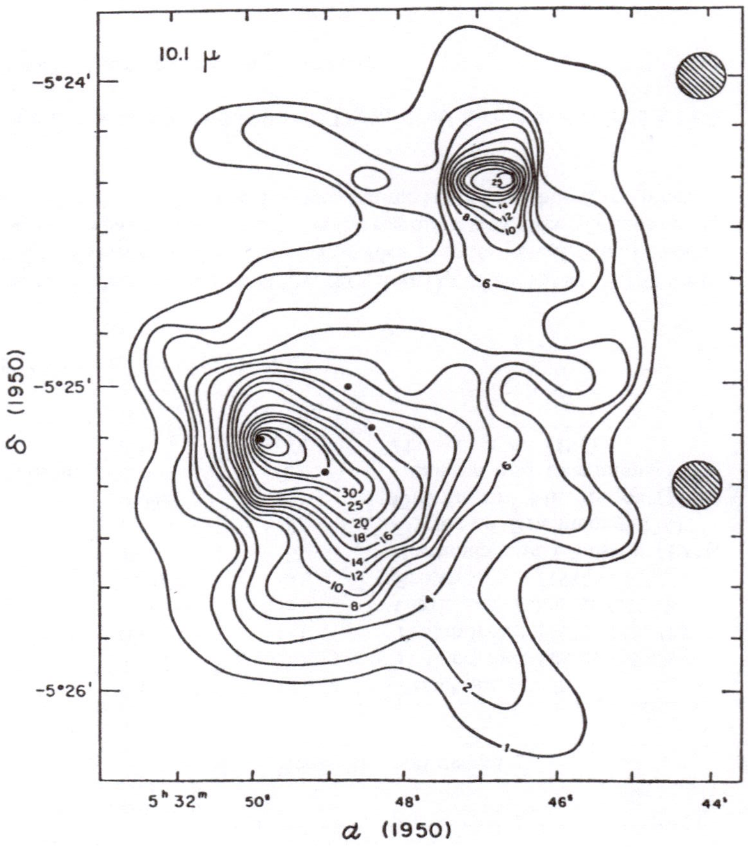

Fig. 3.1.1a.-e.   A sequence of infrared continuum contour maps of the 2' core of Orion (M42) made at 8.6μ, 10.1μ, 11.2μ, 12.3μ, and 13.1μ with 9"
angular resolution. The contour unit is 5×10^{-9} W cm^{-2} μ^{-1} sr^{-1} (after Gehrz et al., 1975). The source around the Trapezium Stars is the Trapezium or
Ney–Allen Nebula and is physically associated with the ionized gas (M42); the source lying 1' NW of Trapezium is the Kleinmann–Low Nebula (KL) and is
physically associated with the Orion Molecular Cloud (OMC 1) which lies on the backside of M42.

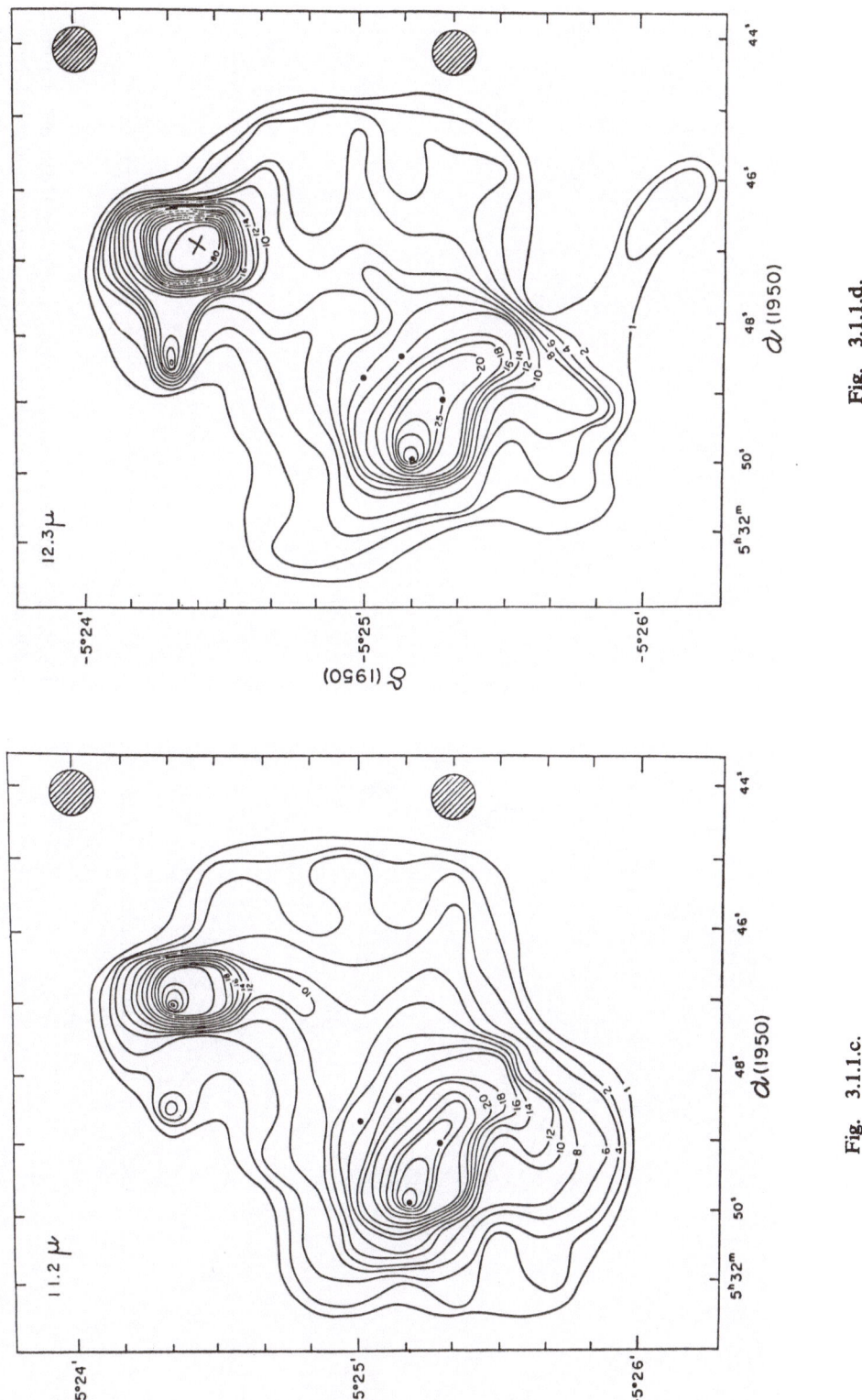

Fig. 3.1.1.d.

Fig. 3.1.1.c.

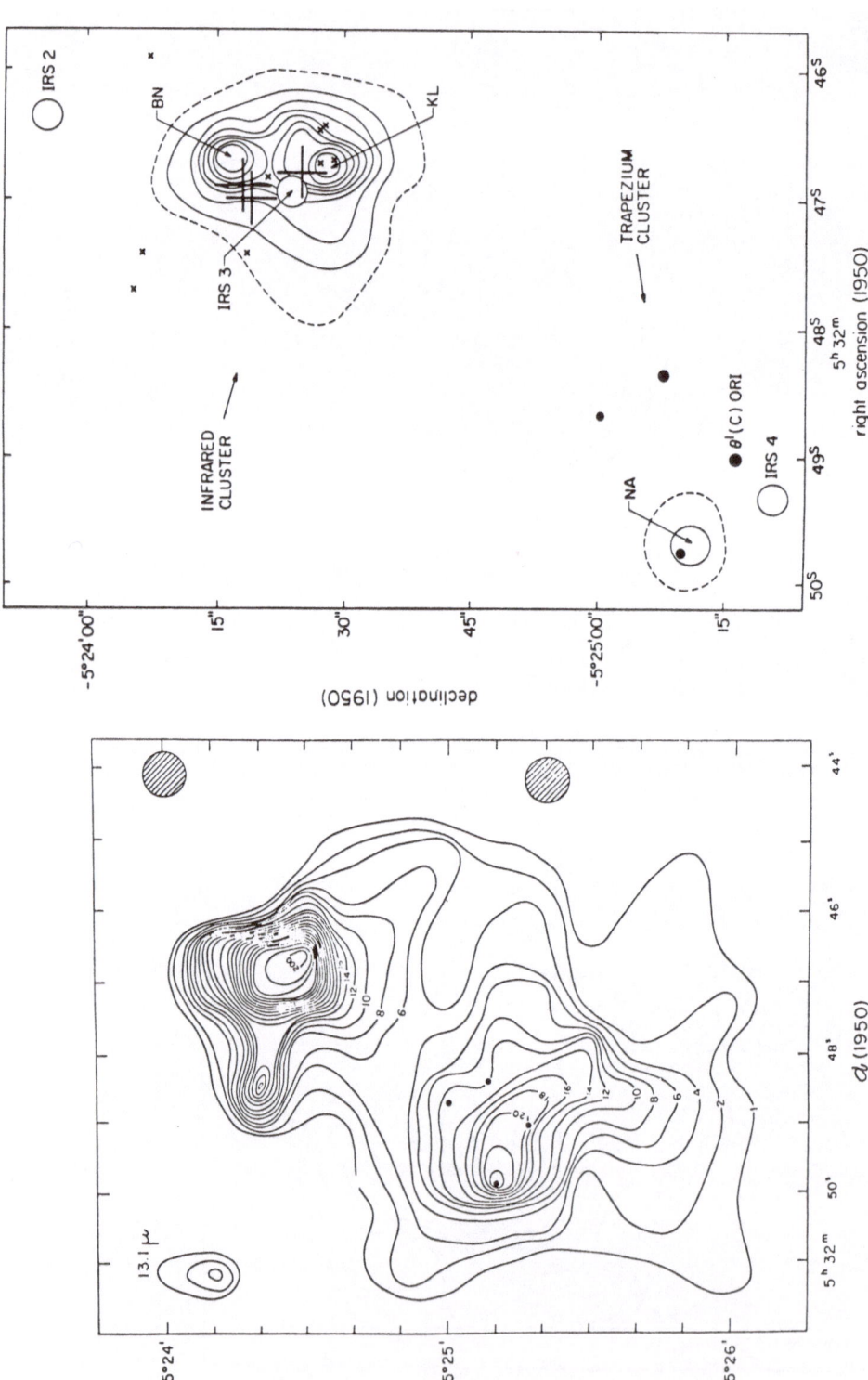

Fig. 3.1.2. Infrared continuum contour map of the 1′ core of Orion made at 20μ with a 5″ angular resolution. Both the Trapezium and the KL Nebula are shown. The contour interval is 3×10^{16} W m^{-2} Hz^{-1} sr^{-1} with the dashed line at half this integral (after Wynn-Williams and Becklin, 1974). Note the close spatial correlation of KL and the sources embedded in it (BN object, IRS 3) with the OH masers (large crosses; Raimond and Eliasson, 1969) and the H$_2$O masers (small crosses; Moran et al., 1973).

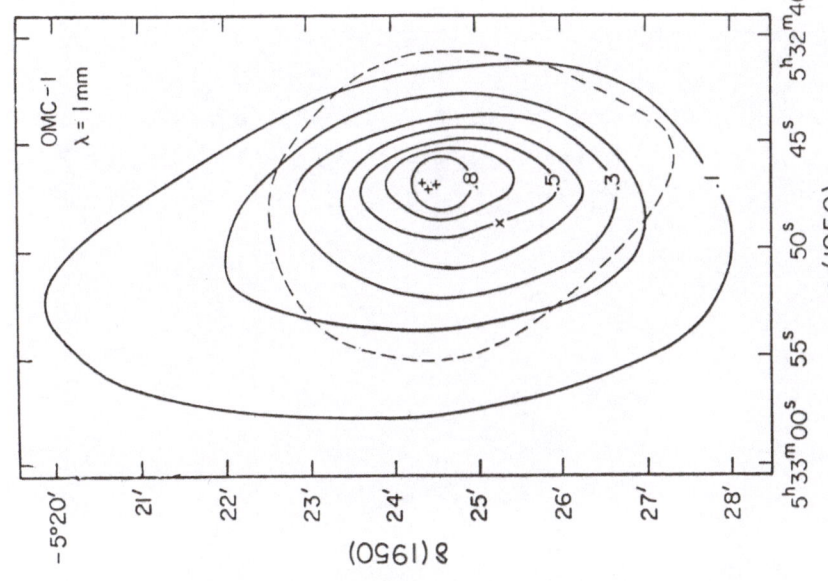

Fig. 3.1.4. Infrared continuum contour map of the Orion Molecular Cloud (OMC 1) made at 1 mm with an angular resolution of 1′. The contours are normalized to the peak flux density which is 215 Jy into a 1′ beam (after Westbrook *et al.*, 1976). Note the similar distribution of the molecular 2 cm H$_2$CO emission (dashed line marking the 10%, contour of the molecular map by Harvey *et al.*, 1974) and the spatial coincidence of the core of the OMC 1, the KL Nebula ($\simeq$40″ core of the infrared map) and the most prominent sources in the infrared cluster (crosses; Wynn-

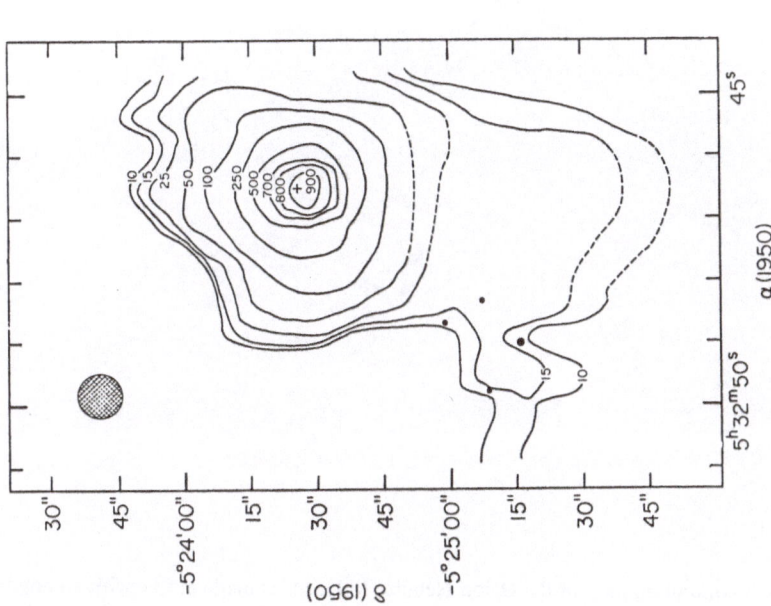

Fig. 3.1.3. Infrared continuum contour map of the 2′ core of the Orion Nebula made at 33 μ with an angular resolution of $\simeq$10″. The contour unit is 8.75 × 10^{-17} W m^{-2} Hz^{-1} sr^{-1} (after Beichman *et al.*, 1973). Note the

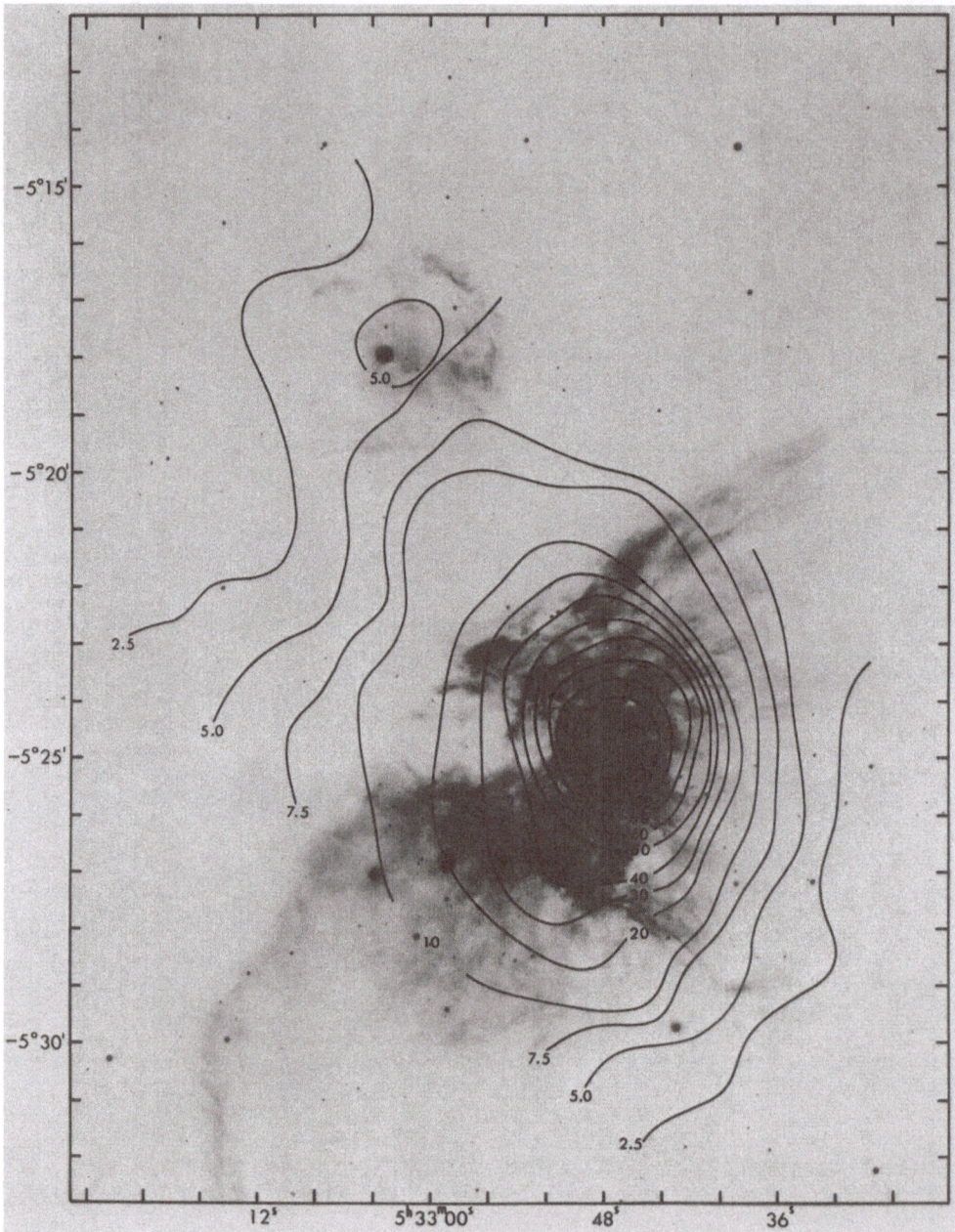

Fig. 3.1.5.   Infrared continuum map of the Orion Nebula (M42, M43) made at 91μ with an angular resolution of $2\overset{.}{.}2$ (after Harper, 1974). The contour units are arbitrary. The map is superimposed on a photograph taken by T. R. Gull at Kitt Peak National Observatory. The central part of the M42 contour map coincides with KL.

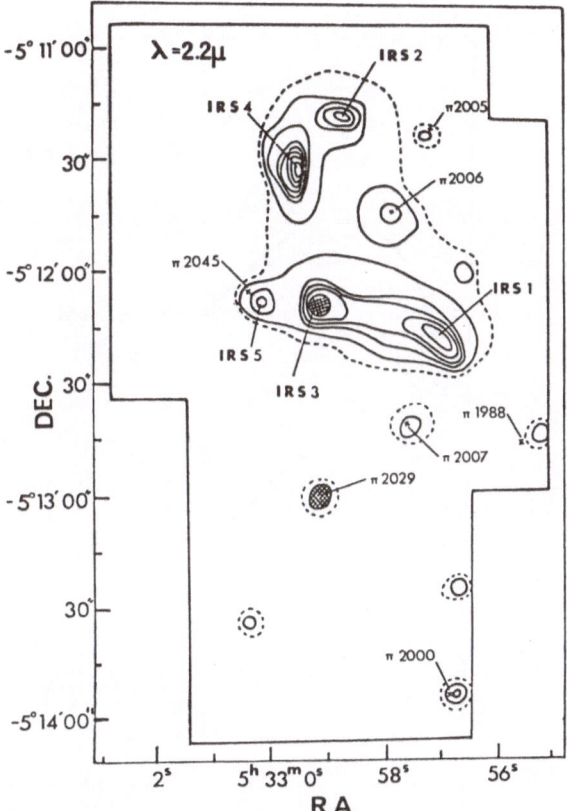

Fig. 3.1.6. Infrared continuum contour map of the core of OMC 2 made at 2.2μ with an angular resolution of 7".5 (after Gatley *et al.*, 1974). The objects marked with π are visible stars identified by Parenago. The five infrared sources marked IRS1, 2, 3, 4, 5 are probably very young stars. Several of these objects have also been photographed at the near infrared (effective wavelength $\simeq$ 9400 Å) by Cohen and Frogel (1977).

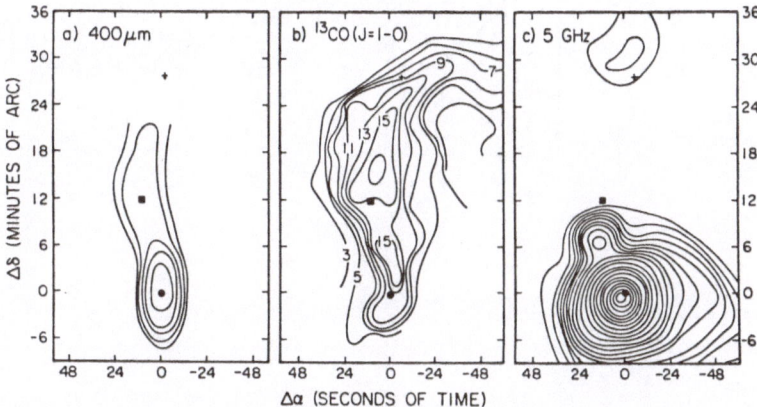

Fig. 3.1.7. Three contour maps of the $\simeq 10' \times 40'$ ridge of the large dust/molecular cloud L1641, showing the emission from dust, molecular gas and ionized gas respectively (after Smith *et al.*, 1979). The infrared continuum contour map was made at 400μ with 3' angular resolution. The ^{13}CO ($J = $ 1–0) map (Kutner *et al.*, 1976) was made with 2'.6 angular resolution. The radio continuum contour map of M42, M43, and NGC 1977 was made at 5 GHz with 4' resolution (Goss and Shaver, 1970). The molecular clouds OMC 1, OMC 2, and OMC 3 marked in all maps with the symbols $\bullet$, $\blacksquare$ and + respectively are probably the outcome of the fragmentation of L1641. Note that all the H II regions involved lie on the edges of the corresponding molecular clouds.

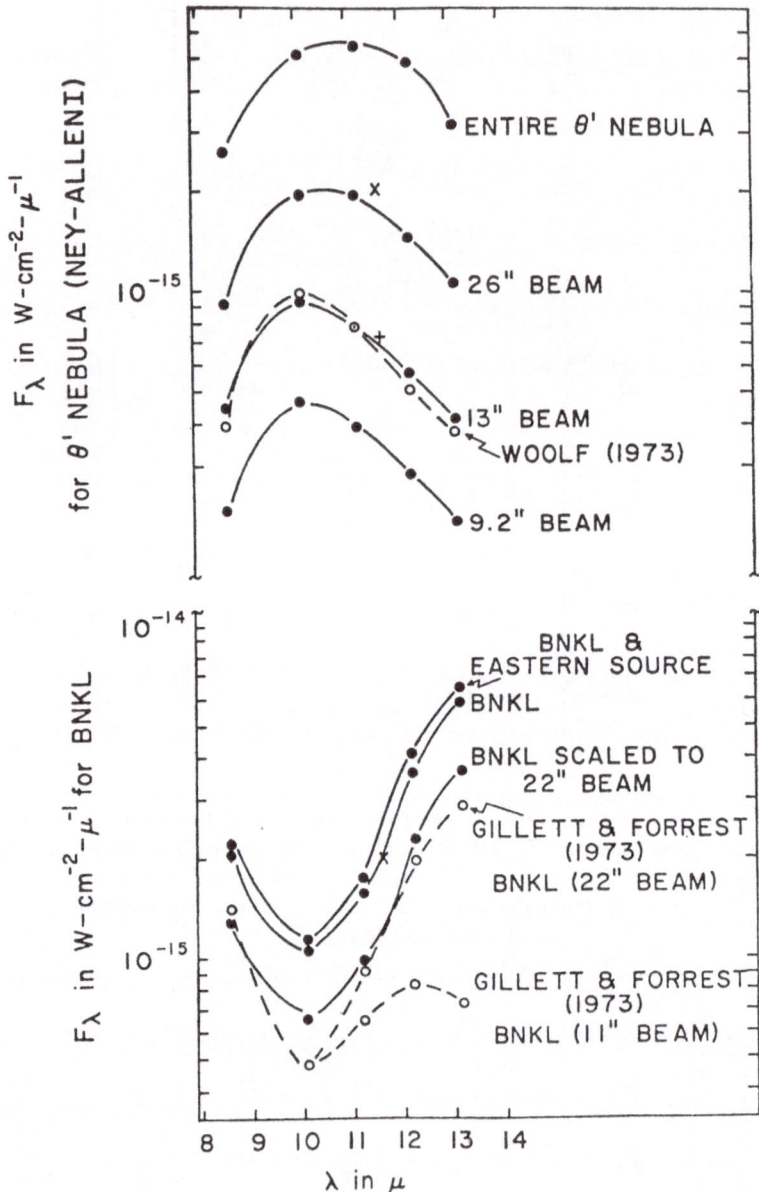

Fig. 3.2.1. The 8–14μ energy spectrum of the Trapezium (θ^1 or Ney–Allen I) and the KL Nebula showing the 10μ silicate band in emission (Trapezium or θ^1 Nebula) and in absorption (KL or BNKL Nebula). Beams and references are denoted on the figure (after Gehrz *et al.*, 1975).

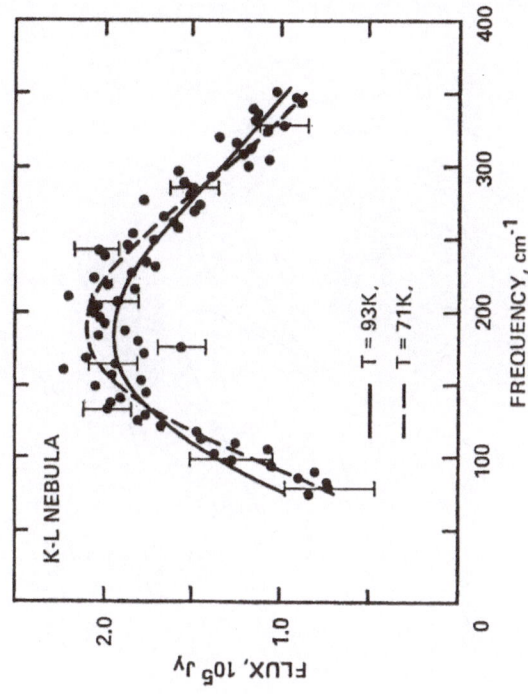

Fig. 3.2.3.   The 30–125 μ energy distribution in the 1′.4 core of KL (after Errickson *et al.*, 1977). The dust temperature T_d is in the range of 70–95 K (after Errickson *et al.*, 1977)

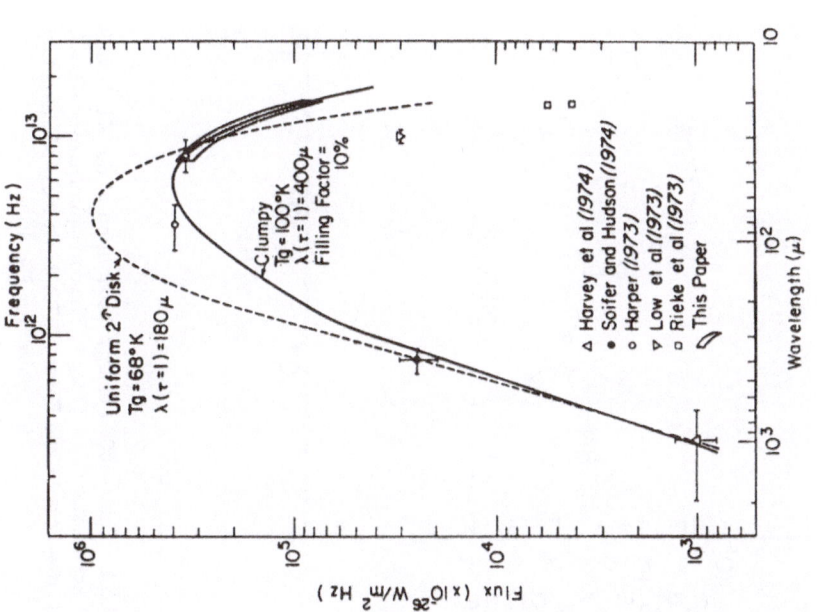

Fig. 3.2.2.   The 10–1000 μ energy distribution in the 2′ core of KL fitted with curves respresenting (1) a uniform distribution of dust characterized by a grain temperature $T_g \simeq 70K$, and (2) a clumpy distribution of dust with $T_g \simeq 100K$ (after Houck *et al.*, 1974).

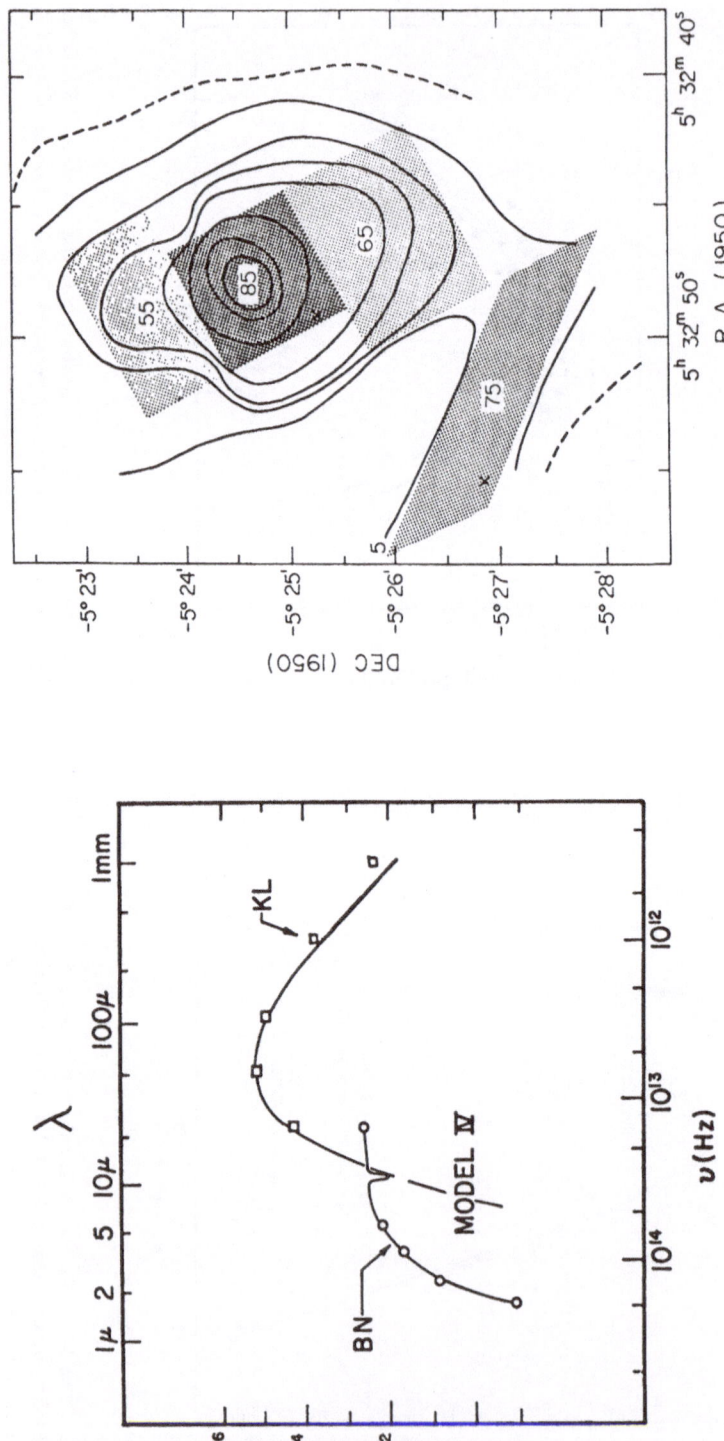

Fig. 3.2.5. The average 50–100μ colour temperature of the dust for several regions of Orion. The contours are from the 100μ map of Figure 2.4.7 (after Werner *et al.*, 1976).

Fig. 3.2.4. The 1–1000μ energy distribution of the KL Nebula and the BN object measured in a 1' beam. The curve plotted over KL is based on a model (after Scoville and Kwan, 1976). Note the predominance of the BN object for $\lambda < 10\mu$.

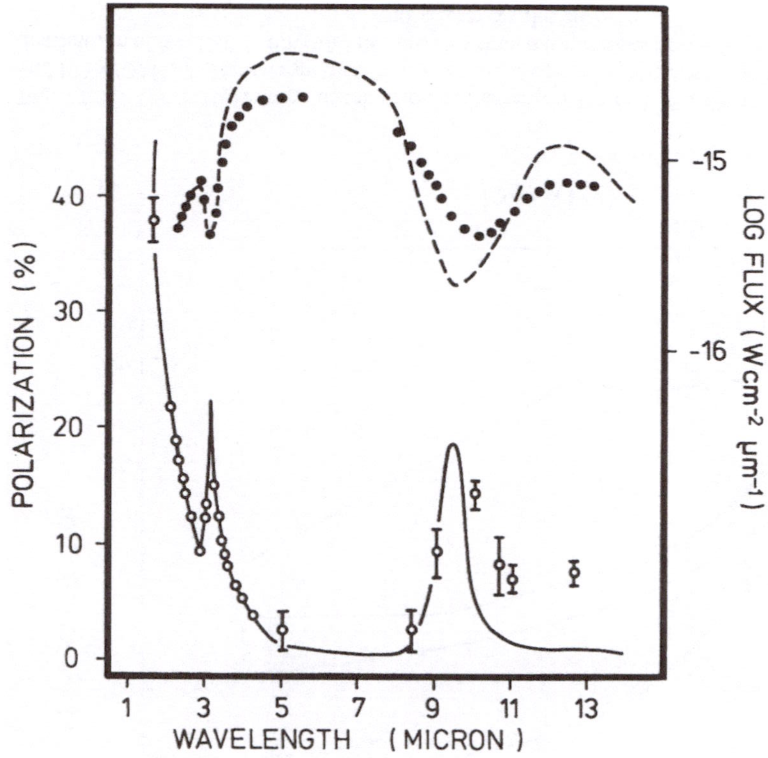

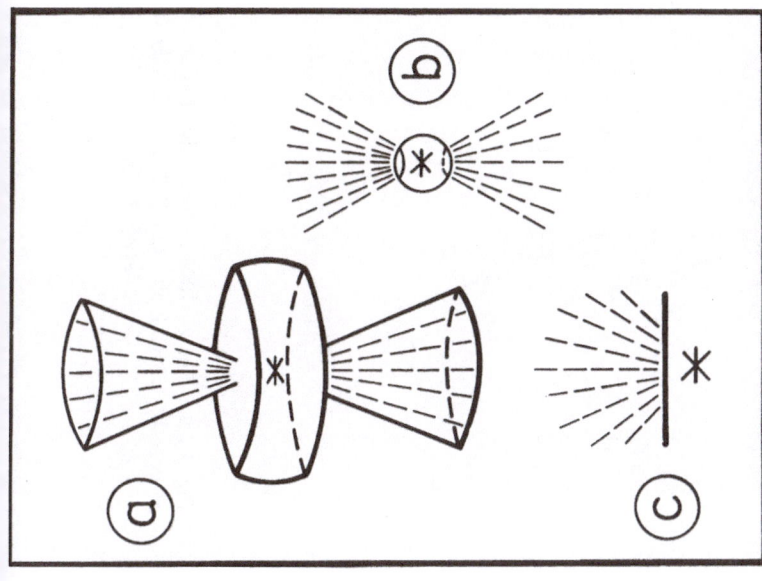

Fig. 3.2.7. Configurations proposed by Elsässer and Staude (1978) to explain the observed linear polarization of the BN object in terms of scattering dust grains in non spherical circumstellar shells: (a) *Bipolar nebula model* (built on a model structurally similar with the S106 'Bipolar Nebula'; Eiroa *et al.*, 1979). The star is embedded in an optically thick disk with optically thin polar lobes (in which the dust density decreases rapidly). An observer on the same plane with the 'disk' will see a heavily reddened star and scattering polarization from the poles. The polarization at the absorption features is due to the strong absorption of the direct light at these wavelengths and hencefore to the predominance of the scattered component. (b) *Disrupted circumstellar shell model*. An alternative configuration of a compact circumstellar

Fig. 3.2.6. Energy distribution (filled circles) and linear polarization (empty circles) of the BN object in the 1–13μ range. The absorption features at 3μ and 10μ indicate the presence of water ice and silicate grains respectively. The curves, drawn to match the observational data, are from a model interpreting the polarization in terms of scattering by grains and electrons in a non-spherical circumstellar shell (after Elsässer and Staude, 1978).

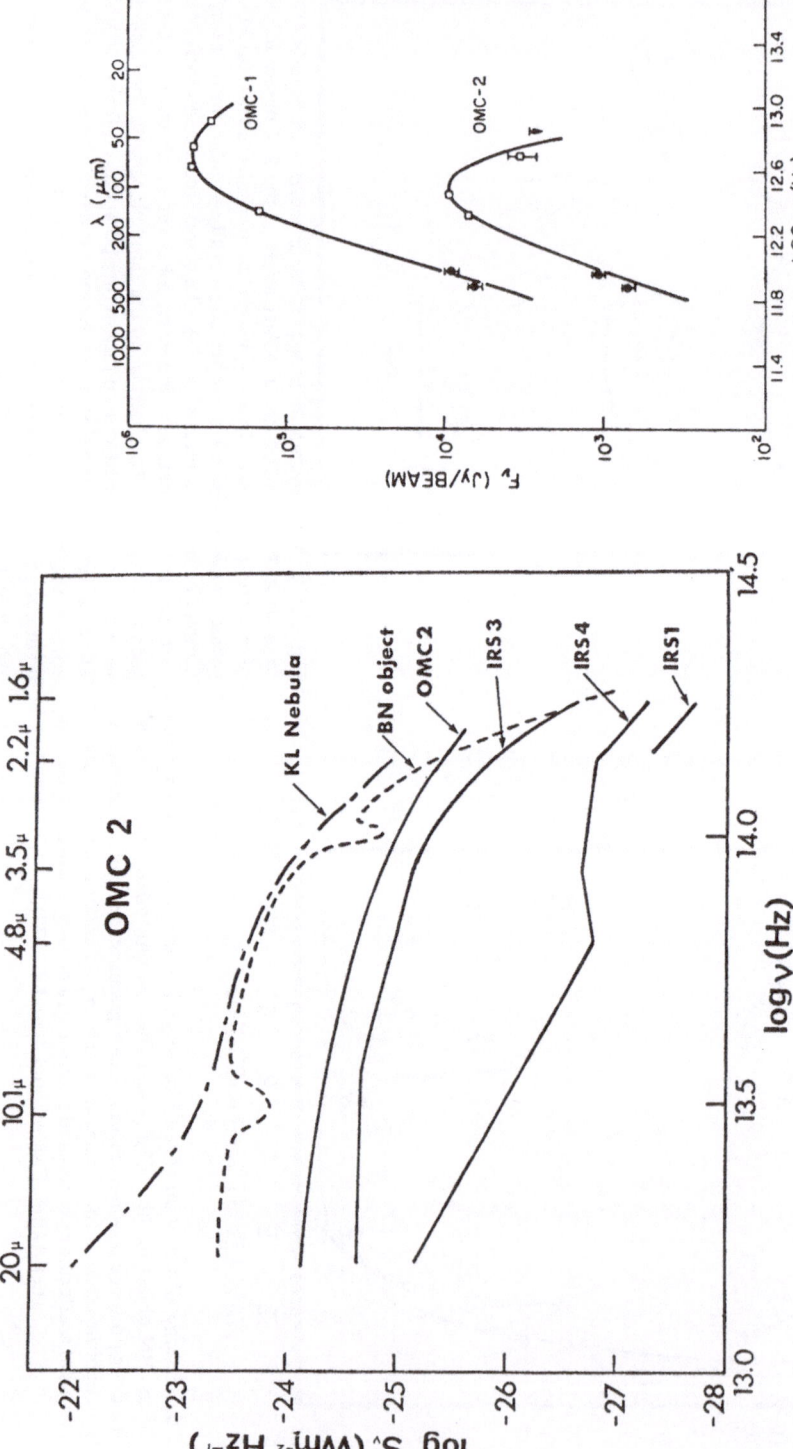

Fig. 3.2.9. The 20–1000μ energy distribution of the 3' core of OMC 1 and OMC 2, respectively (after Smith et al., 1979).

Fig. 3.2.8. The 2–20μ energy distribution of the central 7".5 of IRS 1/OMC 2, IRS 3/OMC 2, and IRS 4/OMC 2. The energy distribution of the 1' core of OMC 2 is also shown. The energy distribution of KL (30" beam) and BN are also shown for comparison (after Gatley et al., 1974, and references therein).

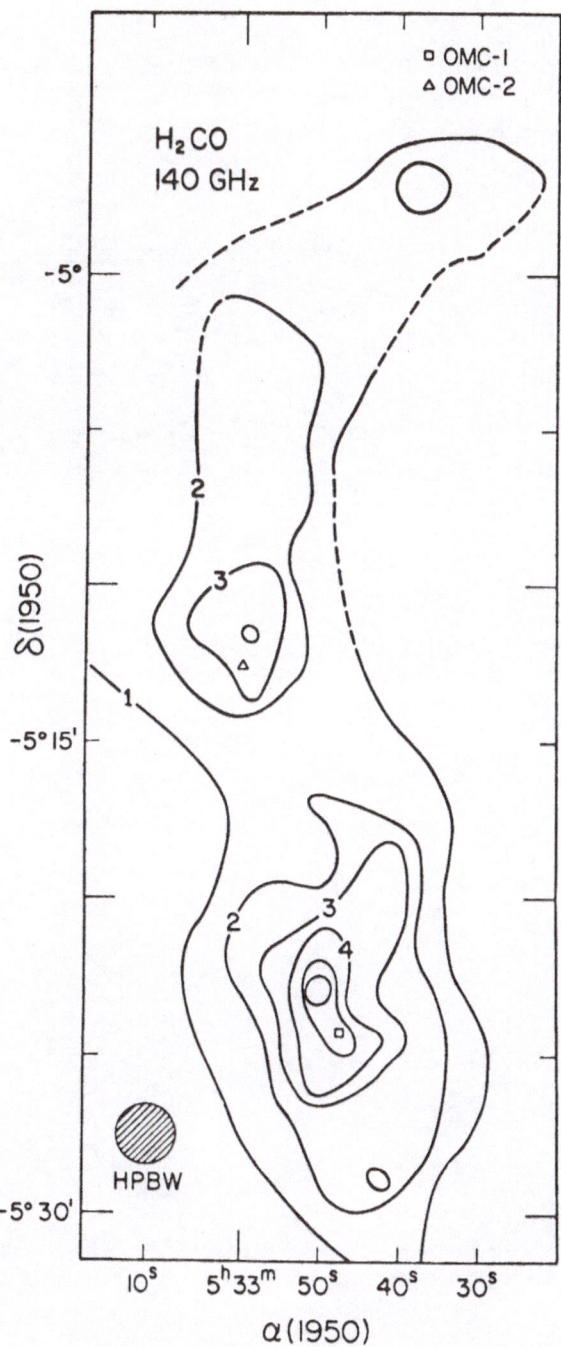

Fig. 3.3.1.   Contour map of the H_2CO ($2_{11} - 2_{12}$) emission from the OMC 1/OMC 2 region made at 140 GHz with a 2′ angular resolution (after Kutner *et al.*, 1976).

Fig. 3.3.2.   Contour map of the ^{13}CO ($J = 1$–0) emission from the OMC 1/OMC 2 region made at 110.2 GHz with a 2ʹ.6 angular resolution, superimposed on a Lick Observatory photograph showing M42/M43 and NGC 1977 (after Kutner *et al.*, 1976).

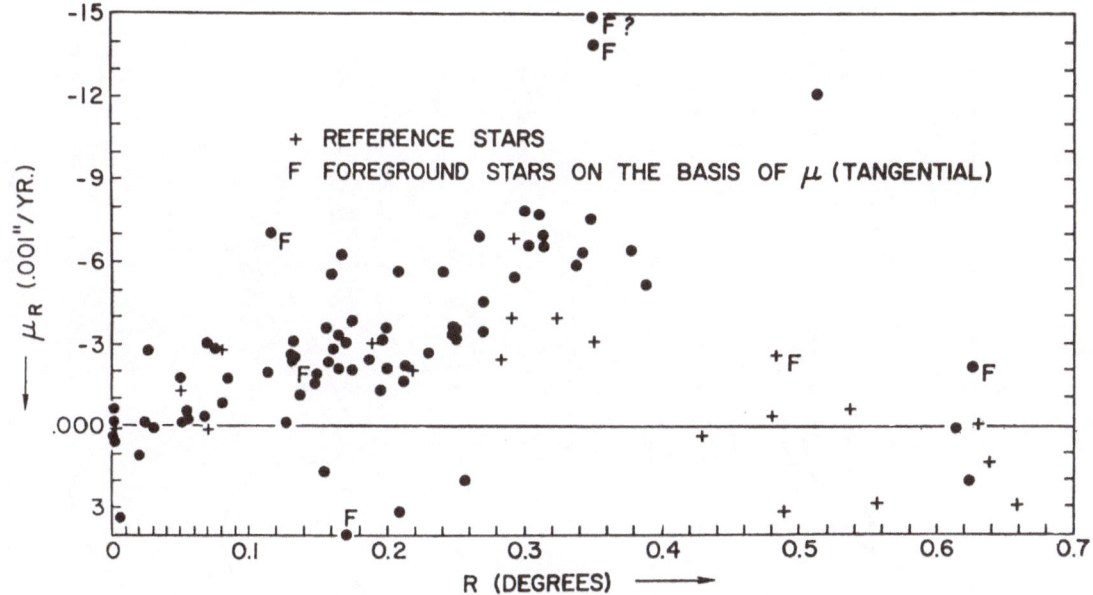

Fig. 3.3.3 Radial component of proper motion of the Orion cluster of stars as a function of radial distance from the Trapezium (after Fallon *et al.*, 1977). Foreground stars (with large perpendicular component of proper motion) are marked *F*. Note the contraction of the cluster with a velocity directly proportional to the distance from the Trapezium stars. The collapse of the cluster, whose mass is less than the critical Jeans mass can be understood if both the cluster and the molecular cloud are within each other and contract together.

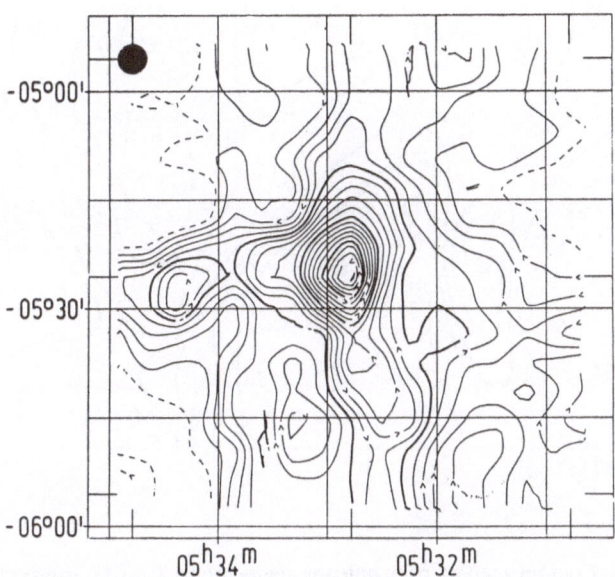

Fig. 3.3.4 Contour map of the ^{12}CO ($J = 1$-0) emission from the OMC 1/OMC 2 region made at 115 GHz with an angular resolution of 4′ (after Gillespie and White, 1980). The contour interval is 2.8 K. The zero contour is dashed. Note the North to South extension of the main molecular complex and a previously undetected feature 24′ to the East of the main peak.

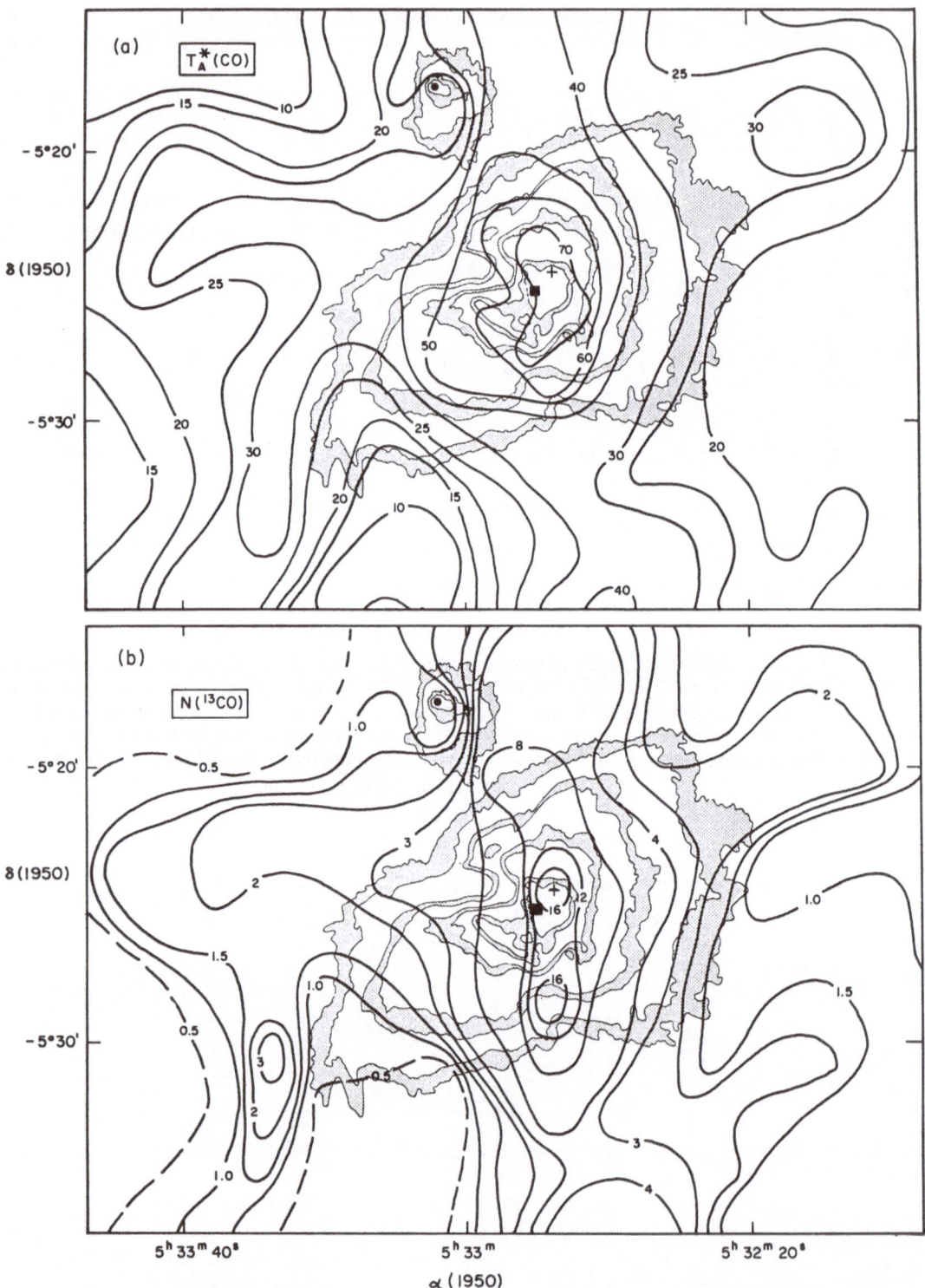

Fig. 3.3.5. Contour map of peak antenna temperature T_A (CO) irrespective of velocity (a) and of column density N (^{13}CO) of OMC 1 (b) made with an angular resolution of 2′ (after Loren, 1979). The contour map of M42, M43 (Hβ isophotes) is from Dopita *et al*. (1975). The KL Nebula is indicated by a cross, the Trapezium by a square and the exciting star of M43 (NU Orionis) by a filled circle. Note the ridge like shape of the central part of the N (^{13}CO) contours.

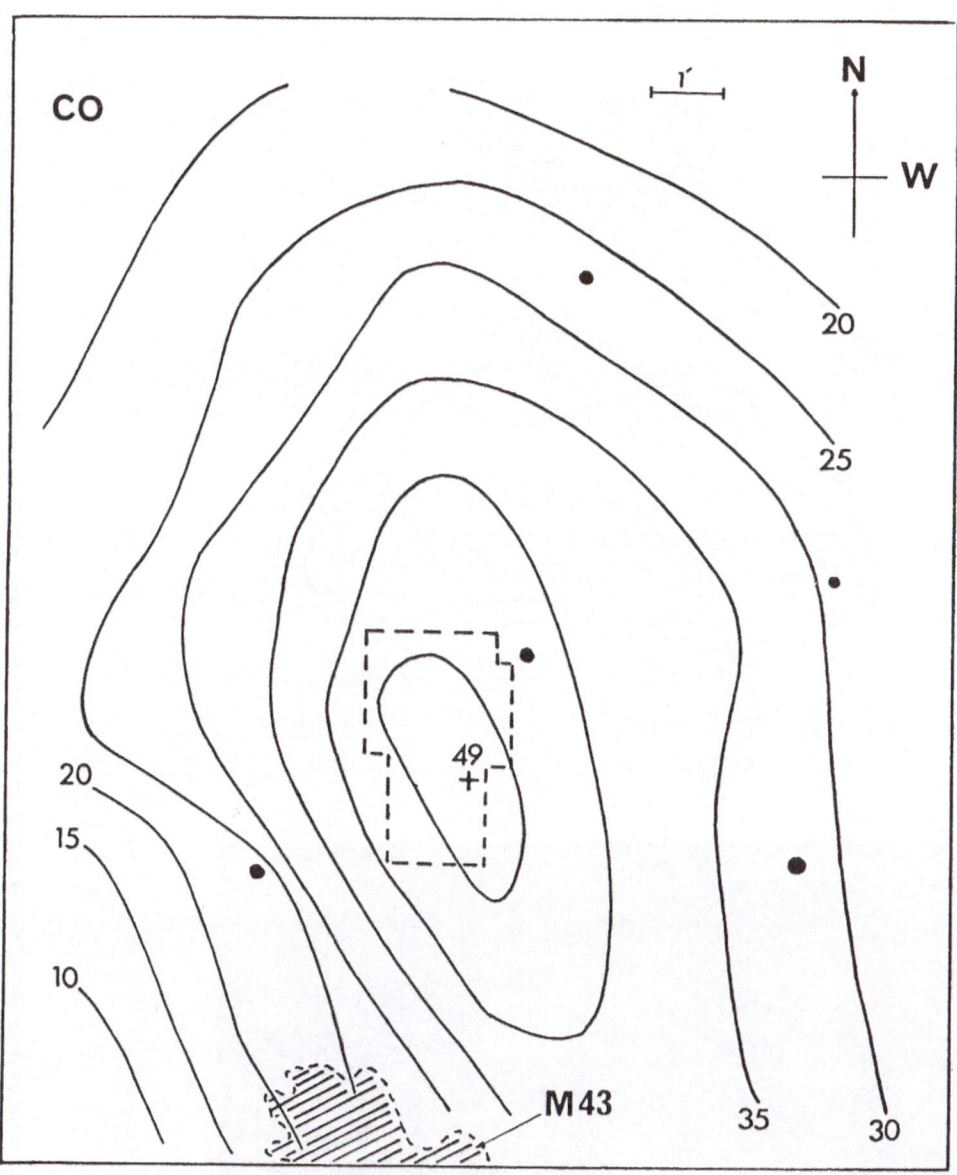

Fig. 3.3.6.   Contour map of the ^{12}CO ($J = 1$–0) emission from OMC 2 made at 115 GHz with a 2′ angular resolution. The insert corresponds to the outline of Figure 3.1.6 (after Gatley *et al.*, 1974).

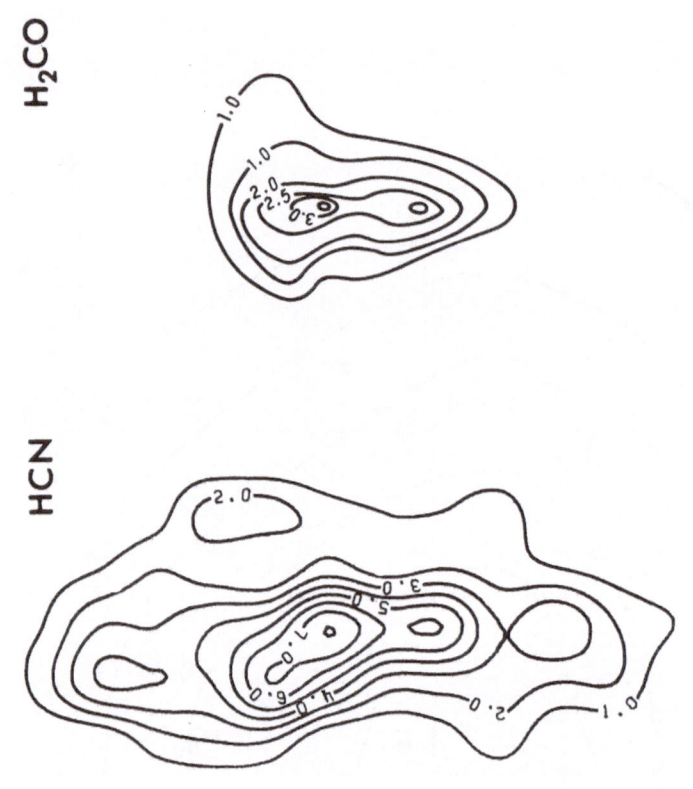

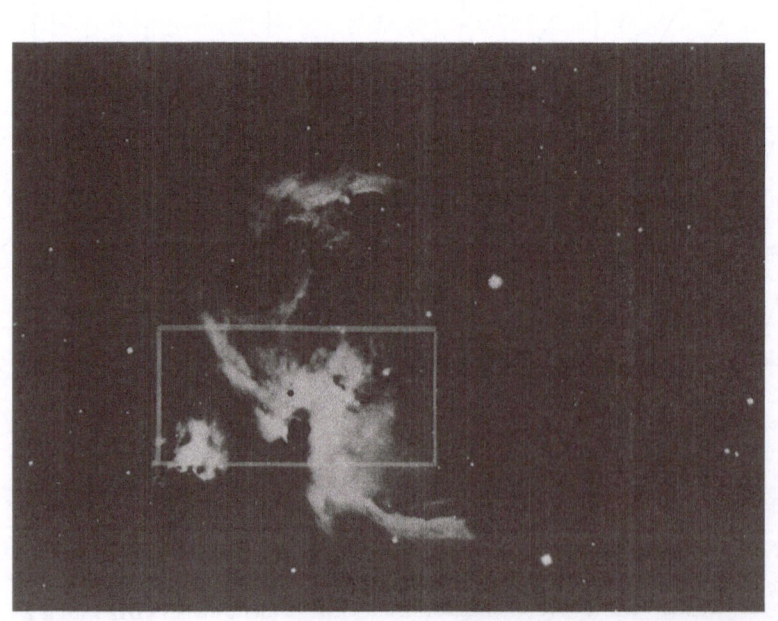

Fig. 3.3.7. Radio brightness contour map of HCN and H₂CO emission from the core of OMC 1 made at 88.6 GHz and 140.8 GHz with angular resolution of 1′.3 and 1′, respectively. The H₂CO map was produced by Thaddeus *et al.* (1971) (after Clark *et al.*, 1974).

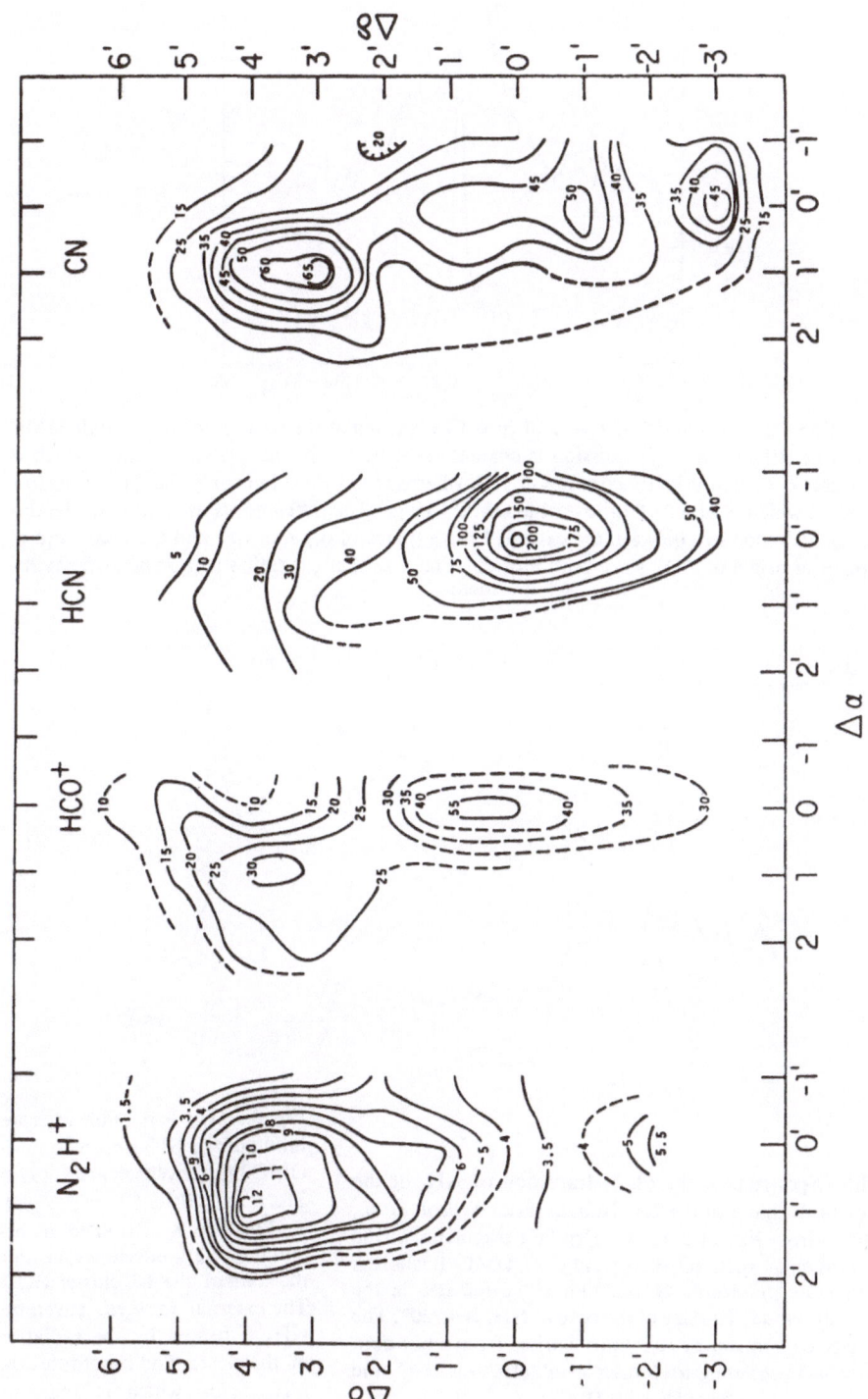

Fig. 3.3.8. Integrated intensity contours of N_2H^+, HCO^+, HCN, and CN for the central part of the OMC 1, made with an angular resolution of 71″, 75″, 75″, and 63″, respectively. The (0, 0) position is that of the KL nebula. The contour units are in Kelvins km s^{-1} (after Turner and Thaddeus, 1977).

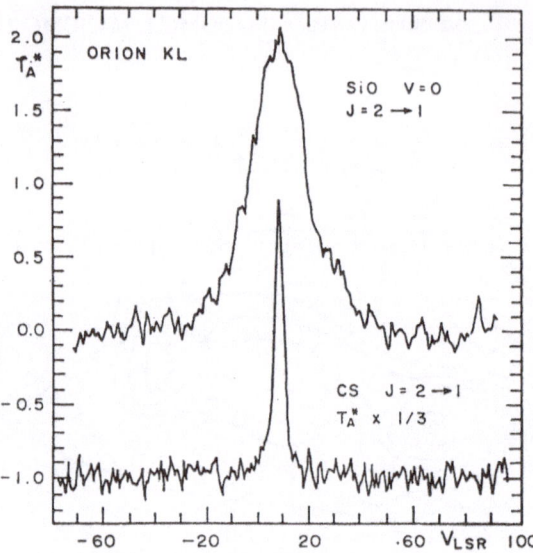

Fig. 3.3.9. Spectra of SiO and CS, $J = 2 - 1$ ($v = 0$) observed at the position of KL Nebula/OMC 1 (after Scoville, 1980). The SiO emission is characterized by the broad 'plateau' feature which is localized in the < 1' high velocity core of OMC 1, whereas the CS exhibits only the typical narrow profile ('spike') which is observed all over the OMC 1 ridge. The difference is attributed to the different chemical abundances between the extended ridge (place of orgin of CS) and the small core of OMC 1 (place of origin of SiO), since both molecules have similar excitation requirements (Scoville, 1980)

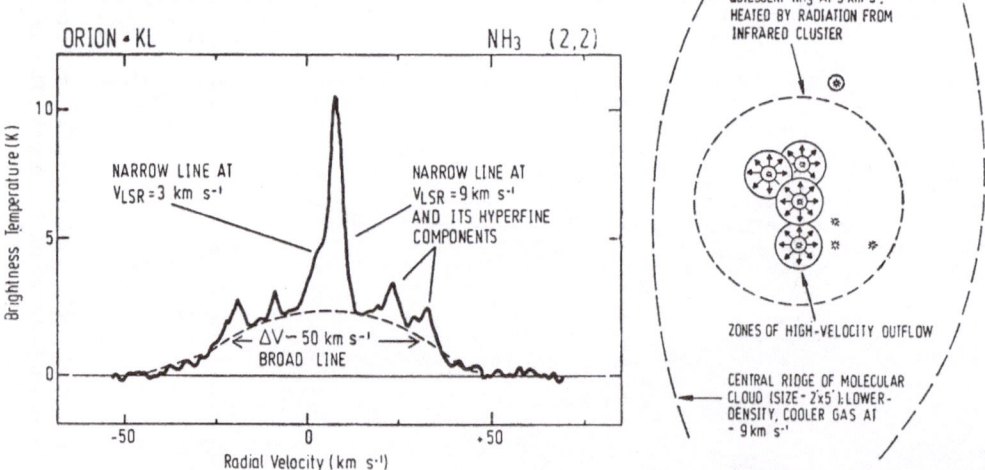

Fig. 3.3.10. Spectrum of the (2, 2) transition of NH₃ at the position of peak signal in the KL Nebula (after Wilson *et al.*, 1979). The narrow line at 9 km s⁻¹ ('spike') originates in the extended molecular material of the ridge of OMC 1 whereas the broad feature ('plateau', $\Delta V \simeq 50$ km s⁻¹) originates in the < 1' core of the cloud. The size of the core in NH₃ is < 20″. The third velocity component at $\simeq 3$ km s⁻¹ arises from a hot condensation which according to Morris *et al.* (1980) is $\simeq 6″$ and coincides with IRc 2.

Fig. 3.3.11. A possible interpretation of the NH₃ observations in OMC 1 (after Wilson *et al.*, 1979). The broad $\Delta V \simeq 50$ km s⁻¹ 'plateau' feature is attributed to expanding NH₃ envelopes around the stars of the IR cluster (KL). The external to these envelopes NH₃ is heated by the radiation of the cluster and constitutes the NH₃ region which is characterized by the narrow line ('spike') at 9 km s⁻¹. This region is in turn embedded in the lower density 'ridge' of OMC 1.

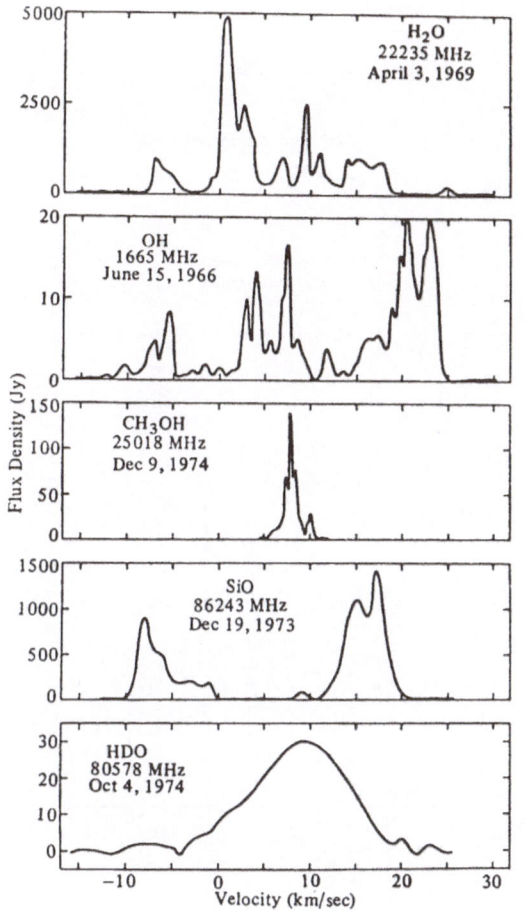

Fig. 3.4.1. The spectra of the four masers detected in KL/OMC 1 (after Moran, 1976). The H_2O spectrum is from Sullivan (1973); the OH spectrum is from Manchester *et al.* (1970) and Menon (1967); the SiO spectrum is from Snyder and Buhl (1974); and the CH_3OH spectrum is from Barrett *et al.* (1975). A non-maser spectrum (HDO from Turner *et al.*, 1975) is included for comparison. Note the numerous narrow spectral features which are one of the characteristics of a masering source (see also Table 3.4.I).

OH SPECTRUM OF ORION A

(March 1977)

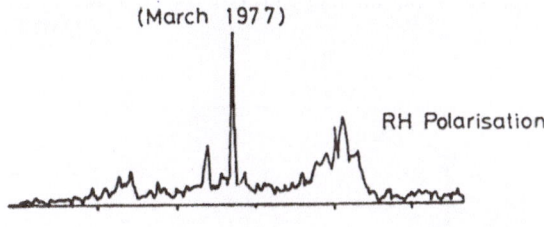

Fig. 3.4.2. Spectra of the OH (1665 MHz) maser in KL/OMC 1 in right hand (RH) and left hand (LH) polarization (after Norris *et al.*, 1980). The dominant features at 7.1 km s^{-1} (RH) and 8.6 km s^{-1} (LH) are components of one line at 7.8 km s^{-1} which is splitted under the influence of a 3 mG magnetic field. The position of the masering spot coincides with that of IRc4/OMC 1 (see also Table 3.4.VI).

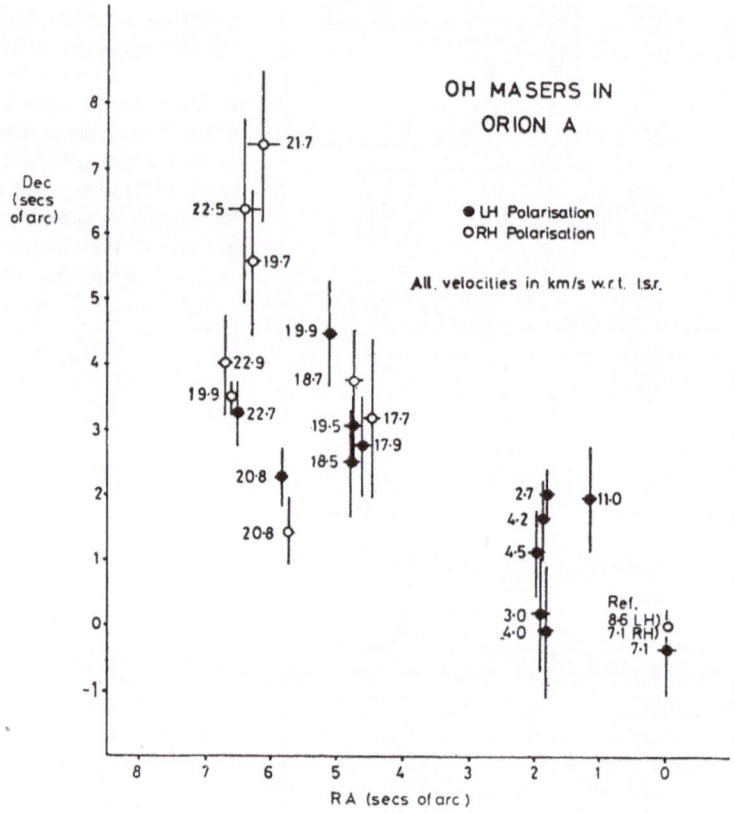

Fig. 3.4.3. Relative positions of the main features of the OH masers in KL/OMC 1 (after Norris *et al.*, 1980). The resolution is 0.40 km s⁻¹. The masering spot at $+7.1/+8.6$ km s⁻¹ and the cluster of features in the range 2–11 km s⁻¹ are probably associated with IRc 4/OMC 1, whereas the cluster of features in the range 17–23 km s⁻¹ is probably associated with IRc 2 (IRS 3)/OMC 1 (see also Table 3.4.VI).

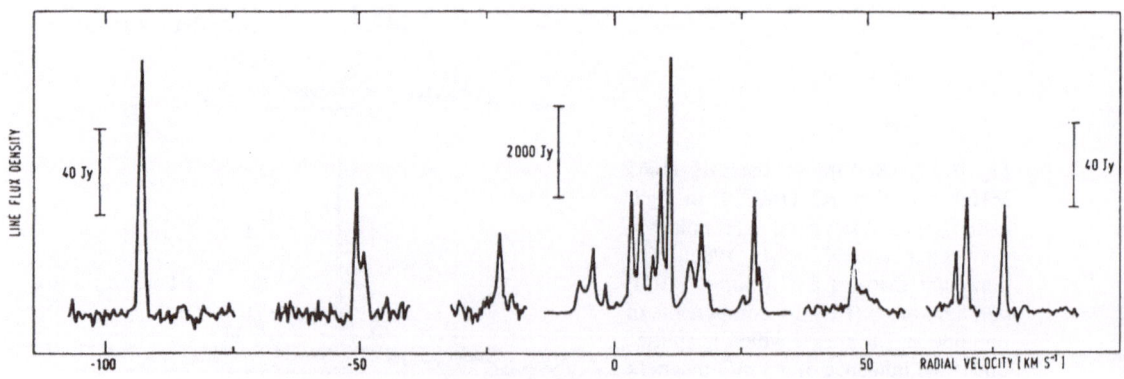

Fig. 3.4.4. Composite spectrum of the H₂O masers in KL/OMC 1. This is a blend of the H₂O emission received in a 40″ beam (after Genzel and Downes, 1977). The resolution is 0.43 km s⁻¹.

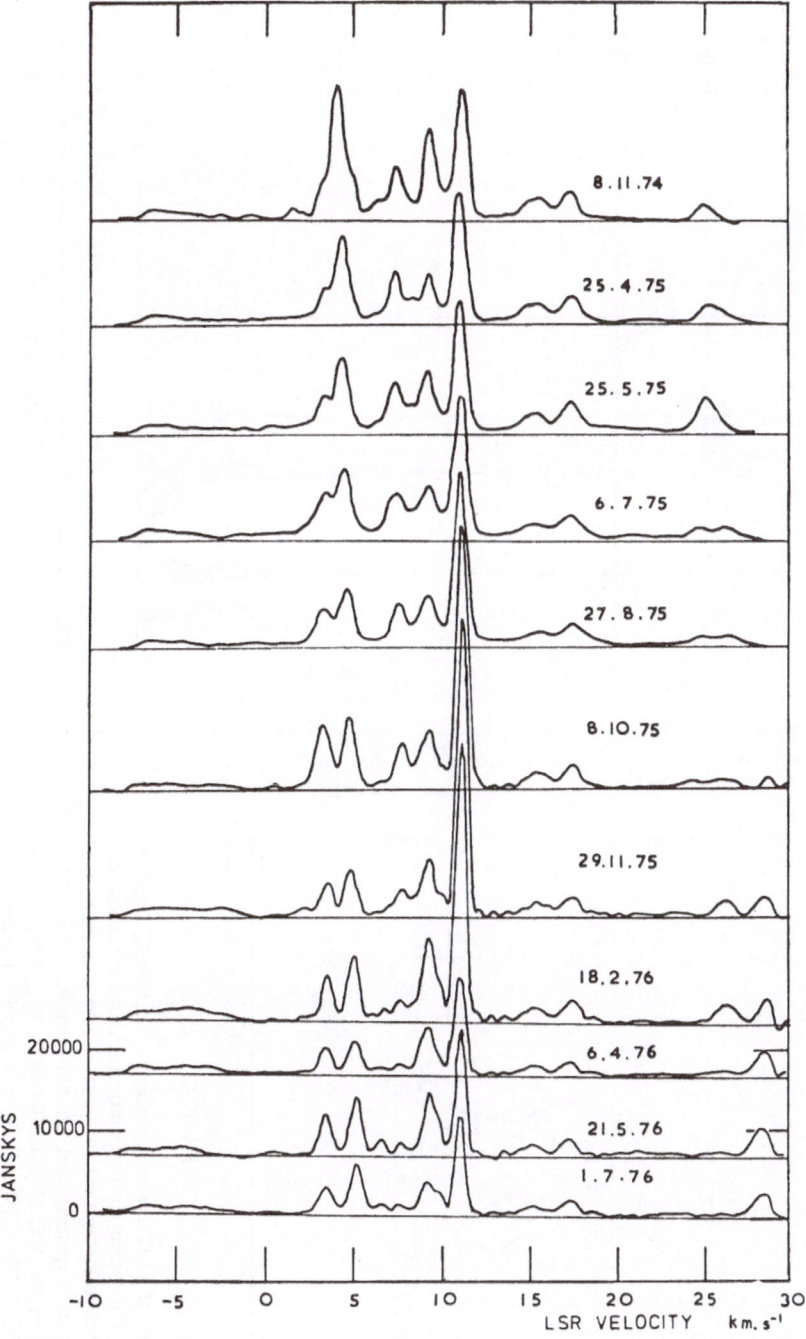

Fig. 3.4.5. Time variations of the H₂O maser in KL/OMC 1 in the range − 10 to + 30 km s⁻¹ (strong, low velocity features). The resolution is 0.40 km s⁻¹ (after Little *et al.*, 1977).

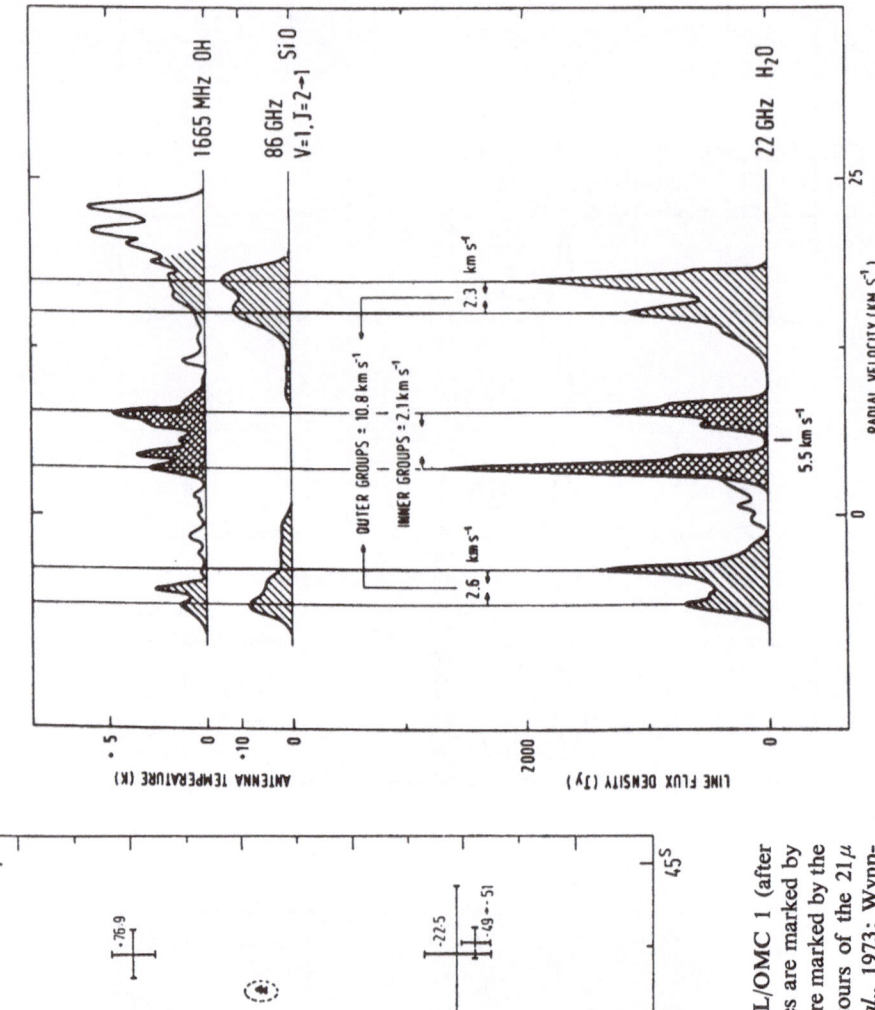

Fig. 3.4.6. Positions of H$_2$O maser sources in KL/OMC 1 (after Genzel *et al.*, 1978). The strong, low velocity features are marked by black dots whereas the weak, high velocity features are marked by the crosses. The masers are superimposed on the contours of the 21μ infrared radiation from the KL Nebula (Rieke *et al.*, 1973; Wynn-Williams and Becklin, 1974). The strong low velocity features are concentrated in clusters and are probably associated with circumstellar shells enveloping the stars of the IR cluster. 'Source A' probably coincides with IRc 4 whereas the H$_2$O cluster just to the NE of 'source A' is associated with IRc 2(IRS 3)/OMC 1. The weak, low velocity features appear to be dispersed over a more extended area around KL and are probably fragments of planetary mass blown out from the cocoons enveloping certain young stars in KL/OMC 1.

Fig. 3.4.7. Comparison of the H$_2$O maser emission spectrum from the core of KL with the SiO and OH maser emission spectra (after Genzel and Downes, 1977). Note the absence of the inner groups (appearing at ±2 km s^{-1}) from the SiO spectrum.

←Fig. 3.4.8. A model explaining the H_2O emission of 'source A' at the core of KL/OMC 1 under the assumption that the H_2O 'shell features' at ±11 km s⁻¹ originate in this source (after Genzel and Downes, 1977). The diagram on the upper right part of the figure shows the predicted variation of number density, kinetic temperature and velocity as a function of distance from the exciting star (Cochran and Ostriker, 1977). The H_2O maser features are embedded in the dust shell which is sandwiched between the shock front and the ionization front and envelope a compact H II region created by the young exciting star in the centre. According to this model the outer group of lines arise from the front and back of the expanding shell whereas the inner groups may originate from the edge of the shell (Cross-hatched area).

Fig. 3.4.9. SiO and H_2O spectra in KL/OMC 1. The shaded features of the H_2O spectrum and the corresponding features of SiO may originate in a shell which coincides with IRc 2 (IRS 3)/OMC 1 (after Genzel et al., 1979); see also Table 3.4.VI.

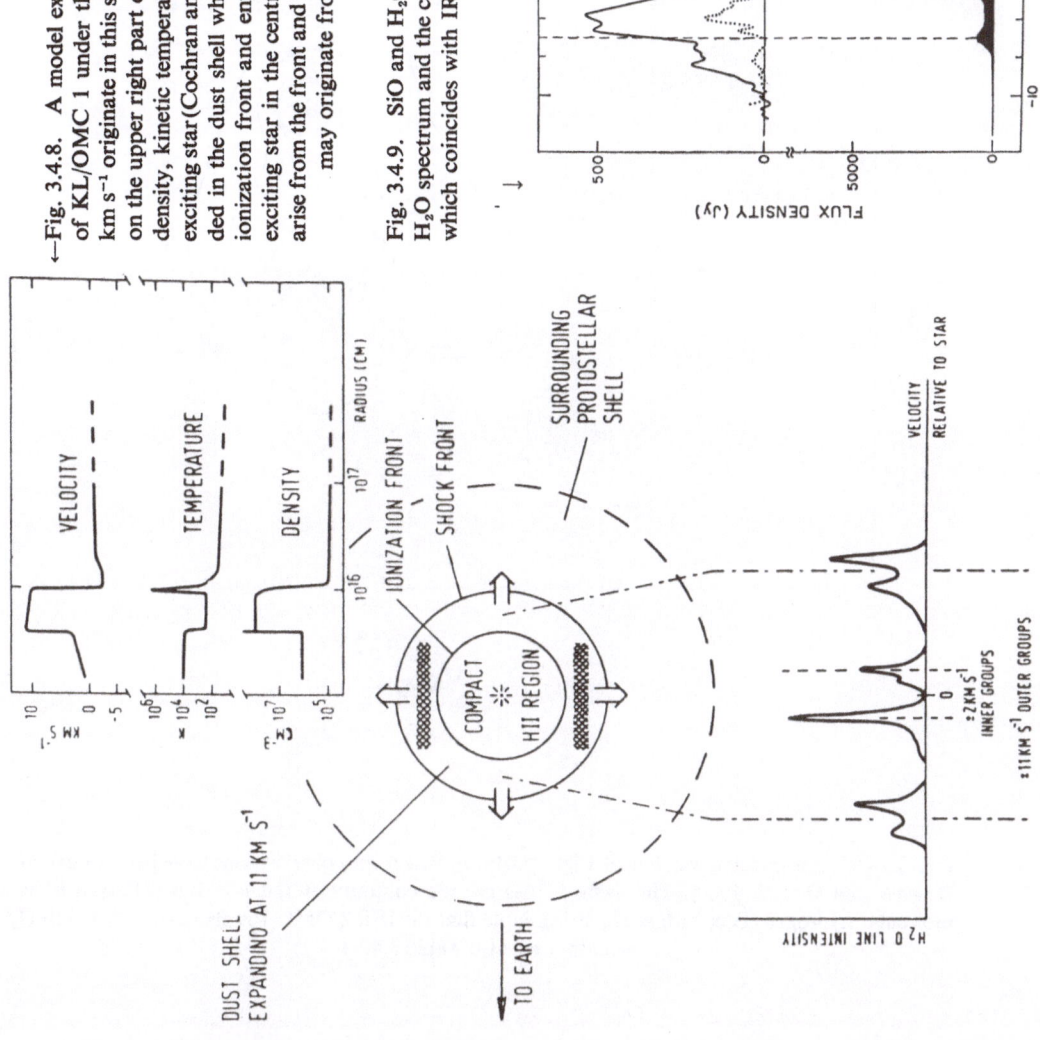

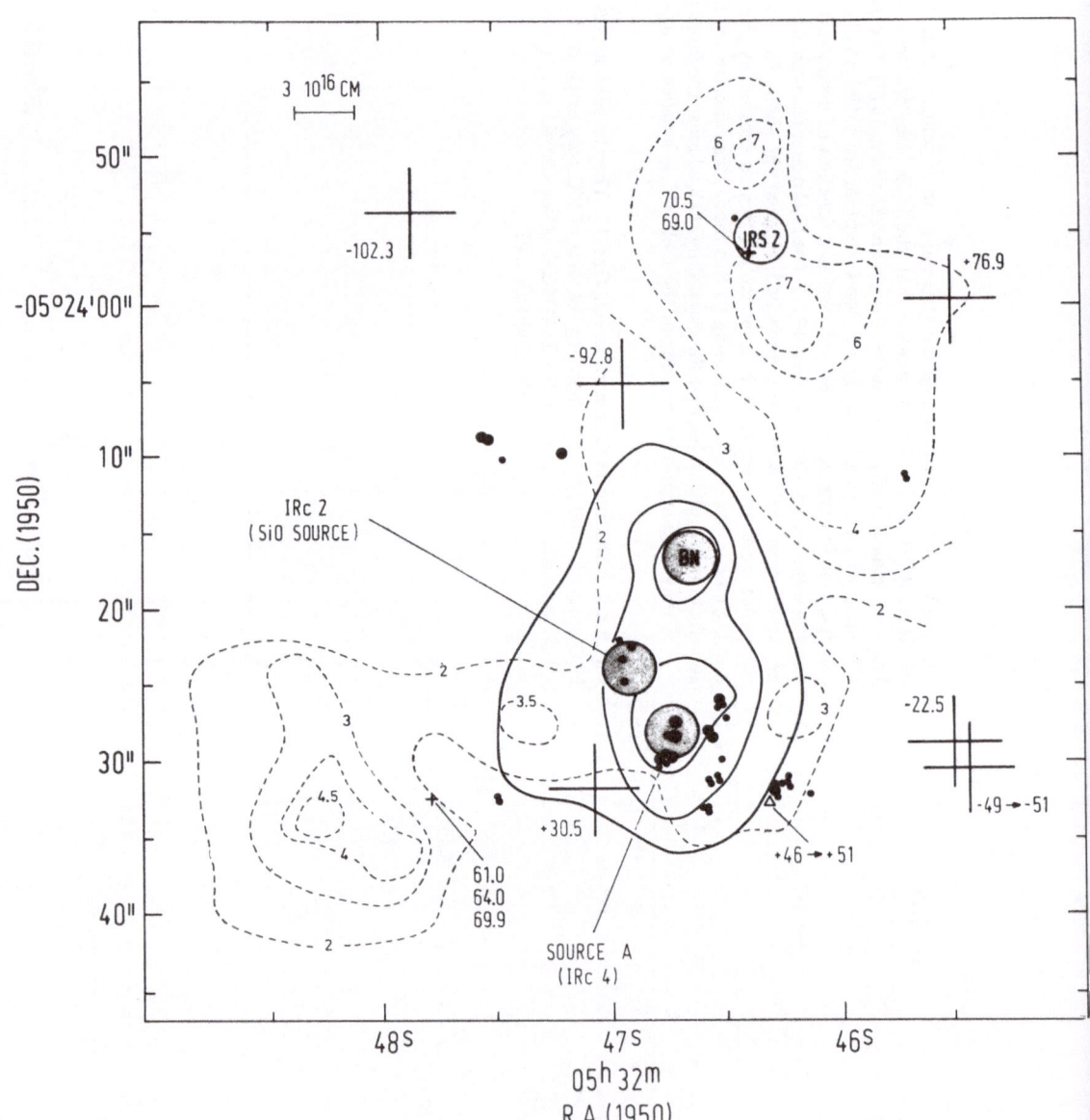

Fig. 3.4.10. An updated version of Figure 3.4.6 in which new observations have been added (after Downes and Genzel, 1980). The dashed lines are the contours of the $v = 1–0$ S(1) transition of molecular hydrogen (Beckwith *et al.*, 1978). Note that the IRS 2/OMC 1 is also associated with H_2O masers see also Table 3.4.VI.

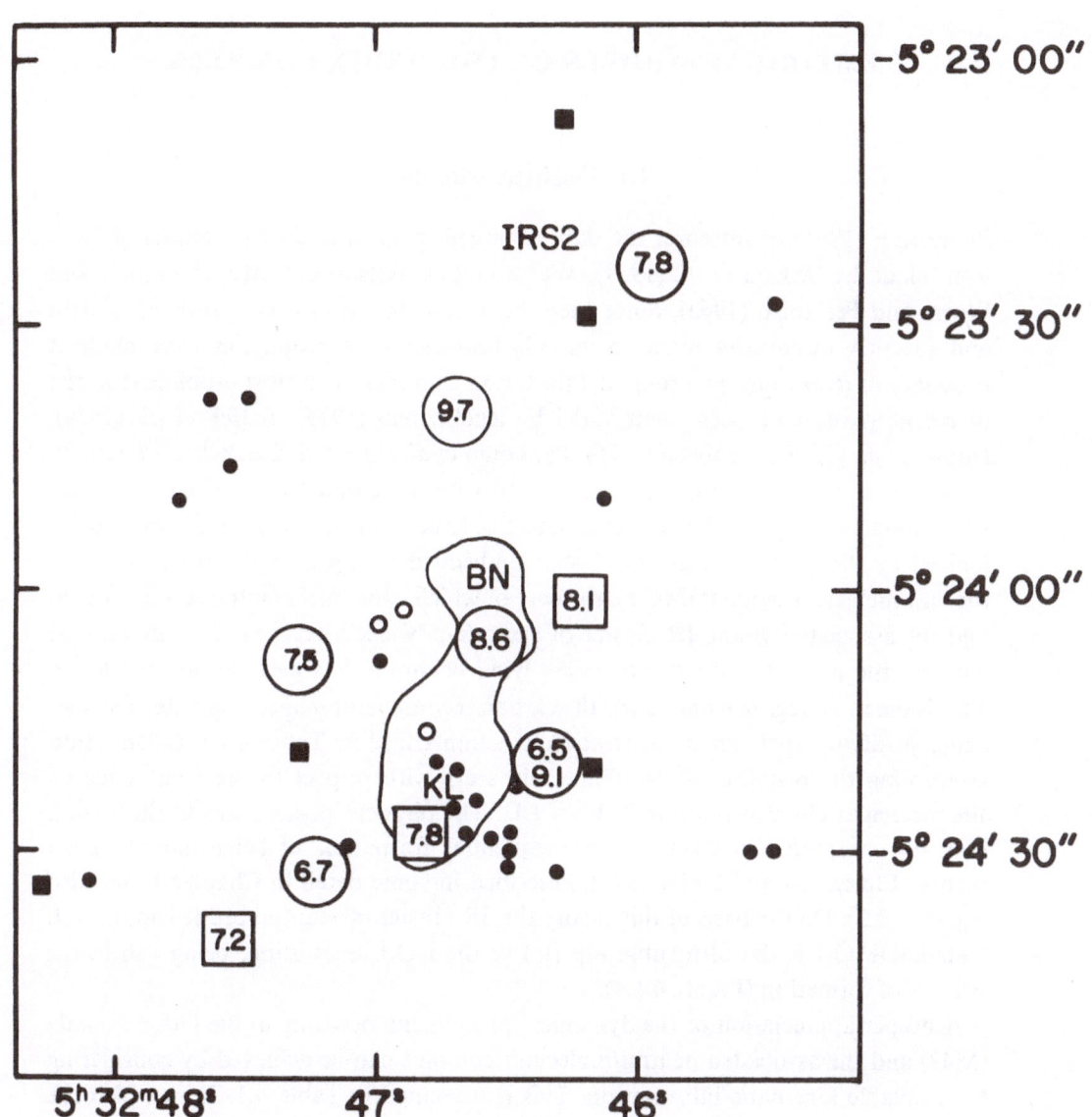

Fig. 3.4.11. Distribution of the various CH$_3$OH masering features over the KL/OMC 1 (after
Matsakis *et al.* 1980). Numbers within circles or squares indicate the V_{LSR} (km s^{-1}) of the methanol
masers at that position. H$_2$O (filled circles), OH (open circles), and H$_2$ (filled squares) positions are
shown for comparison. The contour shows the extent of the infrared emission. The methanol masers
do not coincide with H$_2$O, OH, and SiO masers; they are probably denser clumps embedded in OMC 1.

CHAPTER 4

EMPIRICAL MODELS OF THE ORION COMPLEX

4.1. Empirical Models

Pioneering efforts to interpret the data obtained from optical observations of M42 were made by Wilson *et al.* (1959), Wurm (1961), Wilson and Münch (1962), and Wurm and Perinotto (1965). Since then the accumulation of a vast amount of data and the new discoveries made in various branches of Astrophysics have made it necessary to re-examine and re-model the Orion Complex. The most prominent works of recent years have been contributed by Zuckerman (1973), Balick *et al.* (1974), Dopita *et al.* (1974), Meaburn (1975), Pankonin *et al.* (1979), and Balick *et al.* (1980).

The broad structure of the area associated with the Orion Nebula (M42) is shown in the model of Figure 4.1.1. According to this (Zuckerman, 1973) the Orion Nebula, ionized by the Trapezium group of stars embedded within it, is located in front of the vast molecular cloud (OMC 1), the core of which contains the infrared KL Nebula and the associated young IR cluster of stars. Sandwiched between the fully ionized gas and the neutral molecular complex lying behind it, is a partially ionized layer. This is the main region from which the carbon recombination lines originate. The evidence justifying such an interpretation is summarized in Table 4.1.I. Information concerning the position of the Trapezium stars with respect to the front edge of the molecular cloud is given in Table 4.1.II. The physical phenomena in the region may be interpreted in terms of the sequential formation of subgroups of stars theory (Elmegreen and Lada, 1977), described in some detail in Chapter 1 (see also Figure 1.3.3). On the basis of this theory the IR cluster of stars embedded in the KL Nebula/OMC 1 is the fifth subgroup (Ie) of the I Ori association, being still in the process of formation (Figure 4.1.1).

A deeper appreciation of the dynamical phenomena occuring in the Orion Nebula (M42) and the associated neutral/molecular complex can be achieved by considering the available kinematic information. This is presented in Table 4.1.III in which the kinematics of the various constituents of the complex (molecular cloud, carbon zone, ionized region, Trapezium stars) are summarized. Consideration of the radial velocities of the various features with respect to the molecular cloud (Table 4.1.III) leads to the conclusion that the Trapezium stars are moving into the molecular cloud with a velocity of $+3$ km s^{-1}. The partially ionized interface layer between the H II region and the molecular cloud is also moving into the cloud with roughly the same velocity (~ 2 km s^{-1}). The situation can be expressed in terms of ionizing stars and their associated H II region 'eating' into the molecular cloud. The ionized gas, however, flows away from the molecular cloud, with a velocity of -10 km s^{-1}. This is not surprising since the newly ionized gas at the interface of the H II–H I/molecular region, finds itself in the denser part of the ionized region ($N_e = 10^4$–10^5 cm^{-3}). Since the electron temperature in the ionized region is roughly the same everywhere ($T_e \simeq 10^4$ K), the pressure ($P = 2kN_eT_e$) in the denser part is higher than in the less dense foreground region ($N_e \simeq 10^2$ cm^{-3}). This pressure gradient forces the newly ionized gas to flow away from the place of its origin into the relatively tenuous foreground space. The expelled ionized gas is immediately replenished by new gas, ionized at the H II–H I interface. This mechanism maintains the pressure gradient which forces the newly produced ionized gas to flow continuously away. Thus the picture emerging is of an H II region strongly ionization bounded on the back side (and probably on the northern and eastern sides as well) and freely expanding into the tenuous foreground gas.

This process which is thought to be occuring not only in Orion (M42) but in the majority of H II regions associated with neutral/molecular complexes (Israel, 1978) creates a 'fire cracker' or 'blister' burned at the edge of the molecular complex and can hardly be approximated by the traditional spherically symmetric models (Strömgren spheres). The inadequacy of the classical theoretical models (Strömgren, 1939; Kahn, 1954) to cope with the observational picture has inspired new theoretical modelling of H II regions and has already generated a number of important works (Tenorio-Tagle, 1979; Whitworth, 1979; Tenorio–Tagle et al., 1979; Bodenheimer et al., 1979; Icke et al., 1980). The schematic representation of the evolution of an H II region according to one of these is shown in Figure 4.1.2, while a semi-empirical model coping with the particular kinematic and morphological structure of the Orion Nebula (M42) and the associated neutral/molecular complex is shown in Figure 4.1.3. Details concerning both models are included in the captions of the respective figures.

Finer interpretations of the events occuring in the core of Orion (M42) have to take into account the complicated radial velocity structure observed over the ionized area. This structure is shown in Figure 4.1.4 in which the splitting of the [O III] line over the core of M42 and the distribution of the radial velocities of the H76α lines are presented. Also shown in the same figure are the 'bar', two weaker filaments near the Trapezium stars, the position of the radio continuum peak (which coincides with the peak density position lying $\sim 30''$W of Trapezium) and the positions of the peaks in the brightness distribution of H$_2$CO emission. The distribution of the H76α velocities shows a roughly rounded structure with the outer velocity contours blueshifted towards the observer. This was attributed by Pankonin et al. (1979) to the presence of a 'stellar bubble' created by stellar winds or radiation pressure around the main ionizing star θ^1 Ori C. Their model, a cup shaped ionized cavity moving into the molecular cloud, also explains the displacement of the radio continuum (peak density) position from that of the Trapezium stars as a projection effect: they suggest that the observer views the nebula at an angle relative to the axis of symmetry. Further details concerning this model are included in the legend of Figure 4.1.5.

Another characteristic of the available kinematic data is the dependence of the observed velocities not only on the position in the nebula but also on the state of excitation of the observed line. This was first noted by Kaler (1967) (Table 2.2.XVIII and note (8) in the same table) and more recently by Balick *et al.* (1980) (see however alternative argument by Fehrenbach, 1977). The observed velocities of the forbidden lines [O I], [N II], [S II], and [O III] for a 2′ region near θ^1 Ori C are included in Table 4.1.III. There is substantial evidence supporting the view that certain elements like Fe^+, Fe^{++}, and [O I] with low ionization potentials (7.9, 16.7, and 13.6 eV, respectively) do in fact originate in regions located in or close to the H II–H I/Molecular interface layer. They all have the same low velocity with respect to the molecular cloud (in the range of 0–2 km s^{-1}) which is similar to the velocity of the carbon lines (Table 4.1.III). There is also independent evidence indicating that the O I in the 'bar' originates in the neutral material associated with it (Taylor and Münch, 1974). However, the observed velocities of [N II], [S II], and [O III] do not show a simple pattern and are more difficult to understand. Current explanations of this more complex scheme of velocities attribute the phenomenon to secondary flows interfering with the primary flow of ionized gas which is directed roughly along the line of sight (see Figure 4.1.3). It should be mentioned that the observed velocity is a weighted velocity containing contributions from different flows and different regions along the line of sight. The observed hydrogen lines are superpositions of lines originating in regions near the molecular cloud where densities are high and velocities low, and lines originating in regions far from the molecular cloud where densities are low and velocities high. The end result is a velocity measurement weighted towards higher velocities. This however may not be the case with velocity measurements based on the [S II], [N II], or [O III] lines, if their ionized zones do not participate in the same motion as that of hydrogen.

The flows of ionized gas, employed to explain certain kinematic aspects of the core of the Orion Nebula, do probably characterize the whole nebula, as suggested by Meaburn (1975). These flows are thought to be the cause of the observed splitting of the [O III] line over the core of Orion (Wilson *et al.*, 1959; Figure 4.1.4) and the splitting of the [N II] line observed several minutes of arc away (Deharveng, 1973). An alternative cause of the [O III] splitting may be the presence of partially ionized globules (PIGs) of neutral matter enveloped by ionization fronts (Dopita *et al.*, 1974; Figure 2.3.4). Evidence for the existence of such condensations has been produced recently by Laques and Vidal (1979) (Section 2.3). The detected condensations, however, are within the area defined by the Trapezium stars and do not coincide with the areas where [O III] splitting was observed.

A characteristic feature of the core of Orion, marked in Figure 4.1.4, is the 'bar' which is most probably an ionization front viewed edge on and 'eating' into the neutral material with which it is associated (Section 2.1; see also Figure 2.1.12). A summary of the reasons supporting such an interpretation is given in Table 4.1.IV. Two more filaments, marked in the same figure and seen distinctly in photographs taken in the light of the low excitation lines O I, [S II], and [N II] (Figures 2.1.8, 2.3.5, and 2.3.6, respectively) appear to be associated with areas of [O III] splitting and a peak of H_2CO emission originating in the molecular ridge behind the ionized matter. Such a correlation may imply a generic relationship of the filaments with the molecular cloud. The possibility that they represent ionization fronts seen edge on has already been suggested in the literature (Meaburn, 1975). It should be noted that the positions of the various features seen in the correlation of Figure 4.1.4 may be affected by projection effects. These however are not expected to be important since the depth of the features involved is relatively small (Münch, 1980).

M42 is enveloped to the south by a faint looplike filamentary structure beautifully depicted in Figure 4.1.6. Kinematic information concerning these filaments is not yet available and any suggestion concerning their nature and energetics is necessarily speculative. A morphological model of M42 incorporating the filamentary structure seen in Figure 4.1.6 is presented in Figure 4.1.7.

The companion of the Orion Nebula, the small H II region M43, has not stimulated the interest of the astronomical community to the same extent as M42. There is evidence supporting the view that M43 is not dynamically coupled to M42. Although both are offsprings of molecular fragments originating in the same complex (OMC 1 and OMC 2, respectively) M43 seems to be an ionization bounded spherical H II region embedded in the neutral cloud (Thum *et al.*, 1978) which exhibits a far simpler structure than the Great Orion Nebula M42. The indications supporting the independent existence of M43 as an H II region are summarised in Table 4.1.V.

TABLE 4.1.I

Location of the main constituents of the Orion Complex M42/OMC 1

Location	Evidence supporting it
a. The molecular cloud (OMC 1) lies behind the H II region (M 42)	1. The morphological and kinematic structure of the molecular cloud is distinctly different from that of the H II region. 2. The vast majority of the detected molecular lines are seen in emission; if even a very small fraction of optical molecules (CN, CH, CH⁺) was lying in front of the H II region, their absorption features should have been detected in the visible region.
b. The Kleinmann–Low (KL) nebula and the IR cluster are embedded in the molecular cloud (OMC 1)	1. The KL nebula is not associated with any optical or radio feature of the H II region. 2. The positions of the KL nebula, the IR cluster of stars, the OH, H_2O masers and the core of the molecular cloud are very close to each other. 3. The temperature of the dust in the KL nebula is equal to the molecular gas temperature of the core of OMC 1 (~ 80 K).
c. The carbon recombination lines originate in the H II–H I/molecular interface layer.	1. The morphological and kinematic structure of the carbon emitting region is similar to that of the molecular ridge of OMC 1. 2. The carbon line widths are similar to the widths of the molecular lines detected in the ridge of OMC 1 (very narrow, ~ 4 km s⁻¹, indicating a cool place of origin, ~ 50 K).

TABLE 4.1.II

Distance between the Trapezium stars and the front edge of the molecular cloud OMC 1[1]

Distance (pc)	Derived from arguments based on:	References
0.10	an excitation parameter of $u = 100$ pc cm⁻² for the H II region and an electron density $N_e \sim 3 \times 10^4$ cm⁻³.	Zuckerman (1973)
≤0.26	the scattering of light by dust.	Schiffer and Mathis (1974)
~0.4	a carbon line contour map.	Balick et al. (1974)
~0.3	the extent of the Trapezium (Ney–Allen) infrared nebula.	Harper (1974)

(1) As summarized by Zuckerman (1974).

TABLE 4.1.III

Summary of the kinematics of the Orion Complex M42/OMC 1

Lines	V_{Hel} (km s^{-1})	V_{LSR} (km s^{-1})	V with respect to the molecular cloud (km s^{-1})	Full width ΔV (km s^{-1})
a. Molecular cloud (OMC 1)				
Molecular lines originating in the 'ridge' of OMC 1 }	+26	+8	0	4
b. Partially ionized H II–H I/molecular interface				
Carbon recombination lines	+28	+10	+2	4
c. Trapezium system of stars				
	+29	+11	+3	
d. H II region (M42)				
H recombination lines	+16	−2	−10	27
He recombination lines	+16	−2	−10	24
[O I] lines	+28	+10	+2	15
[O III] lines (near θ^1 Ori C)	+20	+2	−6	18
[N II] lines (near θ^1 Ori C)	+22	+4	−4	20
[S II] lines (near θ^1 Ori C)	+22	+4	−4	22

TABLE 4.1.IV

Main evidence supporting the interpretation of the 'bar' as an ionization front viewed edge on

Evidence	Relevant figures
1. Strong emission from low excitation lines.	2.1.5, 2.1.6, 2.1.8, 2.1.9
2. Association with dust/neutral material lying on the side away from the Trapezium stars (infrared emission, [O I] line emission). }	2.4.7, 2.4.8
3. Association with areas of [O III] splitting lying on the side towards the Trapezium stars. }	2.2.2

TABLE 4.1.V

Evidence supporting the independent existence of M43 as an H II region

Evidence	Relevant figures and tables
1. The radio continuum contour maps of the M42/M43 area show a minimum of emission between M42 and M43 indicating that the dark lane, visible in the optical photographs between the two nebulae, separates them physically. }	Figures 2.1.14, 2.1.15, 2.1.16 a, b, c
2. The linear polarization map of the dust scattered light emitted by the star NU Orionis shows a centrosymmetric pattern indicating that there is no contribution to the ionization of M43 from the stars associated with M42. }	Figure 2.4.5
3. The kinematics of the ionized gas of M43 is distinctly different from that of M42. }	Tables 2.2.XIX, 2.2.XX

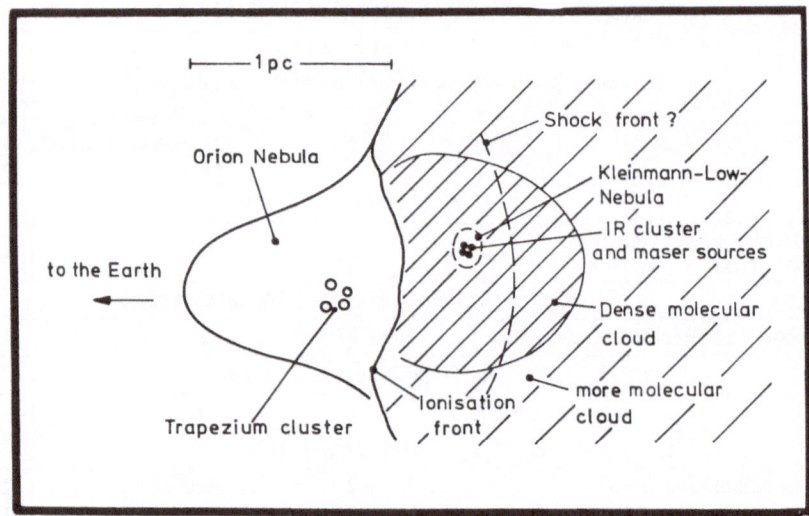

Fig. 4.1.1. Model of the Orion Nebula (M42)/Molecular Cloud proposed by Zuckermann (1973) (after Lemke and Harris, 1979). The Trapezium cluster of stars, formed on the outer edge of the molecular cloud, have ionized part of it, creating the Orion Nebula. The H II region is strongly ionization bounded on the back-side (and probably on the eastern and northern side as well), but expands freely on the front side (towards the observer) where the density of the foreground interstellar matter is very low. On the back side, the interface between the H II region and the neutral/molecular complex consists of a partially ionized layer in which the carbon recombination lines originate. According to Elmegreen and Lada (1977), the ionization front powered by the Trapezium stars, is encroaching into the molecular cloud preceded by a shock front. The matter sandwiched between the two fronts becomes gravitationally unstable. The fragments of this layer are probable birthplaces of new clusters of stars. This may be the case with the IR cluster which is associated with the KL Nebula (see also Figure 1.3.3).

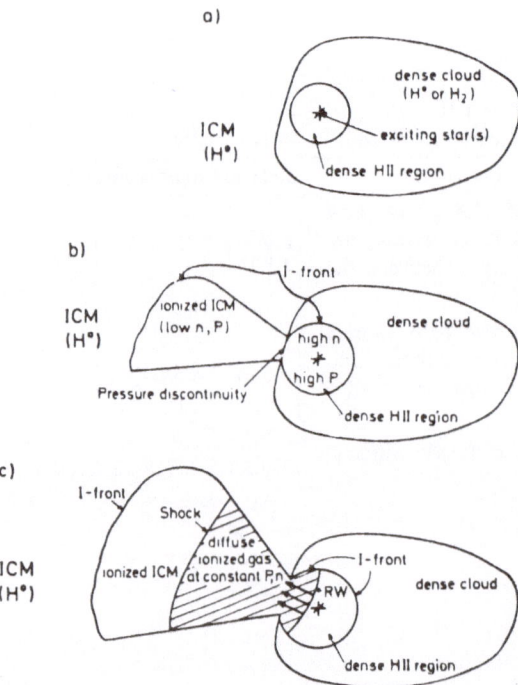

Fig. 4.1.2. The initial stages of the evolution of an H II region created by an O star lying inside and near the edge of a dense neutral/molecular cloud (after Tenorio-Tagle, 1979). ICM stands for the low density ($\simeq 1$ cm^{-3}) intercloud medium surrounding the dense could (C) ($N_{ICM} \ll N_C \simeq 10^4$ cm^{-3}). P is the pressure inside the H II region ($P = 2N_e kT_e$). *Phase (a):* Ultraviolet radiation emitted from an O star forms a dense spherical H II region near the edge of a dense neutral cloud; the H II region is probably optically unobservable. *Phase (b):* The ionization front facing the edge of the cloud, encroaches into it and eventually breaks into the low density intercloud medium. Since the temperature in the H II region does not vary considerably, the difference in density sets a pressure discontinuity between the ionized cloud material and the now ionized intercloud medium; the situation strongly resembles that of a shock tube when the diaphragm is released. *Phase (c):* A strong shock produced by the pressure gradient moves into the ionized intercloud medium while a rarefaction wave (RW) moves towards the star. The ionized cloud material flows supersonically behind the shock and eventually spread over an extended region. At this stage the H II region becomes optically observable.

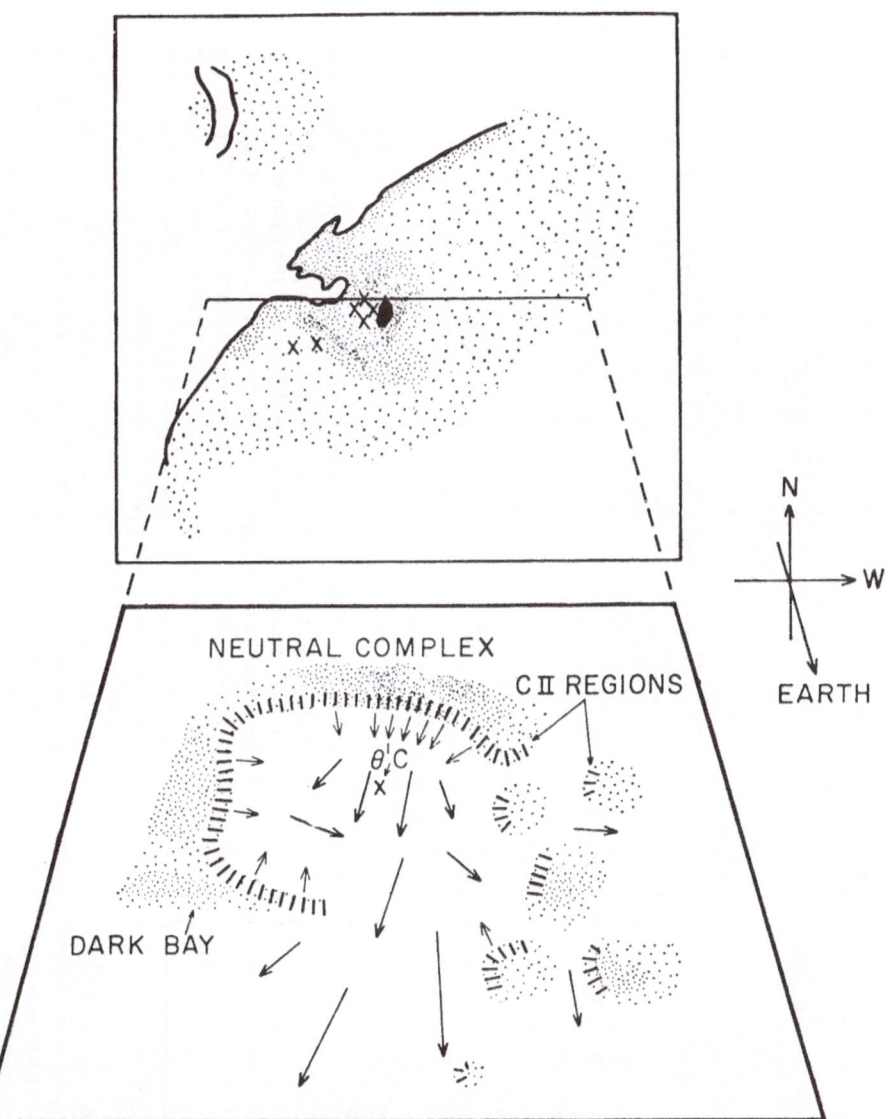

Fig. 4.1.3. A system of flows of ionizing matter originating in the ionization bounded areas of the Orion Nebula (M42) (rear and most probably eastern and northern side) can explain the main kinematic features observed in Orion (after Balick *et al.*, 1974). The lower part of the figure represents a cross section through the Trapezium region in the E-W direction along the line of sight. The partially ionized layer (C II zone), which constitutes the interface between the H II region, and the H I/ Molecular cloud, is shown cross-hatched. The actual, however, structure of the C II layer is quite complicated. Three distinct carbon velocity features detected by Jaffe and Pankonin (1978) appear to originate in different places. The higher velocity component (at 11 km s^{-1}) is localized in the C II shell at the back of the ionized gas, the middle component (at 8.5 km s^{-1}) is concentrated in the dark lane between M42 and M43, whereas the low component (at 6 km s^{-1}) originates in the dark bay which is a foreground protrusion of the molecular cloud enveloping the NE part of the nebula. Probable density fluctuations in the neutral/molecular gas are symbolised by a density variable dot pattern. Ultraviolet radiation from θ^1 Ori C incident on the dense neutral/molecular cloud ionizes the gas in the relatively dense region ($N_e \simeq 10^4$ cm^{-3}) near the Trapezium stars. The ionized gas flows away (towards the observer) under the influence of a pressure gradient set up by the density difference between the dense gas in the Trapezium region ($N_e \simeq 10^4$ cm^{-3}) and the tenuous gas in the foreground ($N_e \leq 10^2$ cm^{-3}). The expelled ionized gas is instantaneously replenished by new gas, ionized by the UV radiation of θ^1 Ori C.

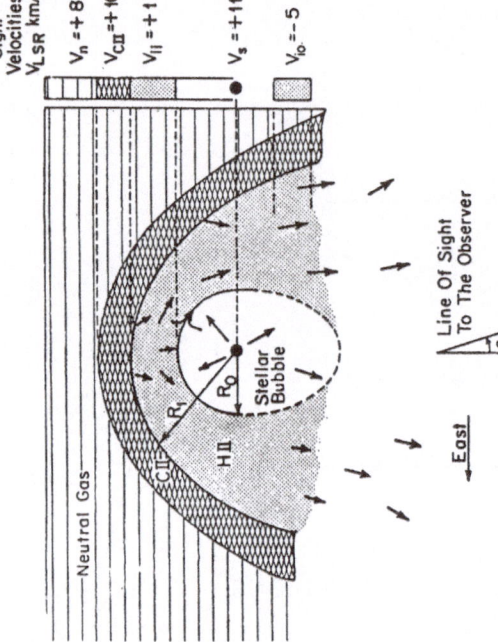

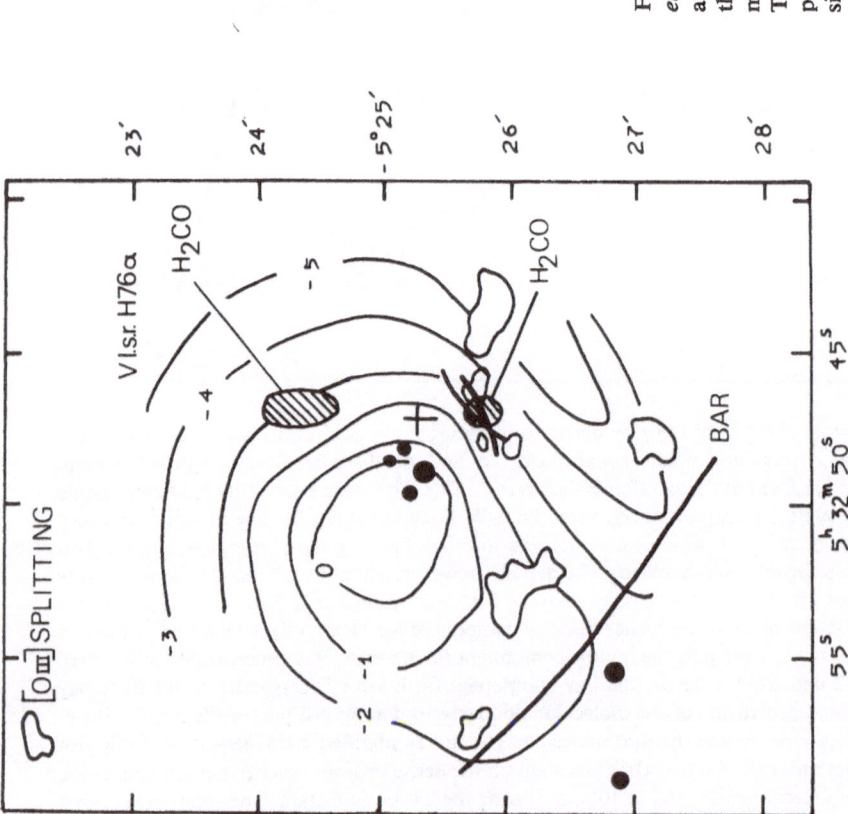

Fig. 4.1.5. Model of the core of the Orion Nebula (after Pankonin *et al.*, 1979). Ultraviolet radiation emitted from the star θ^1 Ori C creates a cup shaped cavity of ionized gas. Marked in the right hand scale are the velocities (with respect to the local standard of rest) of the neutral/molecular gas (V_n), of the carbon recombination lines (C II), of the Trapezium stars (V_s) and of the ionized hydrogen at the inner and outer part of the cavity (V_{ii} and V_{io} respectively). The fundamental idea is similar to that of the model shown in Figure 4.1.3.

The presence of a stellar bubble around the Trapezium stars, evacuated by stellar winds or radiation pressure, creates an obstruction preventing the outflow of the ionized gas along this direction. This can explain the observed velocity distribution of ionized hydrogen as derived from the H76α line (Figure 4.1.4). The 'bluer' velocities in the west than in the east are attributed to the neutral protrusion wrapping the nebula from the east ('bay' area). The offset of the radio continuum peak from the Trapezium stars is also explained as a projection effect
(observer viewing the nebula at an angle θ).

Fig. 4.1.4. The distribution of the H76α radial velocities (V_{LSR} in km s^{-1}) over the core of M42 (Pankonin *et al.*, 1979; angular resolution of 1') compared to the distribution of the areas for which the ([O III] line is split (Wilson *et al.*, 1959) angular resolution of 1''3). Also shown are the radio continuum peak (which roughly coincides with the density peak position), the peaks of the H$_2$CO emission originating in the molecular ridge (Thaddeus *et al.*, 1971; angular resolution of 1'), the 'bar' and two fainter filaments which are associated with areas for which the [O III] line is observed split.

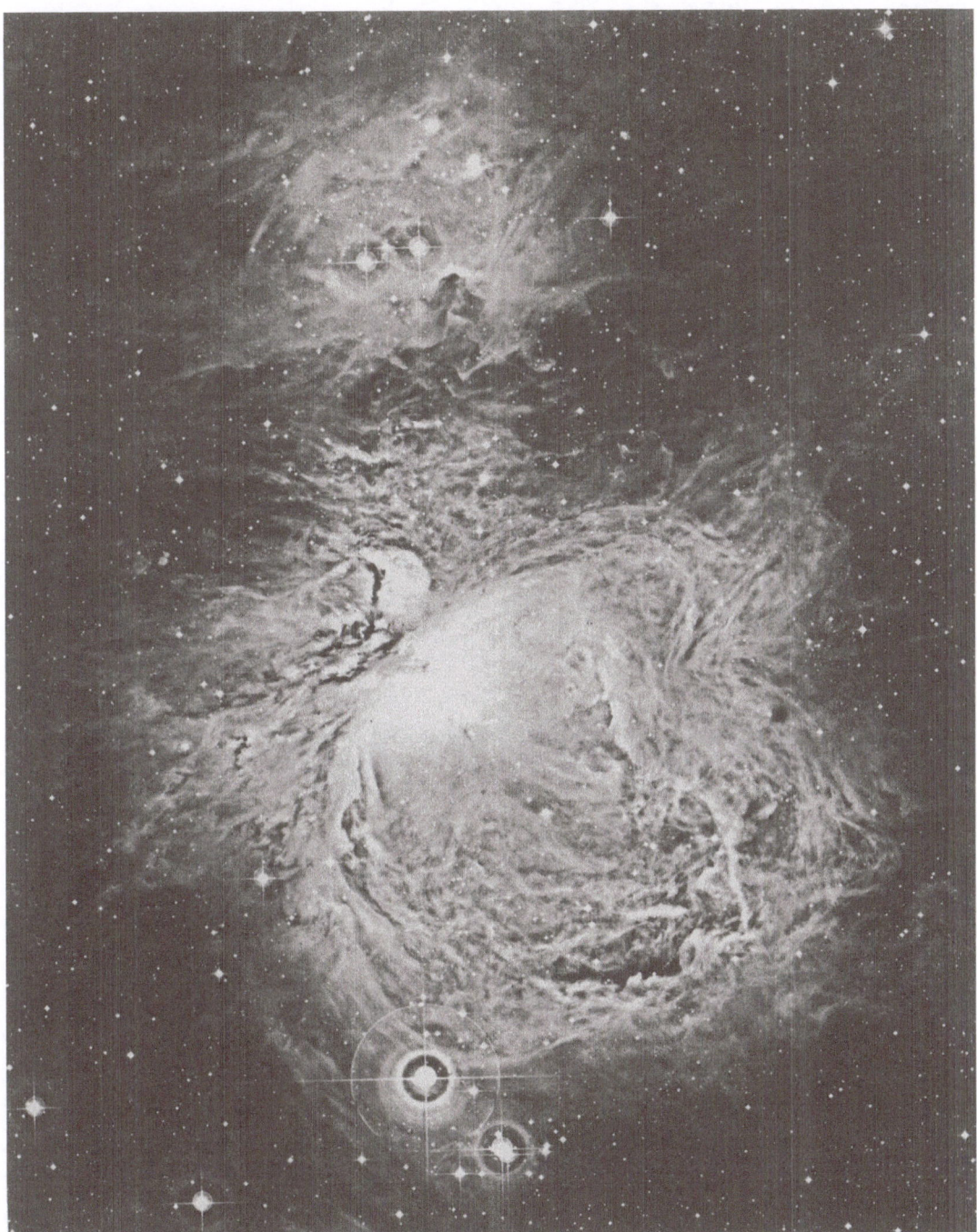

Fig. 4.1.6. A red photograph of the Orion Nebulae (M42, M43) and NGC 1977 (to the north) taken with the 1.2 m UK Schmidt telescope through a Schott RG630 filter (bandpass: 6300–7000 Å); the exposure time was 90 m; the print was made through an unsharp mask (courtesy of Dr D. F. Malin). Note the filamentary looplike structure enveloping M42 to the south.

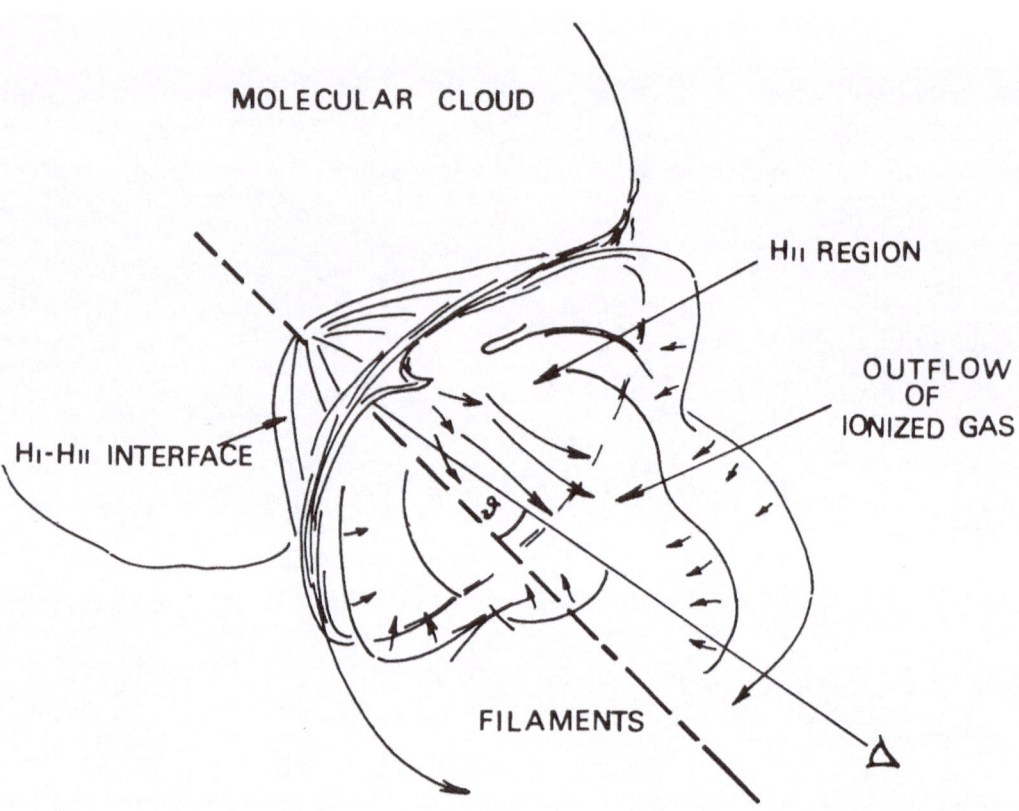

Fig. 4.1.7.   Morphological model of the Orion Nebula (M42). Ionized gas flowing away from its place of origin (H II–H I/molecular interface on the rear, eastern and northern side of the nebula) can produce the depicted loops. Some of them might be collapsing towards M42. Alternative energetic processes sustaining these loops (e.g. stellar winds) cannot be ruled out.

CHAPTER 5

NGC 2024 AND THE ASSOCIATED MOLECULAR COMPLEX

5.1. Optical and Radio Structure of NGC 2024

The second most prominent source in the Orion region is NGC 2024 (W12, Orion B). It lies eastwards of the bright star ζ Orionis (the eastern star in the belt of Orion) and is seen in the sky together with the elongated filamentary nebula IC 434 and the small reflection nebula NGC 2023, although they are probably not physically associated. A photograph of the region containing all these sources is shown in Figure 5.1.1, and an Hα contour map of the same area is presented in Figure 5.1.2. A radio continuum contour map of the fainter area around NGC 2024 (mainly IC 434) is shown in Figure 5.1.3.

A red photograph of NGC 2024 printed through an unsharp mask and showing the fine structure of the source and the characteristic dust lane dividing the nebula is presented in Figure 5.1.4. The radio continuum contour maps of the object made with an intermediate angular resolution (2'–4') show a spherically symmetric structure centred at a point lying behind and in the middle of the dust lane. This implies that the dust lane does not separate the H II region physically, but lies in the foreground. Such a contour map made with a 2' angular resolution is shown in Figure 5.1.5. Another radio continuum contour map made with a higher angular resolution ($\sim 1'$) is presented in Figure 5.1.6a and shows the existence of two ionizing centres of activity. Both of these are located behind the obscuring layer as can be seen distinctly in the superposition of Figure 5.1.6b. A compilation of the published radio continuum maps of the object is given in Table 5.1.I. The thermal nature of the radio continuum is demonstrated in the radio spectrum of NGC 2024 shown in Figure 5.1.7. The distribution of the visual extinction of light derived from the comparison of optical (attenuated by dust) and radio (unattenuated) data is presented in Figure 5.1.8.

The physical parameters of the source derived from observations of the radio continuum emission are presented in Table 5.1.II. Included are the angular size ($\theta\alpha \times \theta\delta$ and a mean size θ), the linear diameter ($2R$) for a distance D from the Earth, the root mean square electron density (rms N_e), the mass of the ionized gas ($M_{\mathrm{H\,II}}$), the emission measure (E), the excitation parameter (u), and the required rate of ionizing Lyman continuum photons to produce such an H II region (Lc). The physical parameters of the nearby nebulocity IC 434 are also included for comparison.

NGC 2024 is a particularly important and untypical source of recombination line emission showing:

(1) weak He line emission both in the radio and optical part of the spectrum;
(2) a strong and narrow ($\Delta V \simeq 5$ km s^{-1}) C recombination line;

(3) a line emission originating in an element (or elements) heavier than carbon (Z line emission) which also exhibits a narrow width ($\Delta V \simeq 5$ km s^{-1}); and

(4) narrow ($\Delta V \simeq 5$ km s^{-1}) hydrogen line emission (in addition to the normal broad hydrogen lines with $\Delta V \simeq 25$ km s^{-1}).

The presence of a weak He line leads to an estimated helium abundance $N(\text{He}^+)/N(H^+) \simeq 0.02$ (Table 5.1.III) which is far lower than the typical value of $\simeq 0.1$. This is not, however, thought to imply a real He deficiency, since NGC 2024 is the only galactic spiral arm H II region observed to have such a deficiency. Such an explanation for the weakness of the observed He lines should be rejected on statistical grounds (Sarazin, 1976).

Alternative explanations (proposed by Cesarsky, 1971) attribute the observed deficiency to:

(1) the low effective temperature of the exciting star (or stars) which results in a weak flow of He ionizing photons; or

(2) the presence of 'ultraviolet absorbing dust' i.e. dust which preferentially absorbs the He ionizing photons.

Further elaboration on these views, which are related to the nature of the ionizing star(s) of the source, will be encountered in section 5.2 which deals with the problem of the excitation of NGC 2024.

The region of origin of the other three lines (C, Z, and narrow H recombination lines), a characteristic spectrum of which is shown in Figure 5.1.9, is thought to be the partially ionized layer constituting the interface between the H II and H I/molecular cloud associated with NGC 2024. This relatively thin layer can be envisaged as a shell (probably incomplete) partly enveloping the H II region.

The association of the C line with the neutral/molecular cloud detected in the direction of NGC 2024 is based on the striking kinematic correlation of C with the atomic hydrogen and the various molecular species (Figure 5.1.10 and Tables 5.1.V and 5.2.V later in the text). In addition the narrow width of the C line implies a cool place of origin ($T \simeq 50$ K; Dupree, 1974) which is compatible with the kinetic energy of the molecular cloud (T_k in the range of 20–50 K; Table 5.2. III). The extent of the C II zone, which does not coincide with the H II source, is shown in Figure 5.1.11 together with the extent of the He II zone which is distinctly smaller than that of H II.

The heavy element (Z) lines probably originate in a region largely overlapping with that of the C line without, however, being entirely coincident (Pankonin, 1980). These lines are also narrow and have been attributed to sulphur (Pankonin et al., 1977).

The 'narrow' hydrogen recombination line emission also originates in the cold, partially ionized medium surrounding the H II zone and blends to a large extent with the broad hydrogen recombination line emission originating in the hot H II region. The origin of the 'narrow' hydrogen lines has been attributed to the outer portion of the ionization front bounding the H II region (Zuckerman and Ball, 1974) or to soft X-rays escaping from the ionized gas (Pankonin et al., 1977; Krügel and Tenorio-Tagle, 1978).

Temperature estimates of the hot H II region ($T_k \simeq 8000$ K) and the relatively cold ionized H I zone ($\simeq 500$–1000 K), derived mainly from the investigation of hydrogen recombination lines, are presented in Table 5.1.IV. Estimates of the turbulence characterizing the H II region are also included in the same table. The distribution of the electron temperature over the H II region is presented in Figure 5.1.12.

Kinematic information concerning both the H II region and the surrounding partially ionized H I zone, derived from investigations of the H, C, and Z lines is presented in Tables 5.1.V and 5.2.VI. The latter table contains the characteristic widths of the various lines in comparison with the widths of the molecular lines which originate in the associated neutral/molecular complex (Section 5.2). The distribution of the radial velocities over the ionized region are shown in Figure 5.1.13.

The dynamical age of the nebula has been estimated to be $(4$–$5) \times 10^5$ yr (Gordon, 1969; Grasdalen, 1974). The magnetic field associated with the source is less than 1.3 mG (Troland and Heiles, 1977).

5.2. Infrared and Molecular Structure of the NGC 2024/Molecular Complex

Mapping of NGC 2024 at two different far infrared wavelength ranges centred on 93μ and 350μ with comparable angular resolution ($\sim 2'$), has revealed completely different structures as can be clearly seen in Figure 5.2.1. The 93μ map shows a similarity to the radio continuum map of Figure 5.1.5 ($2'$ angular resolution) and to the distribution of the carbon recombination line shown in Figure 5.1.11, whereas the 350μ map shows striking correlations with the dust lane and the distribution of emission from the HCN molecule (mapped with an angular resolution of 1.3) (Figure 5.2.2).

These facts led Hudson and Soifer (1976) to suggest that the two maps result from the emission of two different dust components, a hot one ($T \simeq 80$ K) intermixed with the H II region and a colder one (with T probably of the order of 25 K) associated with the molecular cloud detected in the direction of the dust lane of NGC 2024 and thought to be in front of the ionized region. A ^{12}CO map of this cloud showing the spatial coincidence of a 'ridge' of CO emission with the dust lane is shown in Figure 5.2.3.

A compilation of the published infrared maps of the NGC 2024/molecular complex is presented in Table 5.2.I. Maps made in the near infrared (2–10μ) have provided useful information on the possible excitation sources of NGC 2024 and will be discussed in Section 5.3. The physical parameters of the H II region and the associated molecular complex, derived from infrared observations, are presented in Table 5.2.II. The clarification of various points concerning these parameters can be found in the accompanying commentary. The physical parameters of the associated molecular cloud, derived from the observations of the emission of many molecular lines are presented in Tables 5.2.III to 5.2.VI. These include estimates of the kinetic temperature of the molecular gas (Table 5.2.III), its chemical composition (Table 5.2.IV) and information concerning its radial velocities (Table 5.2.V). The widths of molecular lines in comparison with the widths of the lines originating in the ionized H II region and the partially ionized interface between the H II and the H I/molecular cloud are also given (Table 5.2.VI).

It should be remembered that the NGC 2024 complex is probably physically associated with the vast dust cloud L1630/Norther Molecular Complex (Chapter 1.1, Figure 1.1.9). The position of NGC 2024 and other nebulosities with respect to the dust/molecular cloud L1630, based on observations of the OH 1667 MHz line, are shown in Figure 5.2.4. Relevant information concerning the kinematics of the L1630 cloud is given in Table 5.2.VII.

5.3. Excitation of NGC 2024

The excitation requirements of NGC 2024 implied by the excitation parameter derived in Table 5.1.II. ($u \simeq 30$) demand the presence of a single main sequence star of spectral type in the range of O9–O9.5 (Appendix II.1, Figure II.1.2). The rate, however, of the He ionizing photons emitted by such a star is far too high to explain the low $N(H_e^+)$/ $N(H^+)$ observed in NGC 2024 (Table 5.1.III). This discrepancy can be resolved by assuming either a dust population which preferentially absorbs the He ionizing photons (Mathis, 1971; Cesarsky, 1971; Leibowitz, 1973; Mezger *et al.*, 1974; McLeod *et al.*, 1975; Sarazin, 1976), or an exciting star of low effective temperature already evolved from the main sequence and therefore having an overall luminosity greater than that of a main sequence star of the same temperature (Grasdalen, 1974).

The search for such a star or stars has revealed two candidates: a reddened B star, NGC 2024 #1 (Johnson and Mendoza, 1964) lying on the west edge of the dust lane and a bright IR point source lying right in the middle of the dust lane, NGC 2024 #2 (Grasdalen, 1974). The position of these stars can be seen in Figure 5.3.1 in which a 3.5μ contour map of NGC 2024 is superimposed on an infrared photograph. The energy distribution of the star NGC 2024 #2 strongly resembles that of the BN object (Figure 3.2.6).

Grasdalen (1974) has suggested that NGC 2024 #2 is a low temperature ($T_{eff} \simeq$ 23 000 K) star which has already left the main sequence, probably a supergiant with high luminosity ($L \simeq 10^6 L_{\odot}$) and has claimed to be the exciting star of NGC 2024. A more elaborate version of this suggestion was later proposed·by McLeod *et al.* (1975) who suggested that both NGC 2024 #1 and #2 stars are responsible for the excitation of the nebula. Their model is shown in Figure 5.3.2. The main drawback with models using NGC 2024 #2 as an exciting star is its high luminosity ($L \simeq 10^6$ $L_{\odot}$) with respect to that of the H II region ($L_{H\ II} < 10^5 L_{\odot}$, Table 5.2.II).

It is not improbable that NGC 2024 #2 is a much cooler main sequence star lying within the molecular cloud associated with the dust lane and located in front of the H II region (Thum, 1975; and Hudson and Soifer, 1976). In that case NGC 2024 #2 is the star heating the cooler infrared source associated with the molecular cloud, whereas the real exciting star lies far behind and is still undetected (Hudson and Soifer,

1976; Cesarsky, 1977). This arrangement, however, is not without its problems. Such a situation resembles the Orion situation viewed from behind (Frey *et al.*, 1979) with the molecular cloud facing the observer and the H II region expanding away from the molecular cloud. This, however, does not follow from the available kinematic information according to which a part of the ionized gas (SE of the nebula) is nearly at rest with respect to the molecular cloud, whereas another part (NW of the nebula) moves with respect to it towards the observer at -5 km s^{-1} (Figure 5.1.13, Tables 5.1.V and 5.2.V). This could be accomodated by adopting an 'ad hoc' geometry (H II region ionization bounded on the far side as well?) but such an interpretation simply adds more speculation to the problem. Information concerning the nature of the probable ionizing sources of NGC 2024 and the theoretical predictions of the parameters of the exciting stars are given in Table 5.3.I.

TABLE 5.1.I

Radio continuum contour maps of NGC 2024

No.	HPBW	Frequency ν (MHz)	Wavelength λ (cm)	References
1	6ʹ4	5 000	6	Mezger and Henderson (1966)
2	4ʹ2	5 000	6	Gardner and Morimoto (1968)
3	4ʹ2	7 800	3.9	Gordon (1969)
4	4ʹ0	5 000	6	Goss and Shaver (1970)
5	2ʹ8	10 700	2.8	MacLeod and Doherty (1968)
6	2ʹ8	408	73.5	Shaver (1969)
7	2ʹ0	15 400	1.95	Terzian *et al.* (1968), Schraml and Mezger (1969)
8	2ʹ0	1 420	21	Lockhart and Goss (1978)
9	1ʹ9	36 500	0.82	Berulis and Sorochenko (1973)
10	80″	23 400	0.13	Rodriguez and Chaisson (1978)
11	40″ × 90″	4 830	6.2	Fomalont and Weliachew (1973)
12	53″ × 56″	1 415	21	Löbert and Goss (1978)
13	6″ × 32″	2 695	11.1	Turner *et al.* (1974)

TABLE 5.1.III

Helium abundance of NGC 2024 derived from radio recombination lines

No.	Lines used	$N(He^+)/N(H^+)$	Remarks	References
	H85α, He85α	0.015 ± 0.01		Cesarsky (1971, 1977)
	H85α, He85α	0.042 ± 0.02	(1)	MacLeod *et al.* (1975)
	H86α, He86α	< 0.046		Lichten *et al.* (1979)
	H92α, He92α	0.031 ± 0.01	(2)	Chaisson (1973)
	H109α, He109α	< 0.02		Palmer *et al.* (1969)
	H109α, He109α	$\leqslant 0.02$		Schraml and Mezger (1969)
	H109α, He109α	$\leqslant 0.035$		Reifenstein *et al.* (1970)
	H109α, He109α	0.018 ± 0.01		Cesarsky (1971, 1977)
	H109α, He109α	< 0.02		Churchwell *et al.* (1974)

Remarks to Table 5.1.III:

(1) Maximum ratio.

(2) Mean value from two positions.

TABLE 5.1.II

Physical parameters of NGC 2024 and IC 434 derived from radio continuum measurements

No.	Coordinates (1950)	Distance D (kpc)	Angular size $\theta\alpha \times \theta\delta$ (arc min × arc min)	Angular diameter θ_G $\theta_G = \sqrt{\theta\alpha \times \theta\delta}$ (arc min)	Linear diameter $2R$ (pc)	Root mean square density rms N_e (cm^{-3})	Mass of ionized hydrogen $M_{H_{II}}$ ($M_\odot$)	Emission measure E (pc cm^{-6})	Excitation parameter u (pc cm^{-2})	Rate of Lyman continuum photons L_C[1] (photons s^{-1})	Remarks	References
					a. NGC2024 (G206.5−16.4)							
1	$5^h38^m58^s$ −1°54′12″	0.4	4.2×3.9		0.70	746	3.0	3.8×10^5				Mezger and Henderson (1967), Mezger and Höglund (1967)
2	5 39 12 −1 55 54		3.2×2.1		0.48	1350	1.9	8.7×10^5				MacLeod and Doherty (1968)
3	5 39 06 −1 56 06		3.3×2.7					7.4×10^5				Gardner and Morimoto (1968)
4						580		$(2.68\pm0.29)\times10^5$				Gordon (1969)
5			3.5×1.9		0.44	1347	1.5	8×10^5	26.9	7.4×10^{47}		Schraml and Mezger (1969)
6		0.6		3.8	0.70	1052	4.5	7.9×10^5	36.9	1.9×10^{48}		Shaver and Goss (1970b)
7		0.6				1157	1.8		31	1.1×10^{48}		Goss and Shaver (1970)
8	5 39 12 −1 55 42	0.6±0.1	3.6×3.0		0.60	790	6.7	3.7×10^5	38	2.1×10^{48}		Reifenstein et al. (1970)
9		0.4				1930	0.73	12×10^5				Berulis and Sorochenko (1973)
10		0.46			0.40	1300						Harper (1974)
11		0.6	3.0×2.5		0.60	1200			32	1.25×10^{48}		Churchwell et al. (1974)
12	5 39 12 −1 55 42	0.43			0.50	1250	2.1	7.8×10^5	28	8.3×10^{47}		Goudis (1975e)
13		0.5	2.6×3.0			2100		12×10^5			(2)	Churchwell et al. (1978)
14	{5 39 9.8 −1 55 43; 5 39 13.2 −1 56 28}		{1.2; 1.1}		{0.21; 0.19}	{2100; 2600}	{0.22; 0.22}	{8.7×10^5; 12×10^5}	{16; 19}	{1.6×10^{47}; 2.6×10^{47}}	(3)	} Löbert and Goss (1978)
15	5 39 15.0 −1 56 25	0.4	3.7×1.9	2.7		1600	2	12×10^5	32	1.25×10^{48}		Rodriguez and Chaisson (1978)
					b. IC 434 (G 206.9−16.8)							
16				16		6	266		28.5	8.8×10^{47}	(4)	Caswell and Goss (1974)

Remarks to Table 5.1.II:

(1) Calculated from relation (II.1.51) (see Appendix II.1) for $T_e = 8000$ K.
(2) Angular size expressed as $\theta_l \times \theta_b$.
(3) Double source.
(4) The quoted parameters, estimated on the basis of a spherical model, are probably not realistic.

TABLE 5.1.IV

Electron temperature of NGC 2024 derived from radio recombination lines and radio continuum measurements

No.	Method[6]	Electron temperature T_e (K)	Turbulence $\langle V_t^2 \rangle^{1/2}$	Remarks	References
a. H II region					
1	H56α	8400 ± 650			Berulis et al. (1975)
2	H66α	7000 ± 1300			Waltman and Johnston (1973)
3	H66α	7200 ± 500		(4)	Wilson et al. (1979)
4	H85α	9000 ± 600	9.4	(1), (2)	MacLeod et al. (1975)
5	H86α	9330 ± 650			Lichten et al. (1979)
6	H94α	8500	13.1 ± 0.3	(5)	Gordon (1969)
7	H109α		9.6		Meger and Henderson (1967)
8	H109α	7800^{+650}_{-550}			Mezger and Ellis (1968)
9	H109α	7100	10.6		Schraml and Mezger (1969)
10	H109α	7200 ± 890	11 ± 1.6		Reifenstein et al. (1970)
11	H109α	7700			Shaver and Goss (1970)
12	H109α	7510			Churchwell et al. (1978)
13	H126α + H127α	9170^{+1830}_{-1230}			Mezger and Ellis (1968)
14	H158α + H156α	6130^{+1370}_{-920}			
15	Average of H$n\alpha$ lines	7600 ± 800			
16	H108β	9120 ± 700			Lichten et al. (1979)
17	H137β	7560			Churchwell et al. (1978)
18	Continuum	6500^{+800}_{-600}			Shaver (1969)
19	Radio spectrum	4000^{+2000}_{-1000}			Terzian et al. (1968)
20	Continuum	6850			Shaver and Goss (1970)
21	Model (Non LTE)	8810			Churchwell et al. (1978)
22	Model	6800		(4)	Wilson et al. (1979)
b. Partially ionized H I–H II layer					
23	H94α	1150^{+1450}_{-1270}		(3)	Pedlar and Hart (1974)
24	H137α	560^{+2300}_{-980}		(3)	
25	H157α	490^{+1000}_{-640}		(3)	
26	H166α	480^{+350}_{-280}	3 ± 1.5	(3)	

Remarks to Table 5.1.IV:

(1) The temperature of the central part of the nebula remains stable around the value of 9000 K; it varies however monotonically from 8000 K in the west to 11 000 K in the NE (see Figure 5.1.12).

(2) The turbulence has a stable value (within ± 1 km s^{-1}) over the central part of the nebula.

(3) The temperature of the cool partially ionized H I–H II interface was estimated from many narrow hydrogen lines (observed by various authors).

(4) $\alpha = 5^h39^m13^s$, $\delta = -1°56'15''$.

(5) Values refer to the central part of NGC 2024; for variations of temperature and turbulence over the nebula see the original paper.

(6) LTE solution unless otherwise stated.

TABLE 5.1.V
Radial velocities of NGC 2024 derived from recombination and neutral hydrogen lines

Line	HPBW	E= emission A= absorption	Radial velocity V_{LSR} (km s^{-1})	Remarks	References
a. Hydrogen lines					
H$^+$		E	+13.3		Courtès et al. (1967)
H66α	43″	E	+6.4±0.2		Wilson et al. (1979)
H76α		E	+6.9±0.3	(8)	Harwit et al. (1979)
H85α	2.″7	E	+5.5±0.1	(5)	MacLeod et al (1975)
H85α (H I)	2.″7	E	+10.1±0.5	(5), (2)	
H85α		E	+5.5±1		Cesarsky (1977)
H85α(H I)		E	+9.4±3	(2)	
H86α	3.″5	E	+5.8±0.1	(4)	Lichten et al. (1979)
H92α	4′	E	+5.0±0.2		Chaisson (1973)
H94α	4′	E	+5.75±0.2	(6)	Gordon (1969)
H94α	4′	E	+5.4±0.3	(7), (6)	Chaisson (1973)
H109α	6.″4	E	+4.4±1.2		Mezger and Henderson (1967)
H109α	2′	E	+7		Schraml and Mezger (1969)
H109α	6.″5	E	+7.0±0.2		Reifenstein et al. (1970)
H109α		E	+4.7±1		Cesarsky (1977)
H109α	2.″6	E	+5.2±0.1		Churchwell et al. (1978)
H109α (H I)	2.″6	E	+9.4±0.6		
H126α+H127α		E	+4		McGee and Gardner (1968)
H134α		E	+8.7		Wilson et al. (1975)
H157α		E	+3.2±1.0		Ball et al. (1970)
H157α (H I)		E	+8.2±1.0		
H157α		E	+4.3±0.2		Pankonin et al. (1977)
H157α (HI)		E	+8.7±0.2		
H158α		E	+5.9		Lilley et al. (1966)
H166α	21′	E	+4±3		Zuckerman and Ball (1974)
H166α (H I)	21′	E	+8±2	(2)	
H166α	13′	E	+3.7±0.7	(3)	Pedlar and Hart (1974)
H166α (H I)	13′	E	+8.8±0.5	(3), (2)	
H166α		E	+4.1±0.2		Pankonin et al. (1977)
H166α (H I)		E	+9.0±0.1	(2)	
H183α	30′	E	+6.3±5		Zuckerman and Ball (1974)
H83β		E	+6.3±1.0		Wilson et al. (1979)
H108β		E	+6.8±0.3	(4)	Lichten et al. (1979)
H137β		E	+5.4±0.2		Churchwell et al. (1978)
H197β		E	+4.1±0.6		Pankonin et al. (1977)
H148δ		E	+3.7±0.5	(7)	Chaisson (1973)
b. Helium lines					
He85α	3.″2	E	+2.9±0.9	(9)	MacLeod et al. (1975)
He85α		E	+4.5±3		Cesarsky (1977)
He94α		E	+3.1±0.5	(7)	Chaisson (1973)
He109α		E	+3.6±3		Cesarsky (1977)
He109α	2.″6	E	+6.2±0.9		Churchwell et al. (1978)
c. Carbon lines					
C76α		E	+11.2		Rickard et al. (1977)
C76α		E	+10.8±0.3	(8)	Harwit et al. (1979)
C85α		E	+10.2±1		Cesarsky (1971α)
C85α	3.″2	E	+10.6±0.3	(10)	MacLeod et al. (1975)

(continued)

Table 5.1.V (continued)

Line	HPBW	E= emission, A= absorption	Radial velocity V_{LSR} (km s^{-1})	Remarks	References
C85α		E	+10.2±1		Cesarsky (1977)
C86α	3ʹ5	E	+10.7±0.2	(4)	Lichten *et al.* (1979)
C92α	4ʹ	E	+10.5±0.3		Chaisson (1974α)
C92α		E	+10.5		} Cesarsky (1977)
C92α (H I)		E	+7.8	(11)	
C94α	4ʹ2	E	+9.8±0.7		Chaisson (1971b)
C94α		E	+10.5±0.2	(7)	Chaisson (1973)
C109α	6ʹ	E	+10.2		Palmer *et al.* (1967)
C109α	6ʹ	E	+10.5		Palmer (1968)
C109α		E	+10.28		Palmer *et al.* (1969)
C109α		E	+9.4±2		Cesarsky (1971α)
C109α		E	+10.1±1		Cesarsky (1977)
C109α		E	+10.4±0.1		Churchwell *et al.* (1978)
C110α		E	+10.2		Palmer (1968)
C134α	11ʹ	E	+9.4		Zuckerman and Ball (1970)
C134α		E	+9.6		Wilson *et al.* (1975)
C137α	11ʹ	E	+10.3		} Cesarsky (1971b)
C137α (H I)	11ʹ	E	+9.4±3	(11)	
C157α	18ʹ	E	+8.8±0.3		} Ball *et al.* (1970)
C157α (H I)	18ʹ	E	+8.2±1	(11)	
C157α		E	+9.3±0.1		Pankonin *et al.* (1977)
C158α	18ʹ	E	+10.1		Chaisson *et al.* (1972)
C166α	13ʹ	E	+9.3±0.2	(3)	Pedlar and Hart (1974)
C166α	21ʹ	E	+9±2		Zuckerman and Ball (1974)
C166α		E	+9.3±0.1		Pankonin *et al.* (1977)
C183α	30ʹ	E	+9.7±2		Zuckerman and Ball (1974)
C198α	36ʹ	E	+9.5±3		Zuckerman and Ball (1974)
C211α	43ʹ	E	+8±3		Zuckerman and Ball (1974)
C220α	31ʹ	E	+9		Pedlar and Davies (1972)
C137β		E	+9.4±2		Cesarsky (1977)
C137β		E	+10.7±0.2		Churchwell *et al.* (1978)
C197β		E	+9.3±0.3		Pankonin *et al.* (1977)
d. Heavy element lines					
Z92α		E	+1.6		Cesarsky (1977)
Z109α	2ʹ6	E	+11.1±0.5		Churchwell *et al.* (1978)
Z157α		E	+0.7±0.2	(12)	Pankonin *et al.* (1977)
Z166α		E	+0.9±0.5		Pedlar and Hart (1974)
Z166α		E	+0.7±0.2	(12)	Pankonin *et al.* (1977)
e. Neutral hydrogen H I (21 cm) lines					
H I (21 cm)		A	+9±1		Gordon (1970)
		A	+7.5±0.2		}
H I (21 cm)		A	+11.0±0.1		} Radhakrishnan *et al.* (1972b)
		A	+9.7±0.1		}
H I (21 cm)	2ʹ	A	+10.4, +2.4		Lockhart and Goss (1978)

Remarks to Table 5.1.V:

(1) V_{LSR} at the continuum peak ($\alpha = 5^h39^m13^s$, $\delta = -1°56'$); the exact coordinates are often quoted; $V_{Hel} = V_{LSR} + 23.7$ Km s^{-1}.
(2) Narrow line originating in the H II–H I interface layer which partially envelopes the H II region.
(3) At $\alpha = 5^h39^m12^s$, $\delta = -1°56'2''$ (radio continuum peak).
(4) At $\alpha = 5^h39^m13^s$, $\delta = -1°56'13''$ (radio continuum peak).
(5) At $\alpha = 5^h39^m12^s6$, $\delta = -1°56'04''$ (radio continuum peak).
(6) For V_{LSR} variations over the nebula see the original paper.
(7) At $\alpha = 5^h39^m12^s$, $\delta = -1°55'42''$ (radio continuum peak).
(8) At $\alpha = 5^h39^m14^s$, $\delta = -1°57'18''$ ($\sim 1'$ south of the continuum peak).
(9) At $\alpha = 5^h39^m9^s6 \pm 1^s2$, $\delta = -1°55'29'' \pm 18''$ (peak of the He emission).
(10) At $\alpha = 5^h39^m10^s3 \pm 1^s0$, $\delta = -1°56'39'' \pm 14''$ (peak of the C emission).
(11) Narrow C line ($\Delta V \sim 2$ km s^{-1}).
(12) Probably sulphur; see also note (18) to Table 2.2.XIX.

TABLE 5.2.I
Infrared continuum contour maps of NGC 2024

No.	Beam	$\lambda (\mu)$	References
1	2!2	93	Harper (1974)
2	1!6	400	Hudson and Soifer (1976)
3	62″	3.5	Fahrbach *et al.* (1976)
4	62″	2.5	Frey *et al.* (1979)
5	62″	3.5	Frey *et al.* (1979)
6	36″	8.4	Grasdalen (1974)

TABLE 5.2.II

Physical parameters of the dust associated with the NGC 2024/molucular cloud complex derived from infrared continuum measurements

No.	Band-width (μ)	λ (μ)	Airborne (A) or Ground (G) observations	Beam	Beam separation	Flux (Jy)[10] 1 Jy $= 10^{-26}$ W m^{-2} Hz^{-1}	Flux (10^{-14} W cm^{-2})	Dust temperature T_d (K)	Luminosity[10] L ($L_\odot$)	$\dfrac{M_{dust}}{M_{gas}}$	M_{dust} ($M_\odot$)	Remarks	References
1	45–750		A	8′.4±1′			42	70^{+20}_{-15}	2.1×10^4			(1)	Harper and Low (1971)
2	40–350		A	4′.5	7′.5		25						Furniss et al. (1972)
3		21	G	1′	7′.5	1.2×10^3		54				(8), (9)	Lemke and Low (1972)
4	40–350		A	4′.5	45″		32		1.6×10^4			(5)	Emerson et al. (1973)
5	28–40	34	G	25″	45″	$<3\times10^3$							Low et al. (1973)
6	56–500	100	A	5′		5.5×10^4						(2), (3),	Harper (1974)
7	53–82	68	A	5′		7.6×10^4						(4), (12),	
8	110–500	167	A	5′		3.4×10^4		80		0.02	0.036	(14)	
9	45–750	93	A	8′.4		8.8×10^4							
10	30–300		A	5′			$47^{(8)}$		4.6×10^4			(6)	Hudson and Soifer (1976)
11		400	G	1′.6	4′.1	$(5.14\pm1.8)\times10^3$		25?	3×10^3–10^4				
12		39	A	50″	2′.4	$(8.2\pm0.91)\times10^3$						(7), (11),	Thronson et al. (1978)
13		57	A	50″	2′.4	$(10\pm0.85)\times10^3$			5.5×10^3		0.04–0.09	(13)	
14		76	A	50″	2′.4	$(9.7\pm0.8)\times10^3$		60					
15		140	A	50″	2′.4	$(4.9\pm0.82)\times10^3$							
16	1–1000		A				74		4.7×10^4				Frey et al. (1979)

Remarks to Table 5.2.II:

(1) The quoted temperature is the colour temperature estimated from the ratio of fluxes in the 45–750μ and 60–750μ bands.

(2) Observations made at $\alpha = 5^h39^m19^s \pm 6^s$, $\delta = -1°55\overset{'}{.}7 \pm 1\overset{'}{.}5$.

(3) The wavelength in the second column is given by the relation:

$$\lambda_{eff} = \frac{\int\limits_0^\infty B(\lambda, T_c)\, G(\lambda)\, \lambda\, d\lambda}{\int\limits_0^\infty B(\lambda, T_c)\, G(\lambda)\, d\lambda},$$

where $B(\lambda, T_c)$ is the spectral emittance of a blackbody of colour temperature T_c and $G(\lambda)$ is the instrumental response.

(4) The quoted temperature is the temperature of a blackbody curve, fitted to match the energy distribution in the 30–300μ range.

(5) Peak flux density and luminosity.

(6) Integrated flux density over the 3$\overset{'}{.}5$(N–S) × 2$\overset{'}{.}5$(E–W) source associated with the dust lane and the molecular cloud.

(7) The cited luminosity is the total infrared luminosity into a 50″ beam.

(8) Peak flux density.

(9) Colour temperature at 22 μ.

(10) Integrated flux densities and luminosities unless otherwise stated.

(11) The value of M_d refers to a volume of $\simeq 0.1$ pc in diameter.

(12) The values of M_d/M_g and M_d refer to a volume of 0.4 pc in diameter.

(13) The dust mass M_d was estimated on the assumption that the emitting grains consist of a lunar rock core of radius $\alpha = 0.1\mu$ with a water ice mantle 0.01μ thick and are characterized by density $\rho = 2$ g cm^{-3} and infrared emission efficiency Q_{IR} as given by Aannestad (1975) (see Appendix, Section III.2, relation III.2.10). A constant grain temperature T_d was also assumed.

(14) The dust mass M_d was estimated for grain radius $\alpha = 0.1\mu$, grain density $\rho = 1$ g cm^{-3}, infrared emission efficiency $Q_{IR} = 0.005$ and a constant grain temperature T_d (see Appendix, Section III.2).

TABLE 5.2.III

Kinetic temperature of the molecular gas in the NGC 2024/molecular
cloud inferred from ^{12}CO measurements

No.	Kinetic temperature T_k (K)	Remarks	References
1	22	(2)	Goldsmith *et al.* (1975)
2	28	(2)	Phillips *et al* (1979)
3	45	(2)	Plambeck and Williams (1979)
4	41	(1)	Penzias (cited by Cronin *et al.* 1976)

Remarks to Table 5.2.III:

(1) From ^{12}CO, $J = 1$–0 observations.
(2) From ^{12}CO, $J = 2$–1 observations.

TABLE 5.2.IV

Molecular abundances* in the NGC 2024/molecular cloud complex

No.	Molecule	Column density N (molecules cm^{-2})	References
1	H_2	5×10^{22}	Thronson *et al.* (1978)
2	CO	8.2×10^{18}	Wilson *et al.* (1974)
3	CO	1.5×10^{19}	Liszt *et al.* (1975), (cited by Turner and Gammon, 1975)
4	CO	1.2×10^{19}	Hudson and Soifer (1976)
5	CN	1.3×10^{14}	Turner and Gammon (1975)
6	CS	$\geqslant 1 \times 10^{14}$	Liszt and Linke (1975)
7	CS	1.65×10^{14}	Linke and Goldsmith (1980)
8	SO	6×10^{13}	Gottlieb *et al.* (1978)
9	NH_3	5×10^5	Schwartz *et al.* (1977)
10	H_2CO	2.3×10^{13}	Zuckerman *et al.* (1970)
11	H_2CO	2.6×10^{13}	Fomalont and Weliachew (1973)
12	C°	$\geqslant 7 \times 10^{17}$	Phillips *et al.* (1980)

* Carbon abundances are also included for comparison.

TABLE 5.2.V

Radial velocities of the NGC 2024 molecular cloud derived from molecular lines

Molecule	Beam	E = emission A = absorption	Radial velocity V_{LSR} (km s^{-1})	Remarks	References
C I*	2ʺ5	E	+12	(19), (44)	Phillips et al. (1980)
OH		A	+9.5±0.2		Goss (1968), Manchester and Gordon (1971)
OH			+9.4, +12.8		Chaisson (1974)
OH	27′×3′	A	+9.3, +11.2	(1)	
OH	27′×3′	A	+9.5, +11.4, +12.7	(2)	
OH	27′×3′	A	+9.3, +13	(3)	Goss et al. (1976)
OH	27′×3′	A	+9.3, +10.7	(4)	
CH		E	+9.5	(6), (5)	Rydbeck et al. (1973, 1974)
CH	6ʹ4	E	+9.6, +12.6	(7)	
CH	6ʹ4	E	+9.6, +12.6	(8)	Whiteoak et al. (1978)
CH	6ʹ4	E	+9.6, +12.6	(9)	
CH	4′	E	+9.4, +13.2, +7.9	(7), (10)	
CH	4′	E, A	+9.9, +13.3, +8.2, +5.3	(8), (10), (11)	Genzel et al. (1979)
CH	4′	E	+9.9, +12.7	(9), (10)	
CO	70″	E	+11	(12)	
CO	70″	E	+10	(13)	Wilson et al. (1974)
CO	2′	E	+11.2	(12)	
CO	2′	E	+10.4	(13)	Wannier et al. (1976)
CO	2′	E	+10.0	(16)	
CO		E	+11.2	(12), (19)	Scoville and Wannier (1979)
CO	2ʹ25	E	+9	(14)	Goldsmith et al. (1975)
CO	2ʹ25	E	+10	(15)	
CO		E	~ +10	(14), (18)	Watt et al. (1979)
CO		E	~ +11	(15), (18)	
CO		E	+11	(14), (19)	Plambeck and Williams (1979)
CO	4ʹ5	E	+10.5, +5	(14), (20)	White et al. (1980)
CO	4ʹ5	E	+9	(17), (20)	
CN	65″	E	+11.5	(24)	Turner and Gammon (1975)
CS	4ʹ7	E	+10.4±0.5	(21), (19)	
CS	2ʹ5	E	+9.4±0.5	(22), (19)	Liszt and Linke (1975)
CS	1ʹ8	E	+9.5±0.5	(23), (19)	
CS	2ʹ6	E	+10.5	(21), (19), (42)	Linke and Goldsmith (1980)
CS	2ʹ1	E	+10.8	(22), (19)	
SO		E	+9.8	(25)	Gottlieb and Ball (1973)
SO		E	+10.1	(26)	
SO		E	+10.1±0.05	(25), (19)	Gottlieb et al. (1978)
SO		E	+10.4±0.2	(26), (19)	
H$_2$O		E	+11.4, +17.5	(27)	Johnston et al. (1973)
H$_2$O	40″	E	+10	(27), (28)	Genzel and Downes (1977)
HCN		E	+11	(29)	Turner and Thaddeus (1977)
HCN		E	+9	(30), (31)	Baudry et al. (1980)
HCO$^+$		E	+10.9	(33), (29)	Turner and Thaddeus (1977)
HCO$^+$	3ʹ5	E	+9.5	(32), (31)	Baudry et al. (1980)
HNO		E	+8±1	(34), (5)	Ulich et al. (1977)
N$_2$H$^+$		E	+11.4	(35), (29)	Turner and Thaddeus (1977)
NH$_3$		E	+9	(36), (37)	Schwartz et al. (1977)

(continued)

Table 5.2.V (continued)

Molecule	Beam	E= emission, A= absorption	Radial velocity V_{LSR} (km s^{-1})	Remarks	References
H_2CO		A	$+8.65 \pm 1$	(38)	Zuckerman *et al.* (1970)
H_2CO		A	$+9.3$	(38)	Whiteoak and Gardner (1970)
H_2CO	4″2	A	-6.4	(38), (39)	Whiteoak and Gardner (1974)
H_2CO	2″1	A	$+9.0$	(40), (19)	Scoville and Wannier (1979)
H_2CO	4′	A	$+9.5$	(38)	Genzel *et al.* (1979)
H_2CS		E	$+10.5$	(41), (19)	Liszt (1978)
CH_3OH		E	$+10.5 \pm 0.1$	(43), (19)	Gottlieb *et al.* (1979)

* Carbon is included for comparison.

Remarks to Table 5.2.V:

(1) $^2\Pi_{3/2}$, $J = \frac{3}{2}$, $F = 1$–2 at 1612 MHz.

(2) $^2\Pi_{3/2}$, $J = \frac{3}{2}$, $F = 1$–1 at 1665. 401 MHz.

(3) $^2\Pi_{3/2}$, $J = \frac{3}{2}$, $F = 2$–2 at 1667.358 MHz.

(4) $^2\Pi_{3/2}$, $J = \frac{3}{2}$, $F = 2$–1 at 1720.533 MHz.

(5) At $\alpha = 5^h39^m12^s$, $\delta = -1°57'42''$.

(6) $^2\Pi_{1/2}$, $J = \frac{1}{2}$; average of $F = 1$–0 at 3449.193 MHz, $F = 1$–1 at 3335.481 MHz and $F = 0$–1 at 3263.794 MHz.

(7) $^2\Pi_{1/2}$, $J = \frac{1}{2}$, $F = 0$–1 at 3263.794 MHz.

(8) $^2\Pi_{1/2}$, $J = \frac{1}{2}$, $F = 1$–1 at 3335.481 MHz.

(9) $^2\Pi_{1/2}$, $J = \frac{1}{2}$, $F = 1$–0 at 3449.193 MHz.

(10) At $\alpha = 5^h39^m14^s$, $\delta = -1°56'11''.8$.

(11) Weak absorption feature at $+8.2$ km s^{-1}.

(12) ^{12}CO, $J = 1$–0 at 115 271.2 MHz.

(13) ^{13}CO, $J = 1$–0 at 110 201.4 MHz.

(14) ^{12}CO, $J = 2$–1 at 230 537.97 MHz.

(15) ^{13}CO, $J = 2$–1 at 220 398.7 MHz.

(16) $^{12}C^{18}O$, $J = 1$–0.

(17) $^{12}C^{18}O$, $J = 2$–1 at 219 GHz.

(18) At $\alpha = 5^h38^m58^s$, $\delta = -1°40'57''$.

(19) At $\alpha = 5^h39^m12^s$, $\delta = -1°55'42''$.

(20) At $\alpha = 5^h39^m14^s$, $\delta = -1°56'57''$.

(21) $^{12}C^{32}S$, $J = 1$–0 at 48 991 MHz.

(22) $^{12}C^{32}S$, $J = 2$–1 at 97 981 MHz.

(23) $^{12}C^{32}S$, $J = 3$–2 at 146 969 MHz.

(24) CN, $K = 1$–0, $J = 3/2$–1/2 at 113 491 MHz.

(25) ^{32}SO, 3_2–2_1 at 99 300 MHz.

(26) ^{32}SO, 4_3–3_2 at 138 200 MHz.

(27) H_2O, $6_{16} - 5_{23}$ at 22 GHz (maser emission).

(28) $\alpha = 5^h39^m13^s.7$, $\delta = -1°57'30''$.

(29) $\alpha = 5^h39^m13^s.5$, $\delta = -1°55'57''$.

(30) HCN, $J = 1$–0 at 86 340.05 MHz; $F = 2$–1.

(31) At $\alpha = 5^h39^m3^s$, $\delta = -1°51'00''$.

(32) HCO^+, $J = 1$–0.

(33) HCO^+, $J = 3$–2 at 89 188.545 MHz.

(34) HNO, 1_{01}–0_{00}.

(35) N_2H^+ at 93 173.58 MHz ($F = 1$–2, $F = 1$–1, and $F = 1$–0).

(36) NH_3, J, $K = 1$, 1 at 23 694.48 MHz.

(37) $\alpha = 5^h39^m13^s$, $\delta = -1°55'48''$.

(38) H_2CO, $1_{11} - 1_{10}$ at 4829.66 MHz (6 cm).

(39) At $\alpha = 5^h39^m12^s$, $\delta = -1°56'$.

(40) H_2CO, $2_{12} - 2_{11}$ (2 cm).

(41) H_2CS, $3_{13} - 3_{12}$ at 104 620 MHz.

(42) CS size, $2' \times 5'$, $n_{H_1} = 1 \times 10^5$.

(43) CH_3OH, $J = 2$–1 near 96.7 GHz.

(44) C I $(^3P_1 - ^3P_0)$, $J = 1$–0 at 492 GHz.

TABLE 5.2.VI

Full width at half intensity, ΔV(km s^{-1}), of various lines detected
in the NGC 2024 /molecular cloud complex

Lines originating in the H II region		Lines originating in the partially ionized medium (H I–H II interface layer)			Lines originating in the neutral/molecular complex	
H lines (broad)	He lines	C lines	Z lines	H lines (narrow)	H I (21 cm) line	Molecular lines
Hα:	He85α: 12[1]	C85α: 4[1]	Z92α: 4[1]	H66α: 5[4]	H I: 5[12]	OH: 1[16]
H85α: 23[1]	He94α: 15[2]	C92α: 3[1]	Z157α: 2.5[4]	H85α: 5[1]	H I: ~4[13]	CH: 2[15]
H94α: 23[2]	He109α: 13[2]	C92α: 5[11]	Z166α: 3[4]	H157α: 5.5[4]		^{12}CO: 5[19]
H109α: 23[1]		C94α: 5.5[3]				^{13}CO: 3[19]
H157α: 31[3]		C109α: 5[1]				^{18}CO: 3[20]
H157α: 29[4]		C110α: 4[6]				CN: 3.5[14]
H166α: 31[4]		C134α: 5.5[7]				CS: 3[17]
H183α: 35[5]		C137α: 4[8]				SO: 2.5[22]
H197β: 33[4]		C157α: 4[3],[4]				HCO^{+}: 4[14]
H148δ: 20[2]		C158α: 4[9]				HCN: 2[14]
		C166α: 4[5]				HCN: 4[21]
		C183α: 4[5]				N$_2$H^{+}: 2[14]
		C198α: 5.5[5]				H$_2$CO: 2[18]
		C211α: 4[5]				
		C220α: 8[10]				
		C137β: 6[1]				
		C197β: 4[4]				

References to Table 5.2.VI:

(1) Cesarsky (1977).
(2) Chaisson (1973).
(3) Ball et al. (1970).
(4) Pankonin et al. (1977).
(5) Zuckerman and Ball (1974).
(6) Palmer (1968).
(7) Zuckerman and Palmer (1970).
(8) Cesarsky (1971).
(9) Chaisson et al. (1972).
(10) Pedlar and Davies (1972).
(11) Chaisson (1974).

(12) Radakrishnan et al. (1972).
(13) Lockhart and Goss (1978).
(14) Turner and Thaddeus (1977).
(15) Genzel et al. (1979).
(16) Goss et al. (1976).
(17) Linke and Goldsmith (1980).
(18) Whiteoak and Gardner (1970).
(19) Goldsmith et al. (1975).
(20) White et al. (1980).
(21) Baudry et al. (1980).
(22) Gottlieb et al. (1978).

TABLE 5.2.VII

Kinematics of the L1630/northern molecular complex derived from molecular line emission

No.	Molecular line emission	Frequency ν(MHz)	Wavelength λ	HPBW	V_{LSR}[1] (km s^{-1})	ΔV[2] (km s^{-1})	Remarks	References
1	^{12}CO, $J=1$–0	115 271.2	2.6 mm	2ʹ6	+9.6	2.1	(3), (4), (5)	Tucker et al. (1973)
2	^{12}CO, $J=1$–0	115 271.2	2.6 mm	70″	10.5	~3	(7)	Milman et al. (1975)
3	^{13}CO, $J=1$–0	110 201.4	2.7 mm	70″	~+10.5	~2		
4	^{12}CO, $J=1$–0	115 271	2.6 mm	8ʹ	+9.6	2.1	(3), (4), (6)	Kutner et al. (1977)
5	^{13}CO, $J=1$–0	110 201	2.7 mm	8ʹ	+9.8	1.3		
6	^{12}CO, $J=1$–0	115 271	2.6 mm	2ʹ6	+9.6	2.1		
7	OH, $^{2}\Pi_{3/2}$, $J=\frac{3}{2}$, F=1–1, F=2–2	1665,1667		27ʹ × 31ʹ	+9.3	3.5	(3)	Goss et al. (1976)

Remarks to Table 5.2.VII:

(1) Radial velocity with respect to the local standard of rest.
(2) Full width at half intensity.
(3) No evidence for rotation was found.
(4) A second velocity feautre, at $+4$ km s^{-1}, is present in a $1° \times 1°$ area centred at $\alpha = 5^{h}40^{m}$, $\delta = -1°30'$.
(5) The mass of the cloud M_{H_2} is in the range $2.5 \times 10^{4} - 1 \times 10^{5}\ M_{\odot}$ (for an assumed $13 \times 35 \times 13$ pc^{3} volume).
(6) The mass of the cloud is $M_{H_2} = 6 \times 10^{4}\ M_{\odot}$.
(7) $n_{H_2} < 300$ cm^{-3}, for a $20 \times 20 \times 20$ pc^{3} cloud.

TABLE 5.3.I

Physical parameters of stars associated with the NGC 2024/molecular cloud complex

No.	Star	Spectral type	M_V (m)	A_v (m)	Distance D (pc)	Effective temperature T_{eff} (K)	Luminosity L ($L_\odot$)	Remarks	References
1	ζ Ori	O9.5 Ib			350			(1)	Conti and Alschuler (1971) Becker and Fenkart (1963) Mezger and Höglund (1967)
					400				
2	NGC 2024 #1	B0.5Vp	−4.5 or −7.9					(2), (3)	Johnson and Mendoza (1964), Garrison (1968)
		B0 I		8.3±0.2 32±2.6 <30–35	450	23 000 1000–2000	$(1.6\pm0.5)\times10^6$	(7)	Grasdalen (1974) Grasdalen (1974) Thum (1975)
3	NGC 2024 #2	B0.5 I V	−9.3	40	460	23 000±1500 1800–2400 37 000	$(4\pm2)\times10^3$ 3×10^3–1×10^4 6×10^4	(4) (10) (9) (5)	Cesarsky (1977) MacLeod et al. (1975) Hudson and Soifer (1976) Cesarsky (1977) Balick (1974)
4	Undetected exciting star of NGC 2024	O 9.5 V O 8.5–9 V O 6.5–7 III O 6.5 ZAMS O9 V				33 500 35 000 37 500 40 000		(6) (8)	Sarazin (1976) Goudis (1975)

Remarks to Table 5.3.I:

(1) It was considered as exciting star originally by Hubble, 1922; its location, however, (away from the centre of the H II region) excludes it from playing the main role in the ionization of NGC 2024.

(2) $M_V = -4.5$ for $R = A_v/E(B - V) = 5$ while $M_V = -7.9$ for $R = 7$.

(3) NGC 2024 #1 is not sufficiently luminous to excite the H II region (Gordon, 1969). The spectrum of the star shows, however, abnormally broad Balmer lines (Morgan and Keenan, 1973), which occur only in stars associated with H II regions (O'Dell, 1975; Sarazin, 1976).

(4) NGC 2024 #2 is probably not the exciting star of NGC 2024 but the heating source of the cooler infrared source (Hudson and Soifer, 1976) associated with the dust lane and the molecular cloud.

(5) Balick (1974) has also argued that the dust does not play any serious role in the absorption of ionizing radiation.

(6) Estimated from models of NGC 2024 containing ultraviolet absorbing dust.

(7) The estimated unreddened visual magnitude is −1.

(8) Or, alternatively, three B0 stars.

(9) A relatively cool BN-like object embedded in and heating the molecular cloud lying in front of the H II region.

(10) See model of NGC 2024 in Figure 5.3.2.

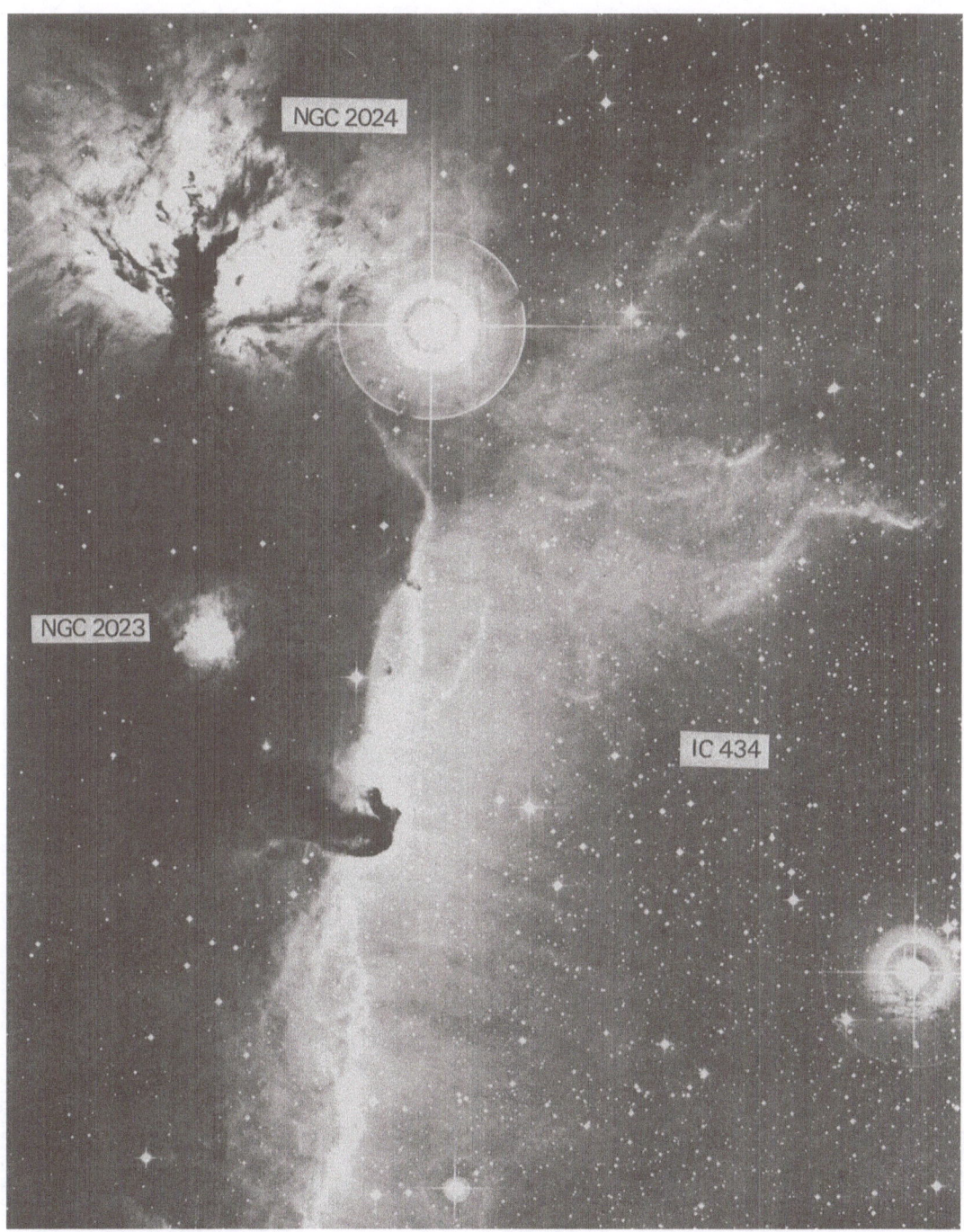

Fig. 5.1.1. A red photograph of the NGC 2024/IC 434 nebulae taken with the 1.2 m UK Schmidt telescope through a RG 630 filter (bandpass 6300–7000 Å). The exposure time was 90 min. The print was made through an unsharp mask (after Malin, 1979). The bright star to the north is ζ Orionis. NGC 2024 is the nebula adjacent to it to the NE. IC 434 is the elongated, filamentary nebula to the south of ζ Orionis.

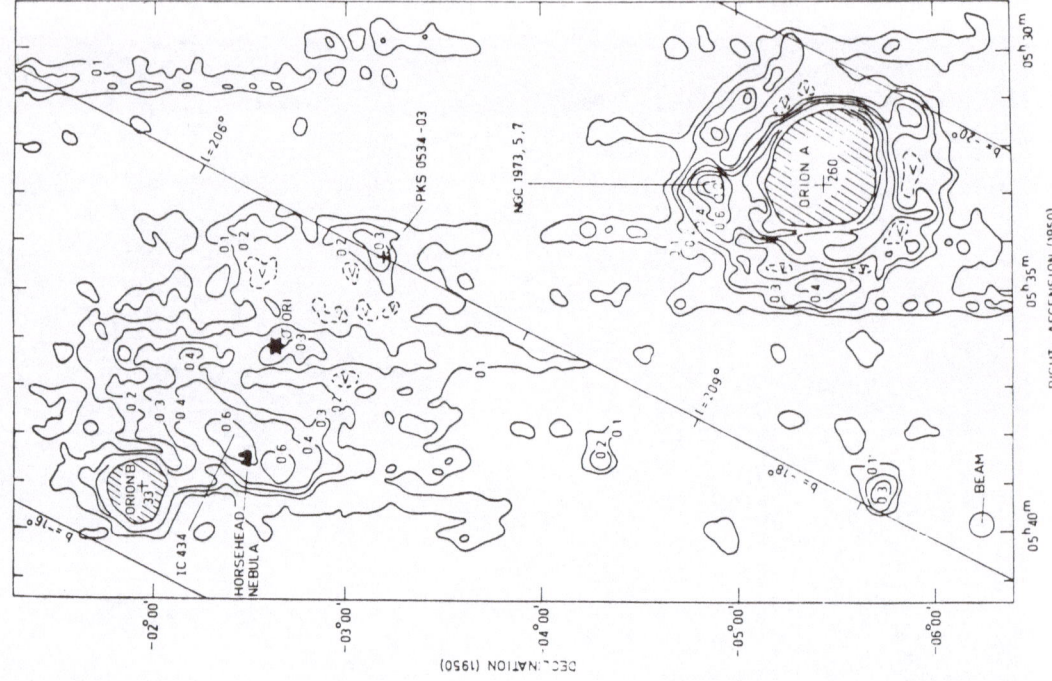

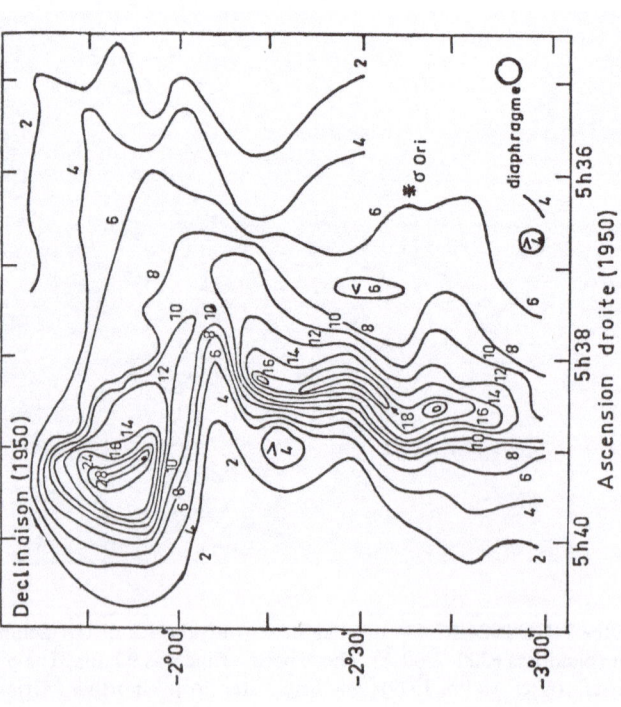

Fig. 5.1.2. Hα contour map of the NGC 2024/IC 434 nebulae made with an angular resolution of 5′ (after Vidal, 1980). The contours are in units 10^{-5} ergs cm^{-2} s^{-1} sr^{-1}.

Fig. 5.1.3. Radio continuum contour map of the area in the vicinity of NGC 2024 (Orion B) made at 2700 MHz with an angular resolution of 8′.2 (after Caswell and Goss, 1974). The contour unit = 1 K (T_b).

Fig. 5.1.4. A red photograph of NGC 2024 (after Malin, 1979). The print was made through an unsharp mask. For technical information concerning this photograph see Figure 5.1.1.

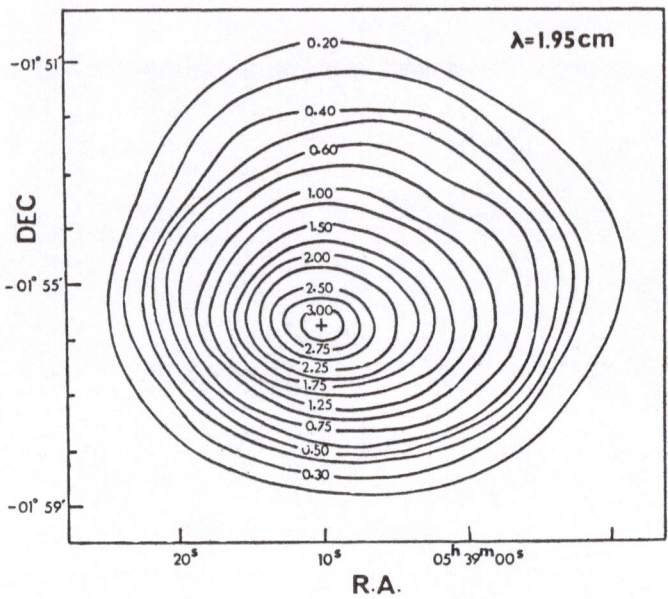

Fig. 5.1.5.   Radio continuum contour map of NGC 2024 made at 15.4 GHz with an angular resolution of 2′ (after Schraml and Mezger, 1969). The contour unit is 1 K (T_A) or 2 K (T_b).

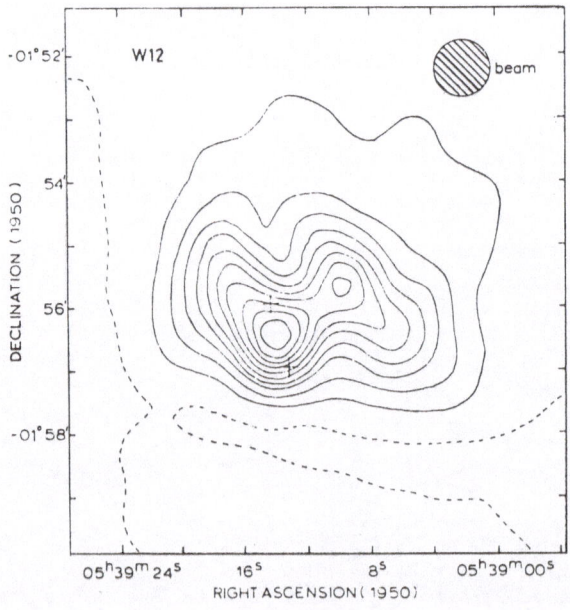

Fig. 5.1.6a.   Radio continuum contour map of NGC 2024 made at 1415 MHz with an angular resolution of $\simeq 1′$ (53″ × 56″) (after Löbert and Goss, 1978). The zero contour is shown dotted. The northern cross marks the position of the IR point source NGC 2024 #2 whereas the southern cross indicates the peak of the 8.4μ extended source detected by Grasdalen (1974).

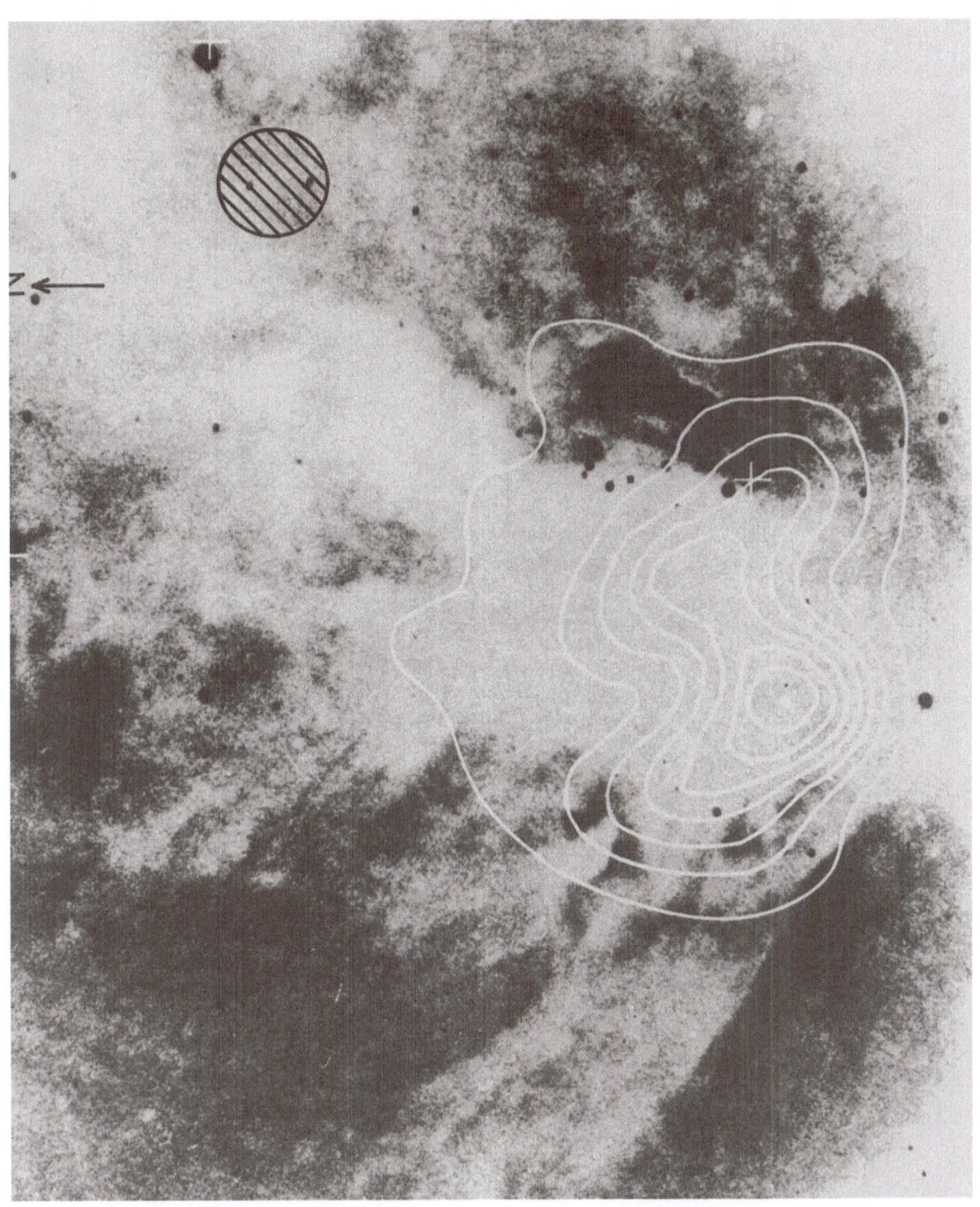

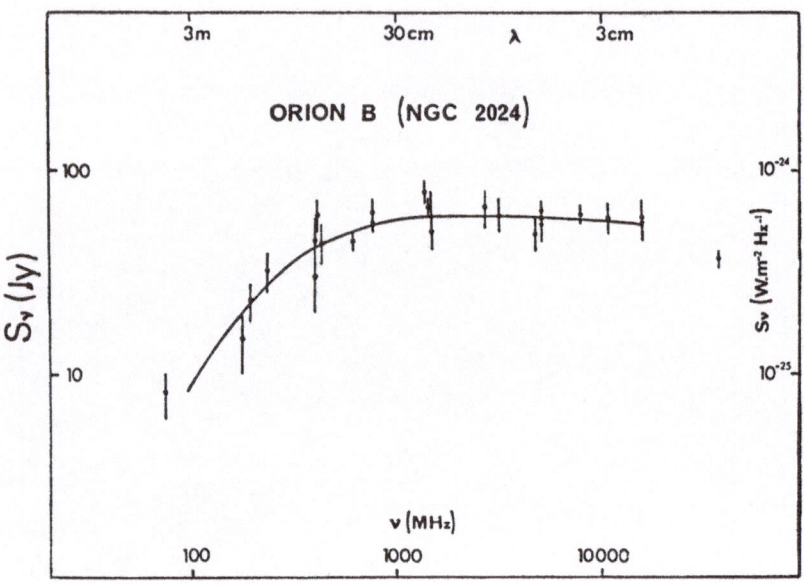

Fig. 5.1.7.   The radio continuum spectrum of NGC 2024 (Goudis, 1975d).

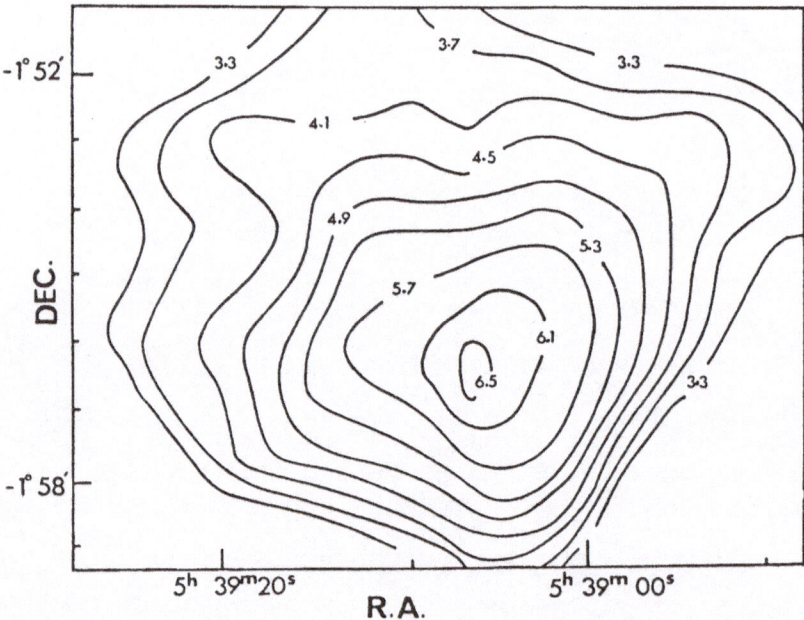

Fig. 5.1.8.   Contour map of the optical depth at Hα of NGC 2024 (after Schmitter, 1971). The map was derived by comparing optical observations made in the light of Hα with radio continuum observations made at 15.4 GHz by Schraml and Mezger (1969, Fig. 5.1.5). The angular resolution of the map is 2'. The contour units are expressed in magnitudes.

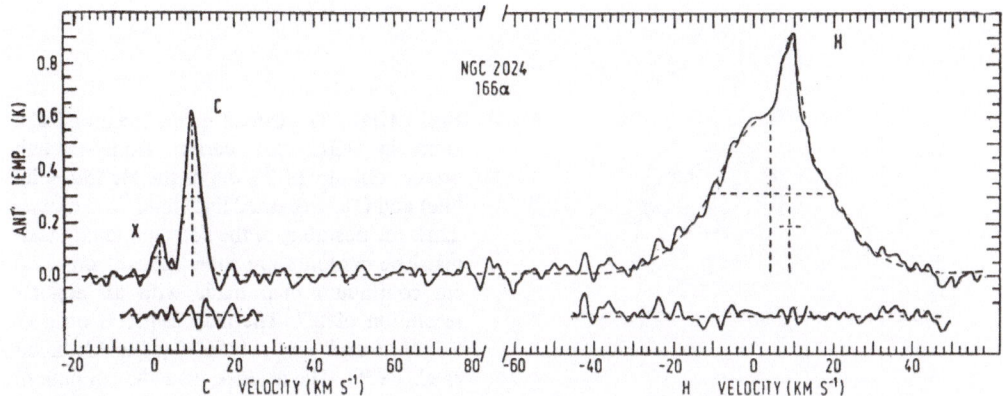

Fig. 5.1.9. The 166α spectrum of NGC 2024 (after Pankonin *et al.*, 1977). The carbon (C), heavy element (X), and 'narrow' hydrogen line (on the top of the broader emission from the ionized region) originate in the partially ionized layer lying between the H II region and the associated neutral/molecular cloud.

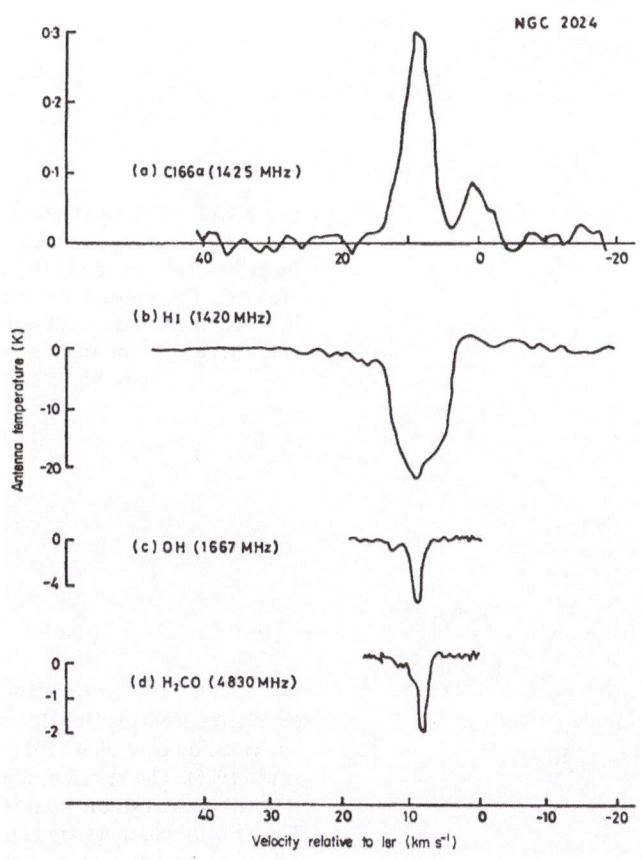

Fig. 5.1.10. The C166α line spectrum of NGC 2024 compared with the spectra of the atomic hydrogen line at 21 cm and various molecular lines (after Hart and Pedlar, 1974). Note the striking agreement of the velocities.

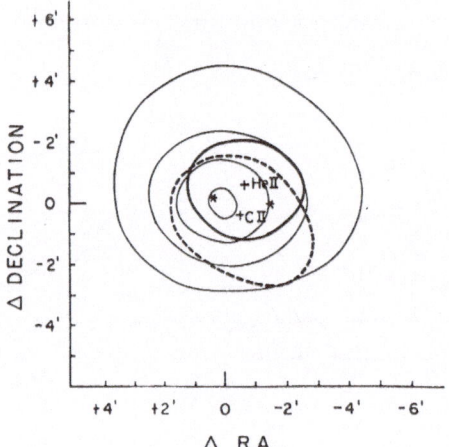

Fig. 5.1.11. The extend of the He II and C II zones in NGC 2024, derived from the half power contours of $T_L \, \varDelta\nu_L$ of the He 85α (solid line) and C 85α (dashed line) lines. The crosses mark the positions of the corresponding maxima. The contours are superimposed on a 2.8 cm continuum map made with an angular resolution of 2.″7. The map centre is on $\alpha =$ 5^{h}39^{m}12.″6 and $\delta = -$ 1°56′4″ (after MacLeod et al., 1975). The asterisk near the continuum maximum marks the position of star NGC 2024 #2 whereas the asterisk to the west marks the position of star NGC 2024 #1.

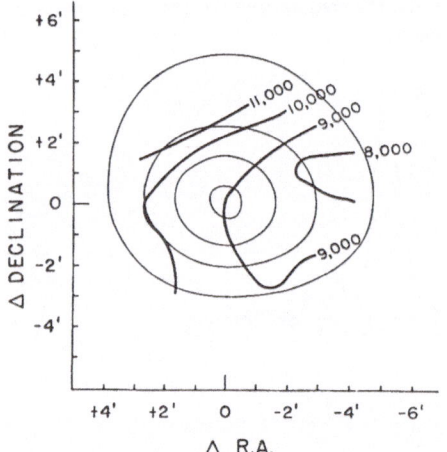

Fig. 5.1.12. The distribution of T_e over NGC 2024, estimated by assuming LTE conditions (after MacLeod et al., 1975). The values of T_e are in K. The temperature contours are superimposed on a 2.8 cm radio continuum contour map made with an angular resolution of 2.″7 (see Fig. 5.1.11).

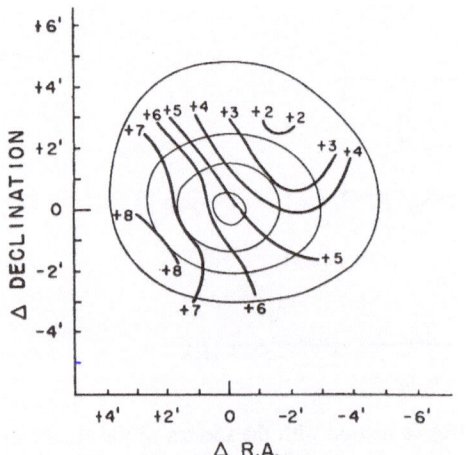

Fig. 5.1.13. The distribution of radial velocities derived from the Doppler shift of the H 85α line emission over NGC 2024 (after MacLeod et al., 1975). The velocities are in units of km s^{-1} (with respect to the local standard of rest). The velocity contours are superimposed on a 2.8 cm radio continuum contour map made with an angular resolution of 2.″7 (see Figure 5.1.11).

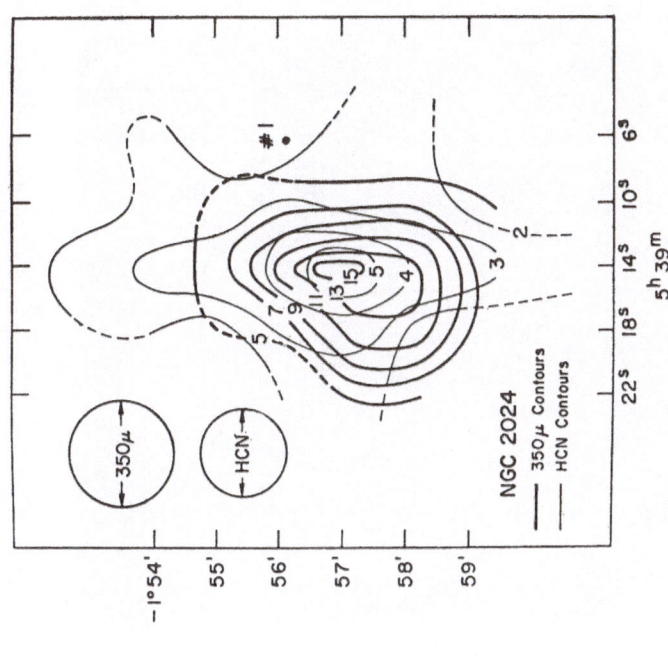

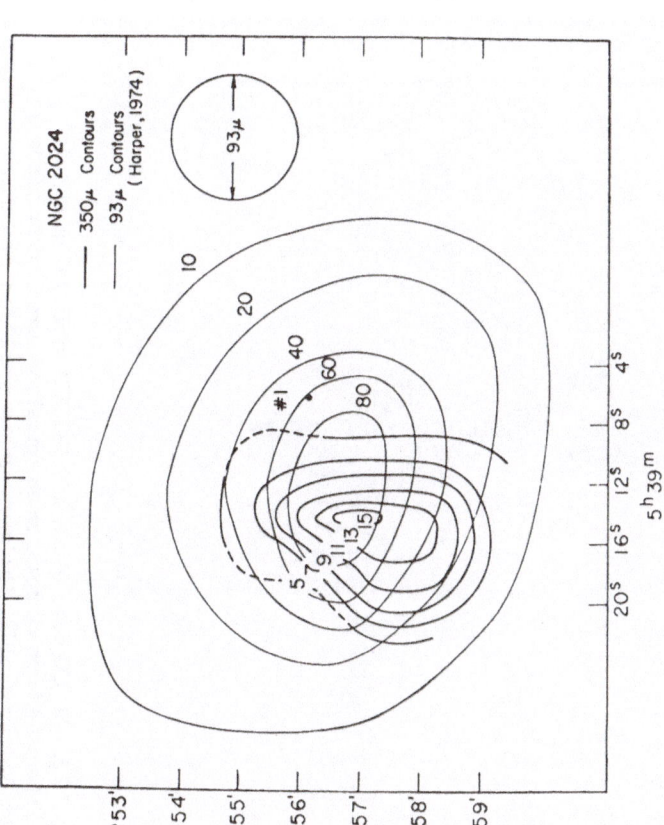

Fig. 5.2.1. Superposition of two infrared maps of NGC 2024 made at 93μ and 350μ with angular resolutions of 2′.2 and 1′.6, respectively (after Hudson and Soifer, 1976). The form of the far-infrared map (93μ; Harper, 1975) resembles that of the ionized gas whereas the form of the submillimeter map (350μ, Hudson and Soifer, 1976) resembles that of the molecular cloud (Figure 5.2.2.) and coincides with the prominent dust lane of NGC 2024.

Fig. 5.2.2. Superposition of a 350μ contour map of NGC 2024 on a HCN map made by Gilmore *et al.* (1975) (after Hudson and Soifer, 1976). The circles mark the beam sizes of the observations. Note the spatial coincidence of the two maps indicating the physical association of the dust lane with the molecular cloud (which probably lies in front of the H II region).

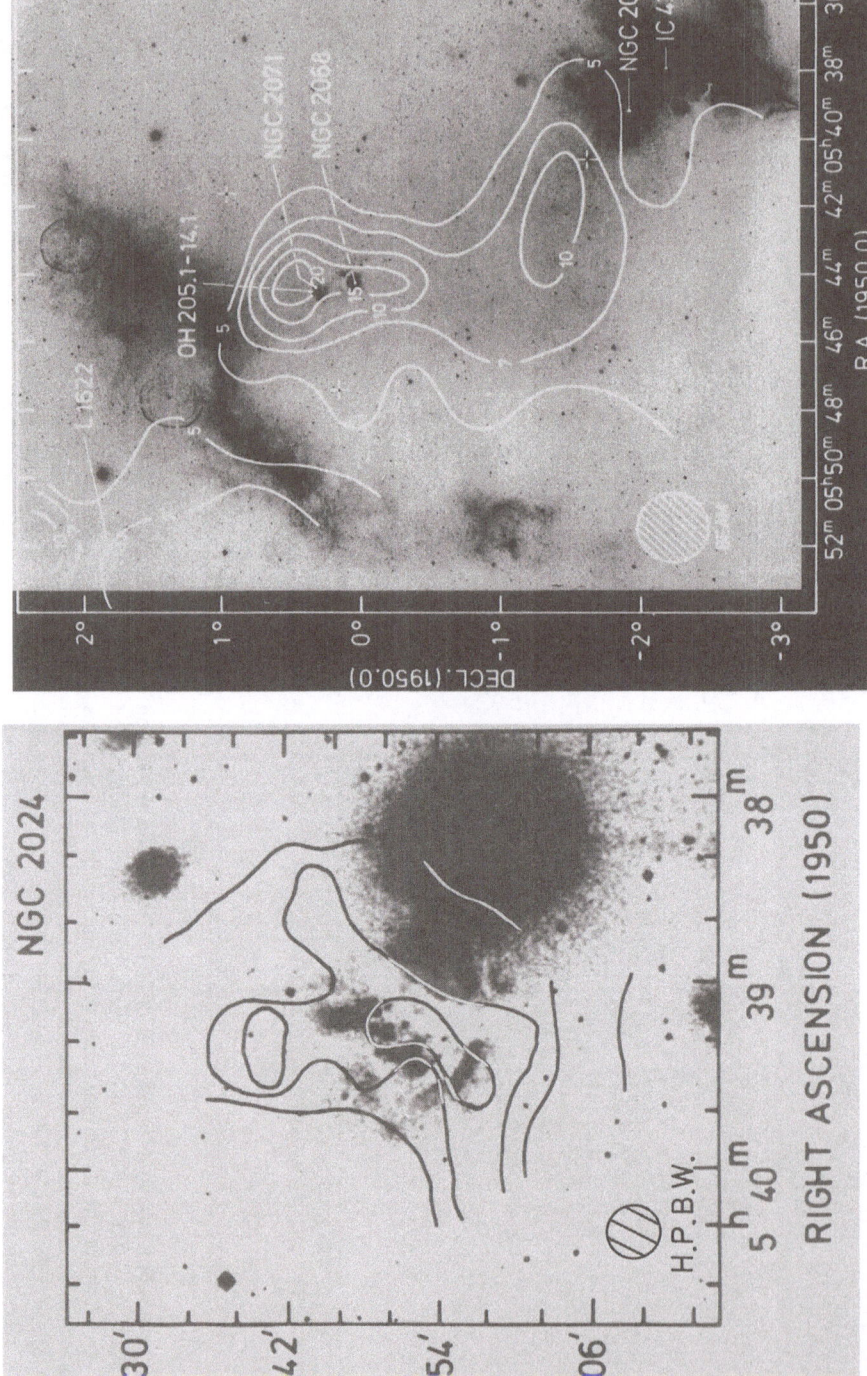

Fig. 5.2.4.   An OH (1667 MHz) contour map of the L1630 dark cloud, made with an angular resolution of 27' × 31', superimposed on the red photograph of the Palomar Sky Survey (after Goss *et al.*, 1976).

Fig. 5.2.3.   A ^{12}CO contour map of NGC 2024 made with an angular resolution of 4'.5 superimposed on an optical photograph of the nebula (after White *et al.*, 1979). The lowest contour is at 20 K and the interval is 5 K (values above the cosmic background of 2.7 K). Note the spatial coincidence of the central molecular 'ridge' with the dust lane.

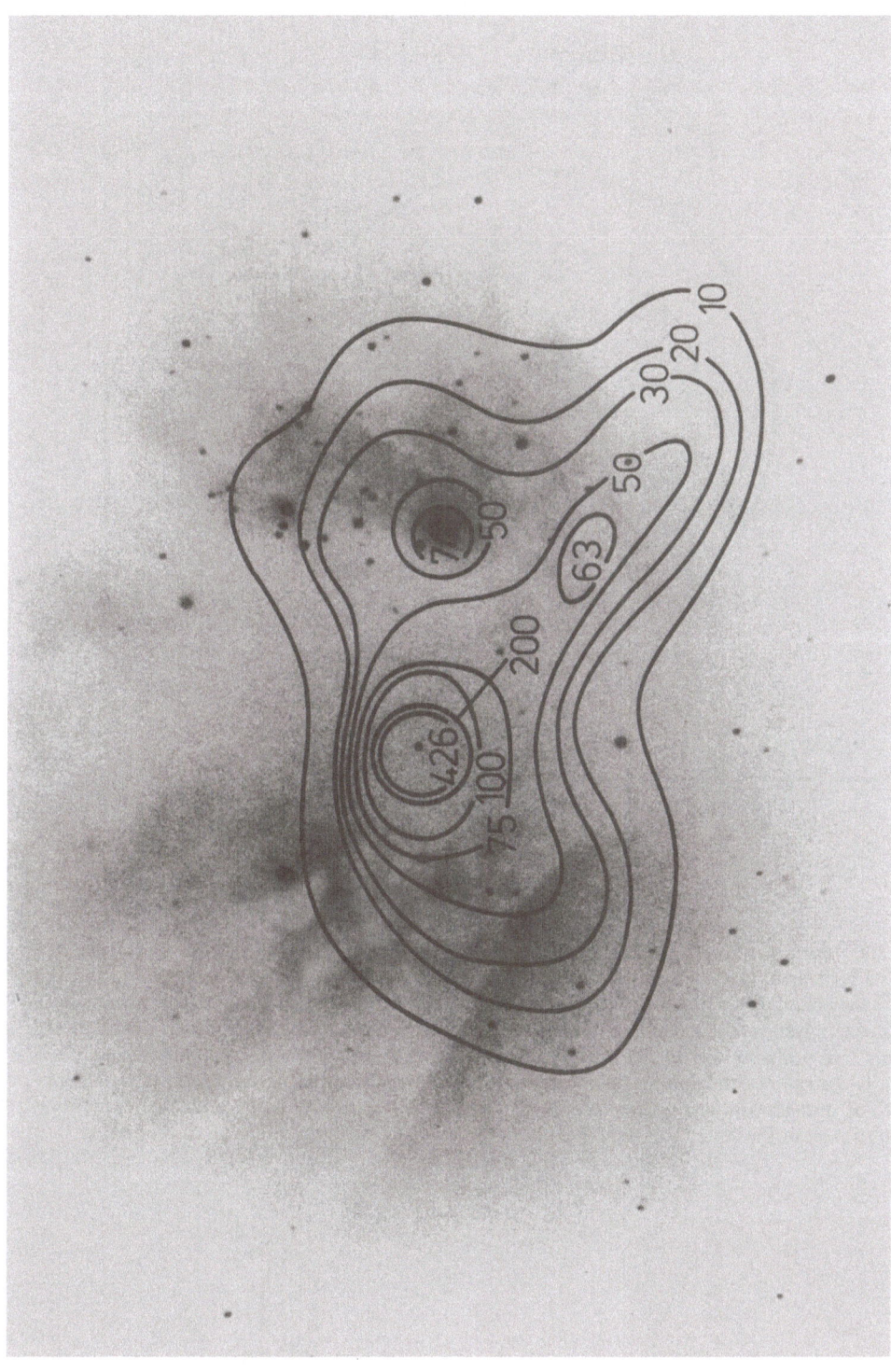

Fig. 5.3.1. Infrared contour map of NGC 2024 made at 3.5μ with 1' angular resolution, superimposed on an I-photograph ($\lambda = 0.92\mu$) of the nebula taken by Dr Birkle with the 1.23 m telescope at Calar Alto (after Frey *et al.*, 1979; and Lemke and Harris, 1979). The peak of the component to the east coincides with NGC 2024 #2 (black dot) whereas the peak of the westerly component occurs at the position of NGC 2024 #1 which is a visible star.

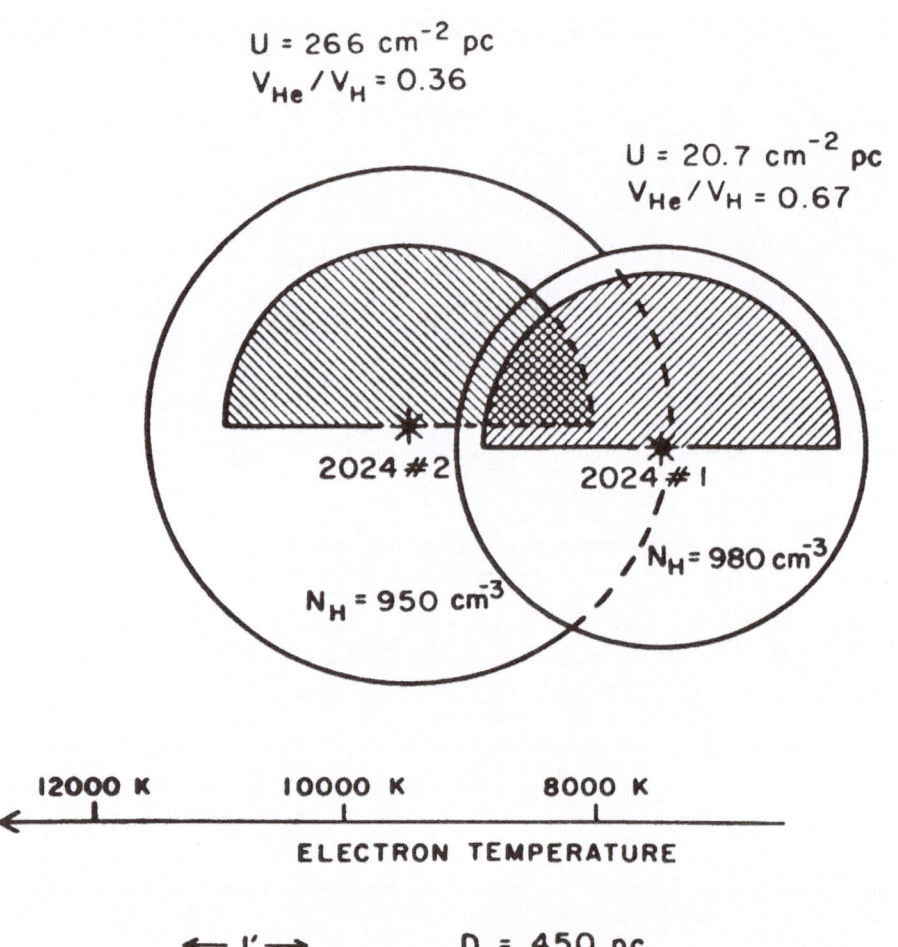

$$U = 266 \ cm^{-2} \ pc$$
$$V_{He} / V_H = 0.36$$

$$U = 20.7 \ cm^{-2} \ pc$$
$$V_{He} / V_H = 0.67$$

2024 #2

2024 #1

$$N_H = 980 \ cm^{-3}$$

$$N_H = 950 \ cm^{-3}$$

12000 K   10000 K   8000 K

ELECTRON TEMPERATURE

← 1' →   D = 450 pc

Fig. 5.3.2.   A model of NGC 2024 assuming that both NGC 2024 #1 and #2 stars have a B0 spectral type and contribute to the ionization of the nebula (after MacLeod *et al.*, 1975). The large circles indicate the boundaries of the hydrogen Strömgren spheres whereas the semi-circles mark the boundaries of the respective helium spheres. The sharply limitted helium ionization zones to the south of the stars, are probably due to the presence of dust absorbing preferentially He-ionizing photons. V_{He}/V_H is the ratio of the volume of the helium sphere (including the part assumed to be absent because of dust absorption) to the volume of the hydrogen sphere. U is the estimated excitation parameters of the stars. This model explains the observed eastwest extension of the nebula (Figure 5.3.1).

APPENDICES

RADIATIVE TRANSFER

I.1. The Various Forms of the Equation of Radiative Transfer

The quantitative interpretation of the radiation emitted from a source is inherently associated with the understanding of the physics of the propagation of radiation through a medium. Consider the situation depicted in Figure I.1.1 where the gas cloud recpresents an H II region, a dust or a molecular cloud which is capable of absorbing radiation passing through it, and generating radiation in it. The monochromatic intensity I_ν, incident on the far side of the elemental volume dV comes out as $I_\nu + dI_\nu$ where dI_ν is the differential intensity contributed from the volume dV, the path length of which is dx. The contributed intensity dI_ν is:

$$dI_\nu = -I_\nu k_\nu \, dx + j_\nu \, dx, \qquad (I.1.1.)$$

where k_ν and j_ν are the linear absorption and emission coefficients respectively. The first term of the sum $(-I_\nu k_\nu \, dx)$ expresses the losses of radiation through absorption whereas the second term $(j_\nu \, dx)$ expresses the gain from emission generated in the differential volume of the gas cloud during the propagation of the incident radiation I_ν through it. Conventionally $k_\nu \, dx$ which depends on two parameters related to the absorbing properties of the differential element of the considered medium (k_ν) and its geometrical depth (dx), is substituted by one variable, the differential optical depth $d\tau_\nu$, given by the relation:

$$d\tau_\nu = k_\nu \, dx. \qquad (I.1.2)$$

Integration of Equation (I.1.1) gives a first form of the equation of radiative transfer, which is:

$$I_\nu = I_0 \, e^{-\tau_\nu} + \int_0^{\tau_\nu} \frac{j_\nu}{k_\nu} e^{-\tau'_\nu} \, d\tau'_\nu, \qquad (I.1.3)$$

where I_0 is the radiation intensity incident on the far side of the cloud and τ_ν is the optical depth at this side (see Figure I.1.1). The term $I_0 e^{-\tau_\nu}$ expresses the attenuation of the background radiation, due to its passage through the cloud. The term $\int_0^{\tau_\nu} (j_\nu/k_\nu) e^{-\tau'_\nu} \, d\tau'_\nu$ expresses the radiation intensity contributed from the gas cloud. In the case where foreground radiation is also present, Equation (I.1.3) should be amended by adding a third term to accomodate the foreground intensity.

Under conditions of thermodynamic equilibrium, characterised by a temperature T, holds:

$$j_\nu = k_\nu B_\nu(T), \qquad (I.1.4)$$

where $B_\nu(T)$ is the Planck function:

$$B_\nu(T) = \frac{2h\nu^3}{c^2} (e^{h\nu/kT} - 1)^{-1}.$$ (I.1.5)

Then, Equation (I.1.3) becomes:

$$I_\nu = I_0 e^{-\tau_\nu} + B_\nu(T)(1 - e^{-\tau_\nu}).$$ (I.1.6)

In the simplified case where no appreciable background radiation is present, Equation (I.1.6) takes the form:

$$I_\nu = B_\nu(T)(1 - e^{-\tau_\nu}).$$ (I.1.7)

Consequently, the radiation originating from an H II region characterized by an electron temperature T_e can be expressed as:

$$I_\nu = B_\nu(T_e)(1 - e^{-\tau_\nu}).$$ (I.1.8)

For radioastronomical observations, for which holds in general $h\nu \ll kT_e$, the Planck function can be approximated by:

$$B_\nu(T_e) = \frac{2k\nu^2}{c^2} T_e.$$ (I.1.9)

In radioastronomy, intensity measurements are usually expressed in terms of brightness temperature, T_b, according to the relationship:

$$I_\nu = \frac{2k\nu^2}{c^2} T_b.$$ (I.1.10)

Then, by taking into account relations (I.1.9) and (I.1.10), Equation (I.1.8) becomes:

$$T_b = T_e (1 - e^{-\tau_\nu}).$$ (I.1.11)

Alternatively, the radiation intensity can be expressed as flux density S_ν, by the relation:

$$S_\nu = \int_{source} I_\nu d\Omega$$ (I.1.12)

(i.e. intensity integrated over the solid angle Ω subtended by the source). Then, because of Equation (I.1.7):

$$S_\nu = \int_{source} B_\nu(T_e)(1 - e^{-\tau_\nu}) \, d\Omega \simeq \Omega B_\nu(T_e)(1 - e^{-\tau_\nu})$$ (I.1.13)

or because of (I.1.10):

$$S_\nu = \frac{2k\nu^2}{c^2} \int_{source} T_b \, d\Omega \simeq \frac{2k\nu^2}{c^2} T_b \Omega.$$ (I.1.14)

For infrared observations of a dust cloud, characterized by dust temperature T_d, the equation of radiative transfer (I.1.7) takes the form:

$$I_{IR_\nu} = B_\nu(T_d)(1 - e^{-\tau_\nu})$$ (I.1.15)

or alternatively:

$$F_{IR_\nu} = \Omega B_\nu(T_d)(1 - e^{-\tau_\nu}),$$ (I.1.16)

where F_{IR} is the flux density at frequency ν and Ω is the solid angle subtended by the source.

For molecular observations, in the simple case of a two-level molecule $(n, m; n > m)$ characterized by excitation temperature T_{nm} and emitting in the line nm, the equation of radiative transfer is expressed by the relation:

$$I_{nm} = B_\nu(T_{nm})(1 - e^{-\tau_\nu}). \tag{I.1.17}$$

For molecules with rotational spectrum in the microwave band of frequencies, for which $h\nu \ll kT_{nm}$, this equation can be expressed in the form:

$$T_b = T_{nm}(1 - e^{-\tau_\nu}) \tag{I.1.18}$$

which is an approximation similar to (I.1.11).

All these relations express the various forms of the equation of the radiative transfer, and will be used in the next sections to derive the physical parameters of an H II region, a dust cloud and a molecular cloud respectively.

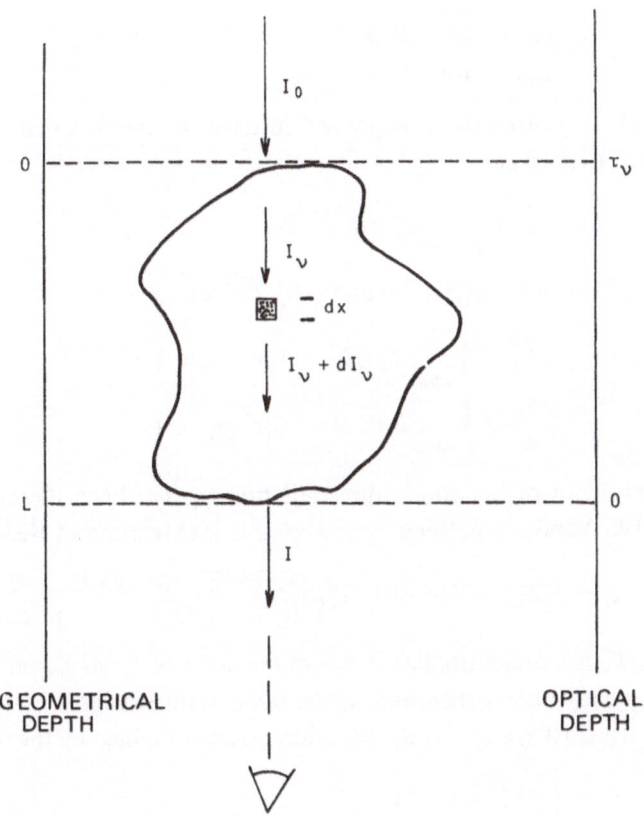

Fig. I.1.1.   Background radiation I_0 travelling through a cloud towards the observer.

PHYSICAL PARAMETERS OF AN H II REGION

II.1. Physical Parameters Derived from Observations
of the Radio Continuum Emission

The brightness temperature of the radio continuum radiation (free-free emission) originating in an H II region can be expressed by the relation (I.1.11) as:

$$T_c = T_e(1 - e^{-\tau_c}),$$ (II.1.1)

where T_c, τ_c are the brightness temperature and the optical depth characterizing the radio continuum radiation. For an isothermal H II region, two limiting cases can be distinguished: (1) an optically thick nebula ($\tau_c \gg 1$) occuring at low frequencies, $\nu < 1$ GHz), and (2) an optically thin nebula ($\tau_c \ll 1$ occuring at high frequencies, $\nu > 1$ GHz). In these cases Equation (II.1.1) takes the forms:

$$T_c = \begin{cases} T_e & \text{for} \quad \tau_c \gg 1 \\ \tau_c T_e & \text{for} \quad \tau_c \ll 1. \end{cases}$$

(II.1.2a)
(II.1.2b)

Equation (II.1.1) can also be expressed in terms of the flux density S_ν (see relations (I.1.10) and (I.1.12)) as:

$$S_\nu = \frac{2k}{c^2} \nu^2 \int_{\text{source}} T_c \, d\Omega$$ (II.1.3)

This relation because of (II.1.2a) and (II.1.2b) can be expressed as:

$$S_\nu = \begin{cases} \dfrac{2k}{c^2} \nu^2 \displaystyle\int_{\text{source}} T_e \, d\Omega & \text{for} \quad \tau_c \gg 1 \\ \dfrac{2k}{c^2} \nu^2 \displaystyle\int_{\text{source}} \tau_c T_e \, d\Omega & \text{for} \quad \tau_c \ll 1. \end{cases}$$

(II.1.4a)
(II.1.4b)

The optical depth of the continuum radiation, τ_c, has been theoretically determined by Oster (1961) and a simplified approximation is (Mezger and Henderson, 1967):

$$\tau_c = 8.235 \times 10^2 \, \alpha(\nu, T_e)\left[\frac{T_e}{\text{K}}\right]^{-1.35}\left[\frac{\nu}{\text{GHz}}\right]^{-2.1}\left[\frac{E}{\text{pc cm}^{-6}}\right],$$ (II.1.5)

where $\alpha(\nu, T_e)$ is a dimensionless factor of the order of 1, whose variation in the region of the frequencies concerned and in the temperature range 5×10^3 K–1.2×10^4 K, does not exceed 10% and E is the emission measure defined by the relation:

$$E = \int_0^L N_e^2 \, dx = N_e^2 L.$$ (II.1.6)

The integration is performed along the path length L of the nebula which is assumed to contain only ionized hydrogen (otherwise $E = \int_0^L N_e N_i \, dx$, where N_e, N_i is the electron and ion number density respectively, with $N_e \neq N_i$). From relations (II.1.4) and (II.1.5) the frequency dependence of the flux density S_ν can be established. This is:

$$S_\nu = \begin{cases} \propto \nu^2 & \text{for} \quad \tau_c \gg 1 \\ \propto \nu^{-0.1} & \text{for} \quad \tau_c \ll 1 \end{cases}$$

(II.1.7a)

(II.1.7b)

and characterizes the radiocontinuum spectrum of every H II region (Figure II.1.1.). The establishment of such a spectrum, from many observations made at various frequencies, leads to the determination of a number of physical parameters.

To appreciate the physical relations currently used to derive the physical parameters of an H II region, a simplistic case will be treated. This involves an H II region, the geometry of which is either cylindrical, with the axis of symmetry lying along the line of sight, or spherical. In both cases a constant electron density is assumed. The knowledge of the radio continuum spectrum, $S_\nu = f(\nu)$, the electron temperature, T_e, and the distance, D, of the H II region from the Earth is also taken for granted. The present treatment will follow the general line of reasoning developed by Mezger (1972). From the shape of the spectrum, the turnover frequency ν_{to} (at which the spectrum starts exhibiting the optically thin pattern) can be established (Figure II.1.1). At this frequency the optical depth τ_c is of the order of unity. From Equation (II.1.5) and $\tau_c = 1$ (sometimes $\tau_c = 1.5$ is also used), the emission measure E can be evaluated. This is given by:

$$E = \text{const. } T_e^{1.35} \nu_{to}^{2.1}$$

(II.1.8)

The next step involves the evaluation of the flux density S_ν for the optically thin case ($\tau_c \ll 1$). From Equations (II. 1.3), (II.1.2b) and (II.1.5) the S_ν can be expressed as:

$$S_\nu = \frac{2k\nu^2}{c^2} \int_{\text{source}} T_e \tau_c \, d\Omega = \text{const. } T_e^{-0.35} \nu^{-0.1} \int_{\text{source}} d\Omega \int_0^L N_e^2 \, dx$$

(II.1.9)

For a cylinder with radius R and axis $L = 2R$ and a sphere with diameter $2R$ hold:

$$\int_{\text{source}} d\Omega \int_0^{L=2R} N_e^2 \, dx = \frac{\pi R^2}{D^2} N_e^2 \, 2R = \begin{cases} V_{\text{cyl}} \dfrac{N_e^2}{D^2} \\ 1.5 V_{\text{sph}} \dfrac{N_e^2}{D^2}. \end{cases}$$

(II.1.10a)

(II.1.10b)

Substitution of (II.1.10a) or (II.1.10b) into (II.1.9) yields:

$$V_{\text{cyl}} \text{ or } V_{\text{sph}} = \text{const. } S_\nu D^2 T_e^{0.35} \nu^{0.1} N_e^{-2}$$

(II.1.11)

(the difference between the two is only in the value of the constant).

Relation (II.1.11) in combination with the analytical expressions

$$V_{\text{cyl}} = \pi R^2 L = 2\pi R^3$$

(II.1.12a)

or

$$V_{\text{sph}} = \tfrac{4}{3}\pi R^3$$

(II.1.12b)

and the relation $E = N_e^2 2R$ (II.1.6) yields the linear size:

$$R = \text{const. } S_\nu^{0.5} D T_e^{0.175} \nu^{0.05} E^{-0.5} \tag{II.1.13}$$

or the angular size ($\theta = 2R/D$):

$$\theta = \text{const. } S_\nu^{0.5} T_e^{0.175} \nu^{0.05} E^{-0.5}. \tag{II.1.14}$$

Then, from (II.1.6) and (II.1.13), the electron density can be estimated by the relation:

$$N_e = \text{const. } S_\nu^{-0.25} D^{-0.5} T_e^{-0.0875} \nu^{-0.025} E^{0.75} \tag{II.1.15}$$

or simply from (II.1.16):

$$N_e = \text{const. } R^{-0.5} E^{0.5} \tag{II.1.16}$$

or, from $\theta = 2R/D$:

$$N_e = \text{const. } \theta^{-0.5} D^{-0.5} E^{0.5}. \tag{II.1.17}$$

The total mass of the ionized hydrogen can be obtained, from the relation:

$$M_{\text{H II}} = V N(\text{H}) m_{\text{H}}, \tag{II.1.18}$$

where V is the total volume of the H II region expressed by (II.1.11), m_{H} is the mass of the hydrogen atom and $N(\text{H})$ is the number density of the hydrogen atoms which for an H II region consisting of pure hydrogen is:

$$N(\text{H}) = N(\text{H}^+) = N_e. \tag{II.1.19}$$

If, however, He is present in the singly ionized stage, with an abundance

$$\frac{N(\text{He}^+)}{N(\text{H}^+)} = \frac{N(\text{He}^+)}{N_{\text{H}}(\text{H})} = \alpha \tag{II.1.20}$$

then,

$$N_e = N(\text{H}^+) + N(\text{He}^+) = N(\text{H}) + N(\text{He}^+) \tag{II.1.21}$$

and

$$N(\text{H}) = N_e - N(\text{He}^+) \tag{II.1.22}$$

or

$$N(\text{H}) = N_e - \alpha N(\text{H}) \tag{II.1.23}$$

from which

$$N(\text{H}) = \frac{N_e}{1 + \alpha} \tag{II.1.24}$$

or

$$N(\text{H}) = N_e \left[1 + \frac{N(\text{He}^+)}{N(\text{H}^+)} \right]^{-1}. \tag{II.1.25}$$

Then, for a nebula containing only hydrogen (Equation (II.1.11), (II.1.18), and (II.1.19)):

$$M_{\text{H II}} = \text{const. } S_\nu D^2 T_e^{0.35} \nu^{0.1} N_e^{-1}, \tag{II.1.26}$$

whereas for a nebula containing both hydrogen and helium (Equations (II.1.11), (II.1.18), and (II.1.25)):

$$M_{\text{H II}} = \text{const. } S_\nu D^2 T_e^{0.35} \nu^{0.1} N_e^{-1} \left[1 + \frac{N(\text{He}^+)}{N(\text{H}^+)}\right]^{-1}. \tag{II.1.27}$$

Another important parameter which can also be derived is the excitation parameter u. This is defined by considering the equilibrium condition of a spherical, dust-free, ionization bounded H II region consisting of pure hydrogen, according to which all the Lyman continuum photons emitted in the unit of time from the surface of the exciting star (Lc) are equal to the electron-proton recombinations occuring within the H II sphere, i.e.:

$$Lc = \tfrac{4}{3}\pi R^3 N_e^2 (\beta - \beta_1), \tag{II.1.28}$$

where $\beta - \beta_1$ is the recombination rate to the excited levels of hydrogen (i.e. minus the first level since recombination on that yields a photon capable to ionize another hydrogen atom and thus is absorbed instantaneously; this is given by:

$$\beta - \beta_1 = 4.10 \times 10^{-10}\, T_e^{-0.8} (\text{cm}^3\ \text{s}^{-1}). \tag{II.1.29}$$

Equation (II.1.28) can also be written as:

$$u^3 = \frac{3}{4\pi(\beta - \beta_1)} Lc, \tag{II.1.30}$$

where u is the excitation parameter, equal to

$$u = R N_e^{2/3}. \tag{II.1.31}$$

The excitation parameter u can be evaluated from (II.1.9). (II.1.10), and (II.1.31) by the relation:

$$u = \text{const. } S_\nu^{0.33} D^{0.67} T_e^{0.117} \nu^{0.033}. \tag{II.1.32}$$

The rate of Lyman-continuum photons Lc which is required to be emitted from the exciting star in order to create such an H II region is also obtained from (II.1.30), (II.1.29), and (II.1.32)

$$Lc = \text{const. } S_\nu D^2 T_e^{-0.45} \nu^{0.1}. \tag{II.1.33}$$

Note that the flux density S_ν is proportional to the rate of the Lyman continuum photons Lc emitted by the exciting star.

The excitation parameter, u, is characteristic of the 'ionization capability' of a star and can be independently estimated for stars of various spectral types and luminosity classes. Such an estimation establishes a relation of the form $u = f$ (spectal type, luminosity class) which, used in combination with the estimation of u of an ionization bounded, dust-free H II region (from observations of the radio continuum emission) provides information about the nature of the exciting star (or stars).

The estimation of the excitation parameter u for a star of a certain spectral type and luminosity class generally demands the knowledge of four parameters: the absolute visual magnitude M_V, the bolometric correction BC, the effective temperature T_{eff}, and the rate of emission of the Lyman continuum photons per unit area of the stellar surface N_L, defined by:

$$N_L = \int_{\nu_L}^{\infty} \frac{\pi F_V}{h\nu} d\nu, \qquad (II.1.34)$$

where ν_L is the frequency at the Lyman limit and πF_V is the emergent flux from the star surface at frequency ν (in ergs cm^{-2} s^{-1}). N_L is a function of the effective temperature of the star (T_{eff}) and the value of gravity (g) and can be evaluated from stellar atmosphere calculations. Then the absolute stellar luminosity L_* (in $L_\odot$), the stellar radius R_* (in $R_\odot$), the emission rate of Lyman continuum photons from the whole stellar surface Lc, (Lyman continuum photon luminosity), and the corresponding excitation parameter u can be estimated according to the formulae (e.g. Panagia, 1973):

$$\log \frac{L_*}{L_\odot} = -0.4(M_V - BC) + 1.888, \qquad (II.1.35)$$

$$\log \frac{R_*}{R_\odot} = 0.5 \log \frac{L_*}{L_\odot} - 2 \log\left(\frac{T_{eff}}{10^4}\right) - 0.473, \qquad (II.1.36)$$

$$Lc = 4\pi R_*^2 N_L \text{ (photons s}^{-1}\text{)}, \qquad (II.1.37)$$

and

$$u = 2.01 \times 10^{-19} \left[\frac{N_L}{\beta - \beta_1}\right]^{1/3} \text{ (pc cm}^{-2}\text{)}, \qquad (II.1.38)$$

where $\beta - \beta_1$ is the recombination rate to the excited levels of hydrogen, (II.1.29).

The Lc and u parameters are characteristic of the ionizing radiation emitted by a star surrounded by a nebulosity. More conveniently, Lc and u are expressed as a function of $N_L/\pi F_V$ (Lyman-visible colour), where πF_V is the emergent flux at the stellar surface in the visual (V) magnitude band. The ratio $N_L/\pi F_V$ is estimated either theoretically from model stellar atmospheres or empirically from observations at optical or radio wavelengths of a nebula excited by the star, from the relations (Morton, 1969; Georgelin et al., 1975):

$$\frac{N_L}{\pi F_V} = 6.93 \times 10^{16} f_{H\alpha} \times 10^{0.4(V - A_v + A_{H\alpha})}, \qquad (II.1.39)$$

where $f_{H\alpha}$ is the Hα flux from the nebula observed at the earth in ergs cm^{-2} s^{-1}:

V is the apparent visual magnitude of the star;

A_v is the total visual interstellar extinction; and

$A_{H\alpha}$ is the total extinction in Hα.

$(A_v = R \times E(B - V)$ where $R \simeq 3$, $A_{H\alpha} = 0.8 A_v$, Whitford (1958), see however discussion for possible anomaly of the extinction law in Orion in Section 2.4), or alternatively:

$$\frac{N_L}{\pi F_V} = 3.76 \times 10^{30} \frac{f_\nu}{g} 10^{0.4(V - A_v)}, \tag{II.1.40}$$

where f_ν is the observed radio continuum flux in ergs cm^{-2} s^{-1} Hz^{-1} from the nebula received at the Earth's surface, and g is the Gaunt factor approximated by

$$g(T_e, \nu) = 1 + 0.130 \log\left[\left(\frac{T_e}{K}\right)^{3/2}\left(\frac{\nu}{Hz}\right)^{-1}\right]. \tag{II.1.41}$$

Both formulae are valid under the assumption than all Lyman continuum photons emitted from the star are used to ionize the hydrogen of the nebula (ionization bounded case). The formulae are inherently dependent on the electron temperature T_e of the associated nebula and were derived for $T_e = 10^4$ K. The above formulae are also dependent on a factor K^{-1} which appears in the conversion of fluxes to photometric magnitude and is determined from the relation

$$f_V = K \times 10^{-0.4V} \text{ ergs cm}^{-2} \text{ s}^{-1}, \tag{II.1.42}$$

where f_V is the flux in the V band observed at the Earth. The above formulae were calculated for $K = 1.03 \times 10^5$ ergs cm^{-2} s^{-1} derived from the calibration of α Lyrae by Oke and Schild (1970). Then the Lyman continuum photon luminocity Lc can be estimated from (Georgelin et al., 1975):

$$Lc = 1.20 \times 10^{40} K \times 10^{-0.4M_V}\left(\frac{N_L}{\pi F_V}\right) \text{ (photons s}^{-1}), \tag{II.1.43}$$

which for $K = 1.03 \times 10^{-5}$ takes the form:

$$Lc = 1.236 \times 10^{35} \times 10^{-0.4M_V}\left(\frac{N_L}{\pi F_V}\right) \text{ (photons s}^{-1}). \tag{II.1.44}$$

The excitation parameter u can then be given by the relation:

$$u = 1.43 \times 10^{-3}\left[\left(\frac{N_L}{\pi F_V}\right) \times 10^{-0.4M_V}\right]^{1/3} \text{ (pc cm}^{-2}). \tag{II.1.45}$$

The excitation parameter, u, as a function of spectral type for stars of ZAMS (zero age main sequence) and luminosity class V (dwarfs), III (giants), and I (supergiants), estimated by various workers, is shown in Figures II.1.2a, b, c, and d. The considerable disagreement of the various models should, however, serve as a reminder of the limitations of the method.

It should also be noted that the excitation parameter u of a nebulosity, estimated from observations at radio wavelengths, can provide information about the nature of the exciting star only in the case of an ionization bounded, dust-free, spherical H II region. For a density bounded, dust bounded or an intermediate more complicated structure (e.g. M42), the excitation parameter u (and the corresponding Lc) represent a lower limit (a stronger ionizing source may be present but not able to create a more extended H II sphere simply because of lack of adequate material in the surroundings or/and absorption of part of the Lc photons by the dust).

Analytical calculations for the derivation of the physical parameters of an H II region can be found in a number of excellent publications on this subject (Pariiskii, 1961; Mezger and Henderson, 1967; Mezger et al., 1967; Terzian, 1968, 1974, 1976). As an example one set of parameters are given (Goudis, 1977) obtained for an ionization bounded, dust-free, spherical H II region of constant density, according to the model introduced by Mezger et al. (1967). These parameters are:

(1) The emission measure at the centre of the source, E, which is given by the relation:

$$\left[\frac{E}{\text{pc cm}^{-6}} \right] = 18.82 \left[\frac{T_e}{K} \right]^{1.35} \left[\frac{\nu_{to}}{\text{GHz}} \right]^{2.1}, \tag{II.1.46}$$

where T_e is the electron temperature of the source and ν_{to} is the turnover frequency of the radio spectrum (for which $\tau_c = 1.5$ was assumed).

(2) The spherical angular diameter θ_{sph} of the source, given by the relation:

$$\left[\frac{\theta_{\text{sph}}}{\text{arc min}} \right] = 94.192 \left[\frac{T_e}{K} \right]^{0.175} \left[\frac{\nu}{\text{GHz}} \right]^{0.05} \left[\frac{E}{\text{pc cm}^{-6}} \right]^{-0.5} \left[\frac{S_\nu}{\text{Jy}} \right]^{0.5}, \tag{II.1.47}$$

where S_ν is the flux density of the source at a frequency ν in which the source is optically thin.

(3) The root mean square electron density, rms N_e, given by

$$\left[\frac{\text{rms } N_e}{\text{cm}^{-3}} \right] = 1.8529 \left[\frac{E}{\text{pc cm}^{-6}} \right]^{0.5} \left[\frac{D}{\text{kpc}} \right]^{-0.5} \left[\frac{\theta_{\text{sph}}}{\text{arc min}} \right]^{-0.5}, \tag{II.1.48}$$

where D is the distance of the source.

(4) The mass of the ionized hydrogen, $M_{\text{H II}}$, given by

$$\left[\frac{M_{\text{H II}}}{M_\odot} \right] = 3.1753 \times 10^{-4} \left[\frac{D}{\text{kpc}} \right]^3 \left[\frac{\theta_{\text{sph}}}{\text{arc min}} \right]^3 \left[\frac{\text{rms } N_e}{\text{cm}^{-3}} \right]. \tag{II.1.49}$$

(5) The excitation parameter u of the object, given by

$$\left[\frac{u}{\text{pc cm}^{-2}} \right] = 0.14549 \left[\frac{\theta_{\text{sph}}}{\text{arc min}} \right] \left[\frac{D}{\text{kpc}} \right] \left[\frac{\text{rms } N_e}{\text{cm}^{-3}} \right]^{2/3}. \tag{II.1.50}$$

(6) The rate of the Lyman continuum photons Lc of the exciting star(s) which is necessary to ionize the source, given by the relation (Rubin, 1968; Mezger, 1972):

$$\left[\frac{Lc}{\text{photons s}^{-1}} \right] = 4.76 \times 10^{48} \left[\frac{S_\nu}{\text{Jy}} \right] \left[\frac{D}{\text{kpc}} \right]^2 \left[\frac{\nu}{\text{GHz}} \right]^{0.1} \left[\frac{T_e}{K} \right]^{-0.45}. \tag{II.1.51}$$

The use of these formulae requires the knowledge of the distance D of the source (usually derived from a velocity-distance relation based on a rotation model of the Galaxy), the electron temperature T_e (which is of the order of 10^4 K), the total flux density of the source S_ν at the optically thin part of the spectrum (at a frequency ν at which the source is optically thin) and the turnover frequency ν_{to} at which the radio spectrum starts showing the optically thin pattern (nearly flat part of the spectrum (see Figure II.1.1). This occurs for optical depths $\tau_c \simeq 1$ (here $\tau_c = 1.5$ was used). The main limitation of this model is in the determination of the turnover frequency ν_{to} which strongly depends on the accuracy with which the continuum spectrum is known.

The method can be applied in Orion (M42), the radio spectrum of which is known to a sufficient accuracy (Figure 2.2.17). The method is also applicable to other H II regions with less well defined spectrum. This is accomplished by using the Orion spectrum for calibration purposes, according to the relation (Hjellming and Davies, 1970):

$$\frac{S_{\nu_{to}}}{S_{\max}} = \left(\frac{S_{\nu_{to}}}{S_{\max}}\right)_{\text{ORION}} = 0.72, \tag{II.1.52}$$

where $S_{\nu_{to}}$ is the flux density at the turnover frequency ν_{to} and $S_{\max}$ is the maximum flux density of the optically thin part of the spectrum. From this relation the $S_{\nu_{to}}$ and consequently the ν_{to} of each source can be determined.

An alternative version of this model, which avoids the uncertainty involved in the estimation of the turnover frequency ν_{to}, utilizes instead the angular size of the source which can be estimated from radio continuum maps. The size which is directly estimable is however the apparent size of the source, θ_α which is the result of the convolution of the true size θ_G and the response pattern of the radio telescope $\theta_{\text{telesc.}}$, i.e.:

$$\theta_\alpha = \theta_G * \theta_{\text{telesc.}}. \tag{II.1.53}$$

A deconvolution is usually performed to establish the true size θ_G (half power width of the Gaussian source). This size is related to the spherical size of the source by the formula:

$$\theta_G = 0.68\theta_{\text{sph.}}. \tag{II.1.54}$$

A density distribution is also assumed which makes the values of N_e, E, and $M_{\text{H II}}$ model dependent. The effects of the presence of helium are also taken into account. The physical parameters derived from such a model which assumes an exponentially tapered distribution of density with a peak density N_0 [$N_e \simeq N_0 \exp\text{-}f(\theta_G^{-2})$], are (Chaisson, 1976):

$$\left[\frac{N_e}{\text{cm}^{-3}}\right] = \frac{35}{\alpha(\nu, T_e)^{0.5}} \left[\frac{\nu}{\text{Hz}}\right]^{0.05} \left[\frac{T_e}{\text{K}}\right]^{0.175} \times$$

$$\times \left[\frac{S_\nu}{\text{Jy}}\right]^{0.5} \left[\frac{D}{\text{kpc}}\right]^{-0.5} \left[\frac{\theta_G}{\text{arc min}}\right]^{-1.5}, \tag{II.1.55}$$

$$\left[\frac{M_{\text{H\,II}}}{M_\odot}\right] = \frac{0.03}{\alpha(\nu, T_e)^{0.5}}\left[\frac{\nu}{\text{Hz}}\right]^{0.05}\left[\frac{T_e}{\text{K}}\right]^{0.175}\left[\frac{S_\nu}{\text{Jy}}\right]^{0.5} \times$$

$$\times \left[\frac{D}{\text{kpc}}\right]^{2.5}\left[\frac{\theta_G}{\text{arc min}}\right]^{1.5}\left[1 + \frac{N(\text{H}_e^+)}{N(\text{H}^+)}\right]^{-1}, \quad \text{(II.1.56)}$$

where $N(\text{He}^+)/N(\text{H}^+)$ is the abundance of the ionized helium and the factor $[1 + (N(\text{He}^+)/N(\text{H}^+))]^{-1}$ is introduced to subtract the contribution of the ionized helium (see relation (II.1.25)),

$$\left[\frac{E}{\text{pc cm}^{-6}}\right] = \frac{503}{\alpha(\nu, T_e)}\left[\frac{\nu}{\text{Hz}}\right]^{0.1}\left[\frac{T_e}{\text{K}}\right]^{0.35}\left[\frac{S_\nu}{\text{Jy}}\right]\left[\frac{\theta_G}{\text{arc min}}\right]^2, \quad \text{(II.1.57)}$$

$$\left[\frac{u}{\text{pc cm}^{-2}}\right] = \frac{2.5}{\alpha(\nu, T_e)^{0.333}}\left[\frac{\nu}{\text{Hz}}\right]^{0.033}\left[\frac{T_e}{\text{K}}\right]^{0.117} \times$$

$$\times \left[\frac{S_\nu}{\text{Jy}}\right]^{0.333}\left[\frac{D}{\text{kpc}}\right]^{0.666}, \quad \text{(II.1.58)}$$

$$\left[\frac{Lc}{\text{photons } s^{-1}}\right] = \frac{6 \times 10^{47}}{\alpha(\nu, T_e)}\left[\frac{S_\nu}{\text{Jy}}\right]\left[\frac{D}{\text{kpc}}\right]^2\left[\frac{\nu}{\text{Hz}}\right]^{0.1}\left[\frac{T_e}{\text{K}}\right]^{-0.45}, \quad \text{(II.1.59)}$$

where $\alpha(\nu, T_e)$ is a dimensionless factor of the order of 1, the variation of which, in the region of frequencies concerned and in the temperature range of 5×10^3–1.2×10^4 K, does not exceed 10%. Computed tables of $\alpha(\nu, T_e)$ can be found in Mezger and Henderson (1967).

A spherical distribution of N_e with constant value within a sphere of size $\theta_{\text{sph.}}$ and zero value outside would alter the estimates of N_e, $M_{\text{H\,II}}$ and E in $0.8N_e$, $0.4M_{\text{H\,II}}$, and $1.4E$, respectively. Cylindrical geometry models with constant density inside are also in use (Mezger and Henderson, 1967).

II.2. Determination of Density from Optical Methods

The observed brightness inhomogeneities in the optical structure of the majority of H II regions (see for example Figure 2.1.3 of the core of M42) is conventionally interpreted as due to density fluctuations within the nebula. Such an interpretation is based on the fact that the brightness of every optically thin emission line is strongly dependent on electron density ($\propto N_e^2$; see also Table II.3.III). The electron density of these 'condensations' or 'clumps' is sometimes referred to as the local density to distinguish it from the root mean square electron density which represents an average density of the whole source.

The local density is estimated from observations made in the light of the optical forbidden lines [O II] and [S II]. The structure of the corresponding O^+ and S^+ ions from which these lines originate has the advantage of possesing two different levels with nearly the same excitation energy (Figure II.2.1.) Therefore, the line emission originating from these levels depends only on the electron density and hardly on temperature. The levels are populated by collisional excitation i.e. from collisions with free electrons which are products of the ionization of nebular hydrogen, caused by the UV radiation emitted from the exciting stars. These electrons quickly establish a Maxwell–Boltzman distribution around an energy peak of a few eV (Böhm and Aller, 1947). The most energetic electrons occasionally collide with ions such as O^+ and S^+, and excite them to their fine structure metastable levels. In the low density limit ($N_e \rightarrow 0$ cm^{-3}), where collisional de-excitation of the ions is insignificant, every eollisional excitation is followed by the emission of a photon resulting in the observed [O II] or [S II] lines. In this case the line ratios of the corresponding ions are (Osterbrock, 1974):

$$[\text{O II}]\ \frac{I(\lambda\ 3729\ \text{Å})}{I(\lambda\ 3726\ \text{Å})} = [\text{S II}]\ \frac{I(\lambda\ 6717\ \text{Å})}{I(\lambda\ 6731\ \text{Å})} = 1.5. \tag{II.2.1}$$

For the high density case ($N_e \rightarrow \infty$) the dominance of collisional processes makes the ratio (Osterbrock, 1974):

$$[\text{O II}]\ \frac{I(\lambda\ 3729\ \text{Å})}{I(\lambda\ 3726\ \text{Å})} = [\text{S II}]\ \frac{I(\lambda\ 6717\ \text{Å})}{I(\lambda\ 6731\ \text{Å})} = 0.35. \tag{II.2.2.}$$

The solution of the equilibrium equations between these limits, is shown in Figure II.2.2 and provides the means of calculating the electron density N_e by estimating the [O II] or [S II] ratio (provided that the observed ratio is always in the range of 0.35–1.5). A summary of the methods, based on the conclusions of Osterbrock's treatment of the problem (Osterbrock, 1974), is given is Table II.2.1.

The electron densities N_e estimated from these methods represent the local densities of the individual condensations which are contained within the more tenuous ionized gas of an H II region. According to Osterbrock and Flather (1959), the condensations are distributed within the volume V of the nebula but occupy only a fraction $\alpha = V_{\text{cond.}}/V$ of it. The total flux density emitted from the whole volume of the optically thin nebula is, however, identical to the flux density emitted from the sum of the condensations. Since $S_\nu \propto \langle N_e^2 \rangle V$ (see relations (II.1.9) and (II.1.10)), one obtains:

$$\langle N_e^2 \rangle V = N_e^2 V_{\text{cond.}}, \tag{II.2.3}$$

where $\langle N_e^2 \rangle$ is the mean square density of the whole nebula and N_e is the local density of the individual condensations. Thus, the condensation factor $a = V_{\text{cond.}}/V$ can be derived from the relation:

$$\alpha = \frac{V_{\text{cond.}}}{V} = \frac{\langle N_e^2 \rangle}{N_e^2}. \tag{II.2.4}$$

In practice, the $\langle N_e^2 \rangle^{1/2}$, which is the rms N_e, is derived from observations of the radio continuum (see section II.1) whereas the N_e is better represented by the local density, determined from the optical methods based on the intensity ratios of the forbidden lines [O II] and [S II].

II.3. Determination of Temperature from Optical and Radio Methods

There are numerous methods to estimate the electron temperature, T_e, of an H II region. These are classified for convenience in 'optical' and 'radio' methods.

The main optical methods, currently in use, derive the electron temperature from:

(1) *The intensity ratio of forbidden lines originating in two different upper levels of the same ion which are characterized by different excitation energies.* Consequently both the collisional excitation and the radiative de-excitation of the ions concerned depend strongly on T_e. The method makes use of the intensity ratio of forbidden lines originating in such ions as O^{++}, N^+, or S^{++}. The energy level diagram for the lowest terms of the O^{++} and N^+ ions is shown in Figure II.2.1. The method is not, however, completely independent of the electron density since for $N_e > 10^4$ cm^{-3} collisional de-excitation becomes important. This contributes more strongly to the depopulation of the middle ($^1D_2$) level than to the depopulation of the upper energy level ($^1S_0$). In addition, collisional excitation of the ^1S from the already excited ^1D level can occur. Both processes result in the weakening of the emission lines originating in the middle level ($^1D \rightarrow {}^3P$) with respect to the line emitted from the upper level ($^1S \rightarrow {}^1D$).

For $N_e \leqslant 10^4$ cm^{-3} collisional de-excitation is not important and the intensity ratio of the [O III] and [N II] forbidden lines can be expressed by the relations (Osterbrock, 1974; Chaisson, 1976):

$$\text{[O III]} \ \frac{I(\lambda\,4959\ \text{Å} + \lambda\,5007\ \text{Å})}{I(\lambda\,4363\ \text{Å})} = 8.32 e^{32\,900/T_e} \tag{II.3.1}$$

and

$$\text{[N II]} \ \frac{I(\lambda\,6548\ \text{Å} + \lambda\,6584\ \text{Å})}{I(\lambda\,5755\ \text{Å})} = 7.53 e^{25\,000/T_e} \tag{II.3.2}$$

These are emissivity ratios and an integration over the nebular path length is required to obtain the corresponding intensity ratios. Since however the H II region is optically thin in forbidden line emission and the path length, over which the integration is performed, is the same for the forbidden lines of the same ion, the ratio of emissivities is equal to that of intensities.

For $N_e > 10^4$ cm^{-3} collisional de-excitation becomes appreciable and a crude estimate of the density N_e is necessary, since the corresponding ratios take the form (Osterbrock, 1974):

$$\text{[O III]} \frac{I(\lambda\,4959\ \text{Å} + \lambda\,5007\ \text{Å})}{I(\lambda 4363\ \text{Å})} = \frac{8.32 e^{32\,900/T_e}}{1 + 4.5 \times 10^{-4}\,N_e T_e^{-1/2}} \tag{II.3.3}$$

and

$$\text{[N II]} \frac{I(\lambda\,6548\ \text{Å} + 6584\ \text{Å})}{I(\lambda\,4363\ \text{Å})} = \frac{7.53 e^{25\,000/T_e}}{1 + 2.7 \times 10^{-3}\,N_e T_e^{-1/2}}. \tag{II.3.4}$$

An analytical treatment of the problem has already been given by Osterbrock (1974).

A graphical representation of the intensity ratios of the [O III], [N II], and the similarly structured [S III] lines as a function of T_e, worked out for $N_e \leqslant 10^4$ cm^{-3}, is shown in Figure II.3.1. A summary of the methods belonging to this category, and of their merits and limitations is presented in Tables II.3.I and II.3.II.

(2) *The ratio of the optical hydrogen recombination continuum* (*bound-free continuum*) *to a hydrogen recombination line* (*bound-bound emission line*). The determination of the intensity ratio of the Balmer continuum, $I(\lambda\ 3646\ \text{Å}_-)$ $I(\lambda\ 3646\ \text{Å}_+)$, (free-bound recombination emission to $\eta = 2$; Figure II.3.2) to the Hβ emission line is a method belonging to this category. This ratio is given by the relation (Aller, 1956; Simpson, 1973):

$$\frac{I_{BC}\,\text{Å}^{-1}}{I_{H\beta}} = 2.118T_e^{-0.673} \tag{II.3.5}$$

from which the T_e can be derived. Another version of the same method is the determination of the ratio of the continuum around Hβ to the Hβ line.

The method is, however, not reliable since the desirable continuum cannot be easily inferred from the observed continuum which is the result of various emission mechanisms. More details, relevant to this method are given in Tables II.3.I and II.3.II.

(3) *The intensity ratios of forbidden lines of trace elements to hydrogen recombination lines.* The method makes use of the different temperature dependence of a recombination and a forbidden line (see Table II.3.III).

A method of this kind, originally proposed by Burbidge *et al.* (1963), derives T_e from the intensity ratio Hα λ 6563 Å/[N II] λ 6584 Å which is given by:

$$\frac{I(\text{H}\alpha\ \lambda\ 6563\ \text{Å})}{I([\text{N II}]\ \lambda 6584\ \text{Å})} = 5.48 \times 10^{-4}\ T_e^{-1/2} \times 10^{9500/T_e} \times$$

$$\times \left[\frac{N(\text{N}^+)}{N(\text{N})}\right]^{-1}\left[\frac{N(\text{N})}{N(\text{H})}\right]^{-1}. \tag{II.3.6}$$

This however requires the knowledge of the total nitrogen abundance $N(\text{N})/N(\text{H})$ and the fraction of the ionized nitrogen $N(\text{N}^+)/N(\text{H}^+)$.

A graphical solution of the determination of T_e from the Hα/[N II] ratio is shown in Figure II.3.3.

A similar method introduced by Pronik (1957) derives temperature from the intensity ratios

$$\frac{[\text{O II}]\ (\lambda\ 3726\ \text{Å} + \lambda\ 3729\ \text{Å})}{\text{H}\alpha\ \lambda\ 6563\ \text{Å}} \quad \text{and} \quad \frac{[\text{O III}]\ (\lambda\ 4959\ \text{Å} + \lambda\ 5007\ \text{Å})}{\text{H}\alpha\ \lambda\ 6563\ \text{Å}}$$

and the assumption that all the oxygen of the H II region is in the form of O$^+$ and O^{++}. This implies that the total oxygen abundance is

$$\frac{N(\text{O})}{N(\text{H})} = \frac{N(\text{O}^+)}{N(\text{H}^+)} + \frac{N(\text{O}^{++})}{N(\text{H}^+)}. \tag{II.3.7}$$

The temperature can be estimated from the relation:

$$\frac{N(\text{O})}{N(\text{H})} = 2.21 \times 10^{-4}\ T_e^{-1/2} \times 10^{16700/T_e}\frac{I_{[\text{O II}]}}{I_{\text{H}\alpha}} +$$

$$+ 5.42 \times 10^{-4}\ T_e^{-1/2} \times 10^{12600/T_e}\ \frac{I_{[\text{O III}]}}{I_{\text{H}\alpha}} \tag{II.3.8}$$

(the Hβ line can also be used intead of the Hα). The method presupposes the knowledge of the oxygen abundance, $N(O)/N(H)$, and the electron density, since for $N_e >$ 100 cm^{-3} collisional de-excitation in the O$^+$ lines becomes important.

Further comments relevant to the methods described above are included in Tables II.3.I and II.3. II.

(4) *The width of the forbidden and recombination line profiles.* The width of a line profile is determined by the motion of the ions characterized by a temperature T_e and by the considerable turbulence $\langle V_t^2 \rangle^{1/2}$ within the nebula. If the turbulence is assumed to be a Gaussian, which is true only under certain conditions (Courtès *et al.*, 1968), the basic shape of the line profile (after the removal of the effects of the instrumental profile by employing a deconvolution process) can be also approximated by a Gaussian:

$$I_{(\lambda)} = I_0 \exp \left\{ - \left[\frac{\lambda_0 - \lambda}{\Delta \lambda_e} \right]^2 \right\} \qquad (II.3.9)$$

with

$$\Delta \lambda_e = \frac{\lambda_0}{c} \left[\frac{2kT_e}{M} + \langle V_t^2 \rangle \right]^{1/2}, \qquad (II.3.10)$$

where

I_0 = the maximum intensity occuring at wavelength λ_0;

$\Delta \lambda_e$ = the $1/e$ half width of the line profile (*e*-folding width, measured at intensity $I = I_0/e$, Figure II.3.4);

M = the mass of emitting atom;

T_e = the electron temperature;

$\langle V_t^2 \rangle$ = the mean square turbulent velocity along the line of sight;

c = the velocity of light; and

k = the Boltzmann constant.

The full width at half-intensity points (i.e. measured at intensity $I = I_0/2$, Figure II.3.4) is given by the relation:

$$\Delta \lambda = \frac{2\lambda_0}{c} \left[\left(\frac{2kT_e}{M} + \langle V_t^2 \rangle \right) \ln 2 \right]^{1/2}. \qquad (II.3.11)$$

The temperature T_e can be derived by measuring the full widths of two lines emitted at wavelengths λ_1 and λ_2 and originating in two different ions of mass M_1 and M_2 respectively, making the assumption that both of them are characterized by the same turbulence and applying twice the relation (II.3.11). This procedure leads to the relation:

$$T_e = \frac{c^2}{8k \ln 2} (\Delta \lambda_1^2 - \Delta \lambda_2^2) \left[\frac{\lambda_1^2}{M_1} - \frac{\lambda_2^2}{M_2} \right]^{-1}. \qquad (II.3.12)$$

Strictly speaking, the derived temperature is the ionic temperature which coincides with the electron temperature T_e under conditions of equipartition of energy.

For lines emitted at essentially the same wavelength, relation (II.3.12) takes the simpler form:

$$T_e = \frac{c^2}{8k \ln 2} \frac{M_1 M_2}{M_2 - M_1} \frac{1}{\lambda_0^2} (\Delta\lambda_1^2 - \Delta\lambda_2^2).$$ (II.3.13)

A commonly used pair of lines is the $H\alpha$ λ 6563 Å and [N II] 6584 Å lines utilising the fact that the nitrogen atom is 14 times more massive than that of hydrogen (note the $M^{-1/2}$ dependence of the thermal broadening in relation (II.3.11).) The $H\alpha$ line exhibits, however, an appreciable natural width due to its fine structure components which should be taken into account (Meaburn, 1970; Dyson and Meaburn, 1971). An additional difficulty is the assumption that both lines originate in areas characterized by the same amount of turbulenc which is probably not true as argued by Dopita et al. (1973) who suggested that the $H\beta$, [O III] lines are more suitable for this purpose (see also Table II.3.I).

The main radioastronomical methods derive the electron temperature T_e from:

(1) *The strength of the radio recombination lines.* For an optically thin plasma under conditions of LTE, in which the optical depth of both radio continuum (τ_c) and radio recombination lines (τ_L) are $\ll 1$, the observed continuum and line temperatures T_c and T_L are given by:

$$T_c = \tau_c T_e$$ (II.1.2b)

and

$$T_L = \tau_L T_e.$$ (II.3.14)

From the above relations, the relation (II.1.5) and its corresponding expression for the optical depth of an α-line formed under conditions of LTE (Mezger, 1972):

$$\tau_L = 1.92 \times 10^3 \, T_e^{-2.5} E \left(\frac{\Delta\nu_L}{kHz}\right)^{-1} \exp\left[\frac{4.5}{T_e}\left(\frac{\nu}{GHz}\right)^{2/3}\right]$$ (II.3.15)

follows that (Kardashev, 1959; Mezger, 1972):

$$(T_{e \, LTE})^{1.15} = \frac{2.33 \times 10^4}{\alpha(\nu, T_e)} \left[\frac{\nu_L}{GHz}\right]^{2.1} \times$$
$$\times \left[\frac{\Delta\nu_L}{kHz} \frac{T_L}{T_c}\right]^{-1} \left[1 + \frac{N(He^+)}{N(H^+)}\right]^{-1}$$ (II.3.16)

where ν_L and $\Delta\nu_L$ are the frequency at the peak and the full width at half intensity of the observed radio recombination α-line, the shape of which is assumed to be Gaussian* (Figure II.3.5). The factor $\alpha(\nu, T_e)$ is of the order of unity and is tabulated

---

* This is a good approximation for radio recombination lines with low principal quantum level *n*, for which electron collisional broadening is negligible. For higher *n* ($n > 150$), however, the observed line width, $\Delta\nu_{obs}$, results from the convolution of a Gaussian (Doppler broadened) with a Lorentzian (collisionally broadened) function; this can be expressed as:

$$\Delta\nu_{obs} = \sqrt{\Delta\nu_D^2 + \Delta\nu_{CB}^2}.$$

The ratio of the Doppler broadened width to that due to electron collisional broadening is given by (Griem, 1967; Chaisson, 1976):

$$\frac{\Delta\nu_D}{\Delta\nu_{CB}} = \frac{(T_e^2 + 4 \times 10^{-5} \, T_e \langle Vt^2 \rangle)^{0.5}}{9.5 \times 10^{-16} \, N_e \, n^7 [\ln(nT_e) - 11.5]}.$$

Line width measurements at high *n* can be used to derive the electron density N_e. The method has yielded for the Orion nebula $N_e = 10^{3.2}$ cm^{-3}, which is consistent with the values derived from radio continuum measurements (Chaisson, 1976).

by Mezger and Henderson (1967) whereas the expression $[1 + \{N(\text{He}^+)/N(\text{H}^+)\}]$ takes into account the He^+ contribution to the radio continuum emission.

Deviations from LTE have as a result the modification of the LTE line intensities. These can be taken into account by using the enhancement factor derived by Goldberg (1966, 1968):

$$\frac{(T_L/T_c)_{\text{Non-LTE}}}{(T_L/T_c)_{\text{LTE}}} = b_n\left[1 + \frac{\tau_c}{2}\frac{kT_e}{h\nu}\frac{d\ln b_n}{dn}\varDelta_n\right], \qquad (\text{II}.3.17)$$

where b_n is a coefficient taking into account the departures form LTE in the population of energy level n and $\varDelta n$ is the difference between the two levels involved in the transition ($\varDelta n = 1$ for transitions between adjacent levels i.e. α-lines). Then, the non-LTE T_e is given by the relation (Mezger, 1968):

$$T_{e\,\text{Non-LTE}} = T_{e\,\text{LTE}}\left[\frac{(T_L/T_c)_{\text{Non-LTE}}}{(T_L/T_c)_{\text{LTE}}}\right]^{-0.87}. \qquad (\text{II}.3.18)$$

Deviations from LTE were invoked to explain the discrepancies between the temperatures derived from optical methods and those derived from radio methods (under the assumption of LTE). This view was however debated by Chaisson, (1976) and Chaisson and Dopita (1977) who argued that such a discrepancy is not sufficiently substantiated by more recent observational data.

(2) *The widths of the recombination line profiles.* The method is similar to the optical method number (3), but utilises instead H and He $n\alpha$ lines. Comments concerning the merits and limitations of the method are included in Table II.3.I, whereas the formula used to derive the temperature is given in Table II.3.II.

(3) *The free-free radio continuum.* The electron temperature, T_e, can be derived directly from observations made at low frequencies. The brightness temperature at a point on the surface of an H II region is given by:

$$T_b = T_e(1 - e^{-\tau_c})\begin{cases} = T_e & \text{for}\quad \tau_c \gg 1 & (\text{II}.1.2\text{a}) \\ = \tau_c T_e & \text{for}\quad \tau_c \ll 1. & (\text{II}.1.2\text{b}) \end{cases}$$

Then, for low frequencies ($\tau_c \gg 1$; source optically thick) the estimated brightness temperature T_b should represent the electron temperature of the ionized gas T_e. However, observations at low frequencies are difficult to interpret, mainly because the source is seen against a hot background radiation due to the galactic synchrotron radiation; in addition the optically thick parts of an H II region do not fill the beam of the radio telescope (electron temperatures diluted by a not accurately estimable factor).

The T_e can be, however, determined from observations made at one high and one lower frequency, ν_H and ν_L respectively. At the high frequency ν_H, the source is optically thin and $T_{b_{\nu_H}} = T_e\tau_{c_{\nu_H}}$(II.1.2b), from which $\tau_{c_{\nu_H}}$ can be estimated for an adopted T_e. From the known frequency dependence of the optical depth (II.1.5) the ratio $\tau_{c_{\nu_H}}/\tau_{c_{\nu_L}}$ can be found and because $\tau_{c_{\nu_H}}$ is known the $\tau_{c_{\nu_L}}$ can also be estimated.

Then, the brightness temperature at the lower frequency should be $T_{b_{\nu_L}} = \tau_{c_{\nu_L}} T_e$. If the estimated $T_{b_{\nu_L}}$ does not agree with the observed one, the adopted electron temperature T_e is not correct and another T_e should be adopted, till a best fit is obtained.

A more elaborate version of this method has also been applied. First, the T_b distribution over the H II region is established from observations made at a high frequency. Then, by adopting an electron temperature T_e, the optical depth τ_c at this frequency and the optical depth at any other frequency can be calculated as above. This leads to the estimation of the distribution of T_b at various frequencies. For each of these frequencies the total flux density S_ν can be derived by summation over the source according to the relation

$$S_\nu = \frac{2k\nu^2}{c^2} \int_\Omega T_b(\alpha, \delta) \, d\Omega, \qquad (I.1.14)$$

where k is the Boltzmann constant, c is the velocity of light and Ω is the solid angle subtended by the source. This leads to the establishment of the radio spectrum of the source, $S_\nu = f(\nu)$, and eventually, by varying the adopted T_e, to a family of radio spectra. The adopted T_e is varied until a best fit between one of the computed spectra and the observed radio spectrum is obtained. An analytical description of the method can be found in Mezger (1968).

Further information related to the various methods of determination of the electron temperature of an H II region is given in Tables II.3.I and II.3.II.

II.4. Chemical Composition

The relative ionic and total abundances of the various elements which are contained in an H II region can be calculated from the determination of the relative strength of their emission lines. This is in principle not complicated because both the recombination and forbidden lines are optically thin; moreover the recombination lines exhibit a weak dependence on temperature

$$\varepsilon = \text{const. } T^{-m}, \qquad (II.4.1)$$

where ε is the emissivity of the line and $m \simeq 1$ over a limited range of temperatures. The forbidden lines, however, show a strong dependence on T:

$$\varepsilon = \text{const. } T^{-1/2} e^{-(\Delta E/kT)}, \qquad (II.4.2)$$

where ε is the emissivity of the line, k is the Boltzmann constant, and ΔE is the energy difference between the ground and the excited level from where the forbidden line originates. This dependence on temperature complicates the estimation of the abundance for those elements whose emission is dominated by forbidden lines.

The simpler case is the estimation of the relative abundance of He^+, $N(He^+)/N(H^+)$. This can be obtained by determining the ratio of an He recombination line, emitted in the optical part of the spectrum, to an H recombination line close in wavelength, to minimize the effect of differential reddening. The $N(He^+)/N(H^+)$ is obtained from the intensity ratio of the lines, according to the expression (Mathis, 1962):

$$\frac{I_{jk}(\text{He})}{I_{lm}(\text{H})} = \frac{b_j}{2b_l} \frac{N(\text{He}^+)}{N(\text{H}^+)} \frac{\lambda_\text{H}^3}{\lambda_\text{He}^3} \frac{g_k}{g_m} \frac{f_{kj}}{f_{lm}} e^{(\Delta E_j - \Delta E_l)/kT_e}, \tag{II.4.3}$$

where b_j and b_l are the parameters indicating the deviation of the population of levels j and l from the corresponding thermodynamic equilibrium population; g_k, and g_m are the statistical weights of levels k and m; f_{kj} and f_{lm} are the oscillator strengths; and ΔE_j, and ΔE_l the energy difference between the levels $j - k$ and $l - m$, respectively.

Applications of this formula for various lines are given in Table II.4.I.

Radio recombination hydrogen and heliums lines of the same n can also be employed. The helium abundance is then obtained by applying twice the relation (II.3.16) which yields the formula:

$$\frac{N(\text{He}^+)}{N(\text{H}^+)} = \frac{T_L(\text{He})}{T_L(\text{H})} \frac{\Delta\nu(\text{He})}{\Delta\nu(\text{H})} \tag{II.4.4}$$

The methods of estimating $N(\text{He}^+)/N(\text{H}^+)$ are only valid if the H^+ and He^+ zones are spatially coexistent which is generally the case with the majority of nebulae (note however exceptions such as NGC 2024 which is a low excitation object with a He^+ sphere distinctly smaller than the H^+; then, the abundance estimate is a lower limit).

For the majority of nebulae where no He^{++} is present the relative abundance of He^+ is approximately equal to the total relative abundance of He, $N(\text{He})/N(\text{H})$.

For the estimation of the relative abundances of various elements whose emission is dominated by forbidden lines, the determination of temperature is required. Then, under the assumption of constant T_e, the relative ionic abundance can be derived from the ratio of a forbidden line of the element to a hydrogen recombination line, as above. Since however the temperature T_e is not uniform over the nebula, a more sophisticated approach is currently used, to take into account the spatial fluctuations t occuring around an average temperature T_0 defined as (Peimbert, 1967; Osterbrock, 1974):

$$T_0 = \frac{\int_0^L T_e N_i N_e \, \mathrm{d}x}{\int_0^L N_i N_e \, \mathrm{d}x}. \tag{II.4.5}$$

The mean square fluctuation is expressed as:

$$t^2 = \left\langle \left(\frac{T_e - T_0}{T_0} \right)^2 \right\rangle = \frac{\int_0^L N_i N_e (T_e - T_0)^2 \, \mathrm{d}x}{T_0^2 \int_0^L N_i N_e \, \mathrm{d}x}, \tag{II.4.6}$$

where N_i, N_e are the ionic and electron number densities respectively and the integration is performed along the path length L of the nebula. The values used for t^2 are in the range of 0.000 (uniform temperature) and 0.055 (strong temperature fluctuations). A mean value $t^2 = 0.035$ constitutes probably a more realistic approximation (Peimbert and Torres-Peimbert, 1977).

The ionic abundances are obtained from relations of the form (Peimbert and Torres–Peimbert, 1977):

$$\frac{N(X^{+p})}{N(H^+)} = \alpha(X^{+p}, N_e, T_{\text{fluct}}) \frac{T_{\text{fluct}}^{0.50}}{T_{H\beta}^{0.84}} \frac{I(X^{+p})}{I(H\beta)} e^{\Delta E/kT_{\text{fluct}}}, \tag{II.4.7}$$

where T_{fluct} and $T_{H\beta}$ are given by Peimbert and Costero (1969) by the relations:

$$T_{\text{fluct}}^{-0.5} e^{-(\Delta E/kT_{\text{fluct}})} =$$

$$= T_0^{-0.50} e^{-(\Delta E/kT_0)} \left[1 + \left[\left(\frac{\Delta E}{kT_0}\right)^2 + 3\left(\frac{\Delta E}{kT_0}\right) + \frac{3}{4} \right] \frac{t^2}{2} \right] \tag{II.4.8}$$

$$T_{H\beta} = T_0(1 - 1.84t^2/2), \tag{II.4.9}$$

where ΔE is the energy difference between the ground and the excited level where the forbidden lines originate.

The total abundance of an element is eventually calculated by the formulae (Peimbert and Costero, 1969; Peimbert and Torres-Peimbert, 1971, 1977):

$$\frac{N(O)}{N(H)} = \frac{N(O^+) + N(O^{++})}{N(H^+)} \tag{II.4.10}$$

$$\frac{N(N)}{N(H)} = \frac{N(N^+)}{N(H^+)} \frac{N(O)}{N(O^+)} \tag{II.4.11}$$

$$\frac{N(Ne)}{N(H)} = \frac{N(Ne^{++})}{N(H^+)} \frac{N(O)}{N(O^{++})} \tag{II.4.12}$$

$$\frac{N(S)}{N(H)} = \frac{N(S^+) + N(S^{++})}{N(H^+)} \frac{N(O)}{N(O^+)} \tag{II.4.13}$$

which are derived by making use of the coincidences in the ionization potentials of various ions (see e.g. Table 2.1.II from which $N(O^+)/N(O) \simeq N(N^+)/N(N)$ can be inferred).

A correction formula for the calculation of $N(He)/N(H)$, taking into account the presence of neutral He in an H II region, has also been derived by Peimbert and Torres-Peimbert (1977):

$$\frac{N(He)}{N(H)} = \frac{N(He^+)}{N(H^+)} \left[1 - 0.35 \frac{N(O^+)}{N(O)} - 0.65 \frac{N(S^+)}{N(S)} \right]^{-1} \tag{II.4.14}$$

II.5. Determination of Radial Velocity and Turbulence

The radial velocity V of an H II region is estimated from the Doppler shift of an observed line, according to the relation

$$V = c \frac{\Delta\lambda_{\text{shift}}}{\lambda_{\text{rest}}}, \tag{II.5.1}$$

where $\Delta\lambda_{\text{shift}}$ is the observed wavelength shift of the peak of the line profile with respect to the wavelength λ_{rest} of the peak of the line, as measured in the laboratory (at rest) and c is the velocity of light ($=2.997\,929 \times 10^5$ km s^{-1}). The radial velocity is usually expressed with respect to the local standard of rest (V_{LSR}), in units of km s^{-1}.

The dispersion of radial velocities along the line of sight (turbulence, $\langle Vt^2 \rangle^{1/2}$) is obtained from the study of optical or radio recombination line profiles. After the estimation of T_e from one of the line profile methods (see Section II.3), the turbulence $\langle Vt^2 \rangle^{1/2}$ can be calculated from the observed line width $\Delta\lambda$ (cleansed from the effects of the instrumental profile), according to the relation (II.3.11):

$$\langle Vt^2 \rangle^{1/2} = \sqrt{\frac{c^2 \Delta\lambda}{4\lambda_0^2 \ln 2} - \frac{2kT_e}{M}}, \tag{II.5.2}$$

where all the symbols are taken with their usual meaning and the turbulence is considered to be Gaussian.

The estimated turbulence $\langle Vt^2 \rangle^{1/2}$ is due to the non thermal motions of small parts of the gas which are randomly distributed within the beam of the observing instrument and moving with most probable velocity Vt.

TABLE II.2.I

Determination of density from optical methods

No.	Method	Merits (M) and limitations (L) of the method	Remarks
1	[O II] $\dfrac{\lambda 3729 \text{ Å}}{\lambda 3726 \text{ Å}}$	M: Advantageous for estimating densities $N_e \sim 10^3$ cm^{-3} L: Lines too close to each other, instrument with high wavelength resolution is required.	(1), (2)
2	[S II] $\dfrac{\lambda 6716 \text{ Å}}{\lambda 6731 \text{ Å}}$	M: Advantageous for estimating densities $N_e \sim 2 \times 10^3$ cm^{-3}. L: Collision strengths not accurately known.	(2), (3)

Remarks to Table II.2.I:

(1) Formula according to Seaton and Osterbrock (1957);

$$[\text{O II}] \frac{I(\lambda 3729 \text{ Å})}{I(\lambda 3726 \text{ Å})} = \left\{ \frac{1 + 0.33\varepsilon + 2.30(1 + 0.75\varepsilon + 0.14\varepsilon^2)x}{1 + 0.40\varepsilon + 9.90(1 + 0.84\varepsilon + 0.17\varepsilon^2)x} \right\},$$

where

$$\varepsilon = \exp(-19\,600/T_e)$$

and

$$x = 10^2 N_e/T_e^{1/2}.$$

(2) See Figure II.2.2 (variations of the [O II] and [S II] intensity ratio versus N_e, for $T_e = 6000$, 8000, 10 000, and 12 000 K).

(3) For the variations of the [S II] ratio versus N_e, see also Pequignot *et al.* (1977).

TABLE II.3.I

Comments on the various methods of determination of electron temperature

No.	Method	Formula (No. in Table II.3.II)	Merits and limitations of the method (M=merits, L=limitations, C=comments, V=variations of the method)
			a. Optical methods
1	[O III] $\dfrac{I(\lambda 4959 \text{ Å} + \lambda 5007 \text{ Å})}{I(\lambda 4363 \text{ Å})}$	1	M: Lines close to each other, no reddening correction is required, photocathodes sensitive in the blue part of the spectrum. L: $\lambda 4363$ Å line weak and contaminated with Hg I $\lambda 4358$ Å line (city lights). C: Appropriate for estimation of temperature in high excitation regions (e.g. nebular cores). V: Other ions with similar structure can also be used (e.g. [S III], see Table II.3.II, No. 2).
2	[N II] $\dfrac{I(\lambda 6548 \text{ Å} + \lambda 6584 \text{ Å})}{I(\lambda 5755 \text{ Å})}$	3	C: Appropriate for estimation of temperature in low excitation regions (e.g. outer parts of H II regions). V: Solved simultaneously with the [O II] [($\lambda 3726$ Å + $\lambda 3729$ Å)/($\lambda 7325$ Å)] ratio to derive T_e and N_e; applicable if the density distribution of O$^+$, N$^+$ is the same; for formula see Table II.3.II, No.4. V: Alternatively the [O II] [($\lambda 3726$ Å + $\lambda 3729$ Å)/($\lambda 7325$ Å)] and [O II] ($\lambda 3729$ Å/$\lambda 3726$ Å) ratios can be solved simultaneously to derive T_e and N_e; strong reddening correction is required.
3	$\dfrac{[\text{N II}] \times [\text{O I}]}{[\text{O II}] \times [\text{N I}]} \times \dfrac{I(\lambda 6548 \text{ Å} + \lambda 6584 \text{ Å}) \times I(\lambda 6300 \text{ Å} + \lambda 6364 \text{ Å})}{I(\lambda 3726 \text{ Å} + \lambda 3729 \text{ Å}) \times I(\lambda 5102 \text{ Å} + \lambda 5198 \text{ Å})}$	5	M: Knowledge of the degree of ionization and abundance is not required since $\dfrac{N(\text{O}^+)}{N(\text{O}^0)} = \dfrac{N(\text{N}^+)}{N(\text{N}^0)}$ L: Reddening correction and knowledge of N_e are required. C: Suitable for estimation of temperature in nebulae characterized by weak ionization.
4	Balmer Continuum $\dfrac{I(\lambda 3646 \text{ Å}_-) - I(\lambda 3646 \text{ Å}_+)}{\text{H}\beta} \cdots \dfrac{}{I(\lambda 4861 \text{ Å})}$	6	L: Continuum weak and affected by weak lines; mixed with continuum emitted by the stars and scattered by the nebula; also mixed with two photon continuum; Balmer continuum flux measured from extrapolation to $\lambda 3646$ Å$_+$ of observations made at longer wavelengths (to avoid interference from the higher Balmer lines); reddening correction is required. V: Measurement of the continuum around $\lambda 4861$ Å instead of the Balmer continuum.

(continued)

Table II.3.I (continued)

No.	Method	Formula (No. in Table II.3.II)	Merits and limitations of the method (M = merits, L = limitations, C = comments, V = variations of the method)
5	$[\text{O II}]$ $\dfrac{I(\lambda 3726\ \text{Å} + \lambda 3729\ \text{Å})}{I(\lambda 6563\ \text{Å})}$, $[\text{O III}]$ $\dfrac{I(\lambda 4959\ \text{Å} + \lambda 5007\ \text{Å})}{I(\lambda 6563\ \text{Å})}$	7	*M*: Bright lines are used; variations of temperature within a nebula can be derived. *L*: The total abundance of oxygen is required ($N(\text{O})/N(\text{H})$ is usually taken as 6×10^{-4}); knowledge of N_e is also necessary since for $N_e > 100$ cm⁻³ collisional de-excitations affect the strength of the [O II] line; reddening correction is required. *V*: Use of Hβ $\lambda 4861$ Å instead of Hα; [N II]/Hα can also be used instead of [O II]/Hα; in both cases the reddening correction is not very important; in the second variation of the method the effect of collisional de-excitations is not important unless the nebula is extremely dense.
6	$\dfrac{\text{H}\alpha}{[\text{N II}]}$ $\dfrac{I(\lambda 6563\ \text{Å})}{I(\lambda 6584\ \text{Å})}$	8	*L*: The total abundance of nitrogen $N(\text{N})/N(\text{H})$ and the fraction of ionized nitrogen $N(\text{N}^+)/N(\text{H})$ are required ($N(\text{N})/N(\text{H})$ is usually taken as 10^{-4}); in regions characterized by high excitation and high density, N⁺ can be transformed to N⁺⁺ and can also suffer collisional de-excitation. *C*: Appropriate for estimation of temperature in low excitation regions (e.g. outer parts of nebulae).
7	Widths of recombination and forbidden lines $\left(\begin{array}{c} \text{H}\alpha\ \lambda 6563\ \text{Å and [N II]}\ \lambda 6584\ \text{Å} \\ \text{or} \\ \text{H}\beta\ \lambda 4861\ \text{Å and [O III]}\ \lambda 5007\ \text{Å} \end{array} \right)$		*C*: The estimated temperature is the ionic kinetic temperature which is equal to the electron temperature under conditions of equipartition of energy. *L*: The turbulent broadening is assumed to be the same for both lines; this is true if the emission of both lines originate in the same volume of gas which is not the case for the Hα and [N II] lines; the fine structure of the Hα line should be considered in the determination of the thermal width of the line; the accuracy of the method is further limited by the dependence of the temperature on the difference of the squares of the line widths (errors in the measurements of the widths are squared; see formula No. 11 in Table II.3.II and relation II.3.13 in Appendix, section II.3). *V*: Hα and He I lines can also be used; see Table II.3.II, No. 11.

(continued)

Table II.3.1 (continued)

No.	Method	Formula (No. in Table II.3.II)	Merits and limitations of the method (M=merits, L=limitations, C=comments, V=variations of the method)
			b. Radio methods
8	Strength of radio recombination lines (LTE case) $\left(\Delta\nu \frac{T_L}{T_C}\right)$	9	M: At high frequencies the determined temperature expresses a true average temperature along the line of sight; the angular resolution of the observations is also better. L: The formation of lines under conditions of Local Thermodynamic Equilibrium is debatable.
9	Strength of radio recombination lines (Non-LTE case) $\left(\Delta\nu \frac{T_L}{T_C}\right)$ Non-LTE	10	L: Method used assuming the emission of the observed lines is enhanced because they are formed under Non-LTE; the deviation, however, from LTE conditions may not be important (Mezger, 1968; Chaisson, 1976; Chaisson and Dopita, 1977).
10	Radio continuum emission at low frequencies		L: The observations are confused from the presence of the strong non thermal galactic background (synchrotron radiation $\sim \nu^{-1}$); the low angular resolution at low frequencies (large beam not filled by the source) results in the derivation of an electron temperature diluted over the beam which cannot be corrected accurately.
11	Radio continuum spectrum	11	M: The observed spectrum is well defined from many observations. L: Method applicable to H II regions with high emission measure; the estimated temperature is the average temperature of the outer layers of the nebula. V: A simplified version of the method derives the temperature from observations made at two frequencies (high and low).
12	Widths of radio recombination lines		C: See comments made for the corresponding optical method. C: Hnα and Henα lines are used; lines with small n must always be used to avoid impact broadening (due to the collision of the bound, far from the atomic nucleus, electrons with the free electrons and ions of the surrounding plasma). L: In addition to the limitations referred in the corresponding optical method, the radio method suffers from the mixing of Henα with the Cnα line which originates from the cool H I–H II layer enveloping the H II region.

TABLE II.3.II

Formulae to derive electron temperature

No.	Method	Analytical expressions	Remarks
1	[O III] ratio	$\dfrac{I(\lambda 4959 \text{ Å} + \lambda 5007 \text{ Å})}{I(\lambda 4363 \text{ Å})} = 8.32 \exp(3.29 \times 10^4 T_e^{-1})$	(1), (2), (3), (4)
2	[S III] ratio	$\dfrac{I(\lambda 9069 \text{ Å} + \lambda 9532 \text{ Å})}{I(\lambda 6312 \text{ Å})} = 0.66 + 4 \exp(2.3 \times 10^{-4} T_e^{-1})$	(2), (5)
3	[N II] ratio	$\dfrac{I(\lambda 6548 \text{ Å} + \lambda 6584 \text{ Å})}{I(\lambda 5755)} = 7.53 \exp(2.50 + 10^4 T_e^{-1})$	(1), (2), (6), (7)
4	[N II] ratio,	$\dfrac{I(\lambda 6548 \text{ Å} + \lambda 6584 \text{ Å})}{I(\lambda 5755 \text{ Å})} = \dfrac{11.9 \times 10^{10900/T_e}}{1 + 0.37x}$, $x = 10^{-2} N_e T_e^{-1/2}$, $\varepsilon = \exp(-19600 T_e^{-1})$	(8)
	[O II] ratio	$\dfrac{I(\lambda 3726 \text{ Å} + \lambda 3729 \text{ Å})}{I(\lambda 7325 \text{ Å})} = \dfrac{13.7}{\varepsilon} \cdot \dfrac{1 + 0.13\varepsilon + 5.3x(1 + 0.60\varepsilon + 0.07\varepsilon^2)}{1 + 23.8x(1 + 0.23\varepsilon) + 61.2x^2(1 + 0.61\varepsilon + 0.07\varepsilon^2)}$	
5	$\dfrac{[\text{N II}]}{[\text{O II}]} \dfrac{[\text{O I}]}{[\text{N I}]}$	$\dfrac{I(\lambda 6548 \text{ Å} + \lambda 6584 \text{ Å}) I(\lambda 6300 \text{ Å} + \lambda 6364 \text{ Å})}{I(\lambda 3726 \text{ Å} + \lambda 3729 \text{ Å}) I(\lambda 5102 \text{ Å} + \lambda 5198 \text{ Å})} = 0.146 \dfrac{1 + 3.4 \times 10^{-4} N_e}{1 + 10^{-3} N_e T_e^{-1/2}} (1 + 1.6 \times 10^{-2} N_e T_e^{-1/2}) \times 10^{-9350/T_e}$	(9)
6	Balmer Cont. Hβ	$\dfrac{I(\lambda 3646 \text{ Å} -) - I(\lambda 3646 \text{ Å} +)}{I(\lambda 4861 \text{ Å})} = 2.118 T_e^{-0.673}$	(10)
7	$\dfrac{[\text{O II}]}{H\alpha}$, $\dfrac{[\text{O III}]}{H\alpha}$	$\dfrac{N(O)}{N(H)} = \dfrac{2.21 \times 10^{-4}}{T_e^{1/2}} \times 10^{(1.67 \times 10^4)/T_e} \dfrac{I(\lambda 3726 \text{ Å} + \lambda 3729 \text{ Å})}{I(\lambda 6563 \text{ Å})} + \dfrac{5.42 \times 10^{-4}}{T_e^{1/2}} \times 10^{(1.26 \times 10^4)/T_e} \dfrac{I(\lambda 4959 \text{ Å} + \lambda 5007 \text{ Å})}{I(\lambda 6563 \text{ Å})}$	(11)
8	$\dfrac{H\alpha}{[\text{N II}]}$	$\dfrac{I(\lambda 6563 \text{ Å})}{I(\lambda 6584 \text{ Å})} = 5.48 \times 10^{-4} T_e^{-\frac{1}{2}} \times 10^{9500/T_e} \left[\dfrac{N(N)}{N(H)}\right]^{-1} \left[\dfrac{N(N^+)}{N(N)}\right]^{-1}$	(12)
9	$\left(\Delta\nu_L \cdot \dfrac{T_L}{T_C}\right)$ (LTE)	$(T_e)_{\text{LTE}} = 26630 \left(\dfrac{\nu_L}{\text{GHz}}\right)^{1.826} \left[\left(\dfrac{\Delta\nu_L}{\text{kHz}} \dfrac{T_L}{T_C}\right)^{-1} \left(1 + \dfrac{N(He^+)}{N(H^+)}\right)^{-1} \dfrac{m-n}{n} f_{nm}\right]$	(13)
10	$\left(\Delta\nu_L \cdot \dfrac{T_L}{T_C}\right)$ (non-LTE)	$(T_e)_{\text{Non-LTE}} = (T_e)_{\text{LTE}} \left[b_n \left(1 + \dfrac{\tau_c}{2} \dfrac{kT_e}{h\nu} \dfrac{d\ln b_n}{dn} (m-n)\right)\right]^{-0.87}$	(14), (15)
11	Line widths (H, He, lines)	$T_k = \dfrac{c^2}{8k \ln 2} \dfrac{M(H) M(He)}{M(He) - M(H)} \left[\left[\dfrac{\Delta\nu_D(H)}{\nu_{mn}(H)}\right]^2 - \left[\dfrac{\Delta\nu_D(He)}{\nu_{mn}(He)}\right]^2\right]$	(16), (17)

Remarks to Table II.3.II:

(1) Osterbrock (1974) and references therein.

(2) Applicable for $N_e < 10^4$ cm^{-3}.

(3) For $N_e > 10^4$ cm^{-3} the formula should be divided by $(1 + 4.5 \times 10^{-4} N_e T_e^{-1/2})$.

(4) Similar formula calculated by Seaton (1975) is:

$$\frac{I(\lambda 4959 \text{ Å} + \lambda 5007 \text{ Å})}{I(\lambda 4363 \text{ Å})} \simeq 7.2 \exp(3.297 \times 10^4 T_e^{-1}) \frac{1 + 5.4 \times 10^{-6} N_e T_e^{-1/2}}{1 + 6.3 \times 10^{-4} N_e T_e^{-1/2}}.$$

(5) Foukal (1974).

(6) For $N_e > 10^4$ cm^{-3} the formula should be divided by $(1 + 2.7 \times 10^{-3} N_e T_e^{-1/2})$.

(7) Similar formula calculated by Seaton (1975) is:

$$\frac{I(\lambda 6548 \text{ Å} + \lambda 6584 \text{ Å})}{I(\lambda 5755 \text{ Å})} \simeq 6.9 \exp(2.5 \times 10^{-4} T_e^{-1}) \frac{1 + 5.7 \times 10^{-6} N_e T_e^{-1/2}}{1 + 3.1 \times 10^{-3} N_e T_e^{-1/2}}.$$

(8) Aller and Liller (1968), Peimbert (1967).

(9) Seaton (1954).

(10) Simpson (1973), $I(\lambda 3646 \text{ Å}-) - I(\lambda 3646 \text{ Å}+)$ is the intensity due to the Balmer continuum, per Ångström.

(11) Pronik (1957), Pottasch (1965).

(12) Burbidge *et al.* (1963), Baudel (1970), Dopita (1972), Goudis (1976d).

(13) Karadashev (1959), f_{mn} is the oscillator strength for the transition mn; $f_{mn}/n = 0.194$, 0.0271, 0.00841, 0.00365, and 0.00191 if $m - n = 1, 2, 3, 4$, and 5, respectively (Goldwire, 1968; Hjellming and Gordon, 1971).

(14) Goldberg (1966).

(15) Calculations of b_n and $d(\ln b_n)/dn$ have been made by a number of workers (e.g., Dyson, 1968; Hayler, 1968; McCaroll and Binh, 1968; Sejnowski and Hjellming, 1969; Brocklehurst, 1970).

(16) Chaisson (1976), see Appendix, Section II.3.

(17) Similar formulae hold true for the optical line widths (see relations (II.3.12) and (II.3.13) in Appendix, Section II.3).

TABLE II.3.III
Emissivities (ε) and intensities (I) of certain recombination and forbidden lines[1]

No.	Line	ε(ergs cm^{-3} s^{-1})	Remarks	ε(ergs cm^{-3} s^{-1}) (estimated for $T_e=10$
1	Hα λ 6563 Å	$0.320 \times 10^{-20}\, N_e^2 T_e^{-1}$	(2)	$3.20 \times 10^{-25}\, N_e^2$
2	Hα λ 6563 Å	$0.362 \times 10^{-20}\, N_e^2 T_e^{-1}(1-350 T_e^{-1})$	(3)	$3.49 \times 10^{-25}\, N_e^2$
3	Hα λ 6563 Å	$0.169 \times 10^{-20}\, N_e^2 T_e^{-0.92}$		$3.53 \times 10^{-25}\, N_e^2$
4	Hα λ 6563 Å	$0.370 \times 10^{-20}\, N_e^2 T_e^{-1}$	(11)	$3.70 \times 10^{-25}\, N_e^2$
5	Hβ λ 4861 Å	$0.028 \times 10^{-20}\, N_e^2 T_e^{-0.84}$	(4)	$1.22 \times 10^{-25}\, N_e^2$
6	Hβ λ 4861 Å	$0.0252 \times 10^{-20}\, N_e^2 T_e^{-0.827}$	(5)	$1.24 \times 10^{-25}\, N_e^2$
7	Hβ λ 4861 Å	$0.03532 \times 10^{-20}\, N_e^2 T_e^{-0.866}$	(12)	$1.21 \times 10^{-25}\, N_e^2$
8	Hβ λ 4861 Å	$0.136 \times 10^{-20}\, N_e^2 T_e^{-1}(1-792\, T_e^{-1})$	(3)	$1.25 \times 10^{-25}\, N_e^2$
9	[N II] λ 6584 Å	$6.75 \times 10^{-18} \times 10^{-9500/T \cdot}\, T_e^{-1/2}\, N(N^+)\, N_e$	(10)	$7.57 \times 10^{-21} N(N^+)N$
10	[N II] λ 6584 Å	$6.82 \times 10^{-18}\, e^{-21855/T \cdot}\, T_e^{-1/2}\, N(N^+)N_e$	(3), (9)	$7.67 \times 10^{-21} N(N^+)N$
11	[O III] λ 5007 Å	$6.62 \times 10^{-18}\, e^{-28737/T \cdot}\, T_e^{-1/2}\, N(O^{++})\, N_e$	(3)	$3.74 \times 10^{-21} N(O^{++})I$
12	[S III] λ 9069 Å			$5.78 \times 10^{-21}\, N(S^{++})\, I$
13	[S III] λ 9531 Å			$1.50 \times 10^{-20}\, N(S^{++})\, I$
14	Balmer Continuum (per Å)	$5.346 \times 10^{-22}\, T_e^{-3/2}\, N_e^2$	(6)	$5.346 \times 10^{-28}\, N_e^2$

No.	$I = \dfrac{1}{4\pi}\int \varepsilon\, dr$ (ergs cm^{-2} s^{-1} sr^{-1})	I (ergs cm^{-2} s^{-1} sr^{-1}) estimated for $T_e = 10^4$ K and[7] $\dfrac{N(N^+)}{N(H^+)} = 1.5 \times 10^{-5}$, $\dfrac{N(O^{++})}{N(H^+)} = 2 \times 10^{-4}$, $\dfrac{N(S^{++})}{N(H^+)} = 1 \times 10^{-5}$	$I = f(E)$ (ergs cm^{-2} s^{-1} sr^{-1}) (estimated for $T_e=10^4$ K, E in pc cm^{-6})
1	$2.55 \times 10^{-22} \int T_e^{-1}\, N_e^2\, dr$	$2.55 \times 10^{-26} \int N_e^2\, dr$	$7.9 \times 10^{-8} \times E$
2	$2.88 \times 10^{-22} \int T_e^{-1}(1-350\, T_e^{-1})N_e^2\, dr$	$2.78 \times 10^{-26} \int N_e^2\, dr$	$8.6 \times 10^{-8} \times E$
3	$1.35 \times 10^{-22} \int T_e^{-0.92}\, N_e^2\, dr$[8]	$2.81 \times 10^{-26} \int N_e^2\, dr$	$8.7 \times 10^{-8} \times E$
4	$2.94 \times 10^{-22} \int T_e^{-1}\, N_e^2\, dr$	$2.94 \times 10^{-26} \int N_e^2\, dr$	$9.1 \times 10^{-8} \times E$
5	$2.23 \times 10^{-23} \int T_e^{-0.84}\, N_e^2\, dr$	$0.97 \times 10^{-26} \int N_e^2\, dr$	$3.0 \times 10^{-8} \times E$
6	$2.00 \times 10^{-23} \int T_e^{-0.827}\, N_e^2\, dr$	$0.99 \times 10^{-26} \int N_e^2\, dr$	$3.1 \times 10^{-8} \times E$
7	$2.81 \times 10^{-23} \int T_e^{-0.866}\, N_e^2\, dr$	$0.96 \times 10^{-26} \int N_e^2\, dr$	$3.0 \times 10^{-8} \times E$
8	$1.08 \times 10^{-22} \int T_e^{-1}(1-792\, T_e^{-1})\, N_e^2\, dr$	$0.99 \times 10^{-26} \int N_e^2\, dr$	$3.1 \times 10^{-8} \times E$
9	$5.37 \times 10^{-19} \dfrac{N(N^+)}{N(H^+)} \int T_e^{-1/2} \times 10^{-9500/T \cdot}\, N_e^2\, dr$	$0.90 \times 10^{-26} \int N_e^2\, dr$	$1.9 \times 10^{-3} \times \dfrac{N(N^+)}{N(N)} \times \dfrac{N(N)}{N(H)}$
10	$5.42 \times 10^{-19} \dfrac{N(N^+)}{N(H^+)} \int T_e^{-1/2}\, e^{-21855/T \cdot}\, N_e^2\, dr$	$0.92 \times 10^{-26} \int N_e^2\, dr$	$1.9 \times 10^{-3} \times \dfrac{N(N^+)}{N(N)} \times \dfrac{N(N)}{N(H)}$
11	$5.27 \times 10^{-19} \dfrac{N(O^{++})}{N(H^+)} \int T_e^{-1/2}\, e^{-28737/T \cdot}\, N_e^2\, dr$	$5.95 \times 10^{-26} \int N_e^2\, dr$	$0.9 \times 10^{-3} \times \dfrac{N(O^{++})}{N(O)} \times \dfrac{N(O)}{N(H)}$
12		$0.46 \times 10^{-26} \int N_e^2\, dr$	$1.4 \times 10^{-3} \times \dfrac{N(S^{++})}{N(S)} \times \dfrac{N(S)}{N(H)},$
13		$1.20 \times 10^{-26} \int N_e^2\, dr$	$3.6 \times 10^{-3} \times \dfrac{N(S^{++})}{N(S)} \times \dfrac{N(S)}{N(H)},$
14	$4.25 \times 10^{-23} \int T_e^{-3/2}\, N_e^2\, dr$	$4.25 \times 10^{-29} \int N_e^2\, dr$	$1.3 \times 10^{-10} \times E$

Remarks to Table II.3.III:

(1) The emissivity ε of a line (rate of emitted energy per unit volume) is given by

$$\varepsilon = h\nu_{mn} A_{mn} N_m \quad (\text{ergs cm}^{-3}\,\text{s}^{-1}),$$

where ν_{mn} is the frequency of the emitted line resulting from the transition from the upper level m t lower level n; A_{mn} is the radiative transition probability; and N_m is the population density in the upper lev which can also be expressed as:

$$N_m = \frac{N_m}{N(X^{+\alpha})} \frac{N(X^{+\alpha})}{N(X)} \frac{N(X)}{N(H)} \frac{N(H)}{N(H^+)} N_e,$$

where $X^{+\alpha}$ is the ion of the element X which is involved in the transition.

The term $N_m/N(X^{+\alpha})$ can be determined by solving the equations of statistical equilibrium which take account all the populating and depopulating processes occuring in the studied ion. Since the level popula of a forbidden line are primarily determined by collisional excitation, the term $N_m/N(X^{+\alpha})$ is propor to N_e. The collisional excitation rate is also proportional to $T_e^{-1/2}e^{-\Delta E/kT_e}$, where ΔE is the energy diffe between the levels. This dependence in combination with (2), makes Equation (1):

$$\varepsilon \propto \frac{N(X^{+\alpha})}{N(X)} \frac{N(X)}{N(H)} N_e^2 \quad (\text{ergs cm}^{-3}\,\text{s}^{-1}),$$

which holds under the assumption $N(H) = N(H^+)$ (H II region consisting of hydrogen); this is also a reaso approximation for an H II region mixed with helium (in that case $N(H)$ is overestimated up to 10%).

The intensity of the same line (rate of emitted energy per unit surface and unit solid angle) is given by:

$$I = \frac{1}{4\pi} \int_0^L \varepsilon \, dx \quad (\text{ergs cm}^{-2}\,\text{s}^{-1}\,\text{sr}^{-1}),$$

where the integration is performed over the path length L of the nebula (expressed in cm). This can al expressed (3) as:

$$I \propto f(T_e) \frac{N(X^{+\alpha})}{N(X)} Z \int_0^L N_e^2 \, dx \quad (\text{ergs cm}^{-2}\,\text{s}^{-1}\,\text{sr}^{-1}),$$

where Z is the abundance $N(X)/N(H)$ of the element H. Since however $\int_0^L N_e^2 \, dx = E$ (II.1.6), Equation (also be expressed as:

$$I \propto 3.1 \times 10^{18} f(T_e) \frac{N(X^{+\alpha})}{N(X)} ZE \quad (\text{ergs cm}^{-2}\,\text{s}^{-1}\,\text{sr}^{-1}),$$

where E is in units of pc cm^{-6}.

For hydrogen recombination lines ($N(X^+)/N(X) = Z = 1$) holds the simpler form:

$$I \propto f(T_e) E \quad (\text{ergs cm}^{-2}\,\text{s}^{-1}\,\text{sr}^{-1}).$$

The table includes emissivity and intensity formulae, derived for various lines. Various approximations of sivity formulae referring to the same line are essentially the same, as is demonstrated in the table. The emi: and intensity at the Balmer continuum are also included for comparison.

(2) Burgess (1958).

(3) Dopita (1973).

(4) Chaisson (1976).

(5) Pengelly (1964), Costero and Peimbert (1970).

(6) Aller (1956).

(7) Typical ionic abundances for the Orion Nebula (see Table 2.2.XV).

(8) Vidal (1976).

(9) A similar formula derived by Seaton (1975) is:

$$\varepsilon = 29.75 \times 10^{-15} N(O^{++}) N_e T_e^{-1/2} e^{-28810/T_e} \cdot f(x),$$

where

$$f(x) = Y_2(T_e) \frac{1 + Z_2(T_e)x}{1 + L(T_e)x + M(T_e)x^2}$$

$x = 10^{-2} N_e T_e^{-0.5}$ and Y_2, Z_2, L, M are coefficients slowly varying with temperature.

(10) Baudel (1970).

(11) Olthof et al. (1978).

(12) Pengelly (1964), Simpson (1973).

TABLE II.4.1
Formulae to derive the relative abundance of singly ionized helium[1]

No.	Formulae (estimated for $T_e = 10^4$ K)	Remarks
1	$\dfrac{N(\text{He}^+)}{N(\text{H}^+)} = 1.00 \dfrac{I(\text{He I } \lambda\, 5876 \text{ Å})}{I(\text{H}\beta\ \lambda\, 4861 \text{ Å})}$	(2), (3)
2	$\dfrac{N(\text{He}^+)}{N(\text{H}^+)} = 2.44 \dfrac{I(\text{He I } \lambda\, 4472 \text{ Å})}{I(\text{H}\beta\ \lambda\, 4861 \text{ Å})}$	(2), (4)
3	$\dfrac{N(\text{He}^+)}{N(\text{H}^+)} = 4.76 \dfrac{I(\text{He I } \lambda\, 4026 \text{ Å})}{I(\text{H}\beta\ \lambda\, 4861 \text{ Å})}$	(2)
4	$\dfrac{N(\text{He}^+)}{N(\text{H}^+)} = 1.78 \dfrac{I(\text{He I } \lambda\, 4026 \text{ Å})}{I(\text{H}\delta\ \lambda\, 4102 \text{ Å} + \text{H}\varepsilon\ \lambda\, 3970 \text{ Å})}$	(5)
5	$\dfrac{N(\text{He}^+)}{N(\text{H}^+)} = 0.97 \dfrac{I(\text{He I } \lambda\, 3820 \text{ Å})}{I(\text{H}\eta\ \lambda\, 3835 \text{ Å} + \text{H}_\theta\ \lambda\, 3798 \text{ Å})}$	(5)
6	$\dfrac{N(\text{H}^+)}{N(\text{He}^+)} = 1.38 \dfrac{I(\text{He I } \lambda\, 3965 \text{ Å})}{I(\text{H}\varepsilon\ \lambda\, 3970 \text{ Å})}$	(5), (6)

Remarks to Table II.4.1:

(1) Estimated for constant electron temperature $T_e = 10^4$ K; in the presence of spatial temperature fluctuations:

$$\frac{N(\text{He}^+)}{N(\text{H}^+)} = K\, T_0^{\,\alpha-\beta}\Big(1 + \frac{1}{2}[\alpha(\alpha - 1) - \beta(\beta - 1)]t^2\Big)\frac{I(\text{He line})}{I(\text{H line})},$$

where K is a constant, T_0 is an average temperature, t^2 is the mean square fluctuation (see relation (II.4.6) in Appendix, Section II.4) and α and β are powers expressing the dependence of the H and He line intensities on temperature (Peimbert and Costero, 1969).

(2) O'Dell *et al.* (1964).

(3) $\dfrac{N(\text{He}^+)}{N(\text{H}^+)} = 0.97 \dfrac{I(\text{He I } \lambda\, 5876 \text{ Å})}{I(\text{H}\beta\ \lambda\, 4861 \text{ Å})}$ Mathis (1962).

(4) $\dfrac{N(\text{He}^+)}{N(\text{H}^+)} = 1.95 \dfrac{I(\text{He } \lambda\, 4472 \text{ Å})}{I(\text{H}\beta\ \lambda\, 4861 \text{ Å})}$ Mathis (1962).

(5) Dopita *et al.* (1974).

(6) For a nebula thick to self absorption in the $1\,^1S - n\,^1P$ series (case B).

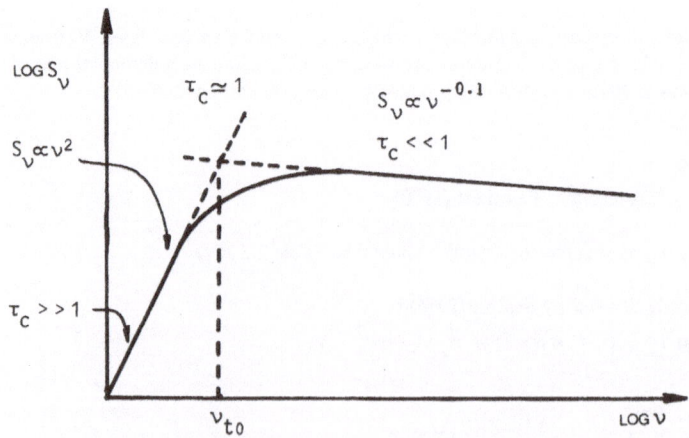

Fig. II.1.1.   A typical radio continuum spectrum of an H II region. The turnover frequency ν_{to} at which $\tau_c \simeq 1$, is also marked.

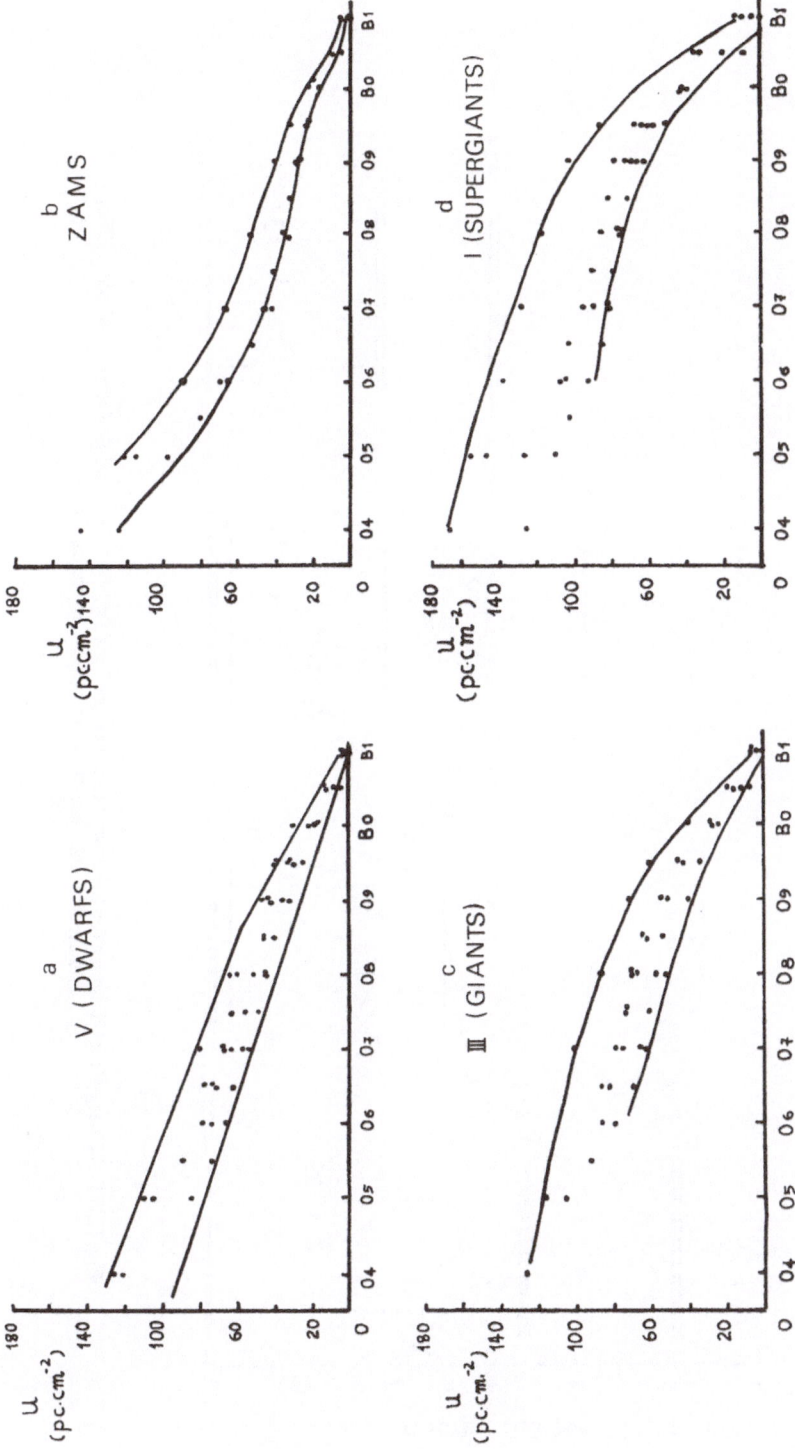

Fig. II.1.2. The excitation parameter *u* of an H II region as a function of the spectral type of the exciting star. Four different cases have been considered: (a) the exciting star is a main sequence dwarf (luminosity class V); (b) the exciting star is on the zero age main sequence (ZAMS); (c) the exciting star is a giant (luminosity class III); (d) the exciting star is a supergiant (luminosity class I). The individual points have been taken from the estimations of the excitation parameter made by Churchwell and Walmsley, (1973); Panagia (1973), Israel *et al.* (1973), Georgelin *et al.* (1975), and Kazès *et al.* (1975).

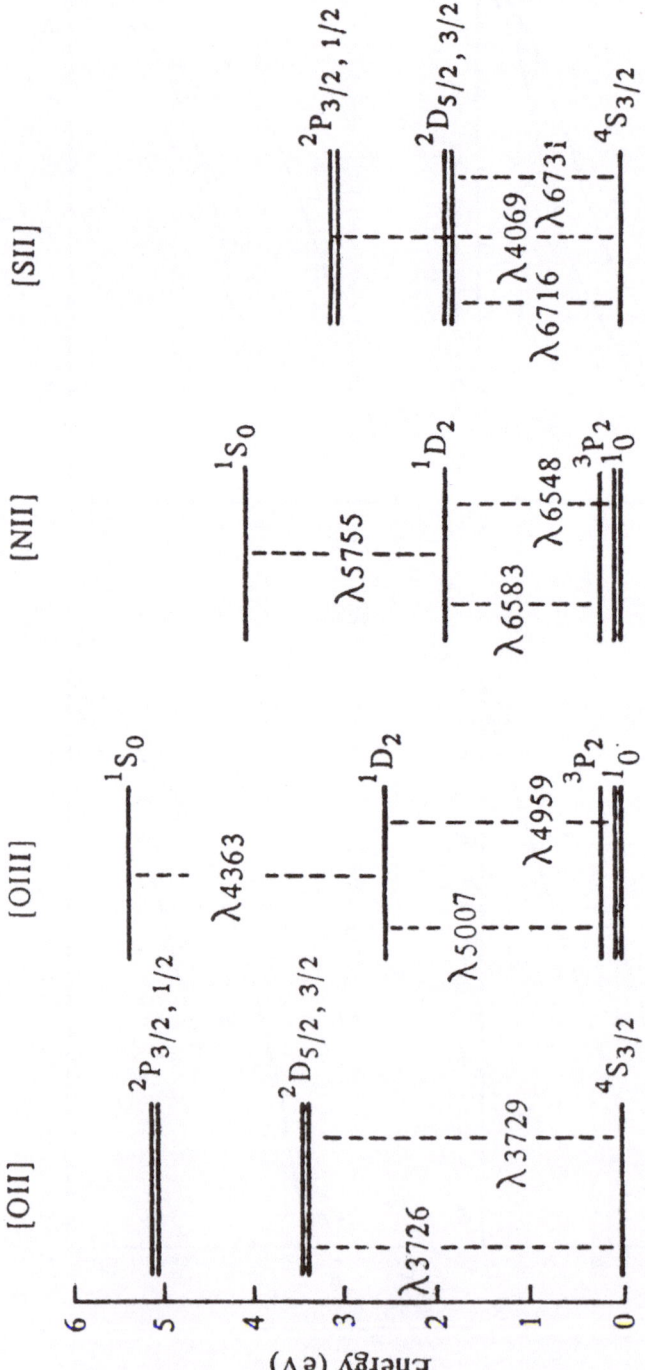

Fig. II.2.1.   Energy level diagram for the lowest terms of the ions O^+, O^{++}, N^+, and S^+ (after Chaisson, 1976). The dashed lines indicate the optical emission lines.

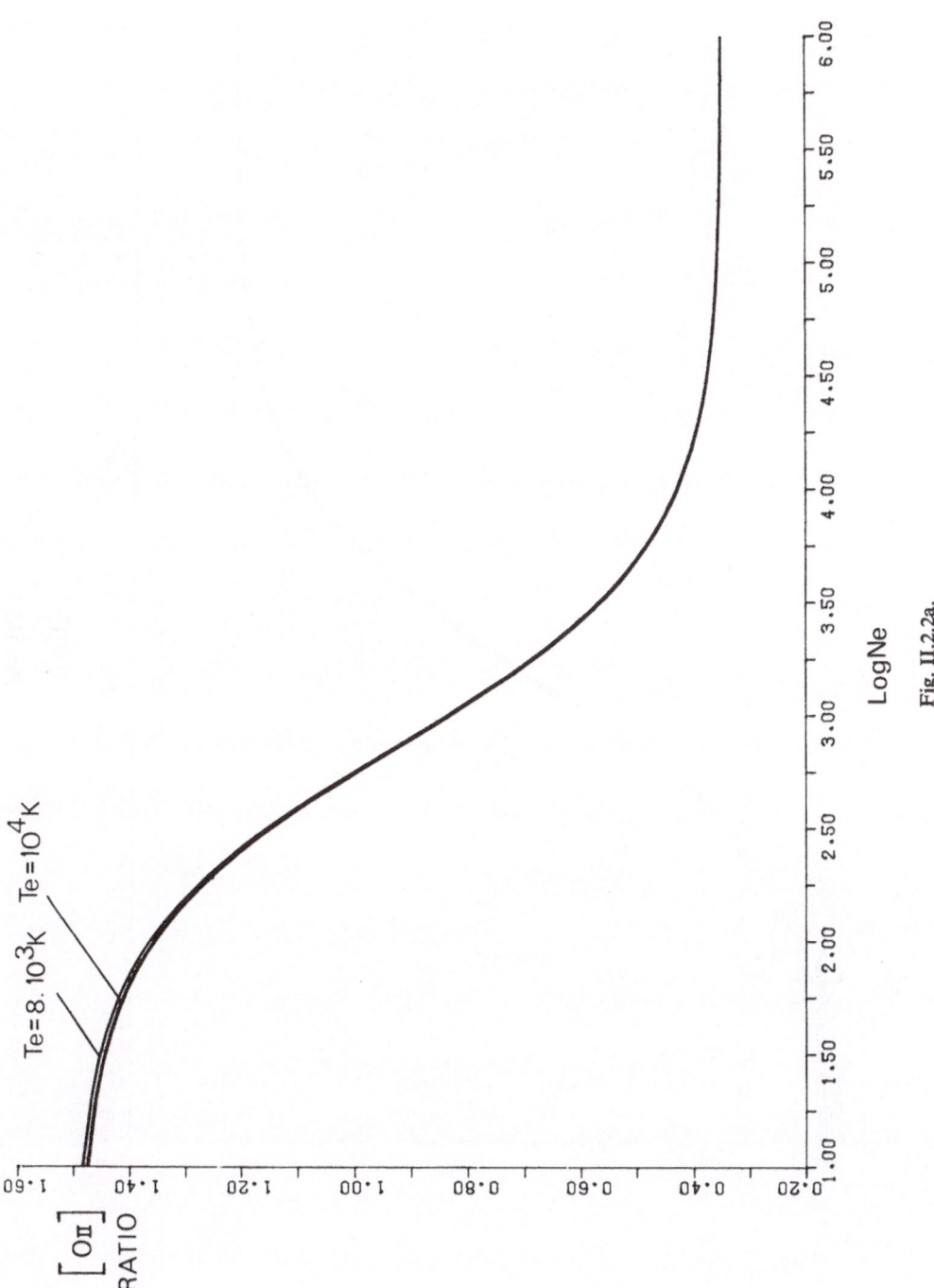

Fig. II.2.2a.

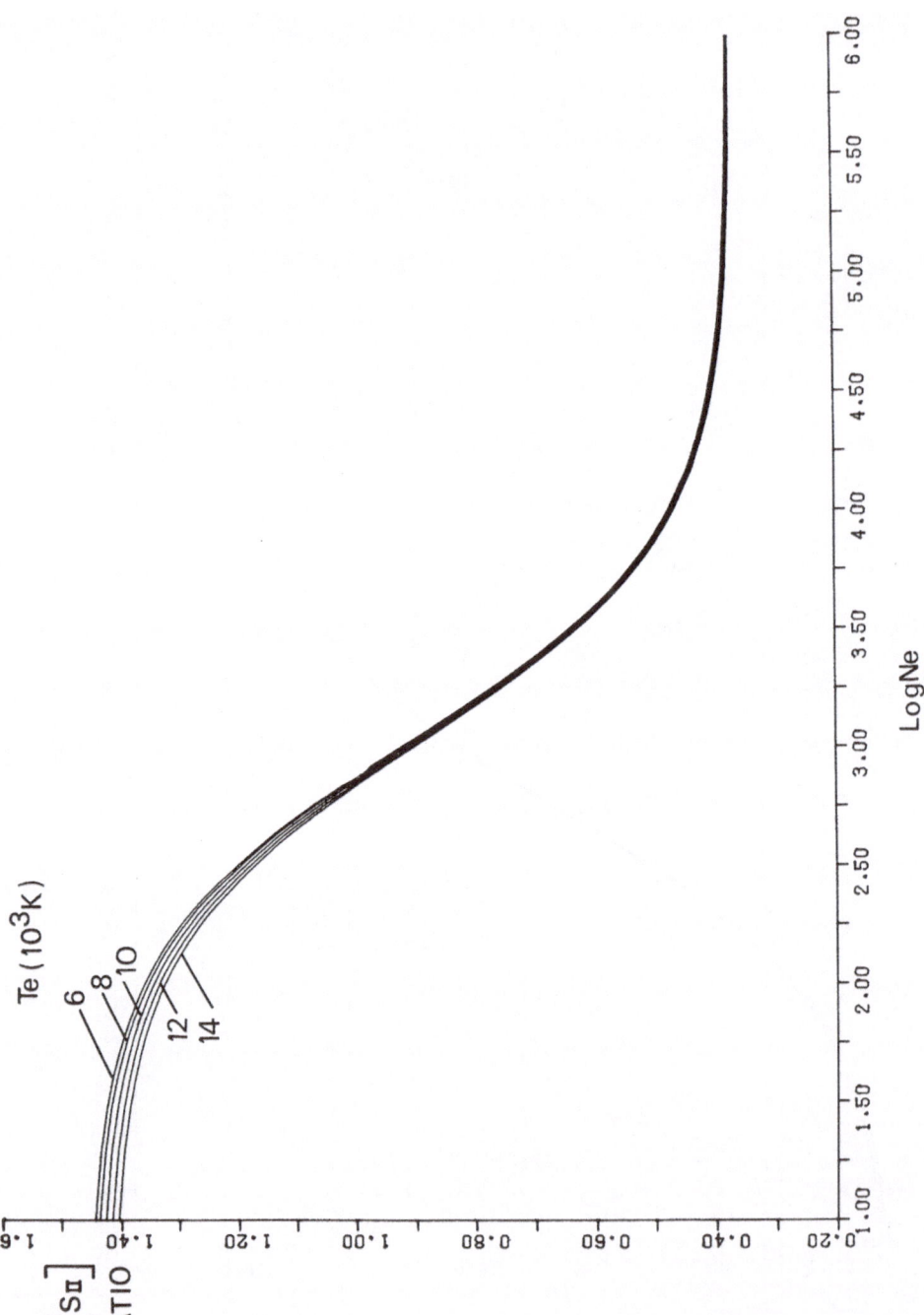

Fig. II.2.2b.

Figs. II.2.2.a-b.   The [O II] (λ 3729 Å/λ 3726 Å) (a), and [S II] (λ 6717 Å/λ 6731 Å) (b) intensity ratios as a function of the logarithm of the electron density N_e (cm^{-3}) for $T_e = 10^4$ K (after Gomez, 1981). For temperatures other than 10^4 K the ratio is expressed as a function of log N_e ($10^4\, T_e^{-1}$)$^{1/2}$. These calculations were made for a five level atom model, by employing collision strengths from Pradhan (1978) and transition probabilities from Garstang (1968).

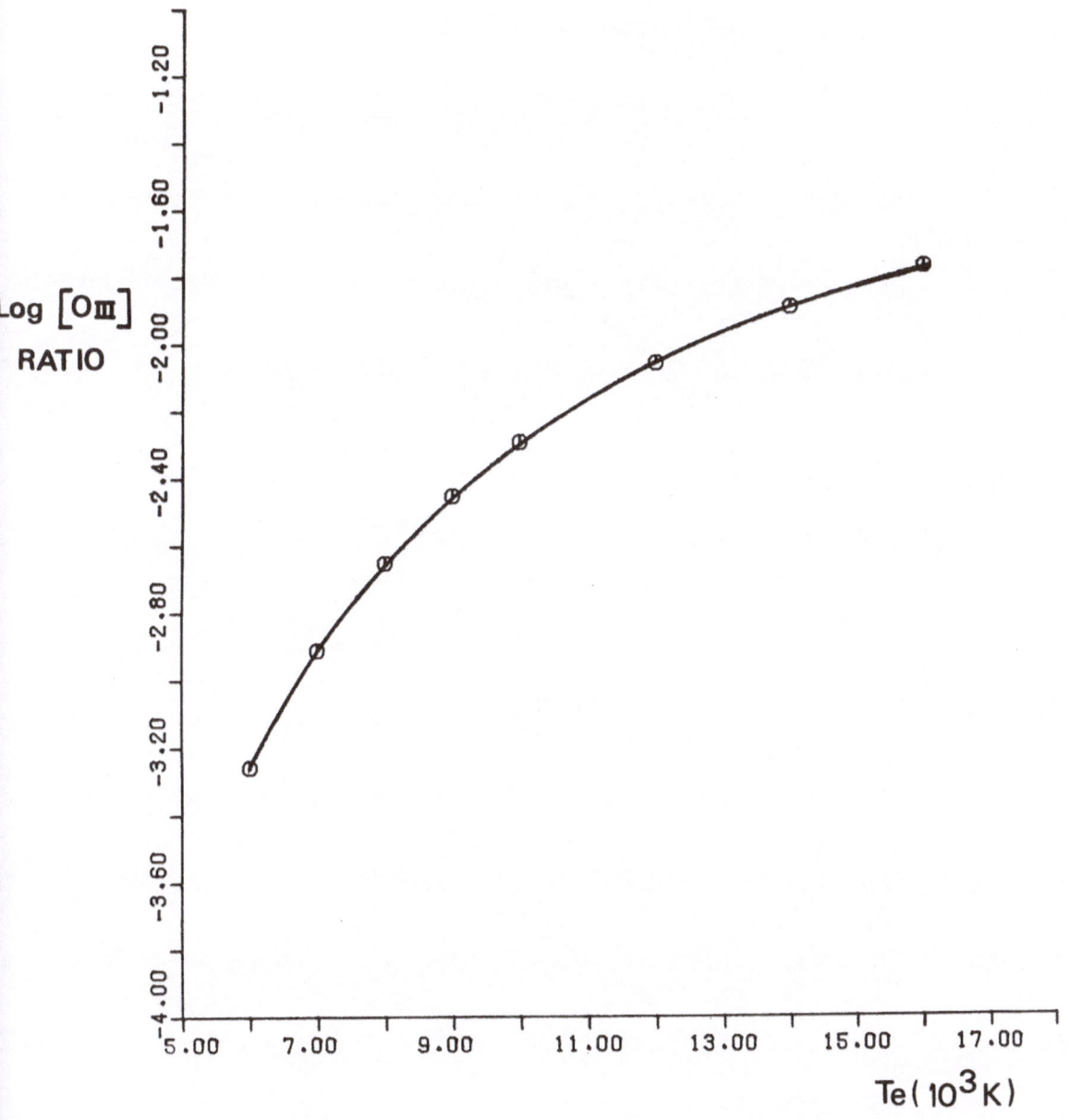

Fig. II.3.1a.

Figs. II.3.1.a-c. The logarithm of the [O III] [λ 4363 Å/(λ 4959 Å + 5007Å)], [N II] [λ 5755 Å/(λ 6548Å + 6584Å)], and [S III] [λ 6312Å/λ 9069Å + λ 9532 Å)] intensity ratios (a), (b), and (c), respectively, as a function of T_e for electron densities $N_e \leq 10^4$ cm^{-3} (after Gomez, 1981). These calculations were made for a five level atom model, by employing collision strengths from Seaton (1975) ([O III], [N II]) and Osterbrock (1974) ([S III]) and transition probabilities from Garstang (1968).

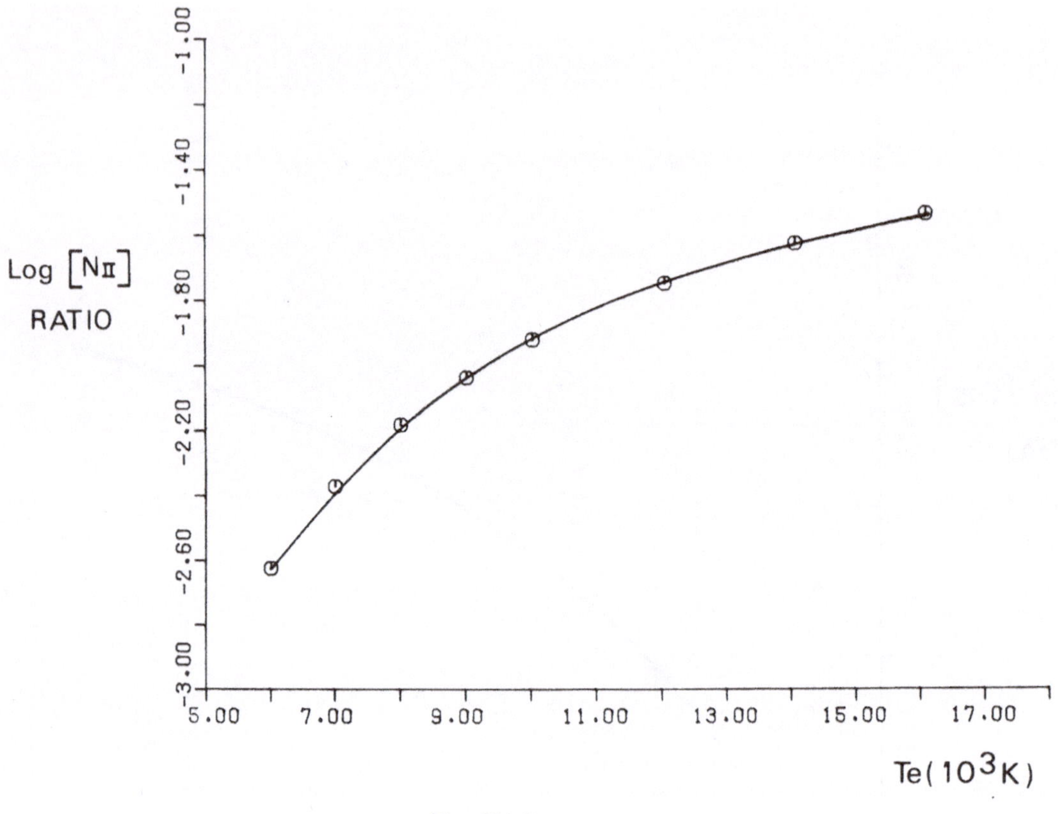

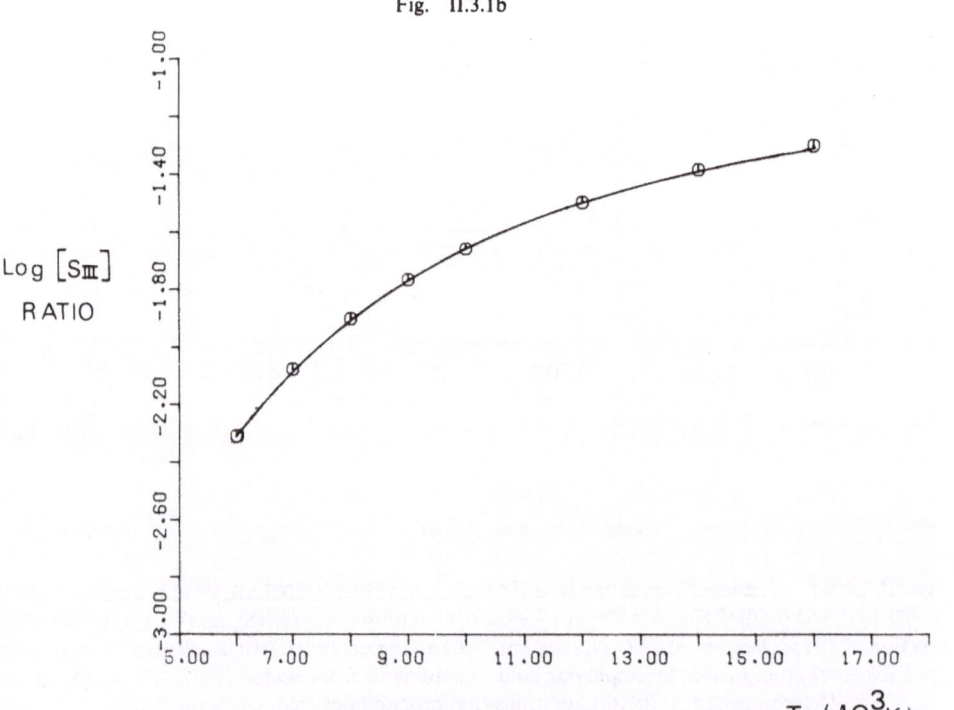

Fig.   II.3.1b

Fig.   II.3.1c

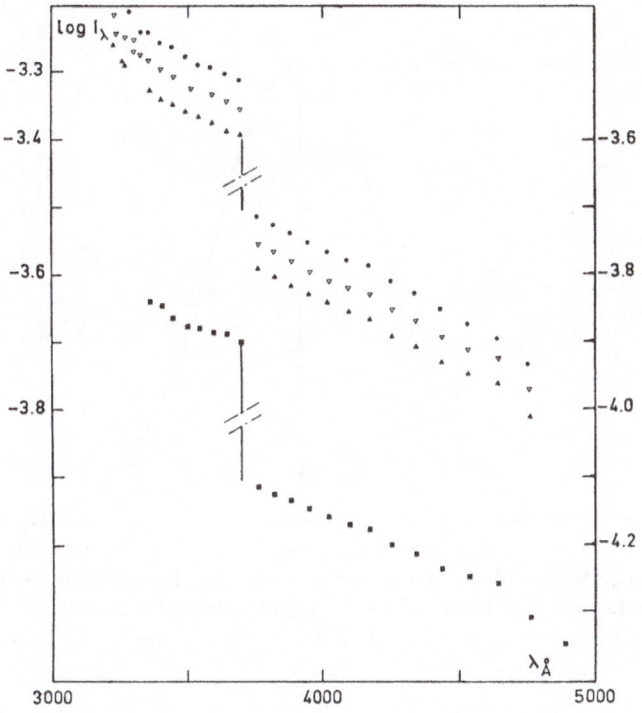

Fig. II.3.2. The observed Balmer discontinuity for four regions of M42 (after Hua, 1974). The observed discontinuity corresponds to

$$D_{obs} = \log \frac{I_{Balmer-} (\lambda < 3646 \text{ Å}) + I_{dust}}{I_{Balmer+} (\lambda > 3646 \text{ Å}) + I_{dust},}$$

where I_{dust} is the continuum due to the scattered by dust starlight. In this case the scattered starlight intensity at the Balmer limit was estimated to be $\simeq 30\%$ of the intensity due to atomic processes. The right-hand scale of $\log I_\lambda$ refers to intensities measured at $\lambda > 3646$ Å whereas the left-hand scale to those at $\lambda < 3646$ Å.

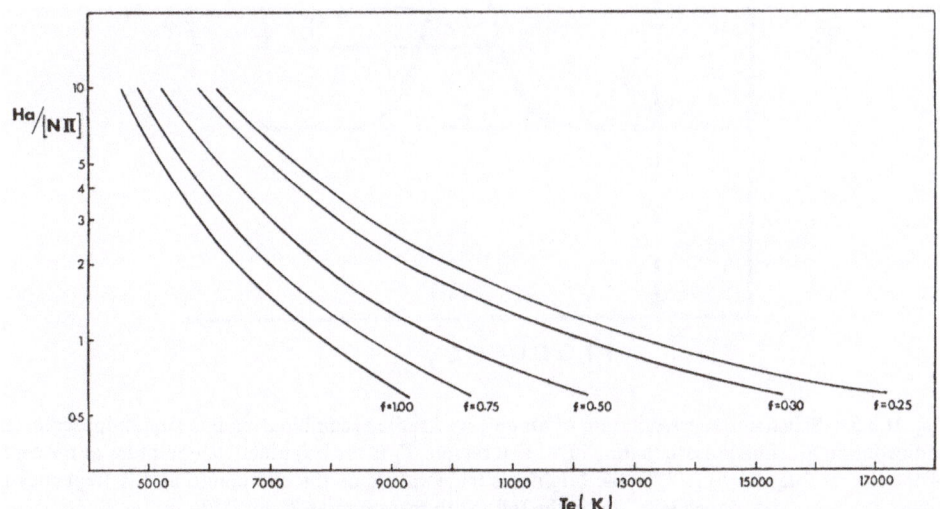

Fig. II.3.3. The $H\alpha/[N \text{ II}]$ intensity ratio as a function of electron temperature T_e for various $f = N(N^+)/N(N)$. The nitrogen abundance $N(N)/N(H)$ is taken equal to 10^{-4} (Goudis, 1976d).

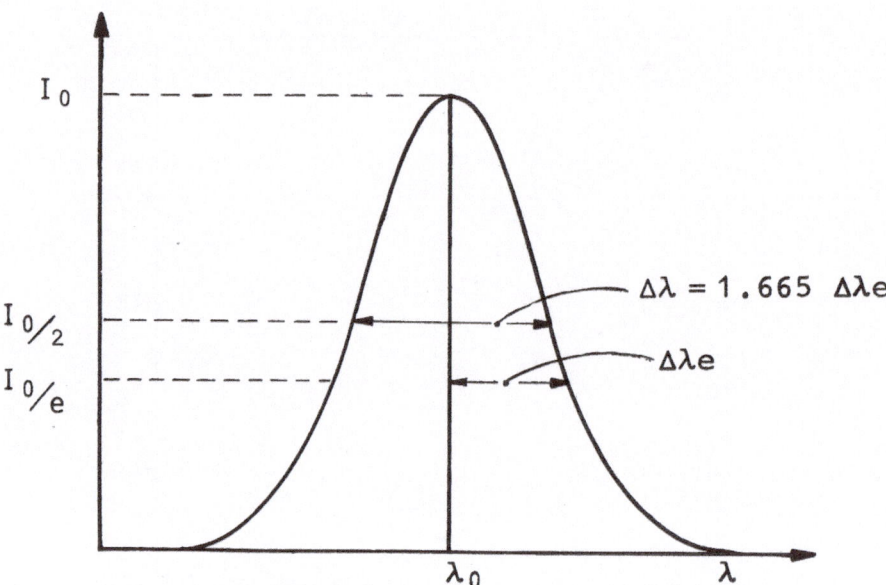

Fig. II.3.4.   Gaussian line profile $I = I_0 \exp\{-[(\lambda_0 - \lambda)/\Delta\lambda_e]^2\}$. The e-folding width $\Delta\lambda_e$, measured at $I = I_0/e$, and the full width $\Delta\lambda$ measured at $I = I_0/2$, are shown. Note that $\Delta\lambda = 2\,(\ln 2)^{1/2}\,\Delta\lambda_e = 1.665\Delta\lambda_e$.

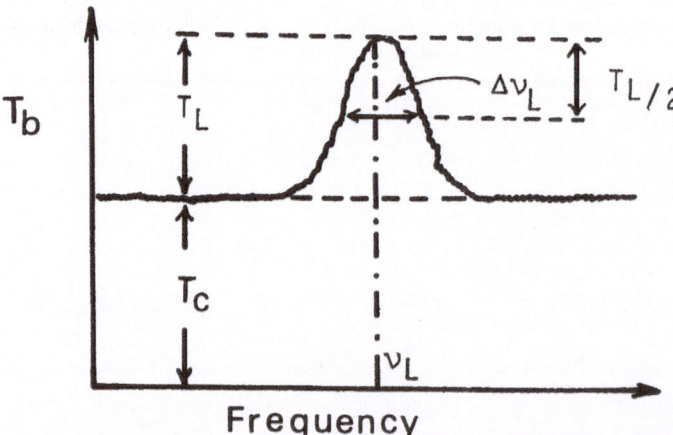

Fig. II.3.5.   Schematic representation of an observed radio recombination line superimposed on the radiocontinuum emission originating in an H II region. T_L is the brightness temperature at the centre of the line at frequency ν_L. T_c is the brightness temperature of the continuum and is frequency independent. $\Delta\nu_L$ is the full width measured at $T_b = T_L/2$.

PHYSICAL PARAMETERS OF THE DUST
ASSOCIATED WITH AN H II REGION

III.1 Visual Extinction Derived from the Comparison of Optical with Radio Data

The visual extinction of light, A_v, caused by dust lying within or in front of an H II region can be estimated by comparing the attenuated optical emission of a hydrogen line (e.g. Hα) with the unaffected by dust radio continuum emission originating in the same area (see e.g., Ishida and Kawajiri, 1968; Goudis, 1976f; Goudis and White, 1980 a, b, c). For a meaningful comparison the data should be obtained with the same angular resolution. At high radio frequencies ($\nu > 1$ GHz) where the nebula is optically thin in the continuum radiation ($\tau_c \ll 1$), holds the relation:

$$T_b = \tau_c T_e. \tag{II.1.2b}$$

This, in combination with (II.1.5) yields:

$$\left[\frac{E}{\text{pc cm}^{-6}} \right] = \frac{12.143}{\alpha(\nu, T_e)} \left[\frac{T_b}{\text{K}} \right] \left[\frac{\nu}{\text{GHz}} \right]^{2.1} \left[\frac{T_e}{\text{K}} \right]^{0.35}, \tag{III.1.1.}$$

where $\alpha(\nu, T_e)$ is of the order of 1 and is tabulated by Mezger and Henderson (1967); ν is the frequency of the radio continuum observations; T_e is the electron temperature of the H II region ($\simeq 10^4$ K); and T_b is the brightness temperature.

Substituting the emission measure E from the above relation (III.1.1) into the relation (see Table II.3.III):

$$\left[\frac{I_{\text{H}\alpha}}{\text{ergs cm}^{-2}\,\text{s}^{-1}\,\text{sr}^{-1}} \right] \simeq 8.5 \times 10^{-8} \left[\frac{E}{\text{pc cm}^{-6}} \right] \tag{III.1.2}$$

and assuming $\alpha(\nu, T_e) = 1$, one obtains:

$$\left[\frac{I_{\text{H}\alpha}}{\text{ergs cm}^{-2}\,\text{s}^{-1}\,\text{sr}^{-1}} \right] \simeq 10^{-6} \left[\frac{T_b}{\text{K}} \right] \left[\frac{T_e}{\text{K}} \right]^{0.35} \left[\frac{\nu}{\text{GHz}} \right]^{2.1}. \tag{III.1.3}$$

This is the intensity of the Hα line which should have been observed in the absence of dust. In the presence of dust the observed Hα intensity, $I_{\text{H}\alpha_{\text{obs}}}$, is always $< I_{\text{H}\alpha}$. The difference between the two is due to the extinction of the Hα line, $A_{\text{H}\alpha}$.
This can be derived by using Pogson's formula from the relation:

$$\left[\frac{A_{\text{H}\alpha}}{\text{mag.}} \right] = 2.5 \log \frac{I_{\text{H}\alpha}}{I_{\text{H}\alpha_{\text{obs}}}}. \tag{III.1.4}$$

For a normal extinction law ($R = 3$) the visual absorption, A_v, can be estimated from

(Whitford, 1958):

$$A_v = 1.28\, A_{H\alpha}.$$ (III.1.5)

The corresponding formulae for the $H\beta$ line are:

$$\left[\frac{I_{H\beta}}{\text{ergs cm}^{-2}\,\text{s}^{-1}\,\text{sr}^{-1}}\right] \simeq 3.7 \times 10^{-7} \left[\frac{T_b}{K}\right]\left[\frac{T_e}{K}\right]^{0.35}\left[\frac{\nu}{\text{GHz}}\right]^{2.1},$$ (III.1.6)

$$\left[\frac{A_{H\beta}}{\text{mag.}}\right] = 2.5 \log \frac{I_{H\beta}}{I_{H\beta_{obs}}},$$ (III.1.7)

$$A_v \simeq 0.90 A_{H\beta}.$$ (III.1.8)

The method cannot however provide information about the location of dust with respect to the HII region. If for example the estimation of the optical depth of the dust at $H\alpha$ is desired, two limiting cases should be discriminated (Mezger, 1972):
(1) the dust lies outside the H II region (Figure III.1.1a);
(2) the dust lies within and is well mixed with the ionized gas (Figure III.1.1b).
In the first case (Figure III.1.1a), holds the relation

$$I_{H\alpha_{obs}} = I_{H\alpha} e^{-\tau_{H\alpha}}$$ (III.1.9)

from which

$$\tau_{H\alpha} = 2.3 \log \frac{I_{H\alpha}}{I_{H\alpha_{obs}}}$$ (III.1.10)

or

$$\tau_{H\alpha} = 0.92 A_{H\alpha} \simeq A_{H\alpha}.$$ (III.1.11)

In the second case (Figure III.1.1b), holds the relation:

$$\frac{I_{H\alpha}}{I_{H\alpha_{obs}}} = \frac{\int_0^L \varepsilon_{H\alpha}\, dx}{\int_0^L \varepsilon_{H\alpha}\, e^{-k_{H\alpha}x}\, dx},$$ (III.1.12)

where $\varepsilon_{H\alpha}$ is the emissivity; $k_{H\alpha}$ the linear adbsorption coefficient; and L the path length of the nebula along the line of sight. The relation (III.1.12) can be written as:

$$\frac{I_{H\alpha}}{I_{H\alpha_{obs}}} = \frac{\varepsilon_{H\alpha} L}{\left(\dfrac{\varepsilon_{H\alpha}}{k_{H\alpha}}\right)(1 - e^{-k_{H\alpha}L})}$$ (III.1.13)

and since $\tau_{H\alpha} = k_{H\alpha} L$ (I.1.2),

$$\frac{I_{H\alpha}}{I_{H\alpha_{obs}}} = \frac{\tau_{H\alpha}}{1 - e^{-\tau_{H\alpha}}}.$$ (III.1.14)

Then, for $\tau_{H\alpha} \gg 1$ one obtains:

$$A_{H\alpha} = 2.5 \log \tau_{H\alpha},$$ (III.1.15)

or

$$\tau_{H\alpha} = 10^{0.4A_{H\alpha}}. \tag{III.1.16}$$

III.2. Physical Parameters Derived from Observations of the Infrared Continuum

The typical energy spectrum of an H II region derived from measurements of the continuum radiation at radio and infrared wavelengths is shown in Figure III.2.1. The energy distribution in the infrared exceeds by far the one predicted from free-free emission originating in the ionized gas. This is probably due to the presence of dust within or in the vicinity of the H II region which absorbs UV radiation emitted from the exciting star(s) and re-radiates it in the infrared part of the spectrum. Such an interpretation is widely accepted today and the reasons supporting it (blackbody spectrum of the emission, absorption features indicating the presence of grains, lack of predominant atomic lines, distribution of the emission etc.) have been summarized by Werner et al. (1977).

The heating of the dust in an H II region was at first attributed to the absorption of the Lα photons emitted by the recombination of the ionized hydrogen and resonantly trapped within the H II region. However, observations of many H II regions have demonstrated that the observed total infrared luminosity L_{IR} is significantly higher than the Lα luminosity. The observed 'infrared excess emission' (IRE) with respect to the emission expected if only Lα photons were absorbed by the dust, is defined as:

$$\text{IRE} = \frac{L_{IR}}{L_{L\alpha}} - 1. \tag{III.2.6}$$

This excess is clearly demonstrated in Figure III.2.2, where the total infrared luminosity is plotted against the Lyman continuum photon flux (Lc) for 46 H II regions and the expected L_{IR} under the assumption that $L_{IR} = L_{L\alpha}$ (Panagia, 1977). This discrepancy led to the assumption that the dust absorbs not only the majority of the resonantly scattered Lα photons but also competes with the gas for the direct absorption of Lc photons*.

Further findings strengthening this assumption are the observed low $N(\text{He}^+)/N(\text{H}^+)$ abundances of many H II regions (lower than 0.10, a value thought to represent the He abundance in our Galaxy) and the inverse proportionality of the abundances with respect to the IRE (Churchwell et al., 1974). This is an indication that the dust absorbs more efficiently He than H ionizing photons.

The IR flux observed at a wavelength λ, F_{IR_λ}, can be expressed by the relation (I.1.16):

$$F_{IR_\lambda} = \Omega B_\lambda(T_d)(1 - e^{-\tau_{IR}}), \tag{III.2.1}$$

where Ω is the solid angle of the field of view; $B_\lambda(T_d)$ is the Planck function, given by

---

* It is now thought that hot dust inside an H II region is mainly heated by resonantly trapped Lα photons while the bulk of the infrared emission is coming from cooler dust outside the H II region (see e.g. Habing and Israel, 1979).

the relation

$$B_\lambda(T_d) = \frac{2hc^2/\lambda^5}{e^{hc/\lambda kT_d} - 1} \text{ (ergs cm}^{-2} \text{ s}^{-1} \text{ sr}^{-1}); \tag{III.2.2}$$

T_d is the blackbody colour temperature which is assumed to express the temperature of the dust grains; this can be derived from the ratio of the observed fluxes F_{λ_1} and F_{λ_2} at two neighbouring wavelengths λ_1 and λ_2, according to the relation:

$$\frac{F_{\lambda_1}}{F_{\lambda_2}} = \left(\frac{\lambda_2}{\lambda_1}\right)^5 \frac{e^{hc/\lambda_2 kT_d} - 1}{e^{hc/\lambda_1 kT_d} - 1}; \tag{III.2.3}$$

and τ_{IR} is the optical depth of the dust at the wavelength λ which is given by the relation (Fazio, 1976):

$$\tau_{IR} = \int_0^L n_d \, Q_{IR} \, \pi\alpha^2 \, dx, \tag{III.2.4}$$

where n_d (cm^{-3}) is the number density of dust grains; $\pi\alpha^2$ (cm^2) is the geometrical area of a dust grain of radius α; L(cm) is the linear dimension of the source along the line of sight; and Q_{IR} is the absorptivity (or emissivity) of the dust which is a function of frequency, grain size and chemical composition of the grains.

An important physical parameter concerning the energetics of the cloud is the total infrared luminosity L_{IR} which is estimated from the relation:

$$L_{IR} = 4\pi D^2 \int_0^\infty F_{IR} \, d\lambda, \tag{III.2.5}$$

where D is the distance to the source. This is always smaller or equal to the total stellar luminosity L_*.

Another important parameter is the mass of dust, M_d, associated with the HII region. For a spherical nebula of radius R containing grains of radius α, density ρ, and number density n_d, the mass of dust is:

$$M_d = \tfrac{4}{3}\pi R^3 n_d \tfrac{4}{3}\pi a^3 \, \rho. \tag{III.2.7}$$

Taking into account the relation (III.2.4) and assuming $L = 2R$ one obtains:

$$\left[\frac{M_d}{\text{gm}}\right] = 2.79 \left[\frac{a}{\text{cm}}\right]\left[\frac{\rho}{\text{gm cm}^{-3}}\right]\left[\frac{R}{\text{cm}}\right]^2 \frac{\tau_{IR}}{Q_{IR}}. \tag{III.2.8}$$

In practice R is the radius of the beam and M_d refers to the mass of dust within the beam. Typical values of the grain radius a are in the range 0.01–0.1μ whereas the value of ρ is ~ 2 gm cm^{-3}. Q_{IR} can be expressed as (Panagia, 1977):

$$Q_{IR} \simeq \frac{2\pi a}{\lambda} \quad \text{for} \quad \lambda \lesssim 100\mu \tag{III.2.9}$$

and (Aannestad, 1975):

$$Q_{IR} \simeq k \left[\frac{2\pi a}{\lambda^p} \right] \quad \text{for} \quad \lambda > 100\mu, \tag{III.2.10}$$

where k and p are constants tabulated for various materials by Aannestad (1975).

The dust to gas mass ratio M_d/M_g can also be derived from relation (III.2.8) and the expression for the ionized gas mass:

$$M_g = \tfrac{4}{3}\pi R^3 \, N_e \, m_H (X + 4Y), \tag{III.2.11}$$

where N_e is the electron density of the ionized gas, m_H is the mass of hydrogen atom ($= 1.6733 \times 10^{-24}$ gm) and X and Y are the fractional abundances by number of hydrogen and helium, usually taken as $X = 1$, $Y = 0.1$. Then:

$$\frac{M_d}{M_g} = 3.65 \times 10^{23} \frac{\tau_{IR}}{Q_{IR}} \left[\frac{a}{\text{cm}} \right] \left[\frac{\rho}{\text{gm cm}^{-3}} \right] \left[\frac{N_e}{\text{cm}^{-3}} \right]^{-1} \left[\frac{R}{\text{cm}} \right]^{-1}. \tag{III.2.12}$$

For typical values of the parameters involved, the M_d/M_g ratio is estimated to be in the order of 10^{-2}.

Another parameter, the dust column density N_D can be estimated from the relation:

$$N_D = (\tfrac{4}{3}\pi a^3) \, \rho n_d \, 2R. \tag{III.2.13}$$

From (III.2.13) and (III.2.4) one obtains:

$$\left[\frac{N_D}{\text{gm cm}^{-2}} \right] = 1.33 \frac{\tau_{IR}}{Q_{IR}} \left[\frac{\rho}{\text{gm cm}^{-3}} \right] \left[\frac{a}{\text{cm}} \right]. \tag{III.2.14}$$

The expected column density of the H_2 gas can then be estimated from the relation:

$$\left[\frac{N_{H_2}}{\text{mol.cm}^{-2}} \right] = \left[\frac{N_D}{\text{gm cm}^{-2}} \right] \left[\frac{m_{H_2}}{\text{gm}} \right]^{-1} \frac{M_g}{M_d} \tag{III.2.15}$$

or for $M_g/M_d \simeq 100$ and $m_{H_2} = 3.35 \times 10^{-24}$ gm,

$$\left[\frac{N_{H_2}}{\text{mol.cm}^{-2}} \right] \simeq 3 \times 10^{25} \left[\frac{N_D}{\text{gm cm}^{-2}} \right]. \tag{III.2.16}$$

The visual extinction A_v can also be estimated from the relation (Wickramasinghe, 1972):

$$\left[\frac{A_v}{\text{mag}} \right] = 1.086 \left[\frac{N_d}{\text{cm}^{-2}} \right] \left[\frac{\pi a^2}{\text{cm}^2} \right] Q_{ext}, \tag{III.2.17}$$

where N_d is the number column density of the dust and Q_{ext} is the extinction efficiency of the grains at visible wavelengths, computed by many authors (e.g. Wickramasinghe, 1972; Greenberg, 1968); the typical value of $Q_{ext} = 1$.

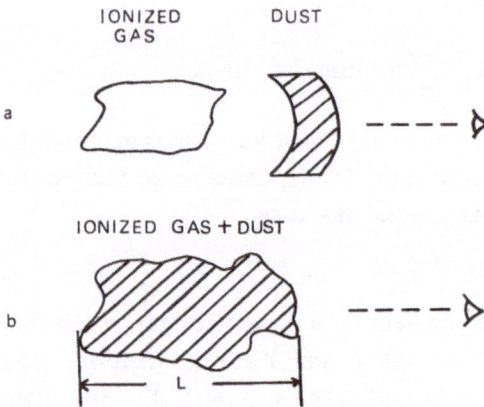

Fig. III.1.1. (a) An H II region obscured by dust lying outside. (b) An H II region mixed with dust.

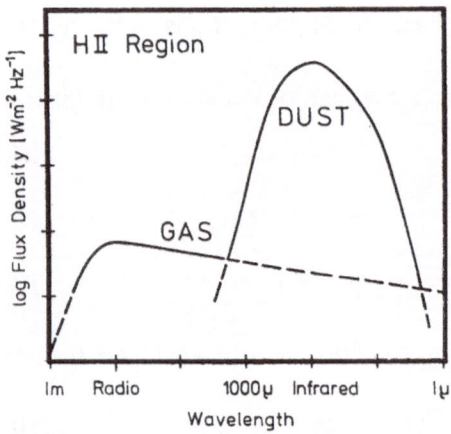

Fig. III.2.1a. Typical spectrum of an H II region mixed with dust (after Lemke and Harris, 1979). The spectrum consists of two components: (1) free-free emission from the ionized gas and (2) thermal emission from the dust grains mixed with the gas.

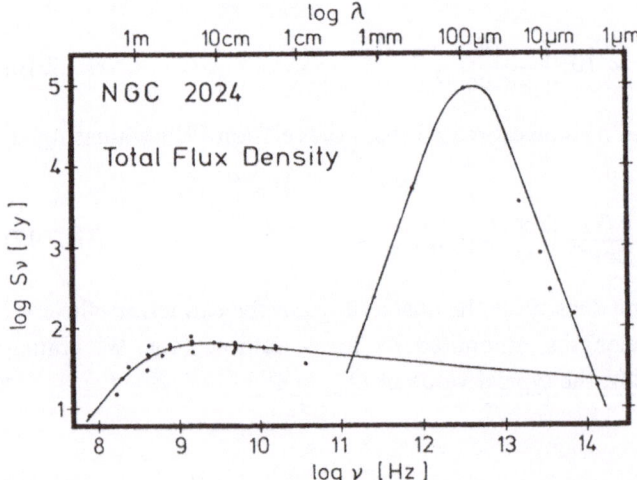

Fig. III.2.1b. Radio and Infrared spectrum of NGC 2024 (after Frey *et al.*, 1979; the radio flux densities were taken from Goudis, 1975).

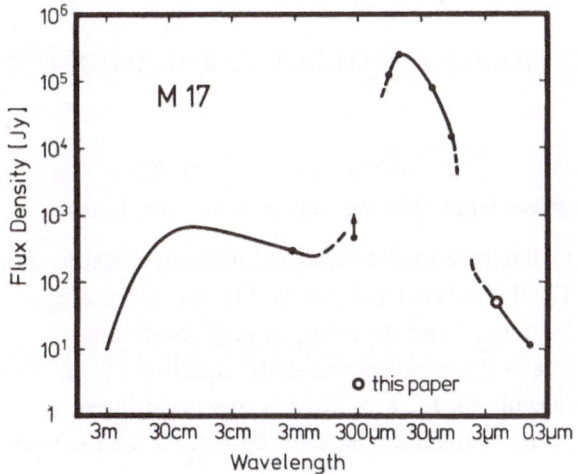

Fig. III.2.1c. Radio and Infrared spectrum of M17 (after Lemke and Harris, 1981; the radio flux densities were taken from Goudis, 1975).

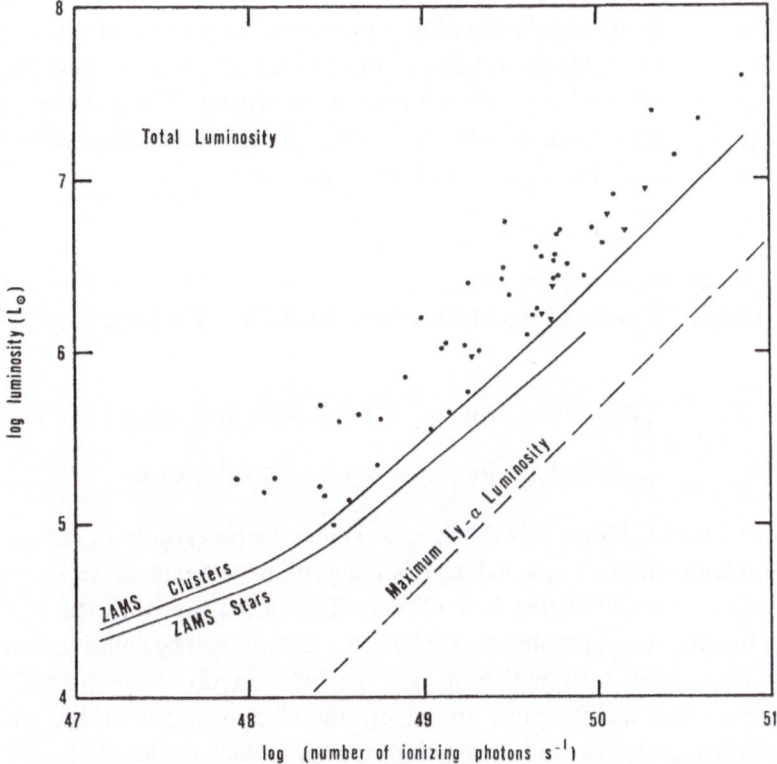

Fig. III.2.2. The total infrared luminosity versus the Lyman-c luminosity (Lc) plotted for 46 H II regions (after Panagia, 1977). The curves of ZAMS clusters, ZAMS stars and the line $L_{IR} = L_{L\alpha}$ are also shown for comparison.

PHYSICAL PARAMETERS OF A MOLECULAR CLOUD

IV.1. Physical Parameters Derived from Observations of Molecular Lines

The most important physical parameters of a molecular cloud, (mainly consisting of H_2) are the kinetic temperature, T_k, the column density of the various molecules, N_{mol} (cm^{-2}), the hydrogen density N_{H_2} (cm^{-3}) and the mass, M_{H_2}, of the cloud.

The starting point for the estimation of these parameters is the equation of radiative transfer. For a simple two-level molecule ($n, m; n > m$), the measured intensity I_{nm} of the emission line originating from the transition nm, after correcting for the background emission, is given by:

$$I_{nm} = B_\nu(T_{nm})(1 - e^{-\tau_\nu}), \tag{I.1.17}$$

where $B_\nu(T_{nm})$ is the Planck function (I.1.5); T_{nm} is the rotational excitation temperature characterizing the population of the upper n and lower m levels; and $\tau_\nu = \mathcal{N}_m \alpha_\nu$ is the optical depth of this line (at frequency ν) where $\mathcal{N}_m$ is the column density of the lower level in cm^{-2} and α_ν is the absorption coefficient. Using the Rayleigh-Jeans approximation which holds true for $h\nu \ll kT_{nm}$ (molecules rotating in the microwave part of the spectrum), Equation (I.1.17) takes the form:

$$T_b = T_{nm}(1 - e^{-\tau_\nu}) \tag{I.1.18}$$

where T_b is the brightness temperature, estimated from the observations. Depending on the optical thickness of the observed molecular line, Equation (I.1.18) is reduced to:

$$T_b = \begin{cases} T_{nm} & \text{for } \tau_\nu \gg 1 \text{ (optically thick case),} & \text{(IV.1.1a)} \\ \mathcal{N}_m \alpha_\nu T_{nm} & \text{for } \tau_\nu \ll 1 \text{ (optically thin case).} & \text{(IV.1.1b)} \end{cases}$$

Thus for a saturated (optically thick, $\tau_\nu \gg 1$) line the observable T_b yields directly the excitation temperature T_{nm} which equals the gas kinetic temperature T_k, if the excitation mechanism is dominated by collisions. This is the case with the ^{12}CO emission lines. In fact the state of molecular excitation is determined by collisions with particles and by interaction with radiation. The particles involved are neutral molecules (mainly H_2 which are the most abundant) and electrons. The kinetic energy of the molecules is acquired by collisions with dust grains which are heated by the UV radiation of stars. The radiation includes the continuum radiation emitted from local stars, and the 2.7 K microwave background radiation. The dominant however excitation mechanism, particularly for CO, is collisions with H_2 (Solomon, 1973). This permits the establishment of lower limits for the H_2 density, n_{H_2} (cm^{-3}), from the mere detection of a molecular line. The presence of such a line requires a density greater than a certain value, which is necessary to make collisional excitations the dominant mechanism.

For an optically thin ($\tau_\nu \ll 1$) molecular line, the observed T_b (IV.1.1b) in combination with a relation expressing the absorption coefficient as a function of frequency ν and excitation temperature T_{nm}, can be used to estimate the column density $\mathcal{N}_m$ of the lower level m according to the relation (Solomon, 1973):

$$\mathcal{N}_m = \frac{3\,k}{8\pi^3\,|\mu_{mn}|^2\,\nu} \int T_b\, dV, \qquad (IV.1.2)$$

where the integration is over velocity (in km s^{-1}) and $|\mu_{mn}|$ is the dipole-moment matrix element. To derive the total column density, N_{tot}, the column density of all levels must be computed. This method is also used to derive the density of H_2. This can be achieved by estimating first the total column density of the optically thin ^{13}CO ($J = 1 \rightarrow 0$) line which is (Solomon, 1973):

$$\left[\frac{N_{^{13}CO}}{cm^{-2}}\right] = 5 \times 10^3 \left[\frac{T_{01}}{K}\right]\int\left[\frac{T_b}{K}\right]\left[\frac{dV}{km\ s^{-1}}\right], \qquad (IV.1.3)$$

where T_{01} is the excitation temperature calculated from observations of the optically thick ^{12}CO ($J = 1$–0) line (IV.1.1a). Formulae, currently used to derive the CO excitation temperature T_{01} and the column density $N_{^{13}CO}$ (Wilson, 1974; Knapp et $al.$, 1977; Elmegreen and Elmegreen, 1978) are given by:

$$T_{01} = T_{01}(^{12}CO) = T_{01}(^{13}CO) = 5.54\left[\ln\left(1 + \frac{5.54}{T(^{12}CO) + 0.817}\right)\right]^{-1} \quad (IV.1.4)$$

$$N_{^{13}CO} = \frac{2.31 \times 10^{14}\ \tau_{^{13}CO}\ \Delta V_{^{13}CO}(T_{01} + 0.91)}{1 - e^{-5.29/T_{01}}}, \qquad (IV.1.5)$$

where

$$\tau_{^{13}CO} = \ln\left[\frac{T_{^{12}CO}}{T_{^{12}CO} - T_{^{13}CO}}\right], \qquad (IV.1.6)$$

$T_{^{12}CO}$, $T_{^{13}CO}$ are the Rayleigh–Jeans equivalent brightness temperatures defined by:

$$T = \frac{h\nu}{k}\ [F(T_B) - F(T_{bb})], \qquad (IV.1.7)$$

where

$$F(T) = (e^{h\nu/kT} - 1)^{-1}, \qquad (IV.1.8)$$

T_B is the temperature of the corresponding line; T_{bb} is the temperature of the microwave background (2.7 K); and $\Delta V_{^{13}CO}$ is the full width at half maximum intensity of the ^{13}CO line, expressed in Km s^{-1}.

These formulae are valid under the assumption that the ^{12}CO line is optically thick,

$T_{01}(^{12}CO) = T_{01}(^{13}CO)$, the antenna beamwidth is smaller than the source and the effective beam efficieny is $\simeq 1$ (Elmegreen and Elmegreen, 1978).

The total column density of molecular hydrogen, N_{H_2}, can then be derived by the relation:

$$N_{H_2} = \tfrac{1}{2} N_{^{13}CO} \frac{N(^{12}C)}{N(^{13}C)} \left(\frac{N(C)}{N(H)} \right)^{-1} f^{-1}, \tag{IV.1.9}$$

where f is the fraction of carbon tied in CO. Then, assuming an isotopic abundance of $N(^{12}CO)/N(^{13}CO) = N(^{12}C)/N(^{13}C) = 89$ (equal to the terrestial value, as found by Vauden Bout and Thaddeus (1971)), all carbon to be in the form of CO ($f = 1$) and using the cosmic abundance $N(C)/N(H) = 3 \times 10^{-4}$, one obtains:

$$\left[\frac{N_{H_2}}{cm^{-2}} \right] \simeq 1.5 \times 10^5 \left[\frac{N_{^{13}CO}}{cm^{-2}} \right]. \tag{IV.1.10}$$

It should be noted that such an estimation is based on a number of assumptions which are still open to question. Thus, the fraction of carbon which is tied in the form of CO may be as low as $f = 0.1$ (Thronson *et al.*, 1978; Goldreich and Kwan, 1974) and the C isotopic abundance may be $N(^{12}CO)/N(^{13}CO) = N(^{12}C)/N(^{13}C) = 40$ (equal to the cosmic value as found by Wannier *et al.* (1976)). Therefore, estimates of N_{H_2} can only be considered as rough approximations.

From (IV.1.10) the H_2 density, n_{H_2} (cm^{-3}), can be derived by dividing the column density N_{H_2} by the cloud's diameter L, i.e.:

$$\left[\frac{n_{H_2}}{cm^{-3}} \right] = \left[\frac{N_{H_2}}{cm^{-2}} \right] \left[\frac{L}{cm} \right]^{-1}. \tag{IV.1.11}$$

Knowledge of the cloud diameter (from the extend of the cloud as seen in molecular maps and the distance from the Earth) yields the total mass of the cloud. This is:

$$M_{cloud} = \tfrac{4}{3} \pi \left(\frac{L}{2} \right)^3 n_{H_2} m_{H_2} \tag{IV.1.12}$$

or

$$\left[\frac{M_{cloud}}{M_\odot} \right] \simeq 2.7 \times 10^{-21} \left[\frac{N_{H_2}}{cm^{-2}} \right] \left[\frac{L}{pc} \right]^2. \tag{IV.1.13}$$

The derivation of the kinematic parameters of a molecular cloud from the observation of a molecular line (radial velocity, line width) is accomplished in a manner similar to that employed for the optical and radio recombination lines.

REFERENCES

Aannestad, P.: 1975, *Astrophys. J.* **200**, 30.

Abbott, D.: 1978, cited as private communication in Reynolds, R. J. and Ogden, P. M.: 1979, *Astrophys. J.* **229**, 942.

Abt, H. A. and Levato, H.: 1977, *Publ. Astron. Soc. Pacific* **89**, 797.

Adams, W.S.: 1944, *Publ. Astron. Soc. Pacific* **56**, 119.

Adams, W.S.: 1949, *Astrophys. J.* **109**, 335.

Ade, P.A.R., Clegg, P.E., and Rather, J.D.G.: 1974, *Astrophys. J.* **189**, L23.

Ahmad, I.A.: 1976, *Astrophys. J.* **205**, 379.

Aikman, G.C.L. and Goldberg, B.A.: 1974, *Roy. Astron. Soc. Canada* **68**, 205.

Aitken, D.K., Roche, P.F., Spenser, P.M., and Jones, B.: 1979, *Astron. Astrophys.* **76**, 60.

Alferova, Z.A., Venger, A.P., Gosachinskii, I.V., Grachev, V.G., Egorova, T.M., Zhelenkov, S.R., Korol'kov, D.V., Mogileva, V.G., Pariiskii, Yu.N., Prozorov, V.A., Ryzhkov, N.F., and Shivris, O.N.: 1979, *Soviet Astron.* **23**, 672.

Aller, L.H.: 1956, *Gaseous Nebulae*, Chapman and Hall, London.

Aller, L.H. and Czyzak, S.J.: 1968, in D.E. Osterbrock and C.R. O'Dell (eds.), 'Planetary Nebulae', *IAU Symp.* **34**, 209.

Aller, L.H. and Liller, W.: 1959, *Astrophys. J.* **130**, 45.

Aller, L.H. and Liller, W.: 1968, *Nebulae and Interstellar Matter*, University of Chicago Press, Chicago, p. 483.

Altenhoff, W.J.: 1972, in N.C. Wickramasinghe, F.D. Kahn, and P.G. Mezger, (eds.), *Interstellar Matter*, edited by the Astronomical Institute, University of Basel, Saas-Fee, p. 190.

Ambartsumian, V.A.: 1955, *Observatory* **75**, 72.

Anderreg, M., Moorwood, A.F.M., Hippelein, H.H., Baluteau, J., Bussoletti, E., and Coron, N.: 1976, in M. Rowan-Robinson (ed.), *Far Infrared Astronomy*, Pergamon Press, p. 171.

Andrillat, Y. and Duchesne, M.: 1974, *Astron. Astrophys.* **35**, 467.

Appenzeller, I.: 1974, *Astron. Astrophys.* **36**, 99.

Baade, W. and Minkowski, R.: 1937, *Astrophys. J.* **86**, 123.

Baars, J.W.M., Mezger, P.G., and Wendker, H.: 1965, *Z. Astrophys.* **61**, 134.

Balick, B.: 1976, *Astrophys. J.* **208**, 75.

Balick, B., Gammon, R.H., and Hjellming, R.M.: 1974, *Publ. Astron. Soc. Pacific* **86**, 616.

Balick, B., Gull, T.R., and Smith, M.G.: 1980, *Publ. Astron. Soc. Pacific* **92**, 22.

Ball, J.A., Cesarsky, D., Dupree, A.K., Goldberg, L., and Lilley, A.E.: 1970, *Astrophys. J.* **162**, L25.

Baluteau, J.-P., Bussoletti, E., Anderegg, M., Moorwood, A.F.M., and Coron, N.: 1976, *Astrophys. J.* **210**, L45.

Barbieri, C., Cosmovici, C.B., Michel, K.W., and Nishimura, T.: 1976, *Astron. Astrophys.* **47**, 255.

Barbon, R., Bernaca, P.L., Tarenghi, M., and Treves, A.: 1972, *Nature Phys. Sci.* **240**, 182.

Barker, T.: 1979, *Astrophys. J.* **227**, 863.

Barnard, E.E.: 1894a, *Astron. Astrophys.* **13**, 811.

Barnard, E.E.: 1894b, *Popular Astronomy* **2**, 151.

Barnard, E.E.: 1903, *Astrophys. J.* **17**, 77.

Barnard, E.E.: 1927, in E.B. Frost and M.R. Calvert (eds.), *A Photographic Atlas of Selected Regions of the Milky Way*, Washington, D.C., Carnegie Institution of Washington.

Barrett, A.H., Schwartz, P.R., and Waters, J.W.: 1971, *Astrophys. J.* **168**, L101.

Barrett, A.H., Ho, P., and Martin, R.N.: 1975, *Astrophys. J.* **198**, L119.

Barrett, A.H., Ho, P.T.P., and Myers, P.C.: 1977, *Astrophys. J.* **211**, L39.

Batchelor, A.S.J. and Brocklehurst, M.: 1972, *Astrophys. Letters* **11**, 129.

Baudry, A., Forster, J.R., and Welch, W.J.: 1974, *Astron. Astrophys.* **36**, 217.

Baudry, A., Combes, F., Perault, M., and Dickman, R.: 1980, *Astron. Astrophys.* **85**, 244

Baudel, L.: 1970, *Astron. Astrophys.* **8**, 65.

Beck, S.C., Lacy, J.H., and Geballe, T.R.: 1979, *Astrophys. J.* **234**, L213.

Becker, W. and Fenkart, R.: 1963, *Z. Astrophys.* **56**, 257.

Becklin, E.E. and Neugebauer, G.: 1967, *Astrophys. J.* **147**, 799.

Becklin, E.E., Neugebauer, G., and Wynn-Williams, C.G.: 1973, *Astrophys. J.* **182**, L7.

Becklin, E.E., Beckwith, S., Gatley, I., Matthews, K., Neugebauer, G., Sarazin, C., and Werner, M.W.: 1976, *Astrophys. J.* **207**, 770.

Beckwith, S., Persson, S.E., Neugebauer, G., and Becklin, E.E.: 1978, *Astrophys. J.* **223**, 464.

Beichman, C.A. and Chaisson, E.J.: 1974, *Astrophys. J.* **190**, L21.

Beichman, C.A., Dyck, H.M., and Simon, T.: 1978, *Astron. Astrophys.* **62**, 261.

Berkhuijsen, E.M.: 1972, *Astron. Astrophys. Suppl.* **5**, 263.

Bertsch, D.L., Fichtel, C.E., and Reames, D.V.: 1972, *Astrophys. J.* **171**, 169.

Berulis, I.I. and Sorochenko, R.L.: 1973, *Soviet Astron.* **17**, 179.

Berulis, I.I., Smirnov, G.I., and Sorochenko, R.L.: 1975, *Soviet Astron. Letters* **1**, 187.

Blaaw, A.: 1956, *Astrophys. J.* **123**, 408.

Blaaw, A.: 1964, *Ann. Rev. Astron. Astrophys.* **2**, 213.

Blaaw, A. and Morgan, W.W.: 1954, *Astrophys. J.* **119**, 625.

Bodenheimer, P., Tenorio-Tagle, G., and Yorke, H.W.: 1979, *Astrophys. J.* **233**, 85.

Böhm, K.H.: 1977, in R. Kippenhahn, J. Rahe, and W. Strohmeier (eds.), *The Interaction of Variable Stars with their Environment*, Bamberg.

Böhm, D. and Aller, L.H.: 1947, *Astrophys. J.* **105**, 131.

Bohlin, R.C. and Stecher, T.P.: 1975, *Bull. Am. Astron. Soc.* **7**, 547.

Bohuski, T.J., Dufour, R.J., and Osterbrock, D.E.: 1974, *Astrophys. J.* **188**, 529.

Boksenberg, A. and Burgess, D.E.: 1975, in J.W. Glaspey and G.A.H. Walker (eds.), *Proc. of Symp. T.V. Sensors*, Univ. of British Columbia, Vancouver, p. 21.

Bologna, J.M., Johnston, K.J., Knowles, S.H., Mango, S.A., and Sloanaker, R.M.: 1975, *Astrophys. J.* **199**, 86.

Boughton, W.L.: 1978, *Astrophys. J.* **222**, 517.

Bradt, H.V. and Kelley, R.L.: 1979, *Astrophys.* **228**, L33.

Brandshaft, D., McLaren, R.A., and Werner, M.W.: 1975, *Astrophys. J.* **199**, L115.

Breger, M.: 1977, *Astrophys. J.* **215**, 119.

Breger, M. and Hardrop, J.: 1973, *Astrophys. J.* **183**, L77.

Brocklehurst, M.: 1970, *Monthly Notices Roy. Astron. Soc.* **148**, 417.

Brocklehurst, M. and Seaton, M.J.: 1971, *Astrophys. Letters* **9**, 139.

Brocklehurst, M. and Seaton, M.J.: 1972, *Monthly Notices Roy. Astron. Soc.* **157**, 179.

Brown, R.L. and Broderick, J.J.: 1973, *Astrophys. J.* **181**, 125.

Brown, R.D., Godfrey, P.D., and Winkler, D.A.: 1980, *Monthly Notices Roy. Astron. Soc.* **190**, 1.

Buhl, D.: 1972, *Mercury* **1**, No. 6, 4.

Buhl, D. and Snyder, L.E.: 1970, *Nature* **228**, 267.

Buhl, D. and Snyder, L.E.: 1973, *Astrophys. J.* **180**, 791.

Buhl. D., Snyder, L.E., Lovas, F.J., and Johnson, D.R.: 1974, *Astrophys. J.* **192**, L97.

Buhl. D., Snyder, L.E., Lovas, F.J., and Johnson, D.R.: 1975, *Astrophys. J.* **201**, L29.

Burbidge, G.R., Gould, R.J., and Pottasch, S.R.: 1963, *Astrophys. J.* **138**, 945.

Burgess, A.: 1958, *Monthly Notices Roy. Astron. Soc.* **118**, 477.

Bussoletti, E. and Zambetta, A.M.: 1976, *Astron. Astrophys. Suppl.* **25**, 549.

Cantó, J., Goudis, C., Johnson, P.G., and Meaburn, J.: 1980, *Astron. Astrophys.* **85**, 128.

Caplan, J.G.: 1972, *Astron. Astrophys.* **18**, 408.

Carruthers, G.R.: 1969, *Astrophys. J.* **157**, L111.

Carruthers, G.R. and Opal, C.B.: 1977a, *Astrophys. J.* **212**, L27.

Carruthers, G.R. and Opal, C.B.: 1977b, *Astrophys. J.* **217**, 95.

Castor, J., McCray, R., and Weaver, R.: 1975, *Astrophys. J.* **200**, L107.

Caswell, J.L. and Goss, W.M.: 1974, *Astron. Astrophys.* **32**, 209.

Cesarsky, C.J., Cesarsky, D.A., Churchwell, E., and Lequeux, J.: 1978, *Astron. Astrophys.* **68**, 33.

Cesarsky, D.A.: 1971a, Ph.D. Thesis, Harvard University.

Cesarsky, D.A.: 1971b, *Astrophys. J.* **167**, L89.

Cesarsky, D.A.: 1977, *Astron. Astrophys.* **54**, 756.

Chaisson, E.J.: 1971, cited in Cesarsky, D.A., 1971a.

Chaisson, E.J.: 1972, Ph.D. Thesis, Harvard University.

Chaisson, E.J.: 1973a, *Astrophys. J.* **182**, 767.

Chaisson, E.J.: 1973b, *Astrophys. J.* **186**, 545.

Chaisson, E.J.: 1974a, *Astrophys. J.* **191**, 411.

Chaisson, E.J.: 1974b, *Astron. J.* **79**, 555.

Chaisson, E.J.: 1976, E.H. Avrett (ed.), *Frontiers of Astrophysics*, Harvard University Press, p. 259.

Chaisson, E.J. and Beichman, C.A.: 1975, *Astrophys. J.* **199**, L39.

Chaisson, E.J. and Dopita, M.A.: 1977, *Astron. Astrophys.* **56**, 385.

Chaisson, E.J., and Lada, C.J.: 1974, *Astrophys. J.* **189**, 227.

Chaisson, E.J. and Vrba, F.J.: 1978, in T. Gehrels, (ed.), *Protostars and Planets*, University of Arizona Press, Tucson, Arizona, p. 189.

Chaisson, E.J., Black, J.H., Dupree, A.K., and Cesarsky, D.A.: 1972, *Astrophys. J.* **173**, L131.

Chevallier, R.A.: 1974, *Astrophys. J.* **188**, 501.

Churchwell, E.: 1970, Ph.D. Thesis, Indiana University.

Churchwell, E. and Edrich, J.: 1970, *Astron. Astrophys.* **6**, 261.

Churchwell, E. and Mezger, P.G.: 1970a, cited in Hjellming, R.M. and Gordon, M.A., 1971.

Churchwell, E. and Mezger, P.G.: 1970b, *Astrophys. Letters* **5**, 227.

Churchwell, E. and Walmsley, C.M.: 1973, *Astron. Astrophys.* **23**, 117.

Churchwell, E., Mezger, P.G., Reifenstein, E., III., Rubin, R., and Turner, B.: 1970, *Astrophys. Letters* **5**, 157.

Churchwell, E., Mezger, P.G., and Huchtmeier, W.: 1974, *Astron. Astrophys.* **32**, 283.

Churchwell, E., Smith, L.F., Mathis, J., Mezger, P.G., and Huchtmeier, W.: 1978, *Astron. Astrophys.* **70**, 719.

Chui, M.F., Cheung, A.C., Matsakis, D., Townes, C.H., and Cardiasmenos, A.G.: 1974, *Astrophys. J.* **187**, L19.

Clark, B.G.: 1965, *Astrophys. J.* **142**, 1398.

Clark, F.O.,: 1975, *Astrophys. J.* **200**, L115.

Clark, F.O. and Johnson, D.R.: 1974 *Astrophys. J.* **191**, L87.

Clark, F.O. and Lovas, F.J.: 1977, *Astrophys. J.* **217**, L47.

Clark, F.O., Buhl, D., and Snyder, L.E.: 1974, *Astrophys. J.* **190**, 545.

Clark, F.O., Brown, R.D., Godfrey, P.D., Storey, J.W.V., and Johnson, D.R.: 1976, *Astrophys. J.* **210**, L139.

Clark, F.O., Lovas, F.J., and Johnson, D.R.: 1979, *Astrophys. J.* **229**, 553.

Cohen, J.G. and Frogel, J.A.: 1977, *Astrophys. J.* **211**, 178.

Cochran, W.D. and Ostriker, J.P.: 1977, *Astrophys. J.* **211**, 392.

Conti, P.S.: 1972, *Astrophys. J.* **174**, L79.

Conti, P.S. and Alschuler, W.R.: 1971, *Astrophys. J.* **170**, 325.

Conti, P.S. and Leep, E.M.: 1974, *Astrophys. J.* **193**, 113.

Cooke, B.A., Fabian, A.C., and Pringle, J.E.: 1978, *Nature* **273**, 645.

Cosmovici, C.B., Strafella, F., and Dirscherl.: 1980, *Astrophys. J.* **236**, 498.

Costero, R. and Peimbert, M.: 1970, *Bol. Obs. Tonantzintla y Tacubaya* **5**, 229.

Courtès, G.: 1960, *Ann. Astrophys.* **23**, 115.

Courtès, G.: 1967, cited in O'Dell *et al.*, 1967.

Courtès, G. and Viton, M.: 1965, *Ann. Astrophys.* **28**, 691.

Courtès, G., Cruvellier, P., and Georgelin, Y.: 1967, *Publ. Obs. Ht. Provence* **8**, 34.

Courtès, G., Georgelin, Y.P., Georgelin, Y.M., Monnet, G., and Pourcelot, A.: 1968a in, Y. Terzian (ed.), *Interstellar Ionized Hydrogen*, W.A. Benjamin, New York, p. 571.

Courtès, G., Louise, R., and Monnet, G.: 1968b, *Ann. Astrophys.* **31**, 493.

Cowie, L.L., Songaila, A., and York, D.G.; 1979, *Astrophys. J.* **230**, 469.

Cronin, N.J., Gillespie, A.R., Huggins, P.J., and Phillips, T.G.: 1976, *Astron. Astrophys.* **46**, 135.

Cruvellier, P.: 1967, *Ann. Astrophys.* **30**, 1059.

Danks, A.C.: 1970, *Astron. Astrophys.* **9**, 175.

Danks, A.C. and Meaburn, J.: 1971, *Astrophys. Space Sci.* **11**, 398.

Davies, R.D.: 1970, cited in Hjellming and Gordon, 1971.

Davies, R.D.: 1971, *Astrophys. J.* **163**, 479.

Davis, J.H., Blair, G.N., Van Till, H., and Thaddeus, P.: 1974, *Astrophys. J.* **190**, L117.

De Boer, J.A., Hin, A.C., Schwarz, U.J., and van Woerden, H.: 1968, *Bull. Astron. Inst. Neth.* **19**, 460.

Deharveng, L.: 1973, *Astron. Astrophys.* **29**, 341.

den Boggende, A.J.F., Mewe, R., Cronenschild, E.H.B.M., Heise, J., and Grindlay, J.E.: 1978, *Astron. Astrophys.* **62**, 1.

Dennison, B., Ward, D.B., Gull, G.E., and Harwit, M.: 1977, *Astron. J.* **82**, 39.

Dickinson, D.F.: 1972, *Astrophys. J.* **175**, L43.

Dickinson, D.F., Gottlieb, C.A., Gottlieb, E.W., and Litvak, M.M.: 1976, *Astrophys. J.* **206**, 79.

Dieter, N.H.: 1967, *Astrophys. J.* **150**, 435.

Doazan, V.: 1976, in A. Sletteback (ed.), 'Be and Shell Stars', *IAU Symp.* **70**, 37.

Doherty, L.H., Higgs, L.A., and MacLeod, J.M.: 1972 *Astrophys. Letters* **12**, 91.

Dopita, M.A.: 1972, *Astron. Astrophys.* **17**, 165.

Dopita, M.A.: 1973, *Astron. Astrophys.* **29**, 387.

Dopita, M.A.: 1974, *Astron. Astrophys.* **32**, 121.

Dopita, M.A.: 1978, *Astrophys. J. Suppl.* **37**, 117.

Dopita, M.A., Gibbons, A.H., and Meaburn, J.: 1973, *Astron. Astrophys.* **22**, 33.

Dopita, M.A., Dyson, J., and Meaburn, J.: 1974a, *Astrophys. Space Sci.* **28**, 61.

Dopita, M.A., Elliott, K.H., and Meaburn, J.: 1974b, *Astrophys. Space Sci.* **28**, 163.

Dopita, M.A., Isobe, S., and Meaburn, J.: 1975, *Astrophys. Space Sci.* **34**, 91.

Downes, D.: 1980, private communication.

Downes, D. and Genzel, R.: 1980, in B. Andrew (ed.), 'Interstellar Molecules', *IAU Symp.* **87**, 565.

Dravskikh, Z.V. and Dravskikh, A.F.: *Astron. Zh.* **44**, 35.

Dufour, R.J.: 1974, Ph.D. Thesis, University of Wisconsin, Madison.

Dufour, R.J. and Mathis, J.S.: 1975, *Publ. Astron. Soc. Pacific* **87**, 345.

Dupree, A.K.: 1974, *Astrophys. J.* **187**, 25.

Dyck, H.M. and Beichman, C.A.: 1974, *Astrophys. J.* **194**, 57.

Dyck, H.M. and Capps, R.W.: 1978, *Astrophys. J.* **220**, L49.

Dyck, H.M., Capps, R.W., Forrest, W.J., and Gillett, F.C.: 1973, *Astrophys. J.* **183**, L99.

Dyson, J.E.: 1968, *Astrophys. J.* **73**, 511.

Dyson, J.E. and Meaburn, J.: 1971, *Astron. Astrophys.* **12**, 219.

Eiroa, C., Elsässer, H., and Lahulla, J.F.: 1979, *Astron. Astrophys.* **74**, 89.

Elliott, K.H. and Meaburn, J.: 1970, *Astrophys. Space Sci.* **7**, 252.

Elliott, K.H. and Meaburn, J.: 1973a, *Nature* **244**, 69.

Elliott, K.H. and Meaburn, J.: 1973b, *Astron. Astrophys.* **27**, 367.

Elliott, K.H. and Meaburn, J.: 1974, *Astrophys. Space Sci.* **28**, 351.

Elliott, K.H., Goudis, C., Meaburn, J., and Pilkington, J.: 1978, *Astrophys. Space Sci.* **55**, 475.

Elmegreen, B.G. and Lada, C.J.: 1977, *Astrophys. J.* **214**, 725.

Elmegreen, D.M. and Elmegreen, B.G.: 1978, *Astrophys. J.* **218**, 510.

Elsässer, H. and Staude, H.J.: 1978, *Astron. Astrophys.* **70**, L3.

Emerson, J.P., Jennings, R.E., and Moorwood, A.F.M.: 1973, *Astrophys .J.* **184**, 401.

Erickson, E.F., Swift, C.D., Witteborn, F.C., Mord, A.J., Augason, G.C., Caroff, L.J., Kunz, L.W., and Giver, L.P.: 1973, *Astrophys. J.* **183**, 535.

Erickson, E.F., Strecker, D.W., Simpson, J.P., Goorvitch, D., Augason, G.C., Scargle, J.D., Caroff, L.H., and Witteborn, F.C.: 1977, *Astrophys. J.* **212**, 696.

Evans, N.J., II., Zuckerman, B., Sato, T., and Morris, G.: 1975, *Astrophys. J.* **199**, 383.

Evans, N.J., II., Plambeck, R.L., and Davis, J.H.; 1979, *Astrophys. J.* **227**, L25.

Fahrbach, U., Hofmann, W., Lemke, D., and Thum, C.: 1976, *Mitteilungen der Astronomischen Gesellschaft* **40**, 199.

Fallon, F.W., Gerda, H., and Sofia, S.: 1977, *Astrophys. J.* **217**, 719.

Fazio, G.G.: 1976, E.H. Avrett (ed.), *Frontiers of Astrophysics*, Harvard University Press, London, p. 203.

Fazio, G.G., Kleinmann, D.E., Noyes, R.W., Wright, E.L., Zeilik, M., II., and Low, F.J.: 1974, *Astrophys. J.* **192**, L23.

Fehrenbach, Ch.: 1977, *Astron. Astrophys. Suppl.* **29**, 71.

Fischel, D. and Feibelman, W.A.: 1973, *Astrophys. J.* **180**, 801.

Flather, E. and Osterbrock, D.E.: 1960, *Astrophys. J.* **132**, 18.

Fomalont, E.B. and Weliachew, L.: 1973, *Astrophys. J.* **181**, 781.

Forman, W., Jones, C., Cominsky, L., Julien, P., Murray, S., Peters, G., Tananbaum, H., and Giacconi, R.: 1978, *Astrophys. J. Suppl.* **38**, 357.

Forrest, W.J. and Soifer, B.T.: 1976 *Astrophys. J.* **208**, L129.

Forrest, W.J., Houck, J.R., and Reed, R.A.; 1976, *Astrophys. J.* **208**, L133.

Forster, J.R., Welch, W.J., Wright, M.C.H., and Baudry, A.: 1978, *Astrophys. J.* **221**, 137.

Foukal, P.V.: 1969, *Astrophys. Space Sci.* **5**, 469.

Foukal, P.V.: 1974, *Publ. Astron. Soc. Pacific* **86**, 211.

Fourikis, N., Takagi, K., and Saito, S.: 1977, *Astrophys. J.* **212**, L33.

Fox, K. and Jennings, D.E.: 1978, *Astrophys. J.* **226**, L47.

Frey, A., Lemke, D., Thum, C., and Fahrbach, V.: 1979, *Astron. Astrophys.* **74**. 133.

Fukui, Y. and Iguchi, T., 1977, *Publ. Astron. Soc. Japan* **29**, 63.

Furniss, I., Jennings, R.E., and Moorwood, A.F.M.: 1972, *Astrophys. J.* **176**, L105.

Gardner, F.F. and McGee, R.X.: 1967, *Nature* **213**, 480.

Gardner, F.F. and Morimoto, M.: 1968, *Australian J. Phys.* **21**, 881.

Garstang, R.H.: 1968, in D.E. Osterbrock and C.R. O'Dell (eds.), 'Planetary Nebulae', *IAU Symp.* **34**, 143.

Gatley, I., Becklin, E.E., Mathews, K., Neugebauer, G., Penston, M.V., and Scoville, N.: 1974, *Astrophys. J.* **191**, L121.

Gehrz, R.D., Hackwell, J.A., and Smith, J.R.: 1975, *Astrophys. J.* **202**, L33.

Genzel, R. and Downes, D.: 1977a, *Astron. Astrophys.* **61**, 117.

Genzel, R. and Downes, D.: 1977b, *Astron. Astrophys. Suppl.* **30**, 145.

Genzel, R. and Downes, D.: 1979 *Astron. Astrophys.* **72**, 234.

Genzel, R., Downes, D., Moran, J.M., Johnston, K.J., Spencer, J.H., Walker, R.C., Haschick, A.D., Matveyenko, L.I., Kogan, L.R., Kostenko, V.I., Rönnäng, B., Rydbeck, O.E.H., and Moiseev, I.G.: 1978, *Astron. Astrophys.* **66**, 13.

Genzel, R., Downes, D., Pauls, T., Wilson, T.L., and Bieging, J.: 1979a, *Astron. Astrophys.* **73**, 253.

Genzel, R., Moran, J.M., Lane, A.P., Predmore, C.R., Ho, P.T.P., Hansen, S.S., and Reid, M.J.: 1979b, *Astrophys. J.* **231**, L73.

Genzel, R., Downes, D., Schwartz, P.R., Spencer, J.H., Pankonin, V., and Baars, J.W.M.: 1980a, *Astrophys. J.* **239**, 519.

Genzel, R., Reid, M.J., Moran, J.M., and Downes, D.: 1980b, *Astrophys., J.* in press.

Georgelin, Y.M., Lortet-Zuckerman, M.C., and Monnet, G.: 1975, *Astron. Astrophys.* **42**, 273.

Gezari, D.Y., Joyce, R.R., Righini, G., and Simon, M.: 1974, *Astrophys. J.* **191**, L33.

Giacconi, R., Murray, S., Gursky, H., Kellogg, E., Schreier, E., and Tananbaum, H.: 1972, *Astrophys. J.* **178**, 281.

Giacconi, R., Murray, S., Gursky, H., Kellogg, E., Schreier, E., Matilsky, T., Koch, D., and Tananbaum, H.: 1974, *Astrophys. J. Suppl.* **27**, 37.

Gibbons, A.H.: 1976, *Monthly Notices Roy. Astron. Soc.* **174**, 105.

Gillespie, A.R. and White, G.J.: 1980, *Astron. Astrophys.*, in press.

Gillett, F.C. and Forrest, W.J.: 1973, *Astrophys. J.* **179**, 483.

Gillett, F.C., Jones, T.W., Merrill, K.M., and Stein, W.A.: 1975, *Astron. Astrophys.* **45**, 77.

Gilmore, W., Morris, M., Palmer, P., Zuckerman, B., and Turner, B.: 1976, cited in Hudson H.S. and Soifer, B.T., 1976.

Gilra, D.P.: 1972, *The Scientific Results from the OAO-2*, NASA SP-310, p. 295.

Gnedin, Yu.N. and Mitrofanov, I.G.: 1975, *Soviet Astron.* **19**, 673.

Goldberg, L.: 1966, *Astrophys. J.* **144**, 1225.

Goldberg, L.: 1968, Y. Terzian (ed.), *Interstellar Ionized Hydrogen*, W.A. Benjamin, New York, p. 373.

Goldreich, P. and Kwan, J.: 1974, *Astrophys. J.* **189**, 441.

Goldsmith, P.F., Plambeck, P.L., and Chiao, R.V.: 1975, *Astrophys. J.* **196**, L39.

Goldwire, H.E. Jr.: 1968, *Astrophys. J. Suppl.* **17**, 445.

Gol'nev, V. Ya., Lipovka, N.M., and Pariiskii, Yu. N.: 1966, *Soviet Astron.* **9**, 690.

Gomez, P.: 1981, Ph.D. Thesis, University of Madrid.

Gordon, C.P.: 1970, *Astron. J.* **75**, 914.

Gordon, M.A.: 1969a, *Astrophys. J.* **158**, 479.

Gordon, M.A.: 1969b, *Australian J. Phys.* **22**, 201.

Gordon, M.A.: 1970, *Astrophys. Letters* **6**, 27.

Gordon, M.A. and Meeks, M.L.: 1967, *Astrophys. J.* **149**, L21.

Gordon, M.A. and Meeks, M.L.: 1968, *Astrophys. J.* **152**, 417.

Goss, W.M.: 1968, *Astrophys. J. Suppl.* **15**, 131.

Goss, W.M. and Shaver, P.A.: 1970, *Australian J. Phys., Astrophys. Suppl.* **14**, 1.

Goss, W.M., Knowles, S.H., Balister, M., Batchelor, R.A., and Wellington, K.J.: 1976a, *Monthly Notices Roy. Astron. Soc.* **174**, 541.

Goss, W.M., Winnberg, A., Johansson, L.E.B., and Fournier, A.: 1976b, *Astron. Astrophys.* **46**, 1.

Gottlieb, C.A. and Ball, J.A.; 1973, *Astrophys. J.* **184**, L59.

Gottlieb, C.A., Lada, C.J., Gottlieb, E.W., Lilley, A.E., and Litvak, M.M.: 1975, *Astrophys. J.* **202**, 655.

Gottlieb, C.A., Gottlieb, E.W., Litvak, M.M., Ball. J.A., and Penfield, H.: 1978, *Astrophys. J.* **219**, 77.

Gottlieb, C.A., Ball, J.A., Gottlieb, E.W., and Dickinson, D.F.: 1979, *Astrophys. J.* **227**, 422.

Goudis, C.: 1975a, *Astrophys. Space Sci.* **35**, 409.

Goudis, C.: 1975b, *Astrophys. Space Sci.* **36**, 79.

Goudis, C.: 1975c, *Astrophys. Space Sci.* **36**, 105.

Goudis, C.: 1975d, *Astrophys. Space Sci.* **37**, 455.

Goudis, C.: 1975e, *Astrophys. Space Sci.* **38**, 13.

Goudis, C.: 1975f, *Astrophys. Space Sci.* **38**, 283.

Goudis, C.: 1976a, *Astrophys. Space Sci.* **39**, 273.

Goudis, C.: 1976b, *Astrophys. Space Sci.* **40**, 281.

Goudis, C.: 1976c, *Astrophys. Space Sci.* **41**, 105.

Goudis, C.: 1976d, *Astrophys. Space Sci.* **43**, 397.

Goudis, C.: 1976e, *Astrophys. Space Sci.* **44**, 281.

Goudis, C.: 1976f, *Astron. Astrophys.* **48**, 145.

Goudis, C.: 1977, *Astrophys. Space Sci.* **47**, 109.

Goudis, C.: 1978, *Astron. Astrophys.* **70**, 635.

Goudis, C.: 1979, *Astrophys. Space Sci.* **61**, 417.

Goudis, C. and Johnson, P.G.: 1978, *Astron. Astrophys.* **63**, 259.

Goudis, C. and Meaburn, J.: 1974, unpublished.

Goudis, C. and Meaburn, J.: 1978, *Astron. Astrophys.* **68**, 189.

Goudis, C. and White, N.: 1980a, *Astron. Astrophys.* **78**, 373.

Goudis, C. and White, N.: 1980b, *Astron. Astrophys.* **83**, 79.

Goudis, C. and White, N.: 1980c, *Astrophys. Space Sci.* **67**, 255.

Goudis, C., Johnson, P.G., and Meaburn, J.: 1977, unpublished.

Gow, C.E., Sanford, M.T., and Honeycutt, R.K.: 1978, *Astron. Astrophys.* **67**, 345.

Grachev, N.I.: 1970, *Astronomical Circular*, published by the Bureau for Astronomical Communications, Academy of Science, U.S.S.R., No. 578, p. 1.

Grandi, S.A.: 1975, *Astrophys. J.* **199**, L43.

Grasdalen, G.L.: 1974, *Astrophys. J.* **193**, 373.

Grasdalen, G.L.: 1976, *Astrophys. J.* **205**, L83.

Greenberg, J.M. and Hong, S.S.: 1974a in, F.J. Kerr and S.C. Simonson (eds.), 'Galactic Radio Astronomy', *IAU Symp.* **60**, 155.

Greenberg, J.M. and Hong, S.S.: 1974b in A.F.M. Moorwood (ed.), H II *Regions and the Galactic Centre*, Proc. 8th ESLAB Symposium, Frascati, Italy, p. 153.

Griem, H.R.: 1967, *Astrophys J.* **148**, 547.

Gudnov, V.M. and Sorochenko, R.L.: 1968, *Astron. Zh.* **11**, 805,

Gull, G.E., Houck, J.R., McCarthy, J.F., Forrest, W.J., and Harwit, M.: 1978, *Astron. J.* **83**, 1440.

Gull, T.R.: 1974a, in A.F.M. Moorwood (ed.), H II *Regions and the Galactic Centre*, Proc. 8th ESLAB Symposium, Frascati, Italy, p. 1.

Gull, T.R.: 1974b, cited in Gull, T.R., 1974a.

Habing, H.J. and Israel, F.P.: 1979, *Ann. Rev. Astron. Astrophys.* **17**, 345.

Hall, D.N.B., Kleinmann, S.G., Ridgway, S.T., and Gillett, F.C.: 1978, *Astrophys. J.* **223**, L47.

Hall, D.S. and Garrison, L.M.: 1969, *Publ. Astron. Soc. Pacific* **81**, 771.

Hansen, S.S., Moran, J.M., Reid, M.J., Johnson, K.J., Spencer, J.H., and Walker, R.C.: 1977, *Astrophys. J.* **218**, L65.

Haro, G.: 1950, *Astrophys. J.* **55**, 72.

Haro, G.: 1952, *Astrophys. J.* **115**, 572.

Harper, D.A.: 1974, *Astrophys. J.* **192**, 557.

Harper, D.A. and Low, F.J.: 1971, *Astrophys. J.* **165**, L9.

Harper, D.A., Low. F.J., Rieke, G.H., and Armstrong, K.R.: 1972, *Astrophys. J.* **177**, L21.

Harvey, P.M., Gatley, I., Werner, M.W., Elias, J.H., Evans II, N.J., Zuckerman, B., Morris, G., Sato, T., and Litvak, M.M.: *Astrophys. J.* **189**, L87.

Harwit, M., Churchwell, E., and Walmsley, M.: 1979, *Astrophys. Space Sci.* **66**, 487.

Haslam, C.G.T., Quigley, M.J.S., and Salter, C.J.: 1970, *Monthly Notices Roy. Astron. Soc.* **147**, 405.

Hawley, S.A.: 1978, *Astrophys. J.* **224**, 417.

Hayler, D.: 1968, *Astrophys. J.* **73**, 518.

Heiles, C.: 1976, *Astrophys. J.* **208**, L137.

Heiles, C. and Habing, H.J.: 1974, *Astron. Astrophys. Suppl.* **14**, 127.

Heiles, C. and Jenkins, H.B.: 1976, *Astron. Astrophys.* **46**, 333.

Herbig, G.H.: 1951, *Astrophys. J.* **113**, 697.

Herbig, G.H.: 1969, L. Detre (ed.), *Non-Periodic Phenomena in Variable Stars*, D. Reidel Publ. Co., Dordrecht, Holland, p. 75.

Hills, R., Janssen, M.A., Thornton, D.D., and Welch, W.J.: 1972, *Astrophys. J.* **175**, L59.

Hills, R., Pankonin, V., and Landecker, T.L.: 1975, *Astron. Astrophys.* **39**, 149.

Hippelein, H. and Münch, G.: 1978, *Astron. Astropnys.* **68**, L7.

Hjellming, R.M. and Churchwell, E.: 1969, *Astrophys. Letters* **4**, 165.

Hjellming, R.M. and Davies, R.D.: 1970, *Astron. Astrophys.* **5**, 53.

Hjellming, R.M. and Gordon, M.A.: 1971, *Astrophys. J.* **164**, 47.

Hjellming, R.M., Andrews, M.H., and Sejnowski, T.J.: 1969, *Astrophys. Letters* **3**, 111.

Ho, P.T.P. and Barrett, A.H.: 1978, *Astrophys. J.* **224**, L23.

Ho, P.T.P., Barrett, A.H., Myers, P.C., Matsakis, D.N., Cheung, A.C., Chui, M.F., Townes, C.H., and Yngvesson, K.S.: 1979, *Astrophys. J.* **234**, 912.

Hocking, W.H., Winnewisser, G., Churchwell, E., and Percival, J.: 1979, *Astron. Astrophys.* **75**, 268.

Houck, J.R., Schaack D.F., and Reed, R.A.: 1974, *Astrophys. J.* **193**, L139.

Hoyle, F. and Wickramasinghe, N.C.: 1962, *Monthly Notices Roy. Astron. Soc.* **124**, 417.

Hua, C.T.: 1974, *Astron. Astrophys.* **32**, 423.

Hubble, E.P.: 1922, *Astrophys. J.* **56**, 162.

Hudson, H.S. and Soifer, B.T.: 1976, *Astrophys. J.* **206**, 100.

Huggins, P.J., Phillips, T.G., Neugebauer, G., Werner, M.W., and Wannier, P.G.: 1977, *Astrophys. J.* **227**, 441.

Icke, V., Gatley, I., and Israel, F.P.: 1980, *Astrophys. J.* **236**, 808.

Ishida, K. and Kawajiri, N.: 1968. *Publ. Astron. Soc. Japan* **20**, 95.

Isobe, S.: 1970, *Publ. Astron. Soc. Japan* **22**, 429.

Iosbe, S.: 1971, *Publ. Astron. Soc. Japan* **23**, 371.

Isobe, S.: 1973, in J.M. Greenberg and H.C. van de Hulst (eds.), 'Interstellar Dust and Related Topics', *IAU Symp.* **52**, 433.

Isobe, S.: 1978, *Publ. Astron. Soc. Japan* **30**, 499.

Isobe, S. and Kurihara, H.: 1970, *Tokyo Astron. Bull.*, No. 203.

Israel, F.P.: 1978, *Astron. Astrophys.* **70**, 769.

Israel, F.P., Habing, H.J., and de Jong, T.: 1973. *Astron. Astrophys.* **27**, 143.

Jaffe, D.T. and Pankonin, V.: 1978, *Astrophys. J.* **226**, 869.

Jefferts, K.B., Penzias, A.A., and Wilson, R.W.: 1970, *Astrophys. J.* **161**, L87.

Jennings, D.E. and Fox, K.: 1979, *Astrophys. J.* **227**, 433.

Johnson, D.R., Lovas, F.J., Gottlieb, C.A., Gottlieb, E.W., Litvak, M.M., Gnelin, M., and Thaddeus, P.: 1977, *Astrophys. J.* **218**, 370.

Johnson, H.L.: 1965, *Astrophys. J.* **141**, 923.

Johnson, H.L.: 1967, *Astrophys. J.* **150**, L39.

Johnson, H.L.: 1968, in B.M. Middlehurst (ed.), *Stars and Stellar Systems*, Vol. 7, University of Chicago Press, Chicago, p. 167.

Johnson, H.L. and Hiltner, W.A.: 1956, *Astrophys. J.* **123**, 267.

Johnson, H.L. and Mendoza, E.E.: 1964, *Bol. Obs. Tonantzintla y Tacubaya* **3**, 331.

Johnson, H.M.: 1953, *Astrophys. J.* **118**, 370.

Johnson, H.M.: 1965, *Astrophys. J.* **142**, 964.

Johnson, P.G.: unpublished.

Johnston, K.J. and Hobbs, R.W.: 1969, *Astrophys. J.* **158**, 145.

Johnston, K.J., Knowles, S.H., Moran, J.M., Burke B.F., Lo., K-Y., Papadopoulos, G.D., Read, R.B., and Hardebeck, E.G.: 1977, *Astron. J.* **82**, 403.

Joyce, R.R., Gezari, D.Y., Scoville, N.Z., and Furenlid, I.: 1978a, *Astrophys. J.* **219**, L29.

Joyce, R.R., Simon, M., and Simon, T.: 1978b, *Astrophys. J.* **220**, 156.

Kahn, F.D.: 1954, *Bull. Astron. Inst. Neth.* **12**, 187.

Kaifu, N., Akabane, K., and Morimoto, M.: 1973, *Publ. Astron. Soc. Japan* **25**, 129.

Kaifu, N., Morimoto, M., Nagane, K., Akabane, K., Iguchi, T., and Takagi, R.: 1974, *Astrophys. J.* **191**, L135.

Kaifu, N., Iguchi, T., and Morimoto, M.: 1975, *Astrophys. J.* **196**, 719.

Kaler, J.B.: 1967a, *Astrophys. J.* **148**, 925.

Kaler, J.B.: 1967b, in Y. Terzian (ed.), *Proc. AIO-NRAO Symposium on* H II *Regions*.

Kaler, J.B.: 1976, *Astrophys. J. Suppl.* **31**, 517.

Kaler, J.B., Aller, L.H., and Bowen, I.S.: 1965, *Astrophys. J.* **141**, 912.

Kaler, J.B., Lien, D.J., and Peck, M.L.: 1979, *Astrophys. J.* **234**, 909.

Kardashev, N.S.: 1959, *Astron. Zh.* **36**, 838.

Kazès, I., Le Squeren, A.M., and Gadéa, F.: 1975, *Astron. Astrophys.* **42**, 9.

Khalesse, B., Pallister, W.S., Warren-Smith, R.F., and Scarrott, S.M.: 1980, *Monthly Notices Roy. Astron. Soc.* **190**, 99.

Kleinmann, D.E. and Low, F.J.: 1967, *Astrophys. J.* **149**, L1.

Klemperer, W.: 1970, *Nature* **227**, 1230.

Knacke, R.F. and Capps, R.W.: 1979, *Astron. J.* **84**, 1705.

Knacke, R.F. and Thompson, R.K.: 1973, *Publ. Astron. Soc. Pacific* **85**, 341.

Knapp, G.R., Kuiper, T.B.H., Knapp, S.L., and Brown, R.L.: 1977, *Astrophys. J.* **214**, 78.

Krishna-Swamy, K.S. and O'Dell, C.R.: 1967, *Astrophys. J.* **147**, 529.

Krügel, E. and Tenorio-Tagle, G.: 1978, *Astron. Astrophys.* **70**, 51.

Ku, W., H.-M. and Chanan, G.A.: 1979, *Astrophys. J.* **234**, L59.

Kuiper, T.B.H. and Evans, N.J., II.: 1978, *Astrophys. J.* **219**, 141.

Kutner, M.L.: 1972, Ph.D. Thesis, Columbia University.

Kutner, M.L. and Thaddeus, P.: 1971, *Astrophys. J.* **168**, L67.

Kutner, M.L., Thaddeus, P., Jefferts, K.B., Penzias, A.A., and Wilson, R.W.: 1971, *Astrophys. J.* **164**, L49.

Kutner, M.L., Thaddeus P., Penzias, A.A., Wilson, R.W., and Jefferts, K.B.: 1973, *Astrophys. J.* **183**, L27.

Kutner, M.L., Evans, N.J., II., and Tucker, K.D.: 1976, *Astrophys. J.* **209**, 452.

Kutner, M.L., Tucker, K.D., Chin, G., and Thaddeus, P.: 1977, *Astrophys. J.* **215**, 521.

Kwan, J. and Scoville, N.: 1976, *Astrophys. J.* **210**, L39.

Lada, C.J., Blitz, L., and Elmegreen, B.G.: 1978a, in T. Gehrels (ed.), *Protostars and Planets*, University of Arizona Press, Tucson, Arizona, p. 341.

Lada, C.J., Oppenheimer, M., and Hartquist, T.W.: 1978b, *Astrophys. J.* **226**, L153.

Lambert, D.L.: 1968, *Monthly Notices Roy. Astron. Soc.* **138**, 143.

Lambert, D.L.: 1978, *Monthly Notices Roy. Astron. Soc.* **182**, 593.

Lambert, D.L. and Warner, B.: 1968, *Monthly Notices Roy. Astron. Soc.* **138**, 181.

Lang, K.: 1978, *Astrophysical Formulae*, Springer-Verlag, p. 183.

Laques, P. and Vidal, J.L.: 1979, *Astron. Astrophys.* **73**, 97.

Lee, P.: 1969, *Astrophys. J.* **157**, L111.

Lee, T.A.: 1968, *Astrophys. J.* **152**, 913.

Leibowitz, E.M.: 1973, *Astrophys. J.* **181**, 369.

Lemke, D.: 1974, in A.F.M. Moorwood (ed.), H II *Regions and the Galactic Centre*, 8th ESLAB Symposium, Frascati, Italy, p. 53.

Lemke, D. and Harris, A.W.: 1979, *Naturwissenschaften* **66**, 73.

Lemke, D. and Harris, A.W.: 1981, *Astron Astrophys.* **99**, 285.

Lemke, D., Low, F.J., and Thum, C.: 1974, *Astron. Astrophys.* **32**, 231.

Lester, D.F., Dinerstein, H.L., and Rank, D.M.: 1979, *Astrophys. J.* **232**, 139.

Levato, H. and Abt. H.A.: 1976, *Publ. Astron. Soc. Pacific* **88**, 712.

Lichten, S.M., Rodriguez, L.F., and Chaisson, E.J.: 1979, *Astrophys. J.* **229**, 524.

Lilley, A.E., Menzel, D.H., Penfield, H., and Zuckerman, B.: 1966, *Nature* **209**, 468.

Linke, R.A. and Goldsmith, P.F.: 1980, *Astrophys. J.* **235**, 437.

Liszt, H.S.: 1978, *Astrophys. J.* **219**, 454.

Liszt, H.S. and Linke, R.A.: 1975, *Astrophys. J.* **196**, 709.

Liszt, H.S., Wilson, W.R., Penzias, A.A., Jefferts, K.B., Wannier, P.G., and Salomon, P.M.: 1974, *Astrophys. J.* **190**, 557.

Little, L.T., White, G.J., and Riley, P.W.: 1977, *Monthly Notices Roy. Astron. Soc.* **180**, 639.

Lockhart, I.A. and Goss, W.M.: 1978, *Astron. Astrophys.* **67**, 355.

Lockman, F.J. and Brown, R.L.: 1975, *Astrophys. J.* **201**, 134.

Löbert, W. and Goss W.M.: 1978, *Monthly Notices Roy. Astron. Soc.* **183**, 119.

Loer, S.J., Allen, D.A., and Dyck, H.M.: 1973, *Astrophys. J.* **183**, L97.

Lohsen, E.: 1975, IAU Information Bull. Variable Stars, No. 988.

Lonsdale, C.J., Dyck, H.M., Capps, R.W., and Wolstencroft, R.D.: 1980, *Astrophys. J.* **238**, L31.

Loren, R.B.: 1979, *Astrophys. J.* **234**, L207.

Lovas, F.J., Johnson, D.R., Buhl, D., and Snyder, L.E.: 1976, *Astrophys. J.* **209**, 770.

Low, F.J.: 1972, in B.T. Lynds (ed.), *Dark Nebulae, Globules and Protostars*, Univ. of Arizona, Tuscon, Arizona, Chapter 11, p. 116.

Low, F.J. and Aumann, H.H.: 1970, *Astrophys. J.* **162**, L79.

Low, F.J., Johnson, H.L., Kleinmann, D.E., Latham, A.S., and Geisel, S.L.: 1970, *Astrophys. J.* **160**, 531.

Low, F.J., Rieke, G.H., and Armstrong, K.R.: 1973, *Astrophys. J.* **183**, L105.

Lowe, R.P., Moorhead, J.M., and Wehlau, W.H.: 1977, *Astrophys. J.* **214**, 712.

Lucas, R., Encrenaz, P.J., and Falgarone, E.G.: 1977, *Astron. Astrophys.* **51**, 469.

Lynds, B.T.: 1962, *Astrophys. J. Suppl.* **7**, 1.

MacLeod, J.M. and Doherty, L.H.: 1968, *Astrophys. J.* **154**, 833.

MacLeod, J.M., Doherty, L.H., and Higgs, L.A.: 1975, *Astron. Astrophys.* **42**, 195.

Malin, D.F.: 1979, *Mercury* **8**, No. 4, 89.

Manchester, R.N. and Gordon, M.A.: 1971, *Astrophys. J.* **169**, 507.

Manchester, R.N., Robinson, B.J., and Goss, W.M.: 1970, *Australian J. Phys.* **23**, 751.

Martin, A.H.M. and Gull, S.F.: 1976, *Monthly Notices Roy. Astron. Soc.* **175**, 235.

Mathews, W.G.: 1967, *Astrophys. J.* **147**, 965.

Mathews, W.G.: 1969, *Astrophys. J.* **157**, 583.

Mathis, J.S.: 1957, *Astrophys. J.* **125**, 318.

Mathis, J.S.: 1962, *Astrophys. J.* **136**, 374.

Mathis, J.S.: 1971, *Astrophys. J.* **167**, 261.

Matsakis, D.N., Cheung, A.C., Wright, M.C.H., Askne, J.I.H., Townes, C.H., and Welch, W.J.: 1980, *Astrophys. J.* **236**, 481.

McCall, M.L.: 1979, *Astrophys. J.* **229**, 962.

McCaroll, R. and Binh, D.H.: 1968, *Ann. Astrophys.* **31**, 123.

McCarthy, J.F., Forrest, W.J., and Houck, J.R.: 1979, *Astrophys. J.* **231**, 711.

McGee, R.X. and Gardner, F.F.: 1967, *Nature* **213**, 579.

McGee, R.X. and Gardner, F.F.: 1968, *Australian J. Phys.* **21**, 149.

Meaburn, J.: 1967, *Z. Astrophys.* **65**, 93.

Meaburn, J.: 1970, *Nature* **228**, 44.

Meaburn, J.: 1971, *Astrophys. Space Sci.* **13**, 110.

Meaburn, J.: 1975, in T.L. Wilson and D. Downes (eds.), H II *Regions and Related Topics*, Springer-Verlag, Heidelberg, New York, p. 222.

Meaburn, J.: 1978, *Astrophys. Space Sci.* **59**, 193.

Melnick, G., Gull, G.E., and Harwit, M.: 1979a, *Astrophys. J.* **227**, L29.

Melnick, G., Gull, G.E., and Harwit, M.: 1979b, *Astrophys. J.* **227**, L35.

Mendez, M.E.: 1967, *Bol. Obs. Tonantzintla y Tacubaya* **4**, 91.

Mendez, M.E.: 1968, *Bol. Obs. Tonantzintla y Tacubaya* **4**, 240.

Menon, T.K.: 1958, *Astrophys. J.* **127**, 28.

Menon, T.K.: 1967, in H. van Woerden (ed.), 'Radio Astronomy and the Galactic System', *IAU Symp.* **31**, 121.

Menon, T.K.: 1970, *Astron. Astrophys.* **5**, 240.

Menon, T.K. and Payne, J.: 1969, *Astrophys. Letters* **3**, 25.

Mezger, P.G.: 1968, in Y. Terzian (ed.), *Interstellar Ionized Hydrogen*, W.A. Benjamin, New York, p. 477.

Mezger, P.G.: 1972, in N.C. Wickramasinghe, F.D. Kahn, and P.G. Mezger (eds.), *Interstellar Matter*, edited by the Astronomical Institute, University of Basel, Saas-Fee, p. 1.

Mezger, P.G. and Ellis, S.A.: 1968, *Astrophys. Letters* **1**, 159.

Mezger, P.G. and Henderson, A.P.: 1967, *Astrophys. J.* **147**, 471.

Mezger, P.G. and Höglund, B.: 1967, *Astrophys. J.* **147**, 490.

Mezger, P.G. and Palmer, P.: 1968, *Science* **160**, 29.

Mezger, P.G., Schraml, J., and Terzian, Y.: 1967, *Astrophys. J.* **150**, 807.

Mezger, P.G., Wilson, T.L., Gardner, F.F., and Milne, D.K.: 1970, *Astrophys. Letters* **6**, 35.

Mezger, P.G., Smith, L.F., and Churchwell, E.: 1974, *Astron. Astrophys.* **32**, 269.

Miller, J.: 1968, *Astrophys. J.* **151**, 473.

Mills, B.Y.: 1967, *IAU Report*, Prague.

Mills, B.Y. and Shaver, P.A.: 1968, *Australian J. Phys.* **21**, 95.

Milman, A.S., Knapp, G.R., Kerr, F.J., Knapp, S.L., and Wilson, W.J.: 1975, *Astron. J.* **80**, 93.

Moran, J.M.: 1980, *CRC Handbook on Laser Science and Technology*, Vol. 1, 'Lasers in All Media', (preprint).

Moran, J.M., Johnston, K.J., Spencer, J.H., and Schwartz, P.R.: 1977, *Astrophys. J.* **217**, 434.

Moran, W.W. and Keenan, P.C.: 1973, *Ann. Rev. Astron. Astrophys.* **11**, 29.

Morris, M. and Knapp, G.R.: 1976, *Astrophys. J.* **204**, 415.

Morris, M., Zuckerman, B., Palmer, P., and Turner, B.E.: 1973, *Astrophys. J.* **186**, 501.

Morris, M., Zuckerman, B., Turner, B.E., and Palmer, P.: 1974, *Astrophys. J.* **192**, L27.

Morris, M., Snell, R.L., and Van den Bout, P.: 1977, *Astrophys. J.* **216**, 738.

Morris, M., Palmer, P., and Zuckerman, B.: 1980, *Astrophys. J.* **237**, 1.

Münch, G.: 1958, *Rev. Mod. Phys.* **30**, L1035.

Münch, G.: 1968, in Y. Terzian (ed.), *Interstellar Ionized Hydrogen*, W.A. Benjamin, New York, p. 507.

Münch, G.: 1977, *Astrophys. J.* **212**, L77.

Münch, G.: 1980, private communication.

Münch, G. and Persson, S.E.: 1971, *Astrophys. J.* **165**, 241.

Münch, G. and Taylor, K.: 1974, *Astrophys. J.* **192**, L93.

Münch, G. and Wilson, O.C.: 1962, *Z. Astrophys.* **56**, 127.

Myers, P.C. and Buxton, R.B.: 1980, *Astrophys. J.* **239**, 515.

Nadeau, D. and Geballe, T.R.: 1979, *Astrophys. J.* **230**, L169.

Naranan, S, Shulman, S., Friedman, H., and Fritz, G.: 1976, *Astrophys. J.* **208**, 718.

Ney, E.P. and Allen, D.A.: 1969, *Astrophys. J.* **155**, L193.

Ney, E.P., Strecker, D.W., and Gehrz, R.D.: *Astrophys. J.* **180**, 809.

Norris, R.P., Booth, R.S., and McLaughlin, W.: 1980, in P.M. Solomon and M.G. Edmunds (eds.), *Giant Molecular Clouds in the Galaxy*, Pergamon Press, p. 193.

Nousek, J.: 1978, Ph.D. Thesis, University of Wisconsin, Madison.

O'Dell, C.R.: 1975, cited in Sarazin, C.L., 1976.

O'Dell, C.R., Peimbert, M., and Kinman, T.D.: 1964, *Astrophys. J.* **140**, 119.

O'Dell, C.R., Hubbard, W.B., and Peimbert, M.: 1966, *Astrophys. J.* **143**, 743.

O'Dell, C.R., York, D.G., and Henize, K.G.: 1967, *Astrophys. J.* **150**, 835.

Ögelman, H.B. and Maran, S.P.: 1976, *Astrophys. J.* **209**, 124.

Öpik, E.: 1953, *Irish Astron. J.* **2**, 219.

Ogden, P.M., Roesler, F.L., Reynolds, R.J., Scherb, F., Larson, H.P., Smith, H.A., and Daehler, M.: 1978, *Astrophys. J.* **226**, L91.

Oke, J.B. and Schild, R.E.: 1970, *Astrophys. J.* **161**, 1015.

Olthof, H. and Pottasch, S.R.: 1975, *Astron. Astrophys.* **43**, 291.

Olthof, H., Cosmovici, C.B., and Canton, G.: 1978, *Astron. Astrophys.* **69**, 219.

Oort, J.H.: 1954, *Bul. Astron. Inst. Neth.* **12**, 177.

Oster, L.: 1961, *Rev. Mod. Phys.* **33**, 525.

Osterbrock, D.E.: 1974 *Astrophysics of Gaseous Nebulae*, W.H. Freeman, San Francisco.

Osterbrock, D.E. and Flather, E.: 1959, *Astrophys. J.* **129**, 26.

Padman, R., Hills, R.E., Cronin, N.J., and Rose, W.B.: 1980, *Monthly Notices Roy. Astron. Soc.* **192**, 87.

Pallister, W.S., Perkins, H.G., Scarrott, S.M., Bingham, R.G., and Pilkington, J.D.H.: 1977, *Monthly Notices Roy. Astron. Soc.* **178**, 93 pp.

Palmer, P.: 1968, Ph.D. Thesis, Harvard University.

Palmer, P. and Zuckerman, B.: 1966, *Nature* **209**, 1118.

Palmer, P., Zuckerman, B., Penfield, H., Lilley, A.E., and Mezger, P.G.: 1967, *Nature* **215**, 40.

Palmer, P., Zuckerman, B., Penfield, H., Lilley, A.E., and Mezger, P.G.: 1969, *Astrophys. J.* **156**, 887.

Panagia, N.: 1973, *Astron. J.* **78**, 929.

Panagia, N.: 1977, in G.G. Fazio (ed.), *Infrared and Submillimeter Astronomy*, D. Reidel Publ. Co., Dordrecht, Holland, p. 43.

Pankonin, V.: 1980, in P.A. Shaver (ed.), *Radio Recombination Lines*, D. Reidel Publ. Co., Dordrecht, Holland, p. 111.

Pankonin, V., Walmsley, C.M., Wilson, T.L., and Thomasson, P.: 1977, *Astron. Astrophys.* **57**, 341.

Pankonin, V., Walmsley, C.M., and Harwit, M.: 1979, *Astron. Astrophys.* **75**, 34.

Pariiskii, Yu., N.: 1961a, *Izv. Gl. Astron. Obs. v Pulkove* **21**, 54.

Pariiskii, Yu., N.: 1961b, *Astron. Zh.* **38**, 798.

Park, W.M., Vickers, D.G., and Clegg, P.E.: 1970, *Astron. Astrophys.* **5**, 325.

Pauls, T. and Wilson, T.L.: 1977. *Astron. Astrophys.* **60**, L31.

Pedlar, A.: 1970, *Nature* **226**, 830.

Pedlar, A. and Davies, R.D.: 1972, *Monthly Notices Roy. Astron. Soc.* **159**, 129.

Pedlar, A. and Hart, L.: 1974, *Monthly Notices Roy. Astron. Soc.* **168**, 577.

Peimbert, M.: 1967, *Astrophys. J.* **150**, 825.

Peimbert, M. and Costero, R.: 1969, *Bol. Obs. Tonantzintla y Tacubaya* **5**, 3.

Peimbert, M. and Goldsmith, D.W.: 1972, *Astron. Astrophys.* **19**, 398.

Peimbert, M. and Torres-Peimbert, S.: 1971, *Astrophys. J.* **168**, 413.

Peimbert, M. and Torres-Peimbert, S.: 1974, *Astrophys. J.* **193**, 327.

Peimbert, M. and Torres-Peimbert, S.: 1977, *Monthly Notices Roy. Astron. Soc.* **179**, 217.

Pengelly, R.M.: 1964, *Monthly Notices Roy. Astron. Soc.* **127**, 145.

Penston, M.V.: 1972, cited in Ney, E.P. *et al.*, 1973.

Penston, M.V.: 1973, *Astrophys. J.* **183**, 505.

Penston, M.V., Allen, D.A., and Hyland, A.R.: 1971, *Astrophys. J.* **170**, L33.

Penston, M.V., Hunter, J.K., and O'Neill, A.: 1975, *Monthly Notices Roy. Astron. Soc.* **171**, 219.

Penzias, A.A.: 1976, cited in Cronin, N.J. *et al.*, 1976.

Penzias, A.A., Wilson, R.W., and Jefferts, K.B.: 1974, *Phys. Rev. Letters* **32**, 701.

Pequignot, D., Aldrovandi, S.M.V., and Stasinka, G.: 1977, *Astron. Astrophys.* **58**, 411.

Perinotto, M. and Patriarchi, P.: 1974, in A.F.M. Moorwood (ed.), H II *Regions and the Galactic Centre*, Proc. 8th ESLAB Symposium, Frascati, Italy, p. 191.

Perinotto, M. and Patriarchi, P.: 1980a, *Astrophys. J.* **235**, L13.

Perinotto, M. and Patriarchi, P.: 1980b, *Astrophys. J.* **238**, 614.

Perrenod, S.C., Schields, G.A., and Chaisson, E.J.: 1977, *Astrophys. J.* **216**, 427.

Petrosian, V.: 1970, *Astrophys. J.* **159**, 833.

Petrosian, V., Silk, J., and Field, G.B.: 1972, *Astrophys. J.* **177**, L69.

Phillips, T.G., Jefferts, K.B., and Wannier, P.G.: 1973, *Astrophys. J.* **186**, L19.

Phillips, T.G., Jefferts, K.B., Wannier, P.G., and Ade, P.A.R.: 1974, *Astrophys. J.* **191**, L31.

Phillips, T.G., Huggins, P.J., Neugebauer, G., and Werner, M.W.: 1977, *Astrophys. J.* **217**, L161.

Phillips, T.G., Scoville, N.Z., Kwan, J., Huggins, P.J., and Wannier, P.G.: 1978, *Astrophys. J.* **222**, L59.

Phillips, T.G., Huggins, P.J., Wannier, P.G., and Scoville, N.Z.: 1979, *Astrophys. J.* **231**, 720.

Phillips, T.G., Huggins, P.J., Kuiper, T.B.H., and Miller, R.E.: 1980, *Astrophys. J.* **238**, L103.

Pipher, J.L., Graeme Duthie, J., and Savedoff, M.P.: 1978, *Astrophys. J.* **219**, 494.

Plambeck, R.L. and Williams, D.R.W.: 1979, *Astrophys. J.* **227**, L43.

Popper, D.M. and Plavec, M.: 1976, *Astrophys. J.* **205**, 462.

Pottasch, S.R.: 1965, in A.Beer (ed.), *Vistas in Astronomy*, Pergamon Press, London, Vol. 6, p. 149.

Pradhan, A.K.: 1978, *Monthly Notices Roy. Astron. Soc.* **184**, 89 pp.

Pronik, V.I.: 1957, *Izv. Krimsk. Astrophys. Obs.* **17**, 14.

Radhakrishnan, V., Brooks, J.W., Goss, W.M., Lockhart, P., Murray, J.D., Schwartz, U.J., and Whittle, R.P.J.: 1972a, *Astrophys. J. Suppl.* **24**, 1.

Radhakrishnan, V., Goss, W.M., Murray, J.D., and Brooks, J.W.: 1972b, *Astrophys. J. Suppl.* **24**, 49.

Raimond, E. and Eliasson, B.: 1969, *Astrophys. J.* **155**, 817.

Reeves, H.: 1978, in T. Gehrels (ed.), *Protostars and Planets*, University of Arizona Press, Tucson, Arizona, p. 399.

Reich, W.: 1978, *Astron. Astrophys.* **64**, 407.

Reifenstein, E.C., III., Wilson, T.L., Burke, B.F., Mezger, P.G., and Altenhoff, W.J.: 1970, *Astron. Astrophys.* **4**, 357.

Reynolds, R.J. and Ogden, P.M.: 1979, *Astrophys. J.* **229**, 942.

Reynolds, R.J., Roesler, F.L., and Scherb, F.: 1974, *Astrophys. J.* **192**, L56.

Rickard, L.J., Zuckerman, B., Palmer, P., and Turner, B.E.: 1977, *Astrophys. J.* **218**, 659.

Rieke, G.H., Low, F.J., and Kleinmann, D.E.: 1973, *Astrophys. J.* **186**, L7.

Risbeth, H.: 1958, *Monthly Notices Roy. Astron. Soc.* **118**, 591.

Rodriguez, L.F. and Chaisson, E.J.: 1978, *Astrophys. J.* **221**, 816.

Russell, R.W., Soifer, B.T., and Merrill, K.M.: 1977a, *Astrophys. J.* **213**, 66.

Russell, R.W., Soifer, B.T., and Puetter, R.C.: 1977b, *Astron. Astrophys.* **54**, 959.

Rydbeck, O.E.H., Elldér, J., Irvine, W.M., Sume, A., and Hjalmarson, A.: 1974, *Astron. Astrophys.* **33**, 315.

Rydbeck, O.E.H., Kollberg, E., Hjalmarson, A., Sume, A., Elldér, J., and Irvine, W.M.: 1975, Research Report No. 120, Research Laboratory of Electronics and Onsala Space Observatory, Chalmers University of Technology, Gothenburg, Sweden.

Rydbeck, O.E.H., Kollberg, E., Hjalmarson, A., Sume, A., Elldér, J., and Irvine, W.M.: 1976, *Astrophys. J. Suppl.* **31**, 333.

Rydbeck, O.E.H., Irvine, W.M., Hjalmarson, A., Rydbeck, G. Elldér, J., and Kollberg, E.: 1980, *Astrophys. J.* **235**, L171.

Salpeter, E.E.: 1977, *Ann. Rev. Astron. Astrophys.* **15**, 267.

Salter, C.J.: 1970, Ph.D. Thesis, University of Manchester.

Sancisi, R.: 1970, *Evolution Stellaire avant la Sequence Principale*, Societé Royal de Liege, No. XIX, p. 313.

Saraph, H.E. and Seaton, M.J.: 1970, *Monthly Notices Roy. Astron. Soc.* **148**, 367.

Sarazin, C.L.: 1976, *Astrophys. J.* **204**, 68.

Schiffer, F.H., III. and Mathis, J.S.: 1974, *Astrophys. J.* **194**, 597.

Schild, R.E. and Chaffee, F.: 1971, *Astrophys. J.* **169**, 529.

Schmitter, E.F.: 1971, *Astron. J.* **76**, 571.

Schraml, J. and Mezger, P.G.: 1969, *Astrophys. J.* **156**, 269.

Schwartz, P.R., Cheung, A.C., Bologna, J.M., Chui, M.F., Waak, J.A., and Matsakis, D.: 1977, *Astrophys. J.* **218**, 671.

Schwartz, R.D.: 1975, *Astrophys. J.* **195**, 631.

Scoville, N.Z.: 1980, in P.M. Solomon and M.G. Edmunds (eds.), *Giant Molecular Clouds in the Galaxy*, Pergamon Press, p. 147.

Scoville, N.Z. and Kwan, J.: 1976, *Astrophys. J.* **206**, 718.

Scoville, N.Z. and Kwan, J.: 1977, in G.G. Fazio (ed.), *Infrared and Submillimeter Astronomy*, D. Reidel Publ. Co., Dordrecht, Holland, p. 77.

Scoville, N.Z. and Wannier, P.G.: 1979, *Astron. Astrophys.* **76**, 140.

Scoville, N.Z., Hall, D.N.B., Kleinmann, S.G., and Ridgway, S.T.: 1979, *Astrophys. J.* **232**, L121.

Seaton, M.J.: 1954, *Monthly Notices Roy. Astron. Soc.* **114**, 154.

Seaton, M.J.: 1975, *Monthly Notices Roy. Astron. Soc.* **170**, 475.

Sejnowski, T.J. and Hjellming, R.M.: 1969, *Astrophys. J.* **156**, 915.

Serkowski, K. and Rieke, G.H.: 1973, *Astrophys. J.* **183**, L101.

Sharpless, S.: 1952, *Astrophys. J.* **116**, 251.

Shaver, P.A.: 1969, *Monthly Notices Roy. Astron. Soc.* **142**, 273.

Shaver, P.A.: 1980, *Astron. Astrophys.*, **90**, 34.

Shaver, P.A. and Goss, W.M.: 1970a, *Australian J. Phys., Astrophys. Suppl.* **14**, 77.

Shaver, P.A. and Goss, W.M.: 1970b, *Australian J. Phys., Astrophys. Suppl.* **14**, 133.

Shcheglov, P.V.: 1968a, in D.E. Osterbrock and C.R. O'Dell (eds.), 'Planetary Nebulae', *IAU Symp.* **34**, 270.

Shcheglov. P.V.: 1968b, *Astrophys. Letters* **1**, 145.

Shklovsky, I.S.: 1968, *Supernovae*, Wiley-Interscience, New York.

Shull, J.M.: 1979, *Astrophys. J.* **233**, 182.

Simpson, J.P.: 1973, *Publ. Astron. Soc. Pacific* **85**, 479.

Sivan, J.P.: 1974, *Astron. Astrophys. Suppl.* **16**, 163.

Smith, H.A., Larson, H.P., and Fink, V.: 1979, *Astrophys. J.* **233**, 132.

Smith, J., Lynch, D.K., Cudaback, D., and Werner, M.W.: 1979, *Astrophys. J.* **234**, 902.

Smith, M.G. and Weedman, D.W.: 1970, *Astrophys. J.* **160**, 65.

Snell, R.L. and Wootten, A.H.: 1977, *Astrophys. J.* **216**, L111.

Snell, R.L. and Wootten, A.H.: 1979, *Astrophys. J.* **228**, 748.

Snow, T.P. and Morton, D.C.: 1976, *Astrophys. J. Suppl.* **32**, 429.

Snyder, L.E. and Buhl, D.: 1971, *Astrophys. Letters* **163**, L47.

Snyder, L.E. and Buhl, D.: 1973, *Astrophys. J.* **185**, L79.

Snyder, L.E. and Buhl, D.: 1974, *Astrophys. J.* **189**, L31.

Snyder, L.E. and Buhl, D.: 1975, *Astrophys. J.* **197**, 329.

Snyder, L.E., Buhl, D., Schwartz, P.R., Clark, F.O., Johnson, D.R., Lovas, F.J., and Giguere, P.T.: 1974, *Astrophys. J.* **191**, L79.

Snyder, L.E., Hollis, J.M., Ulrich, B.L., Lovas, F.J., and Buhl, D.: 1975a, *Bull. Am. Astron. Soc.* **7**, 497.

Snyder, L.E., Hollis, J.M., Ulrich, B.L., Lovas, F.J., Johnson, D.R., and Buhl, D.: 1975b, *Astrophys. J.* **198**, L81.

Snyder, L.E., Dickinson, D.F., Brown, L.W., and Buhl, D.: 1978, *Astrophys. J.* **224**, 512.

Soifer, B.T. and Hudson, H.S.: 1974, *Astrophys. J.* **191**, L83.

Solomon, P.M.: 1973, *Physics Today* **26**, 32.

Sorochenko, R.L. and Berulis, I.I.: 1970, *Astron. Zh.* **47**, 859.

Sorochenko, R.L., Puzanov, V.A., Salomonovich, A.E., and Shteinshleger, V.B.: 1969, *Astrophys. Letters* **3**, 7.

Stein, W.A.: 1975, *Publ. Am. Soc. Pacific* **87**, 5.

Storey, J.W.V., Watson, D.M., and Townes, C.H.: 1979, *Astrophys. J.* **233**, 109.

Strand, K.A.: 1958, *Astrophys. J.* **128**, 14.

Strömgren, B.: 1939, *Astrophys. J.* **89**, 529.

Strom, K.M., Strom, S.E., Carrasco, L., and Vrba, F.J.: 1975a, *Astrophys.* **196**, 480.

Strom, S.E., Strom, K.M., and Grasdalen, G.L.: 1975b, *Ann. Rev. Astron. Astrophys.* **13**, 187.

Sullivan, W.T.: 1973, *Astrophys. J. Suppl.* **25**, 393.

Sweitzer, J.S., Palmers, P., Morris, M., Turner, B.E., and Zuckerman, B.: 1979, *Astrophys. J.* **227**, 415.

Tamura, S.: 1976, *Sci Report*, Tohuku University.

Taylor, K. and Axon, D.J.: 1979, *Monthly Notices Roy. Astron. Soc.* **188**, 687.

Taylor, K. and Münch, G.: 1978, *Astron. Astrophys.* **70**, 359.

Tenorio-Tagle, G.: 1979, *Astron. Astrophys.* **71**, 59.

Tenorio-Tagle, G., Yorke, H.W., and Bodenheimer, P.: 1979, *Astron. Astrophys.* **80**, 110.

Terzian, Y.: 1968, in Y. Terzian (ed.), *Interstellar Ionized Hydrogen*, W.A. Benjamin, New York, p. 283.

Terzian, Y.: 1974, A. Beer (ed.), *Vistas in Astronomy*, Pergamon Press, London, Vol. 16, p. 279.

Terzian, Y., Mezger, P.G., and Schraml, J.: 1968, *Astrophys. Letters* **1**, 153.

Thackeray, A.D.: 1975, *Monthly Notices Roy. Astron. Soc.* **172**, 49p.

Thaddeus, P., Wilson, R.W., Kutner, M., Penzias, A.A., and Jefferts, K.B.: 1971, *Astrophys. J.* **168**, L59.

Thaddeus, P., Kutner, M.L., Penzias, A.A., Wilson, R.W., and Jefferts, K.B.: 1972, *Astrophys. J.* **176**, L73.

Thaddeus, P., Mather, J., Davis, J.H., and Blair, G.N.: 1974, *Astrophys. J.* **192**, L33.

Thomsen, B.: 1975, *Astron. Astrophys.* **43**, 411.

Thronson, H.A., Jr., Harper, D.A., Keene, J., Loewenstein, R.F., Moseley, H., and Telesco, C.M.: 1978, *Astron. J.* **83**, 492.

Thum, C.: 1975, Ph.D. Thesis, University of Heidelberg.

Thum, C., Lemke, D., Fahrbach, V., and Frey, A.: 1978, *Astron. Astrophys.* **65**, 207.

Tolbert, C.W.: 1965, *Nature* **206**, 1304.

Tolbert, C.W.: 1971, *Astron. Astrophys. Suppl.* **3**, 349.

Torres-Peimbert, S., Peimbert, M., and Daltabuit, E.: 1980, *Astrophys. J.* **238**, 133.

Traub, W.A., Carleton, N.P., and Hegyi, D.J.: 1974, *Astrophys. J.* **190**, L81.

Trimble, V.L. and Thorne, K.S.: 1969, *Astrophys. J.* **156**, 1013.

Troland, T.H. and Heiles, C.: 1977a, *Bull. Am. Astron. Soc.* **9**, 611.

Troland, T.H. and Heiles, C.: 1977b, *Astrophys. J.* **214**, 703.

Tucker, K.D., Kutner, M.L., and Thaddeus, P.: 1973, *Astrophys. J.* **186**, L13.

Tucker, K.D., Kutner, M.L., and Thaddeus, P.: 1974, *Astrophys. J.* **193**, L115.

Turner, B.E.: 1974, G.L. Verschuur and K.I. Kellermann (eds.), *Galactic and Extra-Galactic Radio Astronomy*, Springer-Verlag, New York, p. 199.

Turner, B.E. and Gammon, R.H.: 1975, *Astrophys. J.* **198**, 71.

Turner, B.E. and Thaddeus, P.: 1977, *Astrophys. J.* **211**, 755.

Turner, B.E., Zuckerman, B., Palmer, P., and Morris, M.: 1973, *Astrophys. J.* **186**, 123.

Turner, B.E., Balick, B., Cudaback, D.D. Heiles, C., and Boyle, R.J.: 1974, *Astrophys. J.* **194**, 279.

Turner, B.E., Zuckerman, B., Fourikis, N., Morris, M., and Palmer, P.: 1975, *Astrophys. J.* **198**, L125.

Ulich, B.L. and Haas, R.W.: 1976, *Astrophys. J. Suppl.* **30**, 247.

Ulich, B.L., Hollis, J.M., and Suyder, L.E.: 1977, *Astrophys. J.* **217**, L105.

Underhill, A.B.: 1966, *The Early Type Stars*, D. Reidel Publ. Co., Dordrecht, Holland, p. 50.

Unsöld, A.: 1972, cited in Mezger, 1972.

Van den Bout, P. and Thaddeus, P.: 1971, *Astrophys. J.* **170**, 297.

Vandervoort, P.O.: 1964, *Astrophys. J.* **139**, 869.

Van Woerden, H.: 1967, in H. van Woerden (ed.), 'Radio Astronomy and the Galactic System', *IAU Symp.* **31**, 3.

Vaughan, A.H., Jr., 1968, *Astrophys. J.* **154**, 87.

Vaughn, A.E., Large, M.I., and Wielibinski, R.: 1969, *Nature* **222**, 963.

Verschuur, G.L.: 1969, *Nature* **223**, 140.

Verschuur, G.L.: 1970, in H.J. Habing (ed.), 'Interstellar Gas Dynamics', *IAU Symp.* **39**, 150.

Vidal, J.-L., 1980, *Astron. Astrophys. Suppl.* **40**, 33.

Walker, M.F.: 1969, *Astrophys. J.* **155**, 447.

Walsh, J.: 1980, *Astrophys. Space Sci.* **69**, 227.

Waltman, E.B. and Johnston, K.J.: 1973, *Astrophys. J.* **182**, 489.

Wannier, P.G. and Phillips, T.G.: 1977, *Astrophys. J.* **215**, 796.

Wannier, P.G., Penzias, A.A., Linke, R.A., and Wilson, R.W.: 1976, *Astrophys. J.* **204**, 26.

Ward, D.B.: 1975, *Astrophys. J.* **200**, L41.

Ward, D.B. and Harwit, M.: 1974, *Nature* **252**, 27.

Ward, D.B., Dennison, B., Gull, G.E., and Harwit, M.: 1976, *Astrophys. J.* **205**, L75.

Warren, W.H., Jr. and Hesser, J.E.: 1977, *Astrophys. J. Suppl.* **34**, 115.

Waters, J.W., Gunstincic, J.J., Kakar, R.K., Kuiper, T.B.H., Roscoe, H.K., Swanson, P.N., Kuiper, E.N.R., Kerr, A.R., and Thaddeus, P.: 1980, *Astrophys. J.* **235**, 57.

Watt, G.D., White, G.J., Cronin, N.J., and van Vliet, A.H.F.: 1979, *Monthly Notices Roy. Astron. Soc.* **189**, 287.

Webster, W.J., Jr. and Altenhoff, W.J.: 1980, *Astrophys. Letters* **5**, 233.

Werner, M.W., Elias, J.H., Gezari, D.Y., and Westbrook, W.E.: 1974, *Astrophys. J.* **192**, L31.

Werner, M.W., Gatley, I., Harper, D.A., Becklin, E.E., Loewenstein, R.F., Telesco, C.M., and Thronson, H.A.: 1976, *Astrophys. J.* **204**, 420.

Werner, M.W., Becklin, E.E., and Neugebauer, G.: 1977, *Science* **197**, 723.

Westbrook. W.E., Werner, M.W., Elias, J.H., Gezari, D.Y., Hauser, M.G., Lo, K.Y., and Neugebauer, G.: 1976, *Astrophys. J.* **209**, 94.

White, G.J. and Ricketts, M.J.: 1977, *Astrophys. Letters* **18**, 79.

White, G.J., Watt, G.D., Beckman, J.E., Rose, W.B., and van Vliet, A.H.F.: 1980, *Astron. Astrophys.* **84**, 212.

Whiteoak, J.B. and Gardner, F.F.: 1970, *Astrophys. Letters* **5**, 5.

Whiteoak, J.B. and Gardner, F.F.: 1974, *Astron. Astrophys.* **37**, 389.

Whiteoak, J.B., Gardner, F.F., and Sinclair, M.W.: 1978, *Monthly Notices Roy. Astron. Soc* **184**, 235.

Whitford, A.E.: 1958, *Astron. J.* **63**, 201.

Whitworth, A.: 1979, *Monthly Notices Roy. Astron. Soc.* **186**, 59.

Wickramasinghe, N.C.: 1963, *Monthly Notices Roy. Astron. Soc.* **126**, 99.

Wickramasinghe, N.C.: 1967, *Interstellar Grains*, Chapman and Hall Ltd. London.

Wickramasinghe, N.C.: 1972, in N.C. Wickramasinghe, F.D., Kahn, and P.G. Mezger (eds.), *Interstellar Matter*, edited by the Astromomical Institute, University of Basel, Saas-Fee, p. 209.

Wiese, W.L., Smith, M.W., and Glennon, B.M.: 1966, *Atomic Transition Probabilities*, Vol. 1, 'Hydrogen through Helium', Nat. Bur. Stands, Washington, NSRDS-NBS4.

Williams, D.R.W.: 1967, *Astrophys. Letters* **1**, 59.

Williamson, F.D., Sanders, W.T., Kraushaar, W.L., McCammon, D., Borken, R., and Bunner, A.N.: 1974, *Astrophys. J.* **193**, L133.

Wilson, O.C.: 1939, *Publ. Am. Astron. Soc.* **9**, 274.

Wilson, O.C., Münch, G., Flather, E.M., and Coffeen, M.F.: 1959, *Astrophys. J. Suppl.* **4**, 199.

Wilson, R.E.: 1972, *Astrophys. Space Sci.* **19**, 165.

Wilson, T.L. and Pauls, T.: 1979, *Astron. Astrophys.* **73**, L10.

Wilson, T.L., Mezger, P.G., Gardner, F.F., and Milne, D.K.: 1970, *Astron. Astrophys.* **6**, 364.

Wilson, T.L., Thomasson, P., and Gardner, F.F.: 1975, *Astron. Astrophys.* **43**, 167.

Wilson, T.L., Bieging, J., and Wilson, W.E.: 1979a, *Astron. Astrophys.* **71**, 205.

Wilson, T.L., Downes, D., and Bieging, J.: 1979b, *Astron. Astrophys.* **71**, 275.

Wilson, T.L., Fazio, G.G., Jaffe, D., Kleinmann, D., Wright, E.L., and Low, F.J.: 1979c, *Astron. Astrophys.* **76**, 86.

Wilson, W.J., Schwartz, P.R., Epstein, E.E., Johnson, W.A., Etchevery, R.D., Mori, T.T., Berry, G.G., and Dyson, H.B.: 1974, *Astrophys. J.* **191**, 357.

Withbroe, G.L.: 1978, cited in Perinotto, M. and Patriarchi, P. 1980.

Witt, A.N. and Lillie, C.F.: 1978, *Astrophys. J.* **222**, 909.

Wurm, K.: 1961, *Z. Astrophys.* **52**, 149.

Wurm, K. and Perinotto, M.: *Z. Astrophys.* **62**, 30.

Wurm, K. and Rosino, L.: 1956, *Mitteilungen Hamburg Sternwarte Bergedorf* **10**, No. 103.

Wurm, K. and Rosino, L.: 1959, *A Monochromatic Atlas of the Orion Nebula*, Part I, Asiago Observatory, Italy.

Wurm, K, and Rosino, L.: 1965, *A Monochromatic Atlas of the Orion Nebula*, Part II, Asiago Observatory, Italy.

Wynn-Williams, C.G. and Becklin, E.E.: 1974, *Publ. Astron. Soc. Pacific* **86**, 5.

Wyse, A.B.: 1942, *Astrophys. J.* **95**, 356.

York, D.G.: 1974, *Astrophys. J.* **193**, L127.

Zisk, S.H.: 1966, *Science* **153**, 1107.

Zuckerman, B.: 1973, *Astrophys. J.* **183**, 863.

Zuckerman, B. and Ball, J.: 1974, *Astrophys. J.* **190**, 35.

Zuckerman, B. and Palmer, P.: 1968, *Astrophys. J.* **153**, L145.

Zuckerman, B. and Palmer, P.: 1970, *Astron. Astrophys.* **4**, 244.

Zuckerman, B. and Palmer, P.: 1975, *Astrophys. J.* **199**, L35.

Zuckerman, B., Buhl, D., Palmer, P., and Snyder, L.E.: 1970, *Astrophys. J.* **160**, 485.

Zuckerman, B., Palmer, P., and Rickard, L.J.: 1975, *Astrophys. J.* **197**, 571.

Zuckerman, B., Kuiper, T.B.H., and Rodriguez-Kuiper, E.N.: 1976, *Astrophys. J.* **209**, L137.

INDEX